MEDICINAL AND AROMATIC PLANTS

MEDICINAL AND AROMATIC PLANTS

Expanding their Horizons through Omics

Edited by

TARIQ AFTAB

Aligarh Muslim University, Aligarh

KHALID REHMAN HAKEEM

King Abdulaziz University, Jeddah

ELSEVIER

ACADEMIC PRESS

An imprint of Elsevier

Academic Press is an imprint of Elsevier
125 London Wall, London EC2Y 5AS, United Kingdom
525 B Street, Suite 1650, San Diego, CA 92101, United States
50 Hampshire Street, 5th Floor, Cambridge, MA 02139, United States
The Boulevard, Langford Lane, Kidlington, Oxford OX5 1GB, United Kingdom

Library of Congress Cataloging-in-Publication Data

A catalog record for this book is available from the Library of Congress

British Library Cataloguing-in-Publication Data

A catalogue record for this book is available from the British Library

ISBN: 978-0-12-819590-1

For information on all Academic Press publications
visit our website at https://www.elsevier.com/books-and-journals

Publisher: Charlotte Cockle
Editorial Project Manager: Lena Sparks
Production Project Manager: Sreejith Viswanathan
Designer: Christian J. Bilbow

Typeset by Thomson Digital

This book is dedicated to

Sir Syed Ahmad Khan

(October 17, 1817 – March 27, 1898)

Sir Syed Ahmad Khan, one of the architects of modern India was born on October 17, 1817 in Delhi and started his career as a civil servant.

The 1857 revolt was one of the turning points in Syed Ahmed's life. He clearly foresaw the imperative need for Muslims to acquire proficiency in the English language and modern sciences, if the community were to maintain its social and political clout, particularly in Northern India.

He was one of those early pioneers who recognized the critical role of education in the empowerment of poor and backward Muslim community. In more than one way, Sir Syed was one of the greatest social reformers and a great national builder of modern India. He began to prepare the road map for the formation of a Muslim University by starting various schools. He instituted Scientific Society in 1863 to instil a scientific temperament into Muslims and to make the Western knowledge available to Indians in their own language.

The Aligarh Institute Gazette, an organ of the Scientific Society, was launched in March 1866 and succeeded in agitating the minds in the traditional Muslim society. Anyone with a poor level of commitment would have backed off in the face of strong opposition but Sir Syed responded by bringing out another journal, Tehzibul Akhlaq that was rightly named in English as "Mohammedan Social Reformer."

In 1875, Sir Syed founded the Madarsatul Uloom in Aligarh and patterned the MAO College after Oxford and Cambridge universities that he went on a trip to London. His objective was to build a college in line with the British education system but without compromising its Islamic values. He wanted this College to act as a bridge between the old and the new, the East and the West. While he fully appreciated the need and urgency of imparting instruction based on Western learning, he was not oblivious to the value of oriental learning and wanted to preserve and transmit to posterity the rich legacy of the past. Dr. Sir Mohammad Iqbal observes—"The real greatness of Sir Syed consists in the fact that he was the first Indian Muslim who felt the need of a fresh orientation of Islam and worked for it—his sensitive nature was the first to react to modern age."

The aim of Sir Syed was not merely restricted to establishing a college at Aligarh but at spreading a network of Muslim managed educational institutions throughout the length and breadth of the country keeping in view this end, he instituted All India Muslim Educational Conference that revived the spirit of Muslims at national level. The Aligarh Movement motivated the Muslims to help open a number of educational institutions. It was the first of its kind of such Muslim NGO in India, which awakened Muslims from their deep slumber and infused social and political sensibility into them.

Sir Syed contributed many essential elements to the development of the modern society of subcontinent. During Sir Syed's own lifetime, "The Englishman," a renowned British magazine of the 19th century remarked in a commentary on November 17, 1885—Sir Syed's life "strikingly illustrated one of the best phases of modern history." He died on March 27, 1898 and lies buried next to the main mosque at Aligarh Muslim University.

Contents

Contributors

Shakeel Ahmad Dar
Sher-e-Kashmir University of Agricultural Sciences and Technology, Srinagar, Jammu and Kashmir, India

Hilal Ahmad Ganaie
Cytogenetics and Molecular Biology Research Laboratory, Centre of Research for Development (CORD), University of Kashmir, Srinagar, Jammu and Kashmir, India

Jameel R. Al-Obaidi
Kuala Lumpur, Malaysia

Peerzada Arshid Shabir
Department of Botany, Government Degree College, Sopore, Jammu and Kashmir, India

Ifra Ashraf
College of Agricultural Engineering and Technology, Sher-e-Kashmir University of Agricultural Sciences and Technology of Kashmir, Srinagar, Jammu and Kashmir, India

Lubna Azmi
Pharmacognosy and Ethnopharmacology Division, National Botanical Research Institute (CSIR) Lucknow, Uttar Pradesh, India

Hebert Jair Barrales-Cureño
Programa Institucional de Maestría en Ciencias Biológicas. Universidad Michoacana de San Nicolas Hidalgo. Calle de Santiago Tapia 403, Centro, Morelia, Michoacán, México; Natural Science Divsion, Puebla State Intercultural University, Huehuetla, Puebla, México

Ishrat Bashir
Sher-e-Kashmir University of Agricultural Sciences and Technology, Srinagar, Jammu and Kashmir, India

Rouf Ahmad Bhat
Department of Environmental Sciences, School of Life Sciences, Sri Pratap College, Cluster University of Srinagar, Jammu and Kashmir, India; Sher-e-Kashmir University of Agricultural Sciences and Technology, Srinagar, Jammu and Kashmir, India

Javeed I. A. Bhat
Sher-e-Kashmir University of Agricultural Sciences and Technology, Srinagar, Jammu and Kashmir, India

Bilal A. Beigh
Barkatullah University, Bhopal, Madhya Pradesh, India

Juan Antonio Cortés-Ruiz
Instituto Tecnológico de Mazatlán Instituto Tecnológico de Mazatlán. Ingeniería Bioquímica. México Corsario I 203, Urías, Mazatlán, Sin

Arturo Huerta de la Peña
Unit in Development for Research and Technology Transfer in Biological Control, Postgraduate College, San Pedro Cholula, Puebla, México

José Espinoza-Pérez
El Colegio de la Frontera Sur-Unidad San Cristóbal de las Casas. Departamento de Agricultura, Sociedad y Ambiente. Carretera Panamericana y Periférico Sur s/n, Barrio de María Auxiliadora. San Cristóbal de Las Casas, Chiapas – México

Aijaz H Ganie
Department of Botany, University of Kashmir, Srinagar, Jammu and Kashmir, India

Arti Gautam
Pharmacognosy and Ethnopharmacology Division, National Botanical Research Institute (CSIR), Lucknow, Uttar Pradesh, India

Hafiz ul Islam
Barkatullah University, Bhopal, Madhya Pradesh, India

Bushra Jan
Department of Botany, University of Kashmir, Srinagar, Jammu and Kashmir, India

Periyanaina Kesika
Innovation Center for Holistic Health, Nutraceuticals and Cosmeceuticals, Faculty of Pharmacy, Chiang Mai University, Chiang Mai, Thailand

Shamshad A. Khan
Department of Applied Biotechnology, College of Applied Sciences, Ministry of Education, Sur, Oman

Luis Germán López-Valdez
Universidad Autónoma Chapingo. Carretera Federal México-Texcoco Km 38.5, Universidad Autonoma de Chapingo, Texcoco, México

José Lorenzo-Laureano
División de Ciencias Naturales, Universidad Intercultural del Estado de Puebla. Calle Principal a Lipuntahuaca S/N; Lipuntahuaca, Huehuetla, Puebla, Mexico

Gonzalo Guillermo Lucho-Constantino
Universidad Tecnológica de Gutiárrez Zamora, Veracruz. Prolongación Dr. Miguel Patiño s/n, Centro, 93556 Gutiérrez Zamora, Veracruz

Shafat A. Mir
Sher-e-Kashmir University of Agricultural Sciences and Technology, Srinagar, Jammu and Kashmir, India

Himanshu Misra
Greentechnology Department, Ipca Laboratories Ltd., Sejavta, Ratlam, Madhya Pradesh, India

Jorge Montiel-Montoya
Centro Interdisciplinario de Investigación para el Desarrollo Integral Regional. Unidad Sinaloa, Instituto Politécnico Nacional, Guasave, Sinaloa, Mexico

Nighat Mushtaq
Division of Vegetable Science, Sher-e-Kashmir University of Agricultural Sciences and Technology of Kashmir, Srinagar, Jammu and Kashmir, India

Satheeshkumar Nanjappan
Drug Metabolism and Interactions Research Lab, Department of Pharmaceutical Analysis, National Institute of Pharmaceutical Education and Research (NIPER), Hyderabad, Telangana, India

Deepu Pandita
Government Department of School Education, Jammu, Jammu and Kashmir, India

Varsha A. Parasharami
CSIR-National Chemical Laboratory, Division of Biochemical Sciences, Pune, Maharashtra, India

Mani Iyer Prasanth
Age-Related Inflammation and Degeneration Research Unit, Department of Clinical Chemistry, Faculty of Allied Health Sciences, Chulalongkorn University, Bangkok, Thailand

Laiq Ur Rahman
Plant Biotechnology Division, Central Institute of Medicinal and Aromatic Plants (CSIR-CIMAP), Lucknow, Uttar Pradesh, India

Mauji Ram
Greentechnology Department, Ipca Laboratories Ltd., Sejavta, Ratlam, Madhya Pradesh, India

Siva Nageswara Rao Gajula
Drug Metabolism and Interactions Research Lab, Department of Pharmaceutical Analysis, National Institute of Pharmaceutical Education and Research (NIPER), Hyderabad, Telangana, India

Irfan Rashid
Department of Botany, University of Kashmir, Srinagar, Jammu and Kashmir, India

Nowsheeba Rashid
Amity Institute of Food Technology, Amity University, Noida, Uttar Pradesh, India

Gowhar Rashid
Department of Clinical Biochemistry, Sher-e-Kashmir Institute of Medical Sciences, Soura, Jammu and Kashmir, India

Ishfaq Ul Rehman
Department of Botany, University of Kashmir, Srinagar, Jammu and Kashmir, India

Summia Rehman
Department of Botany, University of Kashmir, Srinagar, Jammu and Kashmir, India

Zafar Ah Reshi
Department of Botany, University of Kashmir, Srinagar, Jammu and Kashmir, India

César Reyes
División de Ciencias Naturales, Universidad Intercultural del Estado de Puebla. Calle Principal a Lipuntahuaca S/N; Lipuntahuaca, Huehuetla, Puebla, Mexico

Pulikkottil Stanes Rosmol
Mookken House, Haritha Nagar, Viyyur, Trissur, Kerala, India

Jesús Antonio Salazar-Magallón
Unit in Development for Research and Technology Transfer in Biological Control, Postgraduate College, San Pedro Cholula, Puebla, México

Amrina Shafi
Department of Biotechnology, School of Biological Sciences, University of Kashmir, Srinagar, Jammu and Kashmir, India

Ila Shukla
Pharmacognosy and Ethnopharmacology Division, National Botanical Research Institute (CSIR), Lucknow, Uttar Pradesh, India

Bhagavathi Sundaram Sivamaruthi
Innovation Center for Holistic Health, Nutraceuticals and Cosmeceuticals, Faculty of Pharmacy, Chiang Mai University, Chiang Mai, Thailand

Ashish Srivastava
Pharmaceutics and Pharmacokinetics Division, CSIR- Central Drug Research Institute, Lucknow, Uttar Pradesh, India

Tewin Tencomnao
Age-Related Inflammation and Degeneration Research Unit, Department of Clinical Chemistry, Faculty of Allied Health Sciences, Chulalongkorn University, Bangkok, Thailand

Priyanka Verma
CSIR-National Chemical Laboratory, Division of Biochemical Sciences, Pune, Maharashtra, India

Insha Zahoor
Drug Therapeutics and Neurobiology Lab, Department of Biotechnology and Bioinformatics Centre, University of Kashmir, Srinagar, Jammu and Kashmir, India; Department of Neurology, Henry Ford Hospital, Detroit, MI, United States

Fabiola Zaragoza-Martínez
Centro de Investigación y de Estudios Avanzados del Instituto Politécnico Nacional, Av. Instituto Politécnico Nacional 2508, 07360, Mexico City, Mexico

Preface

Medicinal and aromatic plants (MAPs) are offered in a wide variety of products in the market. The enormous demand in botanicals result in a huge trade from local to international level. Until last century, the production of botanicals relies to a large degree on wild-collection. However, utilization and commerce of wild-plant resources are not detrimental in themselves, For example, the increasing commercial collection, largely unmonitored trade, and habitat loss lead to an incomparably growing pressure on plant populations in the wild. An estimated 9000 medicinal plant species are threatened worldwide. Conservation concepts and measures which have to meet future supply and the provisions of species conservation range from resource management, cultivation, and shifting processing from consumer to source countries, species conservation to trade restrictions. Medicinal and aromatic plants are of high priority for conservation action, as wild crafting will certainly continue to play a significant role in their future trade—the sustainable commercial use of their biological resources may provide a financial instrument for nature conservation. There has been growing preference for organic- and herbal-based products in the world market. Similarly, the world is shifting from crude export to processing of herbal products and essential oils. Continuing advances in Omics methodologies and instrumentation is enhancing the understanding of how plants cope with the dynamic nature of their growing environment. Omics platforms have been only recently extended to cover horticultural crops and MAPs.

The book comprises 15 chapters, most of them being review articles written by experts from around the globe, highlighting the various omic approaches that are being used in the current research on immensely important MAPs and its future prospects. We are hopeful, this volume would furnish the need of all researchers who are working or have interest in this particular field. Undoubtedly, this book will be helpful for general use of research students, teachers, ethnobotanists, oncologists, pharmacologists, and herbal growers, who have interest in MAPs.

We are highly grateful to all our contributors for accepting our invitation for not only sharing their knowledge and research, but for venerably integrating their expertise in dispersed information from diverse fields in composing the chapters and enduring editorial suggestions to finally produce this venture. We also thank Elsevier team for their generous cooperation at every stage of the book production.

Lastly, thanks are also due to well-wishers, research students, and authors' family members for their moral support, blessings, and inspiration in the compilation of this book.

Tariq Aftab
Aligarh Muslim University, India

Khalid Rehman Hakeem
King Abdulaziz University, Saudi Arabia

CHAPTER 1

Review of the active principles of medicinal and aromatic plants and their disease fighting properties

Hilal Ahmad Ganaie
Cytogenetics and Molecular Biology Research Laboratory, Centre of Research for Development (CORD), University of Kashmir, Srinagar, Jammu and Kashmir, India

Introduction

The medicinal plants have been practised for therapeutics from the beginning of human civilization. The herbal system of medicine is not only the oldest form of health care but also an integral part of modern civilization development. Even in the modern world, a vast majority of human population especially from developing countries rely on herbal system of medicine and their products for primary health care needs (Ramawat and Goyal, 2008). Medicinal plants and their extracts have been used by man from prehistoric times to cure various diseases that resulted in the discovery of some important drugs like morphine (analgesic) (Shimomura et al., 1971), reserpine (antihypertensives) (Dzeufiet et al., 2014), taxol, vinca alkaloids (anticancer) (Paul, Gnanam, M Jayadeepa, & Arul, 2013), digitoxin (cardiotonic), codeine (antitussives), quinine and artemisinin (antimalarials) (Adebayo & Krettli, 2011). According to Ramawat (2007) the use of plants as therapeutic agents dates back 2600 BC when the people of Mesopotamia were using oils from cypress, licorice, myrrh, and poppy for the treatment of infectious ailments. He also opined that present age of Mesopotamia civilization is using *Commiphora wightii, Cupressus sempervirens, Cedrus* spp., *Glycirrhiza glabra,* and *Papaver somniferum* for treatment of infectious diseases.

Present scenario of medicinal plants

Herbal medicine is the upshot of therapeutic experiences gained for over centuries by generations of practising physicians of indigenous medicine system. They are known to be the oldest health care products that have

Medicinal and Aromatic Plants: Expanding their Horizons through Omics
http://dx.doi.org/10.1016/B978-0-12-819590-1.00001-X

been used by mankind all over the world in the form of folklore medicines or traditional medicines or ethnic medicines. The therapeutic use of herbal medicines is gaining considerable momentum in the world during the past decade. The World Health Organization (WHO) estimates that herbal medicine is still the mainstay of about 75%–80% of the world population, mainly in the developing countries, for primary health care because of better cultural acceptability, better compatibility with the human body and lesser side effects. In India the value of botanicals-related trade is about US$ 10 billion per annum with annual export of US$1.1 billion. India is sitting on the goldmine of well-recorded and well-practiced knowledge of traditional herbal medicine and it has a rich heritage of medicinal plants (more than 8000 medicinal plant species). China has successfully exploited its herbal medicine knowledge by promoting its use in the developed world medicine system. But, unlike China, India has not been able to capitalize on this herbal wealth due to the non-availability of standardized herbal drugs and their formulations.

Major contributions of medicinal plants in the Himalaya relate to documentation of inventories, which include information on the list of species, part(s) used and distribution range. An exhaustive compendium of medicinal plants of the country and their biological activity is available (Rastogi and Mehrotra, 1991). *In vitro* propagation and conservation of some endangered medicinal species of India have been reviewed (Bhojwani and Arumugam, 1993). In recent years, *in vitro* culture protocols of some Himalayan medicinal plants have been standardized. These include *P. hexandrum, N. jatamansii, Rheum emodi, Valeriana wallichii,* and *A. heterophyllum*. Similarly, induction of somatic embryogenesis and organogenesis is reported in *Bunium persicum*. As per the estimates of Exim Bank, the annual world business of medicinal plants is over US$ 60 billion, with per year growth rate of 7%, out of it 2300 Crore rupees is India's annual turnover (Sharma, 2005). Countries like China, Germany, France, Italy, Japan, Spain, United Kingdom and United States are the leading global trade centers for MAPs (Laird, 1999). The WHO reported that the current global demand of medicinal plant products is approximately US$ 14 billion per year and is expected to go beyond US$ 5 trillion by the year 2050 and the global demand for medicinal plant products is increasing at the rate of 15%–25% per year. The "nutraceutical" sector consisting of herbal medicines, which are added with dietary supplements in order to pass FDA criteria with ease, is now estimated to US$ 5.1 billion. The number of people using medicinal plants in the United States has increased from 2.5% in 1990 to 30% in 2000.

India is one among the 17 mega biodiversity country in the world, with over 45,000 species of plants, of which 7,500 species are estimated to be used by 4,635 communities for human and veterinary purposes. The floral diversity of India is matchless due to existence of diverse climatic conditions across the country. India has a rich heritage of usage of medicinal plants in the Ayurveda, Siddha, and Unani system. India is among the leading exporters of processed products from plants and raw MAPs. The foreign exchange earned by India from plant-based drug exports during 1994–95 prized US $ 53,219 million and US $ 13,250 million from essential oils (Lambert, Srivastava, & Vietmeyer, 1997). India exports herbal material and medicines to tune of Rs. 3,000 crores, which is expected to reach 15,000 crores by 2015. Medicinal plants are a largely unexplored source of drug repository (Dar et al., 2013). There are many reports showing the rising trends of antimutagenic studies with plant extracts (Khader, Bresgen, & Eckl, 2010;Chen et al., 2011; El-Sayed and Hussin, 2013).

Bioactive principles of medicinal and aromatic plants

Historically, natural products are continually being investigated for promising new leads in pharmaceutical development, but are still largely unexplored (Bailly, 2009; Dar et al., 2012, 2013). Many scientists have worked on the antimutagenic potential of plant extracts (Khader et al., 2010; Chen et al., 2011; El-Sayed and Hussin, 2013). It is now well-established that the traditional herbal therapies contain a diverse array of chemopreventive agents (Aruoma, 2003). The use of antimutagens and anticarcinogens in day-to-day life suggested being the most effective procedure for preventing human cancer and genetic diseases (Ferguson, 1994). Lippman, Benner, Hong (1994) opined that medicinal plants have bioactive compounds which act as strategy to block or reverse carcinogenesis at early stages. In addition to their effectiveness, they are considered to be inexpensive and easily applicable to control cancer (Wattemberg, 1985). The widespread medicinal, edible, and herbal plants have been tested for their antimutagenic activity and proved to inhibit the mutagenic and/or carcinogenic effects of some chemical mutagens (Sarkar, Basak, Bishayee, Basak, Chatterjee, 1997; Sripanidkulchai et al., 2002). As the popularity of traditional medicine has spill at global level, in recent years enormous interest is being given in developing new pharmaceutical products from such valuable natural resources (Karekar, Joshi, & Shinde, 2000; Fahmy et al., 1997 Fahmy, Tantawi, & Awad, 1997). A lot of contemporary research work is directed toward unravelling the antimutagenic potential of plants used in folk medicine. Our country, India is sitting on a gold mine of well-recorded and traditionally well-practiced knowledge of herbal medicine. Specially, plants growing at high-altitude Himalayan pastures are time-honored sources of health and general well-being of local inhabitants. The Himalayan plants are the major contributors to the herbal pharmaceutical industry of India and other countries (Dhyani et al., 2007).

The antimutagenic or protective effects have been attributed to many classes of phytocompounds mainly flavonoids and phenolic compounds present in foods. However, such compounds have also been reported to exhibit a wide range of biological activities such as antimicrobial, anti-inflammatory, anti-allergic, antioxidant and free radical scavenging. Natural antimutagens from edible medicinal plants are of particular importance because they may be useful for human cancer prevention and have no undesirable xenobiotic effects on living organisms. Several therapeutic properties of medicinal plants are known in obstetrics and gynecology (Abo, Adeyemi, & Adeite, 2000), respiratory disorders (Neto et al., 2002), skin disorders (Graf, 2000), cardiac diseases (Ankli et al., 2002), and mental health (Ahmad, Mehmood, & Mohammad, 1998). It has been suggested that halving the rate of mutations would delay the onset of most cancers and might be adequate in the lifetime of many individuals (Loeb, Loeb, & Anderson, 2003). There are different classes of secondary metabolites present in the plants exhibiting antimutagenic activities.

Most plants synthesize toxic substances which act as defensive mechanism against insects and herbivores. In addition, the poisonous substances may also affect organisms that feed on them including humans. Therefore, it is reasonable that while some medicinal plants may suppress the effects of mutagens, others may have toxic or mutagenic effects (Vicentini, Camparoto, Teixeira, & Mantovani, 2001). Thus, studies of their mutagenic as well as antimutagenic potential are necessary to establish the safe use of these medicinal plants. *Adhatoda vasica*, also known as Malabar Nut tree belongs to Acanthaceae family, is used widely among Indians for the treatment of inflammation (Chakraborty and Brantner, 2001), cold, cough, chronic bronchitis (Amin and Mehta, 1959), cataract (Patel, Jivani, Malaviya, Gohil, & Bhalodia, 2012), asthma, piles, glandular tumor, and to cure fresh wounds (Palasuwan, Soogarun, Lertlum, Pradniwat, & Wiwanitkit, 2005; Ayyanar and Ignacimuthu, 2008). Padmaja, Srvanthi, & Hemalatha, (2011) suggested that pharmacological activities of *A. vasica* may be due to the presence of vasicine, vasicinone, and vasicinol, which are the major alkaloids. *Carica papaya* (papaya) belongs to the family Caricaceae, well known for its fruits and its parts are used for microbial infections (Sharmeen, Hossain, Rahman, & Foysal, & Miah, 2012), treating burns and wounds, fever, intestinal nematode infection, asthma, and gastric (Starley, Mohammed, Scheinder, & Bickler, 1999; Runnie, Salleh, Mohamed, Head, Abeywardena, 2004Stepek, Behnke, Buttle, & Duce, 2004). Moreover, according to Mazzio and Soliman (2009) the extract of papaya leaves show antitumor activity when tested in Neuro-2A cell lines.

Vitamins have been extensively studied for their antimutagenic potential. Vitamin E not only suppress the mutagenic effect of $Na_2Cr_2O_4$ in Chinese hamster V79 cells but also protects cells from clastogenic and mutagenic effect of chromate compounds, possibly through its ability to scavenge chromium and/or free radicles (Sugiyama, Lin, & Costa, 1991). Dean & Cheesman, (1987) showed that Vitamin E protects proteins against free radical damage in lipid environments. According to Rao, Pulusani, & Chawan (1986), lactic acid bacteria and their fermented food products confer a variety of important nutritional and therapeutic benefits including antimutagenic and/or anticarcinogenic activity. Durnev and Seredenin (1990) found that beta carotene, a natural food colorant and an antioxidant, has been found antimutagenic against cyclophosphamide-induced genotoxicity in mice bone marrow cells and observed a significant reduction in the micronuclei formation in the bone marrow cells. Ames and Gold (1991) described antimutagenic activities of three vitamins E, A, and C against cyclophosphamide-induced genotoxicity in mice. *Curcumin* acts as a protective agent against sysplatin, hydrocortisone, nicotine, lead acetate, ethanol, and irradiation (Antunes & Takahashi, 1998; Ahmad & Afzal, 2004; El-Ashmawyet al., 2006El-Ashmawy, Ashry, El-Nahas, Salama, 2006). Nakamura et al. (2002) tested invitro the fruits of *C. ferra* in Epstein Bar Virus early antigen and showed potent inhibitory activity. Seo and Surh, (2001) found Eupatilin, a pharmacologically active ingredient of the traditional oriental medicinal herb *Artmisia asiatica* (Nakai) reduced viability in cultured human promyelocytic leukemia (HL-60) cells through induced apoptosis in human promyelocytic leukemia cells.

From ancient times, medicinal plants are being used as remedies for various diseases in human. In today's industrialized society, the use of medicinal plants has been traced to the extraction and development of several drugs as they were used traditionally in folk medicine (Shrikumar and Ravi, 2007). Medicinal plants have potent phytoconstituents which are important source of compounds and are responsible for the therapeutic properties (Jeeva, Johnson, Aparna, & Irudayaraj, 2011; Florence, Joselin, & Jeeva, 2012

Florence, Joselin, Brintha, Sukumaran, & jeeva, 2014; Joselin, Brintha, Florence, & Jeeva, 2012, Joselin et al.,2013Joselin, Brintha, Florence, & Jeeva, 2013; Sainkhediya and Ray, 2012; Sumathi and Uthayakumari, 2014). These phytoconstituents endow them with medicinal properties. Many plants possess antioxidant properties because of the presence of phenolic compounds (Brown and Rice-Evans, 1998; Krings and Berger, 2001). These phenolic compounds possess biological properties such as anti-apoptosis, anti-aging, anti-carcinogen, anti-inflammation, anti-atherosclerosis, cardiovascular protection, and improvement of endothelial function, as well as inhibition of angiogenesis and cell proliferation activities (Han, Shen, & Lou, 2007). Tannins bind to proline-rich protein and interfere with protein synthesis. Flavonoids are hydroxylated phenolic substances known to be synthesized by plants in response to microbial infection and they have been found to be antimicrobial substances against wide array of microorganisms in vitro. The activity is probably due to their ability to complex with extracellular and soluble proteins and to complex with bacterial cell wall (Marjorie, 1999). They are also effective antioxidant and show strong anticancer activities (Del-Rio, Obdululio, Casfillo, Main, & Ortuno, 1997). Besides, most of the phytochemicals are known to have therapeutic properties such as insecticidals (Kambu, Di Phenzu, Coune, Wauter, & Angenot, 1982), anti-bacterial, anti-fungal (Lemos et al., 1990), and anti-constipative (Ferdous, Islam, Ahsan, Hassan, & Ahmad, 1992) activities, etc. The plants thus find their medicinal values due to the presence of these phytochemical constituents. The presence of various phytochemicals in the tested plant reveals that this plant may be a good source for production of new drugs for various ailments.

Plant-derived substances have recently become a great interest owing to their versatile applications. Medicinal plants are the richest bioresource of drugs of traditional systems of medicine, modern medicines, nutraceuticals, food supplements, folk medicines, pharmaceutical intermediates, and chemical entities for synthetic drugs (Ncube, Afolayan, & Okoh, 2008). The phytochemical screening of *Ajuga bracteosa* and *Melissa officinalis* showed that their aerial parts were rich in alkaloids, phenol, and tannins. The presence of phenolic compounds in the plants indicates that these plants may be anti-microbial agent. This agreed with the findings of Ofokansi, Esimone, & Anele (2006). Tannins have stringent properties, hasten the healing of wounds and inflamed mucous membranes. Apart from tannin and phenolic compounds, other secondary metabolite constituents detected include the alkaloids, saponin, and flavonoids. Flavonoids, on the other hand are potent water-soluble antioxidants and free radical scavengers, which prevent oxidative cell damage, have strong anticancer activity (Arceusz, Wesolowski, & Ulewicz-Magulska, 2015). Saponin has the property of precipitating and coagulating red blood cells (Sodipo, Akiniyi, & Ogunbamosu, 2000; Okwu, 2001). Therefore, the data generated from these experiments have provided the chemical basis for the wide use of this plant as therapeutic agent for treating various ailments. However, there is need to further carry out advanced hyphenated spectroscopic studies in order to elucidate the structure of these compounds.

Previous reports show that the essential oil of *M. officinalis* is composed of some important compounds like (E)-caryophyllene and caryophyllene oxide in addition to major constituents such as citronellal, neral, and geranial (Sorensen, 2000; van de Berg, Freundl, & Czygan, 1997; Holla, Svajdlenka, Veverkova, & Havranek, 1997Shabby, El-Gengaihi, & Khattab, 1995). Literature reveals that the essential oil of *M. officinalis* subsp. officinalis contains significant amounts of citral and/or citronellal, whereas *M. officinalis* subsp. *altissima* contains only traces (Van den Berg et al., 1997; Dawson, Franich, & Meder, 1988). Van den Berg et al., 1997 identified b-caryophyllene, germacrene-D, sabinene, and b-pinene as the main components in leaf oils of *M. officinalis* subsp. *altissima*. Schnitzler , Schuhmacher, Astani, & Reichling (2008) opined that the chemical composition of Germany *M. officinalis* is slightly in which the major components were acitral (20.13%) b-caryophyllene (17.31%), b-citral (13.58%), citronellal (3.86%). According to Bagˇdat and Cos¸ge, (2006), the essential oil of Turkey *M. officinalis* are also different in which the major components were citronellal (39%) and citral (33%). Singh et al.,(2014) carried out research on Indian *Melissa officinalis* and he found that the major constituents found essential oil were geranial (24.53%), neral (18.8%) and trans-caryophyllene (7.7%). *M. officinalis* is invested with compounds with antioxidant, antinociceptive and antitumoral properties (Sousa et al., 2004; Canadanovic-Brunet et al., 2008; Guginski et al., 2009; Pereira et al., 2009).

Taherpour, Maroofi, Rafie, Larijani, (2012) found that the main components of Iranian *Melissa officinalis* were (E)-citral (37.2%), neral (23.9%) and citronellal (20.3%). However, the age of lemon balm plants affected the concentration of other constituents and the proportions of the following compounds were subject to especially high fluctuations: citronellal (8.7% and 0.4%), geraniol (trace amounts and 0.6%), and geranyl acetate (0.5% and 3.0%), as well as, among others, isogeranial, E-caryophyllene, caryophyllene oxide, germacrene D, and carvacrol (Nurzyńska-Wierdak, Bogucka-Kocka, & Szymczak, 2014). Mothana et al. (2012) carried out research on the phytoconstituents of *Ajuga bracteosa* and found that the main constituents were borneol (20.8%), hexadecanoic acid (16%) and 9,12- octadecadienoic acid (7%). A previous report by Sajjadi and Ghannadi, (2004) on *Ajuga orientalis* grown in northern parts of Iran showed a completely different chemical composition, where sesquiterpene hydrocarbons e.g. germacrene-D (24.2%), β-cubebene (18.3%), and β-caryophyllene (16.9%) predominated.

Biological activity of *Melissa officinalis*

According to Blumenthal, Goldberg, & Brinckmann (2000), lemon balm is used traditionally for different medical purposes such as tonic, antispasmodic, carminative, diaphoretic, surgical dressing for wounds, sedative hypnotic, strengthening the memory and relief of stress induced headache. The distillate (or hydrosol) of lemon balm (known as Arrack) is commonly used as an antidepressant, Ibn Sina (Avicenna), the well-known Iranian scientist, recommended *M. officinalis* for above indications. In addition, other traditional medicines have indicated that lemon balm is useful for seasonal effective disorder when mixed with St. John's wort (Kuhn and Winston, 2000). Furthermore, it is stated that the essential oil of lemon balm is used in aromatherapy and may be beneficial for mild depression. Despite all these reports, no pharmacological study showing antidepressant effect of this plant has been reported. Essential oils of lemon balm are used as potential antitumoral agents for cancer remedy or prevention (Janina, 2003). The volatile oils of lemon balm may also be used as antiviral agents and also contain anti- herpes simplex virus type 2 (HSV-2) substances (Turhan, 2006). Akhondzadeh et al. (2003), carried out the investigation to assess the efficacy and safety of lemon balm extract using a fixed dose (60 drops/day) in patients with mild-to-moderate Alzheimer's disease. Valnet (1990) have suggested that the potential of lemon balm to mitigate the effects of stress indicates *Melissa* for internal use for migraine, indigestion, neuralgic problems, insomnia, and spasms among others. Bolkent, Yanardag, Karabulut-Bulan, & Yesilyaprak (2005), reported that the administration of *M. officinalis* L. extract reduced total cholesterol and total lipid liver tissue, moreover increased glutathione levels in the tissue. As a result, it was suggested that *M. officinalis* L. extract exerted a hypolipidemic effect and showed a protective effect on the liver of hyperlipidemic rats. Sousa et al. (2004) performed the study on antitumoral and antioxidant activities of lemon balm essential oils. The chemical composition of lemon balm essential oils obtained under controlled harvesting and drying conditions were very effective against a series of human cancer cell lines and mouse cell line.

Melissa officinalis L. (Lemon balm) belonging to the family Lamiaceae is a perennial herb that grows wildly in Europe and the Middle Asia and is used as aromatic, culinary, and medical herb (Kato-Noguchi, 2003). The plant is also used for tea and steeped wine manufactures in Taiwan due to its health benefits. Reports indicated that lemon balm had many beneficial effects such as antibacterial, sedative, spasmolytic, mnemonic improvement, and could reduce excitability, anxiety, stress, gastrointestinal disorders, and sleep disturbance (Mantle, Pickering, & Perry, 2000; Perry et al., 1999Perry, Pickering,Wang, Houghton, & Perry, 1999). Many reports indicated that low polar extract of *Melissa officinalis* (lemon balm) leaves, especially its essential oil, had good antioxidant and antitumor activities (Sousa et al., 2004).The ethanolic extract of lemon balm cultivated in Iran could present good antioxidant activity (Dastmalchi et al., 2008). Capecka, Mareczek, & Leja (2005) found that methanolic extracts of lemon balms cultivated in Poland had good free radical scavenging ability. There is no thorough report concerning anti-proliferative activity of the polar extract from lemon balm leaves for cancer cells. Cyclooxygenase-1 (COX-1) is constitutively expressed in most mammalian tissues and plays a role in tissue homeostasis. COX-2, an inducible isoform, could be stimulated by carcinogens, growth factors, inflammatory cytokines and tumour promoters (Shen, Chen, Zhuang, & Wang, 2008). Abnormal or excessive COX-2 expression has been suggested in many pathological conditions such as angiogenesis, inflammation and tumor promotion (Rao, Hirose, Indranie, & Reddy, 2001). An inhibition of the activity or expression of COX-2 was an important target for anti-inflammation or cancer chemoprevention (Shen et al., 2008). According to Marnett & Du Bois, (2002) cyclooxygenase-1 (COX-1) and cyclooxygenase-2 (COX-2) are rate-limiting enzymes in the biosynthesis of prostaglandins.

Biological activity of *Ajuga bracteosa*

The genus Ajuga comprises of about 40–50 species. *Ajuga bracteosa* is distributed in subtropical and temperate regions from Kashmir to Bhutan, Pakistan, Afghanistan, China, and Malaysia. Its distribution is restricted to northern hilly areas of Pakistan, where it is called kori booti owing to its bitter taste (Arfan, Khan, & Ahmad, 1996).The crude extracts of *Ajuga bracteosa* have shown important biological activities like antidiabetic, antioxidant, antibacterial, diuretic, stimulant (Patel, Gulati, & Gokhale, 1962), astringent, rheumatism, febrifuge, and blood purifier effects (Arfan et al., 1996) on different animal models. The compounds responsible for these activities have not been identified from the plant so far. Several compounds such as glycoside, tannin, ceryl alcohol, cerotic acid, have been isolated from leaves of *Ajuga bracteosa*. The aqueous extract of leaves shows diuretic, stimulant action, aperient and febrifugal (Chopra, Nayar, & Chopra, 1986). According to Verma, Mahmood, & Singh (2002), the roots of *Ajuga bracteosa* contains comparatively larger amounts of chromium which may be correlated to its use as remedy for diabetes. The considerably larger amounts of potassium than sodium may have some correlation with the use of the herb in hypertension. The crude extracts from various parts of the *Ajuga bracteosa* are used to treat different disorders in different Indian traditional systems. The whole plant of *Ajuga bracteosa* afforded five compounds including one new clerodane diterpenoid Bracteonin-A. The other compounds identified were 14, 15-dihydroajugapitin, 14-hydro-15- hydroxyajugapitin, β-sitosterol and stigmasterol (Steinhilber, 1999). Budhiraja, Sudhir, & Garg (1984) described the anti-inflammatory activity of 3β-hydroxy-2, 3-dihydrowithanolide F by testing subacute models of inflammation and found it effective as hydrocortisone. Physangulide and 24, 25-epoxywithanolide D have also been reported to produce anti-inflammatory effects. Benton, (1973) found tested Withaferin-A against cytotoxic KB cell cultures. Significant inhibitory activity against sarcoma 180 tumor in mice and walker intramuscular carcinoma 256 in rats was also shown by withaferin A. It exhibited significant growth retardation of Ehrlich ascites carcinoma in mice at a single dose of 25-40 mg/kg 24 h. Besides, withaferin A and withacnistin several other structurally related withanolides have been reported to exhibit cytotoxic activity. Different extracts of from various parts of *Ajuga bracteosa* have been reported to be active against many bacterial strains. Chatterjee and Chakraborti (1980) studies the antibacterial activity of withaferin A and its compound reported that this constituent active against Gram-positive bacteria.

Biological properties of medicinal plants and their products

The aqueous root extract of *Bryonia dioica* were investigated for antiproliferative, apoptotic induction, and acute toxicity (Benarba, Meddah, & Aoues, 2012). The extract was further screened for preliminary phytochemical analysis. The extract induced cell death in a concentration dependent pattern. The IC_{50} value was found to be of 15.63 μg/mL. Staining of the BL41 cell line with propidium iodide (PI) and treatment of *Bryonia dioica* extract (125 μg/mL) for 24h resulted in induction of apoptosis. Upon treatment with *B. dioica* extract (125 μg/mL), the percentage of BL41 cells undergoing apoptosis increased from 2.7% in untreated cells to 15.7% in treated cells. The extract induced loss of mitochondrial membrane potential, which resulted in membrane disruption and activation of caspases and PARP cleavage. Moreover, the preliminary phytochemical analysis revealed the presence of alkaloids, polyphenols, sterols, triterpenes, saponins, and carbohydrates.

Girish, Vasudevaraju, & Raju (2012) investigated the aqueous extract of black gram husk to study the phytochemical components and its effect on oxidative induced damage to DNA and erythrocytes. The total polyphenolic content in BGH extract was found 59 mg of gallic acid equivalents. For identification of different phenolic acids, RP-HPLC technique was used. The aqueous extract exhibited significant antioxidant potential with IC_{50} value of 3.92 μg of gallic acid equivalent against DPPH free radical scavenging. The extract also demonstrated a glucosidase inhibition with IC_{50} value of 2.78 μg of gallic acid equivalent. Black gram husk aqueous extract inhibited oxidative hemolysis induced by H_2O_2 in dose-dependent pattern with IC_{50} value of 11.5 μg of gallic acid equivalent. Black gram husk aqueous extract protected erythrocyte morphological changes induced by H_2O_2. Deng et al. (2011) isolated two alkaloids (stephanine and crebanine) from *Stephania dielsiana* by using activity directed isolation method. Ten animal pathogenic bacterial strains and eight plant pathogenic fungal strains were used to evaluate the in vitro antimicrobial activity of methanolic extract. Methanolic extract demonstrated highest activity against five Gram-positive bacterial strains and four Gram-negative bacterial strains with MIC value ranging from 0.62 to 7.50 mg/mL. Stephanine and crebanine showed very low activity against Gram-negative bacterial strains but showed high activity against Gram-positive bacterial strains with MIC value ranging from 0.078 to 0.312 mg/mL. Methanolic extract of stephanine and crebanine have shown inhibition against plant pathogenic fungal strains *Cercospora kaki, Pyricularia oryzae, Gymnosporangium haraeanum, Rhizoctonia solani* and *Colletotrichum graminicola*.

Azadmehr et al. (2011) investigated *Scrophularia megalantha* extract for immunomodulatory and anticancer activity under *in vitro* and *in vivo* conditions. Microculture tetrazolium assay (MTT assay) was used to study the toxicity effects of *Scrophularia megalantha* extract on human Jurkat (lymphoblast-like) cell line. Furthermore, hemagglutination and dihydrotestosterone tests (DHT) in mice were employed to evaluate the effects of *Scrophularia megalantha* extract on cellular and humoral responses. Jurkat cells exposed to different concentrations of *Scrophularia megalantha* (0.05, 0.1, and 0.2 mg/mL) significantly suppressed their growth in a dose-dependent manner. Moreover, the production of specific antibody to SRBC antigen in immunized mice significantly increased by different concentrations of *S. megalantha* extract. The methanolic extracts and the fractions of some South African medicinal plants were tested for anticancer activity based on MTT assay (Bisi-Johnson et al., 2011). The methanolic extract of *Eucomis autumnalis* showed the highest *in-vitro* inhibition of proliferation of hepatocarcinoma (Huh-7) cell line with IC_{50} value of 7.8 μg/mL. Butanol and ethyl acetate fractions of *Hypoxis latifolia, Lantana camara* and *Eucomis autumnalis* exhibited lowest anti-proliferation effect with IC_{50} values ranging from 24.8 to 44.1 μg/mL. After 72 hours of treatment, all the fractions of *Aloe striatula* and *Aloe arborescens* have exhibited no or insignificant cytotoxicity.

The ethanolic and aqueous leaf extracts of *Bridelia ferruginea* were investigated by Adetutu, Morgan, & Corcoran (2011) for evaluation of antibacterial activity against bacterial species causing a number of infectious diseases: *Bacillus subtilis, Staphylococcus aureus, Pseudomonas aeruginosa,* and *Escherichia coli.* The ethanolic extract was further studied to assess antioxidant potential and its effect on growth of human dermal fibroblasts (FS5). Both ethanolic and aqueous leaf extracts showed weak antibacterial activity against all bacterial pathogens with MIC value more than 470 μg/mL. Ethanolic extract showed significant increased growth in FS5 fibroblasts at the concentration of 5 μg/mL and above which extract showed toxicity. Ethanolic extract showed significant protection of FS5 cells against hydrogen peroxide induced damage and the results were comparable with catalase (82% at 250 μg/mL concentration). The ethanolic leaf extract also showed high DPPH free radical inhibition with $IC_{50}$of 12.5 μg/mL compared to ascorbic acid (7.3 μg/mL). Mazumder, Das, Das, & Das (2010) first time evaluated the two indigenous medicinal plants of India (*Berberis aristata* and *Hemidesmus indicus*) for in vitro cytotoxicty against human breast cancer cell line (MFC7). The methanolic stem extract of *Berberis aristata* and methanolic rhizome extract of *Hemidesmus indicus* at different test concentrations (12.5–1000 μg/mL) were investigated for antiproliferation activity against MCF7 cell line by trypan blue exclusion test. Methanolic stem extract of *Berberis aristata* showed 89% inhibition of cell growth at 1000 μg/mL concentration with IC_{50} value of 50 μg/mL. While as *Hemidesmus indicus* rhizome extract showed 87% inhibition of cell growth at 1000 μg/mL concentration with IC_{50} value of 48 μg/mL.

Shokeen, Bala, & Tandon (2009) evaluated antibacterial activity of (50%) ethanolic extracts from different parts of 16 medicinal plants against different clinical and WHO strains of *Neisseria gonorrhoeae* including multidrug resistant (MDR) strains. They found that the extracts exhibited better inhibition of MDR strains and differential inhibition against *Neisseria gonorrhoeae.* Among all the extracts tested against *Neisseria gonorrhoeae*, 60% showed higher activity, 20% showed moderate activity and 20% showed weak activity. Chomnawang, Surassmo, Wonsariya, & Bunyapraphatsara (2009) studied antimicrobial activity of 17 Thai medicinal plants against methicillin resistant *Staphylococcus aureus* by using disc diffusion method and broth dilution assay. They found that the extracts *of Barleria lupulina, Psidium guajava, Garcinia mangostana, Tagetes erecta, Hibiscus sabdariffa, Senna alata, Eupatorium odoratum* and *Lawsonia inermis* inhibited the growth of standard *Staphylococcus aureus* bacterial strain. Among these *Garcinia mangostana* extract was found as most potent with highest zone of inhibition (11.3 ± 0.60) and MIC value of 39 μg/mL.

Essential oil was extracted from leaves, flowers, stem, and whole aerial part of *Tamarix boveana* by hydrodistillation method (Saidana et al., 2008). After extraction, the oils were further studied for chemical composition and evaluation of antimicrobial activity. A total of 62 components were identified from whole aerial parts. The major components identified were hexadecanoic acid (18.14%), docosane (13.34%), germacrene D (7.68%), fenchyl acetate (7.34%) and benzyl benzoate (4.11%). The essential oil did not show any activity against fungal strains but showed significant antibacterial activity against all bacterial strains except *Pseudomonas aeruginosa.* Garg et al. (2007) evaluated in vitro anticancer activity of 21 medicinal plant extracts used in Indian traditional medicine against different ailments. The cytotoxicity was evaluated against 6 human cancer cell lines by MTT colorimetric and clonogenic assay. From 21 extracts, 7 extracts showed significant cytotoxicity. *Plectranthus urticioides* and *Garcinea morella* ethanolic leaf and stem extracts showed highest activity against all 6 cancer cell lines.

Natural antimutagenic agents

In last few decades, extensive research was done in order to detect and characterize the antimutagenic compounds from edible, nonedible medicinal plants, and marine organisms. It has been suggested that the natural antimutagens belong to major class of compounds out of which major emphasis has been laid on phenolics, flavonoids, tannins, saponins, carotenoids, terpenoids, and anthraquinones and several other secondary metabolites. According to Boone, Kelloff, & Malone (1990), more than 500 compounds belonging to at least 25 chemical classes have been recognized to possess antimutagenic/protective effects. Some of the major classes of antimutagenic compounds are briefly described further.

Vitamins

Vitamins have been extensively studied for their antimutagenic potential. Antunes and Takahashi, (1998) showed that vitamin C and E possess antimutagenic against doxorubicin induced chromosomal aberrations. Tavan, Maziere, Narbonne, & Cassand, (1997) found that vitamin A, C, and E to be antimutagenic toward Methyl Azoxy Methanol (MAM) induced mutagenesis in *Salmonella typhimurium* strain TA100. Vitamin C (ascorbic acid) was administered concurrently with a pesticide and showed significant decrease in the frequency of pesticide induced mutations. Vitamins B1, B6 and B12 also possess antimutagenic properties against methyl-N-nitro-N-nitrosoguanidine (MNNG) and ethyl-N-nitro-N'- nitrosoguanidine (ENNG) induced mutagenicity both in vitro and in vivo with Ames test (Arriaga-Alba et al., 2013).

Flavonoids

In addition to a wide range of biological activities, flavonoids also possess antimutagenic properties. Flavonoids present an important class of antimutagens and anticarcinogens with high potential. Distinct structure activity relationship was detected when 56 flavonoids, 32 coumarins, 5 naphthoquinones, and 12 anthraquinones were tested for their antimutagenic potencies, with respect to mutagenesis induced by 2-nitrofluoro 3-nitro fluoranthene and 1-nitropyrene in *S. typhimurium* TA98. According to Edenharder and Tang (1997) all flavonoids do not possess antimutagenic activity but all flavones and many flavonoids with phenolic hydroxyl group like leuteolin, kaempherol, etc. exerted antimutagenicity. A number of known flavonoids including flavonoid glycosides and isoflavones were reported to possess significant antimutagenic activity. Calomme, Pieters, Vlietinck, & Vanden (1996) described that flavonoids of citrus juice possess anticarcinogenic and antimutagenic properties. Heo, Yu, Kim, Kim, & Au (1992) tested 14 flavonoids including flavones and flavonol derivatives for their antimutagenic effect against benzo[α]pyrene (Bap) induced micronuclei in mice. Manikumar et al. (1989) isolated two new isoflavones, fremontin, and fremontone, from the root of *Psorothamnus fremontii* which are highly active in the inhibition of mutagenicity of Ethyl Methane Sulfonate (EMS) at all tested concentrations. Antimutagenic effect of hispidulin and hortensin, the flavonoids from *Millingtonia hortensis* was seen when tested against 2-amino anthracene, aflatoxin B1 induced mutation (Chulasiri, Bunyapraphatsara, & Moongkarndi, 1992).

Phenolic compounds

This is a major class of secondary metabolites which are possessing antimutagenic or anticarcinogenic activity besides other biological activities. According to Loarca-Pina, Kuzmicky, de Mejia, Kado, & Hsieh (1996), ellagic acid found in straw berries, rasp berries, grapes, walnuts possess antimutagenic activity. The antigenotoxic potential of tea leaf extracts were studied in Salmonella test (Ohe, Marutani, & Nakase, 2001). Lee et al. (1997) investigated the antimutagenic effect of green tea against smoke-induced mutations in humans and found that green tea can block the cigarette smoking-induced increase in sister chromatid exchange frequency. Soudamini, Unnikrishnan, Sukumaran, & Kuttan (1995) found that curcumin and eugenol found in turmeric and clove, respectively inhibit the mutagenicity induced by direct acting mutagens using *S. typhimuricem* strains TA100 and TA1535 and tobacco-induced mutagenesis in Ames test.

Tannins

The mutagenic activity of various mutagens has been found to be reduced by several tannins. The antimutagenic and anticarcinogenic potential of tannins has been related to their antioxidative property, which is important in protecting cellular oxidative damage including lipid peroxidation (Chung, Wong, Wei, Huang, & Lin, 1998). Sasaki et al. (1990) studied in vivo antimutagenic effect of tannic acid and found that the frequency of micronucleus induction by mitomycin C, ethyl nitrosourea or 4-nitroquinoline-1-oxide decreased by oral administration of tannic acid 6 h before the mutagen in mice. Toering, Gentile, & Gentile, (1996) also reported that catechins, ellagic acid and gallic acid have antimutagenic effect against known mutagens. The anticarcinogenic and antimutagenic potentials of tannins may be related to their antioxidative property in protecting cellular components from oxidative damages, including lipid peroxidation, single-strand DNA breakage, and formation of 8-hydroxydeoxyguanosine. For example, quercetin had been demonstrated to inhibit the generation of superoxide anions by neutrophils (Busse, Kopp, & Elliott, 1984).

Saponins

Saponins are secondary metabolites widely distributed in higher plants but also found in some animal sources like marine invertebrates. Saponins are natural glycosides which possess a wide range of pharmacological properties including cytotoxic activity. Elias et al. (1990) isolated and identified thirteen saponins from *Calendula officinalis*, *Convolvulus arvensis,* and *Hedera helix.* The saponins from *C. arvensis* (four) and *H. helix* (three) showed antimutagenic activity against benzo[a]pyrene and a mutagenic concentrate from a smoker with a dose response relationship in modified liquid incubation technique of the Salmonella assay. Ginseng saponin metabolites introduced by human intestinal bacteria were found antigenotoxic against benzo[a] pyrene-induced clastogenecity (Lee et al., 1998). The anticancer properties of ardisiacrispin (A + B), a mixture of triterpene saponins ardisiacrispin A and B from *Ardisia crenata,* against a number of human cancer cell lines were evaluated. The highest antiproliferative effect of the tested mixture was observed on human hepatoma Bel-7402 cells. Moreover, the mixture induced apoptosis in Bel-7402 cells, which was observed by the changes of the mitochondrial membrane depolarization, membrane permeability enhancement and nuclear condensation, and the effects were dose-dependent (Li et al., 2008).

Carotenoids

Carotenoids are natural fat-soluble pigments that provide bright coloration to plants and animals. Dietary intake of carotenoids is inversely associated with the risk of a variety of cancers in different tissues. They are the pigments that give fruits and vegetables such as carrots, cantaloupe, and sweet potato their vibrant orange, yellow, and green colors. Beta-carotene, lycopene, and lutein are all different varieties of carotenoids. They all act as antioxidants with strong cancer-fighting properties. Antioxidants protect cells from free radicals, substances that work to destroy cell membranes and DNA. Smokers tend to have higher concentrations of free radicals in the blood due to the chemicals they inhale. So, it is no surprise that studies haveconfirmed that antioxidants lower the risk of lung cancer for smokers (Ruano-Ravina, Figueiras, & Barros-Dios, 2000). Several studies have shown that carotenoids affect the activation of promutagens. When water insoluble residues of some carotenoid rich fruits and vegetables was sequentially extracted with several solvents and tested for inhibition of cyclophosphamide (CP)-induced mutagenicity, frequency of CP-induced micronuclei in PCEs in bone-marrow of mice was reduced significantly by the carotenoids, namely, lycopene, canthxanthin, lutein and β-cryptoxanthin (Rauscher, Edenharder, & Platt, 1998). Carotenoids singly or in combination could lower cancer risk due to their antimutagenic properties and ability to scavenge free radicals, to protect against tumor development, and to improve immune response (Chew and Park, 2004; Kawashima, 2011).

Anthraquinones

Anthraquinones are group of functionally diverse aromatic chemicals, structurally related to anthracene, with parent structure 9, 10- dioxoanthracene. Li, Wei, Li, Wu, & Guo (2004) reported antiseptic, diuretic, diarrhoeal, antioxidant, and antimutagenic activities of *Cassia obtusifolia*. Shankel et al. (2000) reported the antimutagenic activity of anthraquinones (aloe-emodin-anthraquinone) isolated from *Aloe barborescence*. Anthrone, one of the structurally related compounds to anthraquinone, exhibit antimutagenic activity and being the most potent one. Edenharder and Tang, (1997) described that although all naphthaquinones were potent antimutagens but plumbagin and 2-methyl-5-hydroxy naphthoquinone showed exceptional antimutagenicity.

Diterpenoids

Diterpenes such as phorbol esters from Croton species have been used in many tumor initiation studies and at low concentrations these compounds are also being explored for antitumor properties (Goel, Makkar, Francis, & Becker, 2007Islam, Rahman, Rahman, Qayum, & Alam, 2010). Connolly, Kitahara, Overtion, & Yoshikoshi (1965) isolated erythroxydiol from *Aquillaria agallocha* and demonstrated that it possesses antimutagenic as well as antitumor activity.

Hormonal steroids

It has been reported that hormonal steroids provide protection against various mutagens. Steroid molecules present in bile acids have been found to be antimutagenic against both direct and indirect acting mutagens in Ames test. Synthetic derivatives of β-estradiol (ethinyloestradiol and mestranol), which are used in contraceptives strongly inhibit the mutagenicity even at nanomolar concentrations (Wilpart, Speder, Ninane, & Roberfroid, 1986). Fahrig (1996) conducted experiments in yeast without an external metabolic activation system and found that the hormones testosterone, β-estradiol, and dicthylstilbesterol were antimutagenic and co-recombinogenic.

Marine products

The secondary metabolites are not found only in plants but are also found in marine organisms. Shankel et al. (2000) found that elatol and obtusol isolated from the extracts of sea hare were antimutagenic as these compounds are halogenated forms of bromine and chlorine. Red and brown algae also contain active halogenated compounds. Wall et al. (1989) isolated cymobarbatol and 4-isocymobarbatol from green algae, *Cymopolia barbata* and studied their antimutagenic activity.

Food products as antimutagens

A wide range of activities are being exhibited by dietary products that can affect mutagenesis. It has been shown in laboratory experiments that naturally occurring substances in food serve as dietary antimutagens. In the process of carcinogenesis, the dietary desmutagens act as tumor growth suppressors. Ferguson, (1994) carried out extensive work on the antimutagenic and anticarcinogenic potential of some common spices and vegetables like turmeric, mustard, green leafy, and *Allium* species. DeMarni (1998) demonstrated that supplementation of diet with vegetables containing carotenoids like carrot, tomato, and spinach significantly decreased the lymphocyte DNA damage. According to Choudhury, Das, Sharma, & Talukder (1997), dietary supplements of garlic and mustard oil showed antimutagenic activity against the clastogenic activity of sodium arsenite. In another study, Ishikawa et al. (1996) found that garlic extract inhibits the mutagenicity of direct acting mutagens like N-methyl-N'-nitro-N-nitrosoguanidine and sodium azide using TA100 and TA1535 strains of *S. typhimuium*. The organosulfur constituents of garlic were found to responsible for its antimutagenic activity. Van Boekel, Goeptae, & Alink, (1997) demonstrated that casein showed a strong antimutagenic activity both in vivo and ex vivo in the DNA repair host-mediated assay and liquid suspension assay, respectively. Bakalinsky, Nadathur, Carney, & Gould (1996) reported that a fermented milk product (yogurt) has antimutagenic activity. Antimutagenic effects of *Psidium guajava* (guava) were reported by Grover and Bala, (1993). Grüter, Friederich, & Würgler (1990) gave the mechanism of antimutagenic activity of mushrooms and found that mushrooms act by direct chemical interaction with the mutagens or inhibition of the activation process in promutagens.

Disease fighting properties of medicinal and aromatic plants

Important sources of new bioactive agents are the natural products. These natural products are obtained from medicinal herbs which are not only being used world-wide for the treatment of various diseases but also have great potential for providing novel drug leads with novel mechanism of action (Dar et al., 2013). According to Lohman, Gentile, Gentile, & Ferguson

(2001), the biomarkers are important in understanding the role of both carcinogens and anticarcinogens in human cancer. For the positive use of natural products as therapeutic and chemopreventive agents, it is necessary to explore more thoroughly their real antimutagenic potential in vivo in connection with other antimutagenic factors. Though, the antagonistic role of any natural compound against the genotoxic compounds was waiting to be unravelled and the present study fills the lacunae in this field. According to Mitscher, Telikepalli, McGhee, & Shankel (1996), majority of higher plants contain a number of agents or phytoconstituents that are capable of causing mitigating effects to a number of mutagens. As the antimutagenic studies are getting credence, there are many studies to show the rising trends of antimutagenicity. The plant *Arabidopsis thaliana* was utilized to screen the effects of various antimutagens (e.g., thiourea, cysteine, 9-hydroxyellipticine, phenolic agents) against chemically induced embryonic and chlorophyll mutation (Gichner, Velemínský, Pospíšil, 1985; Gichner and Velemínský, 1986). Vinitketkumnuen et al. (1994) found antimutagenic effects of *Cymbopogon citratus* Stapf (Lemon grass) against known mutagens in *Salmonella typhimurium* strains. Denadai et al. (1998) described the protective effects of mushroom, *Agaricus blazei* teas, in vivo against cyclophosphamide-induced mutagenicity in mice and found that three different tea extracts of mushroom, significantly reduced the frequencies of MN in polychromatic erythrocytes andin reticulocytes. Marnewick, Gelderblom, & Joubert (2000) found that the aqueous extracts of fermented and unfermented rooibos tea (*Aspalathuslinearis*) and honey-bush tea (*Cyclopia intermedia*) possess antimutagenic activity against 2-acetylaminofluorine and aflatoxin B_1. Vitamin C and E also significantly reduced the CA frequency in mouse bone marrow cells against rifampicin, an anti-tuberculosis drug, (Aly & Donia, 2002).

According to Kaur, Kalotra, Walia, & Handa (2013), the phytoconstituents from *Terminalia arjuna* suppressed the mutagenic effect of the aromatic amine, that is, 2-aminofluorene (2-AF). The observed activity caused the inhibition of the metabolic activation of promutagens. Hong, Cho, Jang, & Lyu (2011) found that the extracts of *Acanthopanaxdivaricatus* were able to rapidly eliminate the mutagenic compounds from the cells before they induce the DNA damage. In a similar study, Nardemir et al. (2015) observed that the methanol extracts of the lichens have antimutagenic effects against sodium azide. Durnova and Kurchatova (2014) also demonstrated the antimutagenic properties of plant extracts against cyclophosphamide-induced mutagenicity in mice. In another study, Prakash, Hosetti, & Dhananjaya (2014) found that the different extracts of *Dioscorea pentaphylla* significantly inhibited the effects of methyl methanesulfonate (MMS)-induced mutagenicity. They also found that the methanolic extract was highly mutagenic in comparison to Petroleum ether and chloroform. Entezari, Dabaghian, & Hasemi (2014) compared the antimutagenic and anticancer activities of *Echinophora platyloba* DC on acute promyelocytic leukemia cancer cells and found that the methanolic extract of this plant prevented the reverted mutations and the hindrance was 93.4% in antimutagenic test. Akinboro, Mohamed, Asmawi & Yekeen (2014) utilized the leaves of *Myristica fragrans* (Houtt.) for antimutagenic activity against benzo[a]pyrene and cyclophosphamide induced mutagenicity in *Salmonella typhimurium* and *Mus musculus* and found that the aqueous extract significantly suppressed more than 50 % of the mutations in all the tested concentrations. Sarac (2015) utilized an edible wild plant, *Tragopogon longirostis* for the evaluation of antioxidant, mutagenic and antimutagenic properties and found that the ethanolic extract of its leaves exhibited antimutagenic properties at 2.5, 0.25, and 0.025 mg/plate concentrations. Habibi et al. (2014) found that the ethanolic extract of *Origanum vulgare* reduced the frequency of MN PCR from 10.52 ± 1.07 for CP to 2.17 ± 0.6 for the synergic test of CP and the ethanolic extract.

The results of the present study clearly showed that all the extracts of both the two plants, *Ajuga bracteosa* and *Melissa officinalis*, had an antimutagenic and anticlastogenic potential against the EMS-induced mutagenicity in mice. Among the two plants, *Melissa officinalis* showed highest antimutagenic activity against EMS-induced mutagenicity in mice. All the extracts suppressed the action of EMS as measured in CA and MN tests. The simultaneous treatment showed protective activity against two end points. The antimutagenic effect of extracts in the CA and MN could be due to reduced induction of damage or increased repair. Menoli et al. (2001) get positive results while applying three experimental designs (simultaneous, pre- and post-treatment) against MMS-induced mutagenicity for a better investigation of possible antimutagenic mechanisms of *Agaricus blazei* extracts on Chinese hamster V79 cells by comet assay and MN test. The posttreatment could show its antimutagenic potential by playing a role in optimization of DNA repair. The conclusion obtained in the pretreatment could reflect the effects on the prevention of DNA damage by affecting metabolic pathways, being antioxidant or acting on DNA replication. These action mechanisms, occurring in both pre- and post-treatments, could be called bio-antimutagenicity (Morita, Hara, & Kada, 1978). The use of a simultaneous treatment appeared to identify mechanisms with direct action on the mutagen by inactivating it, which may be classified as desmutagenicity effect (Kada, 1983).

Since inpresent investigations the mice were simultaneously treated with different extracts of *Ajuga bracteosa* and *Melissa officinalis* extract(s) and EMS that may be simply preventing the uptake of EMS and reducing its subsequent toxicity. Waters, Brady, Stack, & Brockman (1990) reported that some drugs, dietary components and endogenous biochemicals can function as antimutagen by altering the rates of mutagen absorption and uptake.

However, the present experimental design does not rule out the possibility of indirect effects of the different extract(s) of *Ajuga bracteosa* and *Melissa officinalis* against EMS-induced mutagenicity in mice. There are various antimutagenic agents that act through multiple mechanisms to provide protection against diverse mutagens. It is worth mentioning that the ability of compounds to affect mutagens simultaneously in several different ways significantly increases antimutagenic effectiveness. Hence, searching for such multifunctionally acting antimutagens is of great importance. The GC-MS analysis of methanolic extracts of *Ajuga bracteosa* and *Melissa officinalis* showed that both the plants contain three major bioactive constituents: 2, 3-dihydro-3, 5-dihydroxy-6-methyl-4H-pyran-4-one (DDMP), 5-(hydroxyl- methyl)-2-furancarboxaldehyde (5HMF), and hexadecanoic acid, methyl ester. All these compounds are likely to possess potent antimutagenic activity. The major constituent DDMP present in the methanolic extracts of *Ajuga bracteosa* and *Melissa officinalis* is a Millard reaction product of glucose and glycine (Hiramoto, Nasuhara, Michikoshi, Kato, & Kikugawa, 1997), having antimutagenic activity against arylamine (Berhow, Wagner, Vaughn, & Plewa, 2000; Xie et al., 2010). This DDMP isolated from onion in one of the previous studies have modulated the activity of NF-κB (nuclear factor kappa-light-chain enhancer of activated B cells) thereby inducing the apoptotic cell death of cancer cells (Ban et al., 2007). Recently, it was isolated and identified as a potent antioxidant from *Pyruspyrifolia* Nakai (Hwang et al., 2013), supporting some recent studies which showed a strong correlation between antioxidant and anticancer activity (Sengottuvelan, Deeptha, Nalini, 2009; Aqil et al., 2012; Del-Toro-Sánchez et al., 2014; Sarac, 2015).

According to Odin (1997), the mechanism of antimutagenicity of many phytoconstituents and vitamins is mainly connected with their free radical scavenging property and to a lesser extent with nucleophilicity. Many of the substances that are antimutagens are also having antioxidant activity. It is not clear whether oxidative reactions activate nonmutagenic or benign chemicals to mutagenic. Antioxidants interfere with these actions either by interfering with the activation or oxidation of the chemical, or by inhibiting the movement of electrons or the oxygen radical. Nevertheless, the oxidative processes are necessary for the health and survival of an organism; there may be a fine line between preventing the oxidative activation of a mutagen and permitting the essential flow of electrons in a healthy cell (Zeiger, 2000). Antioxidants are very effective in preventing oxidation because they are free-radical scavengers. The free radicals are extremely active chemical species and are produced in diverse processes. These free radicals not only react with vital molecules of the cell but also may cross link with DNA and thus alter its structure (Grover and Bala, 1993; Gajowik & Dobrzyńsk, 2014). An important compound 5- (hydroxymethyl)-2-furan carboxaldehyde (5-HMF) was also isolated from various plants and was tested for various important biological activities like antioxidant, uterotonic, antiplatelet aggregation and radical scavenging activity (Sewram, Raynor, Mulholland, & Raidoo, 2001Pyo, Jin, Koo, & Yun-Choi, 2004; Fu, Wang, & Cai, 2008Luo, Zhao, Yang, Shen, & Rao 2009; Zhao et al., 2013; Liu et al., 2014). The new derivatives of furan including 5-HMF have been reported to possess anticancer properties by inhibition of tubulin polymerization (Uckun and Shyi-Tai Jan, 2003). According to Ying, Kevin, Weihan, Xiaoqiang, & Jianrong (2005) 5-HMF and its derivatives potentially inhibit tumor necrosis factor alpha or interleukin-1 beta expression, thus strongly suggesting that 5-HMF might have an exciting antimutagenic potential.

Hexadecanoic acid, methyl ester was also present in both the methanolic extracts of *Ajuga bracteosa* and *Melissa officinalis* in the minor peaks. It has been reported that this compound possess antiinflammatory, antioxidant, hypocholesterolemic, 5-alpha reductase inhibitor, nematicide, antibacterial, antifungal, antiandrogenic, antifibrinolytic, and antialopecic activity (Kumar, Kumaravel, & Lalita, 2010; Othman et al., 2015Othman, Abdullah, Ahmad, Ismail, & Zakaria, 2015Shah, Gnanaraj, Khan, & Iqbal, 2015). Some studies have shown that methyl ester derivatives possess antitumor activity and exhibit potent cytotoxicity in the human cancer screening program (Jin, You, & Ahn, 2001; Whelan and Ryan, 2003; Yu et al., 2005; Hajdu et al., 2014). When the methyl ester compound was applied to human gastric cancer cells, the inhibitory and apoptosis rates were significantly increased (Yu et al., 2005). Hexadecanoic acid, methyl ester isolated in this study is a low-molecular weight polymer and is expected to confer a relatively high hydrophilicity to molecules, one factor that might be responsible for the enhancement of cytotoxic effect on tumor cells, justifying its role as a potent antimutagen. The other compounds identified by GC-MS analysis in the methanolic fractions of both plants were β-sitosterol and stigmasterol-acetate. The β-sitosterol is known to be effective against a number of cancers like human breast cancer (Awad, Downie, & Fink, 2000), colon carcinoma (Manayi et al., 2013), and prostatic cancer (Jourdain, Tenca, Deguercy, Troplin, & Poelman, 2006). The β-sitosterol alsoinhibits the proliferation of breast cancer cells in a dose dependent manner (Chai, Kuppusamy, & Kanthimathi, 2008). The authors revealed that a higher caspase activity after adding β-sitosterol to the cell line resulted in caspase-induced apoptosis. Besides, the compound also showed antiproliferativeand apoptosis activities in human leukemic cells by activating caspase-3 and Bax/Bcl-2 ratio (Park et al., 2007). Stigmasterol is known to possess many important biological activities like antihypercholesterolemic, antimutagenic, antileishmanial, antimalarial, antitrypanosomal, platelet aggregation inhibitor, and antiviral (Barnes, Anderson, & Phillipson, 2007Zhou, Xie, & Yan, 2011; Ahmed, Abd-Rabou, Hassan, & Kotob, 2014). Previous in vitro studies confirm that plant extracts rich in stigmasterol and β-sitosterol are cytotoxic against HepG2 (liver), Caco-2 (colon) and MDA-MB-231 (breast) cancer cell lines (Rahmat et al., 2006; Yaacob et al., 2010), thus suggesting that these compounds are effective drugs for carcinogenesis.

The significant reduction in the MN and CA in the methanolic treated group in the study might be due to different mechanism of action of the abovementioned compound, or due to the synergetic effect of various compounds. Many studies support that phenolic compounds are capable of protecting biological systems in various ways (Edenharder, Sager, Glatt, Muckel, & Platt, 2002; Kelly, Jewell, & O'Brien, 2003). According to Szaefer, Cichocki, Brauze, & Baer-Dubowska (2004), the phenolic compounds have a dual effect on phase I and phase II enzymes. They can repress some mainly in phase I while stimulate others mainly in phaseII. Doostdar, Burke, & Mayer (2000) found that a flavonoid, hesperetin, caninhibit human cytochrome P450, thereby reducing the absorption or elimination of toxic compounds. Kelly et al. (2003) suggested that phenolic compounds, limonoids, stimulates the detoxifying enzyme, gluthatione S-transferase, thereby facilitating the elimination of toxic compounds, resulting in decrease the effect of toxic chemicals. Franke, Pra, Erdtmann, Henriques, & da silva (2005), with the aim of evaluating whether orange juice could reduce DNA damage induced by EMS in mice, showed that, under their experimental conditions, this really occurred. The authors showed that the components of orange juice are biologically effective, including in the role of targets for toxicants and inmodulating metabolization/detoxification routes. It is likely that phenolic compounds can be methylated by alkylating agents, instead of conjugation enzymes, thereby protecting reducing DNA from alkylation. EMS can methylate nucleophilic regions of DNA, as well as amino acid molecules, especially in nitrogen atoms. EMS-induced genotoxicity is mediated by base modifications, which weaken the N-glycosylic bond, thereby leading to depurination/depyrimidination of DNA strands and the appearance of alkali-labile basic sites (AP sites). The removal of AP sites by AP endonucleases cleaves the DNA adjacent to these sites, thereby generating DNA strand breaks (Horváthová et al., 1998; Boiteux and Guillet, 2004; Franke et al., 2005).

References

Abo, K. A., Adeyemi, A. A., & Adeite, D. A. (2000). Ethnobotanical survey of plants used in the treatment of infertility and sexually transmitted diseases in southwest Nigeria. *African Journal of Medicine and Medical Sciences*, *29*(3–4), 325–327.

Adebayo, J. O., & Krettli, A. U. (2011). Potential antimalarials from Nigerian plants: a review. *Journal of Ethnopharmacology*, *133*(2), 289–302.

Adetutu, A., Morgan, W. A., & Corcoran, O. (2011). Antibacterial, antioxidant and fibroblast growth stimulation activity of crude extracts of Bridelia ferruginea leaf, a wound-healing plant of Nigeria. *Journal of Ethnopharmacology*, *133*(1), 116–119.

Ahmad, I., Mehmood, Z., & Mohammad, F. (1998). Screening of some Indian medicinal plants for their antimicrobial properties. *Journal of Ethnopharmacology*, *62*(2), 183–193.

Ahmad, M. S., & Afzal, M. (2004). Amelioration of genotoxic damage by certain phytoproducts in human lymphocyte cultures. *Chemico-Biological Interactions*, *149*(2), 107–115.

Ahmed, H. H., Abd-Rabou, A. A., Hassan, A. Z., & Kotob, S. E. (2014). Phytochemical analysis and anti-cancer investigation of *Bswellia serrata* bioactive constituents in vitro. *Asian Pacific Journal of Cancer Prevention*, *16*(16), 7179–7188.

Akhondzadeh, S., Noroozian, M., Mohammadi, M., Ohadinia, S., Jamshidi, A. H., & Khani, M. (2003). Melissa officinalis extract in the treatment of patients with mild to moderate Alzheimer's disease: a double blind, randomised, placebo controlled trial. *Journal of Neurology, Neurosurgery and Psychiatry*, *74*(7), 863–866.

Akinboro, A., Mohamed, K. B., Asmawi, M. Z., & Yekeen, T. A. (2014). Antimutagenic effects of aqueous fraction of Myristica fragrans (Houtt.) leaves on *Salmonella typhimurium* and Mus musculus. *Acta Biochimica Polonica*, *61*(4), 779–785.

Aleryani, S. L., Cluette-Brown, J. E., Khan, Z. A., Hasaba, H., de Heredia, L. L., & Laposata, M. (2005). Fatty acid methyl esters are detectable in the plasma and their presence correlates with liver dysfunction. *Clinica Chimica Acta*, *359*(1), 141–149.

Aly, F. A., & Donya, S. M. (2002). In vivo antimutagenic effect of vitamins C and E against rifampicin-induced chromosome aberrations in mouse bone-marrow cells. *Mutation Research/Genetic Toxicology and Environmental Mutagenesis*, *518*(1), 1–7.

Ames, B. N., & Gold, L. S. (1991). Endogenous mutagens and the causes of aging and cancer. *Mutation Research/Fundamental and Molecular Mechanisms of Mutagenesis*, *250*(1), 3–16.

Amin, A.H. and Mehta, D.R. (1959).A bronchodilator alkaloid (vasicinone) from *Adhatoda vasica* Nees.

Ankli, A., Heinrich, M., Bork, P., Wolfram, L., Bauerfeind, P., Brun, R., & Wasescha, M. (2002). Yucatec Mayan medicinal plants: evaluation based on indigenous uses. *Journal of Ethnopharmacology*, *79*(1), 43–52.

Antunes, L. M. G., & Takahashi, C. S. (1998). Effects of high doses of vitamins C and E against doxorubicin-induced chromosomal damage in Wistar rat bone marrow cells. *Mutation Research/Genetic Toxicology and Environmental Mutagenesis*, *419*(1), 137–143.

Aqil, F., Gupta, A., Munagala, R., Jeyabalan, J., Kausar, H., Sharma, R. J., & Gupta, R. C. (2012). Antioxidant and antiproliferative activities of anthocyanin/ellagitannin enriched extracts from *Syzygium cumini* L. (Jamun, the Indian Blackberry). *Nutrition and Cancer, 64*(3), 428–438.

Arceusz, A., Wesolowski, M., & Ulewicz-Magulska, B. (2015). Flavonoids and Phenolic Acids in Methanolic Extracts, Infusions and Tinctures from Commercial Samples of Lemon Balm. *Natural Product Communications, 10*(6), 977–981.

Arfan, M., Khan, G. A., & Ahmad, N. (1996). Steroids and terpenoids of the genus Ajuga. *Journal of the Chemical Society of Pakistan, 18*(2), 170–174.

Arriaga-Alba, M., Ruiz-Pérez, N. J., Sánchez-Navarrete, J., de Angel, B. L., Flores-Lozada, J., & Blasco, J. L. (2013). Antimutagenic evaluation of vitamins B1, B6 and B12 in vitro and in vivo with the Ames test. *Food and Chemical Toxicology, 53*, 228–234.

Aruoma, O. I. (2003). Methodological considerations for characterizing potential antioxidant actions of bioactive components in plant foods. *Mutation Research/Fundamental and Molecular Mechanisms of Mutagenesis, 523*, 9–20.

Awad, A. B., Downie, A. C., & Fink, C. S. (2000). Inhibition of growth and stimulation of apoptosis by beta-sitosterol treatment of MDA-MB-231 human breast cancer cells in culture. *International Journal of Molecular Medicine, 5*(5), 541–546.

Ayyanar, M., & Ignacimuthu, S. (2008). Medicinal uses and pharmacological actions of five commonly used Indian medicinal plants: A mini-review. *Iranian Journal of Pharmacology and Therapeutics*, 7(1), 107–114.

Azadmehr, A., Hajiaghaee, R., Afshari, A., Amirghofran, Z., Refieian-Kopaei, M., Yousofi Darani, H., & Shirzad, H. (2011). Evaluation of in vivo immune response activity and in vitro anti-cancer effect by Scrophularia megalantha. *J Med Plants Res., 5*(11), 2365–2368.

Bağdat, R. B., & Cos¸ge, B. (2006). The essential oil of lemon balm (*Melissa officinalis* L.), its components and using fields. *Journal of Faculty of Agriculture, OMU, 21*(1), 116–121.

Bailly, C. (2009). Ready for a comeback of natural products in oncology. *Biochemical Pharmacology*, 77(9), 1447–1457.

Bakalinsky, A. T., Nadathur, S. R., Carney, J. R., & Gould, S. J. (1996). Antimutagenicity of yogurt. *Mutation Research/Fundamental and Molecular Mechanisms of Mutagenesis, 350*(1), 199–200.

Ban, J. O., Hwang, I. G., Kim, T. M., Hwang, B.Y., Lee, U. S., Jeong, H. S., & Hong, J. T. (2007). Anti-proliferate and pro-apoptotic effects of 2, 3-dihydro-3, 5-dihydroxy-6methyl-4H-pyranone through inactivation of NF-κB in human colon cancer cells. *Archives of Pharmacal Research, 30*(11), 1455–1463.

Barnes, J., Anderson, L. A., & Phillipson, J. D. (2007). *Herbal Medicines*. London, UK: Published by the Pharmaceutical Press, RPS Publishing.

Benarba, B., Meddah, B., & Aoues, A. (2012). Bryonia dioica aqueous extract induces apoptosis through mitochondrial intrinsic pathway in BL41 Burkitt's lymphoma cells. *Journal of Ethnopharmacology, 141*(1), 510–516.

Benton, W. (1973). *The new Encyclopedia Britannica., 74*, 6.

Berhow, M. A., Wagner, E. D., Vaughn, S. F., & Plewa, M. J. (2000). Characterization and antimutagenic activity of soybean saponins. *Mutation Research/Fundamental and Molecular Mechanisms of Mutagenesis, 448*(1), 11–22.

Bhojwani, S.S. and Arumugam, N. (1993). Advances in Developmental Biology and Biotechnology of Higher Plants (eds Soh et al.), The Korean Society of Plant Tissue Culture, Korea, pp. 110-127.

Bisi-Johnson, M. A., Obi, C. L., Hattori, T., Oshima, Y., Li, S., Kambizi, L., & Vasaikar, S. D. (2011). Evaluation of the antibacterial and anticancer activities of some South African medicinal plants. *BMC Complementary and Alternative Medicine, 11*(1), 14.

Blumenthal, M., Goldberg, A., & Brinckmann, J. (2000). Herbal medicine. expanded commission E monographs. *Integrative Medicine Communications* pp. xiii + 519 pp.

Boiteux, S., & Guillet, M. (2004). Abasic sites in DNA: repair and biological consequences in *Saccharomyces cerevisiae*. *DNA Repair*, *3*(1), 1–12.
Bolkent, S., Yanardag, R., Karabulut-Bulan, O., & Yesilyaprak, B. (2005). Protective role of *Melissa officinalis* L. extract on liver of hyperlipidemic rats: a morphological and biochemical study. *Journal of Ethnopharmacology*, *99*(3), 391–398.
Boone, C. W., Kelloff, G. J., & Malone, W. E. (1990). Identification of candidate cancer chemopreventive agents and their evaluation in animal models and human clinical trials: a review. *Cancer Research*, *50*(1), 2–9.
Brown, J. E., & Rice-Evans, C. A. (1998). Luteolin-rich artichoke extract protects low density lipoprotein from oxidation in vitro. *Free Radical Research*, *29*(3), 247–255.
Budhiraja, R. D., Sudhir, S., & Garg, K. N. (1984). Antiinflammatory activity of 3 βHydroxy-2, 3-dihydro-withanolide F. *Planta Medica*, *50*(2), 134–136.
Busse, W. W., Kopp, D. E., & Elliott, M. (1984). Flavonoid modulation of human neutrophil function. *Journal of Allergy and Clinical Immunology*, *73*(6), 801–809.
Calomme, M., Pieters, L., Vlietinck, A., & Vanden, B. D. (1996). Inhibition of bacterial mutagenesis by Citrus flavonoids. *Planta Medica*, *62*(3), 222–226.
Canadanovic-Brunet, J., Cetkovic, G., Djilas, S., Tumbas, V., Bogdanovic, G., Mandic, A., & Canadanovic, V. (2008). Radical scavenging, antibacterial, and antiproliferative activities of *Melissa officinalis* L. extracts. *Journal of Medicinal Food*, *11*(1), 133–143.
Capecka, E., Mareczek, A., & Leja, M. (2005). Antioxidant activity of fresh and dry herbs of some Lamiaceae species. *Food Chemistry*, *93*(2), 223–226.
Chai, J. W., Kuppusamy, U. R., & Kanthimathi, M. S. (2008). Beta-sitosterol induces apoptosis in MCF-7 cells. *Malaysian Journal of Biochemistry and Molecular Biology*, *16*(2), 28–30.
Chakraborty, A., & Brantner, A. H. (2001). Study of alkaloids from Adhatoda vasica Nees on their anti-inflammatory activity. *Phytotherapy Research*, *15*(6), 532–534.
Chatterjee, S., & Chakraborti, S. K. (1980). Antimicrobial activities of some antineoplastic and other withanolides. *Antonie van Leeuwenhoek*, *46*(1), 59–63.
Chen, H. H., Chiang, W., Chang, J. Y., Chien, Y. L., Lee, C. K., Liu, K. J., & Kuo, C. C. (2011). Antimutagenic constituents of adlay (Coix lachryma-jobi L. var. ma-yuen Stapf) with potential cancer chemopreventive activity. *Journal of Agricultural and Food Chemistry*, *59*(12), 6444–6452.
Chew, B. P., & Park, J. S. (2004). Carotenoid action on the immune response. *The Journal of Nutrition*, *134*(1), 257–261.
Chomnawang, M. T., Surassmo, S., Wongsariya, K., & Bunyapraphatsara, N. (2009). Antibacterial activity of Thai medicinal plants against methicillin-resistant Staphylococcus aureus. *Fitoterapia*, *80*(2), 102–104.
Chopra, R.N., Nayar, S.L. and Chopra, I.C. (1986). Glossary of Indian Medicinal Plants. Council of Scientific and Industrial Research: New Delhi.
Choudhury, A. R., Das, T., Sharma, A., & Talukder, G. (1997). Inhibition of clastogenic effects of arsenic through continued oral administration of garlic extract in mice in vivo. *Mutation Research/Genetic Toxicology and Environmental Mutagenesis*, *392*(3), 237–242.
Chulasiri, M., Bunyapraphatsara, N., & Moongkarndi, P. (1992). Mutagenicity and antimutagenicity of hispidulin and hortensin, the flavonoids from *Millingtonia hortensis* L. *Environmental and Molecular Mutagenesis*, *20*(4), 307–312.
Chung, K. T., Wong, T. Y., Wei, C. I., Huang, Y. W., & Lin, Y. (1998). Tannins and human health: a review. *Critical Reviews in Food Science and Nutrition*, *38*(6), 421–464.
Connolly, J. D., Kitahara, K. H., Overtion, K. H., & Yoshikoshi, A. (1965). A direct correlation of dolabradiene and erythroxydiolY. *Chemical and Pharmaceutical Bulletin*, *13*(5), 603–605.
Dar, S. A., Ganai, F. A., Yousuf, A. R., Balkhi, M. U. H., Bhat, T. M., & Sharma, P. (2013). Pharmacological and toxicological evaluation of Urtica dioica. *Pharmaceutical Biology*, *51*(2), 170–180.

Dar, S. A., Yousuf, A. R., Ganai, F. A., Sharma, P., Kumar, N., & Singh, R. (2012). Bioassay guided isolation and identification of anti-inflammatory and anti-microbial compounds from Urtica dioica L. *(Urticaceae) leaves. African Journal of Biotechnology*, *11*(65), 12910–12920.

Dastmalchi, K., Dorman, H. D., Oinonen, P. P., Darwis, Y., Laakso, I., & Hiltunen, R. (2008). Chemical composition and in vitro antioxidative activity of a lemon balm (Melissa officinalis L.) extract. *LWT-Food Science and Technology*, *41*(3), 391–400.

Dawson, B. S., Franich, R. A., & Meder, R. (1988). Essential oil of *Melissa officinalis* L. subsp. altissima (Sibthr. et Smith) Arcang. *Flavour and Fragrance Journal*, *3*(4), 167–170.

Dean, R. T., & Cheeseman, K. H. (1987). Vitamin E protects proteins against free radical damage in lipid environments. *Biochemical and Biophysical Research Communications*, *148*(3), 1277–1282.

Del-Rio, A., Obdululio, B. G., Casfillo, J., Main, F. G., & Ortuno, A. (1997). Uses and properties of citrus flavonoids. *J. Agric. Food Chem*, *45*, 4505–4515.

Del-Toro-Sánchez, C. L., Bautista-Bautista, N., Blasco-Cabal, J. L., Gonzalez-Ávila, M., Gutiérrez-Lomelí, M., & Arriaga-Alba, M. (2014). Antimutagenicity of methanolic extracts from Anemopsis californica in relation to their antioxidant activity. *EvidenceBased Complementary and Alternative Medicine*, 2014.

DeMarini, D. M. (1998). Dietary interventions of human carcinogenesis. *Mutation Research/ Fundamental and Molecular Mechanisms of Mutagenesis*, *400*(1), 457–465.

Denadai, R., Alves de Lima, P.L., Salvadore, D., Eira, A.F., Bazo, A.P.,& Ribeiro, L.R. (1998). The protective effect of mushroom (*Agaricus blazei*) teas on the genotoxity induced by cyclophosphamide. In Congresso Latinoamericano De Mutagênese E Teratogênese Ambiental (Vol. 5, p. 247).

Deng, Y., Yu, Y., Luo, H., Zhang, M., Qin, X., & Li, L. (2011). Antimicrobial activity of extract and two alkaloids from traditional Chinese medicinal plant Stephania dielsiana. *Food Chemistry*, *124*(4), 1556–1560.

Dhuley, J. N. (1999). Antitussive effect of Adhatoda vasica extract on mechanical or chemical stimulation-induced coughing in animals. *Journal of Ethnopharmacology*, *67*(3), 361–365.

Dhyani, D., Maikhuri, R. K., Rao, K. S., Kumar, L., Purohit, V. K., Sundriyal, M., & Saxena, K. G. (2007). Basic nutritional attributes of Hippophae rhamnoides (Seabuckthorn) populations from Uttarakhand Himalaya India. *Current Science*, *92*(8), 1148–1152.

Doostdar, H., Burke, M. D., & Mayer, R. T. (2000). Bioflavonoids: selective substrates and inhibitors for cytochrome P450 CYP1A and CYP1B1. *Toxicology*, *144*(1), 31–38.

Durnev, A. D., & Seredenin, S. B. (1990). Antioxidants as means of protecting the genetic apparatus. *Pharmaceutical Chemistry Journal*, *24*(2), 71–82.

Durnova, N. A., & Kurchatova, M. N. (2014). The effect of plant extracts on the cyclophosphamide induction of micronucleus in red blood cells of outbred whiteman mice. *Tsitologiia*, *57*(6), 452–458.

Dzeufiet, P. D. D., Mogueo, A., Bilanda, D. C., Aboubakar, B. F. O., Tédong, L., Dimo, T., & Kamtchouing, P. (2014). Antihypertensive potential of the aqueous extract which combine leaf of Persea americana Mill. (Lauraceae), stems and leaf of Cymbopogon citratus (DC) Stapf. (Poaceae), fruits of Citrus medical L. (Rutaceae) as well as honey in ethanol and sucrose experimental model. *BMC Complementary and Alternative Medicine*, *14*(1), 1–12.

Edenharder, R., & Tang, X. (1997). Inhibition of the mutagenicity of 2-nitrofluorene, 3nitrofluoranthene and 1-nitropyrene by flavonoids, coumarins, quinones and other phenolic compounds. *Food and Chemical Toxicology*, *35*(3), 357–372.

Edenharder, R., Sager, J. W., Glatt, H., Muckel, E., & Platt, K. L. (2002). Protection by beverages, fruits, vegetables, herbs, and flavonoids against genotoxicity of 2 acetylaminofluorene and 2-amino-1-methyl-6-phenylimidazo [4, 5-b] pyridine (PhIP) in metabolically competent V79 cells. *Mutation Research/Genetic Toxicology and Environmental Mutagenesis, 521*(1), 57–72.

El-Ashmawy, I. M., Ashry, K. M., El-Nahas, A. F., & Salama, O. M. (2006). Protection by turmeric and myrrh against liver oxidative damage and genotoxicity induced by lead acetate in mice. *Basic and Clinical Pharmacology and Toxicology, 98*(1), 32–37.

Elias, R., De Meo, M., Vidal-Ollivier, E., Laget, M., Balansard, G., & Dumenil, G. (1990). Antimutagenic activity of some saponins isolated from *Calendula officinalis* L., *C. arvensis* L. and *Hedera helix* L. *Mutagenesis, 5*(4), 327–332.

El-Sayed, W. M., & Hussin, W. A. (2013). Antimutagenic and antioxidant activity of novel 4-substituted phenyl-2, 2'-bichalcophenes and aza-analogs. *Journal of Drug Design, Development and Therapy, 7*, 73–81.

Entezari, M., Dabaghian, F. H., & Hashemi, M. (2014). The comparison of antimutagenicity and anticancer activities of *Echinophora platyloba* DC on acute promyelocytic leukemia cancer cells. *Journal of Cancer Research and Therapeutics, 10*(4), 1004–1007.

Fahmy, K., Tantawi, T. A., & Awad, A. A. M. (1997). A genetic assay for detecting of aneuploidy in the female germ-line cells of Drosophila. *Annals of Agricultural Science (Egypt)., 42*(2), 513–523.

Fahrig, R. (1996). Anti-mutagenic agents are also co-recombinogenic and can be converted into co-mutagens. *Mutation Research/Fundamental and Molecular Mechanisms of Mutagenesis, 350*(1), 59–67.

Ferdous, A. J., Islam, S. M., Ahsan, M., Hassan, C. M., & Ahmad, Z. V. (1992). In vitro antibacterial activity of the volatile oil of Nigella sativa seeds against multiple drug resistant isolates of Shigella spp. and isolates of *Vibrio cholerae* and *Escherichia coli*. *Phytotherapy Research, 6*, 137–140.

Ferguson, L. R. (1994). Antimutagens as cancer chemopreventive agents in the diet. *Mutation Research/Fundamental and Molecular Mechanisms of Mutagenesis, 307*(1), 395–410.

Florence, A. R., Joselin, J., & Jeeva, S. (2012). Intraspecific variation of bioactive principles in select members of the genus *Clerodendrum* L. *Journal of Chemical and Pharmaceutical Research, 11*, 4908–4914.

Florence, A. R., Joselin, J., Brintha, T. S. S., Sukumaran, S., & Jeeva, S. (2014). Preliminary phytochemical studies of select members of the family Annonaceae for bioactive constituents. *Bioscience Discovery, 5*(1), 85–96.

Franke, S. I. R., Prá, D., Erdtmann, B., Henriques, J. A. P., & da Silva, J. (2005). Influence of orange juice over the genotoxicity induced by alkylating agents: an in vivo analysis. *Mutagenesis, 20*(4), 279–283.

Fu, Z. Q., Wang, M. Y., & Cai, B. C. (2008). Discussion of 5-hydroxymethylfurfural (5HMF) in Chinese native medicine research present situation. *Chinese Archives of Traditional Chinese Medicine, 26*, 508–510.

Gajowik, A., & Dobrzyńska, M. M. (2014). Lycopene-antioxidant with radioprotective and anticancer properties: a review. *Roczniki Państwowego Zakładu Higieny, 65*(4).

Garg, A., Darokar, M. P., Sundaresan, V., Faridi, U., Luqman, S. R., & Khanuja, S. P. S. (2007). Anticancer activity of some medicinal plants from high altitude evergreen elements of Indian Western Ghats. *Journal of Research and Education in Indian Medicine, 13*(3), 1–6.

Gichner, T., & Veleminský, J. (1986). Organic solvents inhibit the mutagenicity of promutagens dimethylnitrosamine and methylbutylnitrosamine in a higher plant *Arabidopsis thaliana*. *Mutagenesis, 1*(2), 107–109.

Gichner, T., Veleminský, J., & Pospíšil, F. (1985). Screening of compounds for antimutagenic properties towards dimethylnitrosamine-induced mutagenicity in *Arabidopsis thaliana*. *Biologia Plantarum, 27*(6), 417–423.

Girish, T. K., Vasudevaraju, P., & Rao, U. J. P. (2012). Protection of DNA and erythrocytes from free radical induced oxidative damage by black gram (*Vigna mungo* L.) husk extract. *Food and Chemical Toxicology, 50*(5), 1690–1696.

Goel, G., Makkar, H. P., Francis, G., & Becker, K. (2007). Phorbol esters: structure, biological activity, and toxicity in animals. *International Journal of Toxicology, 26*(4), 279–288.

Graf, J. (2000). Herbal anti-inflammatory agents for skin disease. *Skin Therapy Lett, 5*(4), 35.

Grover, I. S., & Bala, S. (1993). Studies on antimutagenic effects of guava (*Psidium guajava*) in *Salmonella typhimurium*. *Mutation Research/Genetic Toxicology, 300*(1), 1–3.

Grüter, A., Friederich, U., & Würgler, F. E. (1990). Antimutagenic effects of mushrooms. *Mutation Research/Fundamental and Molecular Mechanisms of Mutagenesis, 231*(2), 243–249.

Guginski, G., Luiz, A. P., Silva, M. D., Massaro, M., Martins, D. F., Chaves, J., & Santos, A. R. (2009). Mechanisms involved in the antinociception caused by ethanolic extract obtained from the leaves of *Melissa officinalis* (lemon balm) in mice. *Pharmacology Biochemistry and Behavior, 93*(1), 10–16.

Habibi, E., Shokrzadeh, M., Ahmadi, A., Chabra, A., Naghshvar, F., & Keshavarz-Maleki, R. (2014). Genoprotective effects of Origanum vulgare ethanolic extract against cyclophosphamide-induced genotoxicity in mouse bone marrow cells. *Pharmaceutical Biology, 53*(1), 92–97.

Hajdu, Z., Hohmann, J., Forgo, P., Máthé, I., Molnár, J., & Zupkó, I. (2014). Antiproliferative activity of artemisia asiatica extract and its constituents on human tumor cell lines. *Planta Medica, 80*(18), 1692–1697.

Han, X., Shen, T., & Lou, H. (2007). Dietry polyphenols and their biological significance. *International Journal of Molecular Science*, 950–988.

Heo, M. Y., Yu, K. S., Kim, K. H., Kim, H. P., & Au, W. W. (1992). Anticlastogenic effect of flavonoids against mutagen-induced micronuclei in mice. *Mutation Research/Fundamental and Molecular Mechanisms of Mutagenesis, 284*(2), 243–249.

Hiramoto, K., Nasuhara, A., Michikoshi, K., Kato, T., & Kikugawa, K. (1997). DNA strand-breaking activity and mutagenicity of 2, 3-dihydro-3, 5-dihydroxy-6-methyl-4Hpyran-4-one (DDMP), a Maillard reaction product of glucose and glycine. *Mutation Research/Genetic Toxicology and Environmental Mutagenesis, 395*(1), 47–56.

Holla, M., Svajdlenka, E., Tekel, J., Veverkova, S., & Havranek, E. (1997). *Melissa officinalis* subsp. altissima: characteristics of a possible adulteration of lemon balm. *Journal of Essential Oil Research, 9*, 481–484.

Hong, C. E., Cho, M. C., Jang, H. A., & Lyu, S. Y. (2011). Mutagenicity and antimutagenicity of *Acanthopanax divaricatus* var. albeofructus. *The Journal of Toxicological Sciences, 36*(5), 661–668.

Horváthová, E., Slameňová, D., Hlinčiková, L., Mandal, T. K., Gábelová, A., & Collins, A. R. (1998). The nature and origin of DNA single-strand breaks determined with the comet assay. *Mutation Research/DNA Repair, 409*(3), 163–171.

Hwang, I. G., Kim, H. Y., Woo, K. S., Lee, S. H., Lee, J., & Jeong, H. S. (2013). Isolation and identification of the antioxidant DDMP from heated pear (*Pyrus pyrifolia* Nakai). *Preventive Nutrition and Food Science, 18*(1), 76–79.

Ishikawa, K., Naganawa, R., Yoshida, H., Iwata, N., Fukuda, H., Fujino, T., & Suzuki, A. (1996). Antimutagenic effects of ajoene, an organosulfur compound derived from garlic. *Bioscience, Biotechnology and Biochemistry, 60*(12), 2086–2088.

Islam, M. S., Rahman, M. M., Rahman, M. A., Qayum, M. A., & Alam, M. F. (2010). In vitro evaluation of *Croton bonplandianum Baill.* as potential antitumor properties using *Agrobacterium tumefaciens*. *Journal of Agricultural Technology, 6*(1), 79–86.

Janina, M. S. (2003). Melissa officinalis. *The International Journal of Aromatherapy, 10*, 132–139.

Jeeva, S., & Johnson, M. (2012). Antibacteriial and phytochemical studies on *Begonia flaccifera* Bedd.flower. *Asian Pacific Journal of Tropical Biomedicine, 1*(1), 151–154.

Jeeva, S., Johnson, M., Aparna, J. S., & Irudayaraj, V. (2011). Preliminary phytochemical and antibacterial studies on flowers of selected medicinal plants. *International Journal of Medicinal and Aromatic Plants, 1*(2), 107–114.

Jin, G. Z., You, Y. J., & Ahn, B. Z. (2001). Esters of 2-(1-hydroxyalkyl)-1, 4-dihydroxy-9, 10-anthraquinones with melphalan as multifunctional anticancer agents. *Bioorganic and Medicinal Chemistry Letters, 11*(11), 1473–1476.

Joselin, J., Brintha, T. S. S., Florence, A. R., & Jeeva, S. (2012). Screening of select ornamental flowers of the family Apocyanaceae for phytochemical constituents. *Asian Pacific Journal of Tropical Disease, 2*, 1–6.

Joselin, J., Brintha, T. S. S., Florence, A. R., & Jeeva, S. (2013). Phytochemical evaluation of Bignonaceae flowers. *Journal of Chemical and Pharmaceutical Research, 5*(4), 106–111.

Jourdain, C., Tenca, G., Deguercy, A., Troplin, P., & Poelman, D. (2006). In-vitro effects of polyphenols from cocoa and β-sitosterol on the growth of human prostate cancer and normal cells. *European Journal of Cancer Prevention, 15*(4), 353–361.

Kada, T. (1983).Environmental and biological factors suppressing induction of mutagens. In Toxicology Forum (Vol. 6, pp. 580-589).

Kambu, K., Di Phenzu, N., Coune, C., Wauter, JN. and Angenot, L. (1982). Plants Medicine ET Phytotherapie, 34.

Karekar, V., Joshi, S., & Shinde, S. L. (2000). Antimutagenic profile of three antioxidants in the Ames assay and the Drosophila wing spot test. *Mutation Research/Genetic Toxicology and Environmental Mutagenesis, 468*(2), 183–194.

Kato-Noguchi, H. (2003). Assessment of allelopathic potential of shoot powder of lemon balm. *Scientia Horticulturae, 97*(3), 419–423.

Kaur, H., Kalotra, R., Walia, G. K., & Handa, D. (2013). Genotoxic effects of dyeing industry effluent on a freshwater fish, Cirrhinus mrigala by chromosomal aberration test. *International Journal of Pharmacy and Biological Sciences, 3*, 423–431.

Kawashima, T. (2011). A marine carotenoid, fucoxanthin, induces regulatory T cells and inhibits Th17 cell differentiation in vitro. *Bioscience, Biotechnology and Biochemistry, 75*(10), 2066–2069.

Kelly, C., Jewell, C., & O'Brien, N. M. (2003). The effect of dietary supplementation with the citrus limonoids, limonin and nomilin on xenobiotic-metabolizing enzymes in the liver and small intestine of the rat. *Nutrition Research, 23*(5), 681–690.

Khader, M., Bresgen, N., & Eckl, P. M. (2010). Antimutagenic effects of ethanolic extracts from selected Palestinian medicinal plants. *Journal of Ethnopharmacology, 127*(2), 319–324.

Krings, U., & Berger, R. G. (2001). Antioxidant activity of roasted foods. *Food Chem, 72*, 23229.

Kuhn, M. A., & Winston, D. (2000). *Herbal therapy and supplements: a scientific and traditional approach*. Lippincott Williams and Wilkins.

Kumar, P. P., Kumaravel, S., & Lalita, C. (2010). Screening of antioxidant activity, total phenolics and GC-MS study ofVitex negundo. *African Journal of Biochemistry Research, 4*(7), 191–195.

Laird, S. A. (1999). The botanical medicine industry. Pp: 78-116. In K. ten Kate, & S. A. Laird (Eds.), *The commercial use of biodiversity*. Earth scan: London.

Lambert, J., Srivastava, J., & Vietmeyer, N. (1997). *Medicinal plants. Rescuing a global heritage*. Washington DC: World Bank (World Bank Technical Paper 355).

Lee, B. H., Lee, S. J., Hur, J. H., Lee, S., Sung, J. H., Huh, J. D., & Hui, J. H. (1998). In vitro antigenotoxic activity of novel ginseng saponin metabolites formed by intestinal bacteria. *Planta Medica, 64*(6), 500–503.

Lee, I. P., Kim, Y. H., Kang, M. H., Roberts, C., Shim, J. S., & Roh, J. K. (1997). Chemopreventive effect of green tea (*Camellia sinensis*) against cigarette smoke-induced mutations (SCE) in humans. *Journal of Cellular Biochemistry, 67*(27), 68–75.

Lemos, T. L. G., Matos, F. J. A., Alencar, J. W., Crareiro, A. A., Clark, A. M., & Chesnary, J. D. (1990). Antimicrobial activity of essential oils of Brazilian plants: Phytopther. *Res, 4*, 82–84.

Li, C. H., Wei, X.Y., Li, X. E., Wu, P., & Guo, B. J. (2004). A new anthranquinone glycoside from the seeds of *Cassia obtusifolia. Chinese Chemical Letters, 15*(12), 1448–1450.

Li, M., Wei, S.Y., Xu, B., Guo, W., Liu, D. L., Cui, J. R., & Yao, X. S. (2008). Proapoptotic and microtubule-disassembly effects of ardisiacrispin (A+ B), triterpenoid saponins from Ardisia crenata on human hepatoma Bel-7402 cells. *Journal of Asian Natural Products Research, 10*(8), 729–736.

Lippman, S. M., Benner, S. E., & Hong, W. K. (1994). Cancer Chemoprevention. *Journal of Clinical Oncology, 12*, 851–873.

Loarca-Piña, G., Kuzmicky, P. A., de Mejía, E. G., Kado, N.Y., & Hsieh, D. P. (1996). Antimutagenicity of ellagic acid against aflatoxin B 1 in the Salmonella microsuspension assay. *Mutation Research/Environmental Mutagenesis and Related Subjects, 360*(1), 15–21.

Loeb, L. A., Loeb, K. R., & Anderson, J. P. (2003). Multiple mutations and cancer. *Proceedings of the National Academy of Sciences, 100*(3), 776–781.

Lohman, P. H., Gentile, J. M., Gentile, G., & Ferguson, L. R. (2001). Antimutagenesis/ anticarcinogenesis 2001: screening, methods and biomarkers. *Mutation Research/Genetic Toxicology and Environmental Mutagenesis, 496*(1), 1–4.

Luo, W., Zhao, M., Yang, B., Shen, G., & Rao, G. (2009). Identification of bioactive compounds in *Phyllenthus emblica* L. fruit and their free radical scavenging activities. *Food Chemistry, 114*(2), 499–504.

Manayi, A., Saeidnia, S., Ostad, S. N., Hadjiakhoondi, A., Ardekani, M. R. S., Vazirian, M., & Khanavi, M. (2013). Chemical constituents and cytotoxic effect of the main compounds of *Lythrum salicaria* L. *Zeitschrift für Naturforschung C, 68*(9–10), 367–375.

Manikumar, G., Gaetano, K., Wani, M. C., Taylor, H., Hughes, T. J., Warner, J., & Wall, M. E. (1989). Plant antimutagenic agents, 5.Isolation and structure of two new isoflavones, fremontin and fremontone from Psorothamnus fremontii. *Journal of Natural Products, 52*(4), 769–773.

Mantle, D., Pickering, A. T., & Perry, E. K. (2000). Medicinal plant extracts for the treatment of dementia. *CNS Drugs, 13*(3), 201–213.

Marjorie, M. C. (1999). Plant products as antimicrobial agents. *Clinical Microbiology Review, 12*(4), 564–582.

Marnett, L. J., & DuBois, R. N. (2002). COX-2: a target for colon cancer prevention. *Annual Review of Pharmacology and Toxicology, 42*(1), 55–80.

Marnewick, J. L., Gelderblom, W. C., & Joubert, E. (2000). An investigation on the antimutagenic properties of South African herbal teas. *Mutation Research/Genetic Toxicology and Environmental Mutagenesis, 471*(1), 157–166.

Mazumder, P. M., Das, S., Das, S., & Das, M. K. (2010). Cytotoxic activity of methanolic extracts of *Berberis aristata* DC and *Hemidesmus indicus* R. Br. in MCF7 cell line. *Journal of Current Pharmaceutical Research, 1*, 12–15.

Mazzio, E. A., & Soliman, K. F. (2009). In vitro screening for the tumoricidal properties of international medicinal herbs. *Phytotherapy Research, 23*(3), 385–398.

Menoli, R. C. R. N., Mantovani, M. S., Ribeiro, L. R., Speit, G., & Jordão, B. Q. (2001). Antimutagenic effects of the mushroom *Agaricus blazei* Murrill extracts onV79 cells. *Mutation Research/Genetic Toxicology and Environmental Mutagenesis, 496*(1), 5–13.

Mitscher, L. A., Telikepalli, H., McGhee, E. & Shankel, D. M. (1996). Natural antimutagenic agents. Mutation Research/Fundamental and Molecular Mechanisms of Mutagenesis, 350(1): 143-152.

Morita, K., Hara, M., & Kada, T. (1978). Studies on natural desmutagens: screening for vegetable and fruit factors active in inactivation of mutagenic pyrolysis products from amino acids. *Agricultural and Biological Chemistry, 42*(6), 1235–1238.

Mothana, R. A., Alsaid, M. S., Hasoon, S. S., Al-Mosaiyb, N. M., Al-Rehaily, A. J., & AlYahya, M. A. (2012). Antimicrobial and antioxidant activities and gas chromatography mass spectrometry (GC/MS) analysis of the essential oils of *Ajuga bracteosa* Wall. ex Benth. and

Lavandula dentata L. growing wild in Yemen. *Journal of Medicinal Plants Research, 6*(15), 3066–3071.
Nakamura, E. S., Kurosaki, F., Arisawa, M., Mukainaka, T., Okuda, M., Tokuda, H., & Pastore, F. (2002). Cancer chemopreventive effects of constituents of *Caesalpinia ferrea* and related compounds. *Cancer Letters, 177*(2), 119–124.
Nardemir, G., Yanmis, D., Alpsoy, L., Gulluce, M., Agar, G., & Aslan, A. (2015). Genotoxic, antigenotoxic and antioxidant properties of methanol extracts obtained from *Peltigera horizontalis* and *Peltigera praetextata*. *Toxicology and Industrial Health, 31*(7), 602–613.
Ncube, N. S., Afolayan, A. J., & Okoh, A. I. (2008). Assessment techniques of antimicrobial properties of natural compounds of plant origin: current methods and future trends. *African Journal of Biotechnology*, 7(12), 1797–1806.
Neto, C. C., Owens, C. W., Langfield, R. D., Comeau, A. B., Onge, J. S., Vaisberg, A. J., & Hammond, G. B. (2002). Antibacterial activity of some Peruvian medicinal plants from the Callejon de Huaylas. *Journal of Ethnopharmacology, 79*(1), 133–138.
Nurzyńska-Wierdak, R., Bogucka-Kocka, A., & Szymczak, G. (2014). Volatile constituents of *Melissa officinalis* leaves determined by plant age. *Natural Product Communications, 9*(5), 703–706.
Odin, A. P. (1997). Vitamins as antimutagens: advantages and some possible mechanisms of antimutagenic action. *Mutation Research/Reviews in Mutation Research, 386*(1), 3967.
Ofokansi, K. C., Esimone, C. O., & Anele, C. R. (2006). Evaluation of the in vitro combined antibacterial effect of the leaf extracts of *Bryophyllum pinnatum* (Fam: crassulaceae) and *Ocimum gratissimum* (Fam: labiatae). *Plant Products Research Journal, 9*(1), 2327.
Ohe, T., Marutani, K., & Nakase, S. (2001). Catechins are not major components responsible for anti-genotoxic effects of tea extracts against nitroarenes. *Mutation Research/Genetic Toxicology and Environmental Mutagenesis, 496*(1), 75–81.
Okwu, D. E. (2001). Evaluation of chemical composition of medicinal plants belonging to Euphorbiaceae. *Pakistan Veternary Journal, 14*, 160–162.
Othman, A. R., Abdullah, N., Ahmad, S., Ismail, I. S., & Zakaria, M. P. (2015). Elucidation of in-vitro anti-inflammatory bioactive compounds isolated from *Jatropha curcas* L. plant root. *BMC Complementary and Alternative Medicine, 15*(1), 1–10.
Padmaja, M., Sravanthi, M., & Hemalatha, K. P. J. (2011). Evaluation of antioxidant activity of two Indian medicinal plants. *Journal of Phytology, 3*(3), 86–91.
Palasuwan, A., Soogarun, S., Lertlum, T., Pradniwat, P., & Wiwanitkit, V. (2005). Inhibition of heinz body induction in an in vitro model and total antioxidant activity of medicinal Thai plants. *Asian Pacific Journal of Cancer Prevention, 6*(4), 458–463.
Park, C., Moon, D. O., Rhu, C. H., Choi, B. T., Lee, W. H., Kim, G. Y., & Choi, Y. H. (2007). β-sitosterol induces anti-proliferation and apoptosis in human leukemic U937 cells through activation of caspase-3 and induction of Bax/Bcl-2 Ratio. *Biological and Pharmaceutical Bulletin, 30*(7), 1317–1323.
Patel, D. G., Gulati, O. D., & Gokhale, S. D. (1962). Positive inotropic action of an alkaloidal fraction from *Ajuga bracteosa* Well ex Benth. *Indian Journal of Physiology and Pharmcology, 6*, 224–230.
Patel, P. M., Jivani, N., Malaviya, S., Gohil, T., & Bhalodia, Y. (2012). Cataract: A major secondary diabetic complication. *International Current Pharmaceutical Journal, 1*(7), 180–185.
Paul, J., Gnanam, R., M Jayadeepa, R., & Arul, L. (2013). Anti-cancer activity on Graviola, an exciting medicinal plant extract vs various cancer cell lines and a detailed computational study on its potent anti-cancerous leads. *Current Topics in Medicinal Chemistry, 13*(14), 1666–1673.
Pereira, R. P., Fachinetto, R., de Souza Prestes, A., Puntel, R. L., da Silva, G. N. S., Heinzmann, B. M., & Morsch, V. M. (2009). Antioxidant effects of different extracts from *Melissa officinalis, Matricaria recutita* and *Cymbopogon citratus*. *Neurochemical Research, 34*(5), 973–983.

Perry, E. K., Pickering, A. T., Wang, W. W., Houghton, P. J., & Perry, N. S. (1999). Medicinal plants and alzheimer's disease: from ethnobotany to phytotherapy. *Journal of Pharmacy and Pharmacology*, *51*(5), 527–534.

Prakash, G., Hosetti, B. B., & Dhananjaya, B. L. (2014). Antimutagenic effect of *Dioscorea pentaphylla* on genotoxic effect induced by methyl methanesulfonate in the drosophila wing spot test. *Toxicology International*, *21*(3), 258–263.

Pyo, M. K., Jin, J. L., Koo, Y. K., & Yun-Choi, H. S. (2004). Phenolic and furan type compounds isolated from Gastrodia elata and their anti-platelet effects. *Archives of Pharmacal Research*, *27*(4), 381–385.

Rahmat, A., Edrini, S., Ismail, P., Yap, T., Hin, Y., & Bakar, M. A. (2006). Chemical constituents, antioxidant activity and cytotoxic effects of essential oil from *Strobilanthes crispus* and *Lawsonia inermis*. *Journal of Biological Sciences*, *6*(6), 1005–1010.

Ramawat, K. G., & Goyal, S. (2008). The Indian herbal drugs scenario in global perspectives. In *Bioactive Molecules and Medicinal Plants*. Springer Berlin Heidelberg (pp. 325-347).

Ramawat, K. G. (2007). Secondary metabolites in nature. In K. G. Ramawat, & J. M. Merillon (Eds.), *Biotechnology: Secondary Metabolites*. Enfield, CT: Science Publishers pp: 21.

Rao, C.V., Hirose, Y., Indranie, C., & Reddy, B. S. (2001). Modulation of experimental colon tumorigenesis by types and amounts of dietary fatty acids. *Cancer Research*, *61*(5), 1927–1933.

Rao, D.R., Pulusani, S.R., & Chawan, C.B. (1986). Natural inhibitors of carcinogenesis: fermented milk products. In: B. S. Reddy and L. A. Cohen. Pp 63-75.

Rastogi, R.P. and Mehrotra, B.N. (1991). Compendium of Indian Medicinal Plants, CDRI Lucknow and Publ & Information Directorate Lucknow, vol. I, p. 497.

Rauscher, R., Edenharder, R., & Platt, K. L. (1998). In vitro antimutagenic and in vivo anticlastogenic effects of carotenoids and solvent extracts from fruits and vegetables rich in carotenoids. *Mutation Research/Genetic Toxicology and Environmental Mutagenesis*, *413*(2), 129–142.

Ruano-Ravina, A., Figueiras, A., & Barros-Dios, J. M. (2000). Diet and lung cancer: a new approach. *European Journal of Cancer Prevention*, *9*(6), 395–400.

Runnie, I., Salleh, M. N., Mohamed, S., Head, R. J., & Abeywardena, M. Y. (2004). Vaso relaxation induced by common edible tropical plant extracts in isolated rat aorta and mesenteric vascular bed. *Journal of Ethnopharmacology*, *92*(2), 311–316.

Saidana, D., Mahjoub, M. A., Boussaada, O., Chriaa, J., Chéraif, I., Daami, M., & Helal, A. N. (2008). Chemical composition and antimicrobial activity of volatile compounds of *Tamarix boveana* (Tamaricaceae). *Microbiological Research*, *163*(4), 445–455.

Sainkhediya, J., & Ray, S. (2012). Preliminary study of flowering plant diversity of Nimar region. *Bioscience Discovery*, *3*(1), 70–72.

Salah, N., Miller, N. J., Pagange, G., Tijburg, L., Bolwell, G. P., Rice, E., & Evans, C. (1995). Polyphenolic flavonoids as scavenger of aqueous phase radicals as chai breaking antioxidant. *Archives of Biochemistry and Biophysics*, *2*, 339–346.

Sarac, N. (2015). Antioxidant, mutagenic, and antimutagenic activities of *Tragopogon longirostis* var. longirostis, an edible wild plant in Turkey. *Indian Journal of Pharmacology*, *47*(4), 414–418.

Sarkar, A., Basak, R., Bishayee, A., Basak, J., & Chatterjee, M. (1997). Beta-carotene inhibits rat liver chromosomal aberrations and DNA chain break after a single injection of diethylnitrosamine. *British Journal of Cancer*, *76*(7), 855–861.

Sasaki, Y., Matsumoto, K., Imanishi, H., Watanabe, M., Ohta, T., Shirasu, Y., & Tutikawa, K. (1990). In vivo anticlastogenic and antimutagenic effects of tannic acid in mice. *Mutation Research Letters*, *244*(1), 43–47.

Schnitzler, P., Schuhmacher, A., Astani, A., & Reichling, J. (2008). *Melissa officinalis* oil affects infectivity of enveloped herpesviruses. *Phytomedicine*, *15*(9), 734–740.

Sengottuvelan, M., Deeptha, K., & Nalini, N. (2009). Resveratrol ameliorates DNA damage, prooxidant and antioxidant imbalance in 1, 2-dimethylhydrazine induced rat colon carcinogenesis. *Chemico-Biological Interactions, 181*(2), 193–201.

Seo, H. J., & Surh, Y. J. (2001). Eupatilin, a pharmacologically active flavone derived from Artemisia plants, induces apoptosis in human promyelocytic leukemia cells. *Mutation Research/Genetic Toxicology and Environmental Mutagenesis, 496*(1), 191–198.

Sewram, V., Raynor, M. W., Mulholland, D. A., & Raidoo, D. M. (2001). Supercritical fluid extraction and analysis of compounds from *Clivia miniata* for uterotonic activity. *Planta Medica, 67*(5), 451–455.

Shabby, A. S., El-Gengaihi, S., & Khattab, M. (1995). Oil of *Melissa officinalis* L., as affected by storage and herb drying. *Journal of Essential Oil Research*, 7(6), 667–669.

Shah, M. D., Gnanaraj, C., Khan, M. S., & Iqbal, M. (2015). *Dillenia suffruticosa* L. impedes carbon tetrachloride– induced hepatic damage by modulating oxidative stress and inflammatory markers in rats. *Journal of Environmental Pathology, Toxicology and Oncology, 34*(2), 133–152.

Shankel, D. M., Pillai, S. P., Telikepalli, H., Menon, S. R., Pillai, C. A., & Mitscher, L. A. (2000). Role of antimutagens/anticarcinogens in cancer prevention. *Biofactors, 12*(1–4), 113–121.

Sharma, N. (2005). Micropropagation of Bacopa monneiri L. Penn- An important medicinal plant. Dissertation submitted to department of Biotechnology and E. Sciences, Thapar Inst. of Engineering and technology Patiala, India.

Sharmeen, R., Hossain, M. N., Rahman, M. M., Foysal, M. J., & Miah, M. F. (2012). Invitro antibacterial activity of herbal aqueous extract against multi-drug resistant *Klebsiella* sp. isolated from human clinical samples. *International Current Pharmaceutical Journal, 1*(6), 133–137.

Shen, Y. C., Chen, S. L., Zhuang, S. R., & Wang, C. K. (2008). Contribution of Tomato Phenolics to Suppression of COX-2 Expression in KB Cells. *Journal of Food Science, 73*(1), 1–10.

Shimomura, K., Kamata, O., Ueki, S., Ida, S., Oguri, K., Yoshimura, H., & Tsukamoto, H. (1971). Analgesic effect of morphine glucuronides. *The Tohoku Journal of Experimental Medicine, 105*(1), 45–52.

Shokeen, P., Bala, M., & Tandon, V. (2009). Evaluation of the activity of 16 medicinal plants against *Neisseria gonorrhoeae. International Journal of Antimicrobial Agents, 33*(1), 86–91.

Shrikumar, S., & Ravi, T. K. (2007). Approaches towards development and promotion of herbal drugs. *Pharmacognosy Reviews, 1*(1), 180–184.

Singh, S., Haider, S. Z., Chauhan, N. K., Lohani, H., Sah, S., & Yadav, R. K. (2014). Effect of time of harvesting on yield and quality of *Melissa Officinalis* L. in Doon Valley, India. *Indian Journal of Pharmaceutical Sciences, 76*(5), 449–452.

Sodipo, O. A., Akiniyi, J. A., & Ogunbamosu, J. U. (2000). Studies on certain on certain characteristics of extracts of bark of *Pansinystalia macruceras* (K schemp) picrre Exbeille. *Global Journal of Pure Applied Science, 6*, 83–87.

Sorensen, J. M. (2000). Melissa officinalis, essential oil authenticity, production and pharmacological activity. *International Journal of Aromatherapy, 10*(1), 7–15.

Soudamini, K. K., Unnikrishnan, M. C., Sukumaran, K., & Kuttan, R. (1995). Mutagenicity and anti-mutagenicity of selected spices. *Indian Journal of Physiology and Pharmacology, 39*, 347–353.

Sousa, A. C., Gattass, C. R., Alviano, D. S., Alviano, C. S., Blank, A. F., & Alves, P. B. (2004). *Melissa officinalis* L. essential oil: antitumoral and antioxidant activities. *Journal of Pharmacy and Pharmacology, 56*(5), 677–681.

Sripanidkulchai, B., Tattawasart, U., Laupatarakasem, P., Vinitketkumneun, U., Sripanidkulchai, K., Furihata, C., & Matsushima, T. (2002). Antimutagenic and anticarcinogenic effects of Phyllanthus amarus. *Phytomedicine, 9*(1), 26–32.

Starley, I. F., Mohammed, P., Schneider, G., & Bickler, S.W. (1999). The treatment of paediatric burns using topical papaya. *Burns*, *25*(7), 636–639.

Steinhilber, D. (1999). 5-Lipoxygenase: a target for antiinflammatory drugs revisited. *Current Medicinal Chemistry*, *6*(1), 71–85.

Stepek, G., Behnke, J. M., Buttle, D. J., & Duce, I. R. (2004). Natural plant cysteine proteinases as anthelmintics? *Trends in Parasitology*, *20*(7), 322–327.

Sugiyama, M., Lin, X., & Costa, M. (1991). Protective effect of vitamin E against chromosomal aberrations and mutation induced by sodium chromate in Chinese hamster V79 cells. *Mutation Research/Genetic Toxicology*, *260*(1), 19–23.

Sumathi, B. M., & Uthayakumari, F. (2014). GC MS analysis of Leaves of Jatropha maheswarii Subram & Nayar. *Science Research Reporter*, *4*(1), 24–30.

Szaefer, H., Cichocki, M., Brauze, D., & Baer-Dubowska, W. (2004). Alteration in phase I and II enzyme activities and polycyclic aromatic hydrocarbons-DNA adduct formation by plant phenolics in mouse epidermis. *Nutrition and Cancer*, *48*(1), 70–77.

Taherpour, A., Maroofi, H., Rafie, Z., & Larijani, K. (2012). Chemical composition analysis of the essential oil of *Melissa officinalis* L. from Kurdistan, Iran by HS/SPME method and calculation of the biophysicochemical coefficients of the components. *Natural Product Research*, *26*(2), 152–160.

Tavan, E., Maziere, S., Narbonne, J. F., & Cassand, P. (1997). Effects of vitamins A and E on methylazoxymethanol-induced mutagenesis in Salmonella typhimurium strain TA100. *Mutation Research/Fundamental and Molecular Mechanisms of Mutagenesis*, *377*(2), 231–237.

Toering, S. J., Gentile, G. J., & Gentile, J. M. (1996). Mechanism of antimutagenic action of (+)-catechin against the plant-activated aromatic amine 4-nitro-ophenylenediamine. *Mutation Research/Environmental Mutagenesis and Related Subjects*, *361*(2), 81–87.

Turhan, M. (2006). Hand book of herbal plants, chapter 4. *Melissa officinalis*, *3*, 184–245.

Uckun, F.M. and Shyi-Tai Jan, M. (2003). US Patent. 6, 258, 841 B1.

Valnet, J. (1990). Aromatherapy. 11th Edn. France: Maloine, Paris, 11: 242-246.

Van Boekel, M. A. J. S., Goeptar, A. R., & Alink, G. M. (1997). Antimutagenic activity of casein against MNNG in the E. coli DNA repair host-mediated assay. *Cancer Letters*, *114*(1), 85–87.

Van den Berg, T., Freundl, E., & Czygan, F. C. (1997). *Melissa officinalis* subsp. altissima: characteristics of a possible adulteration of lemon balm. *Pharmazie*, *52*(10), 802–808.

Verma, V., Mahmood, U., & Singh, B. (2002). Clerodane diterpenoids from Ajuga bracteosa Wall. *Natural Product Letters*, *16*(4), 255–259.

Vicentini, V. E. P., Camparoto, M. L., Teixeira, R. O., & Mantovani, M. S. (2001). In: *Averrhoacarambola* L., *Syzygiumcumini* (L.) *Skeelsannd Cis-sussicoydes* L.: Medicinal herbal tea effects on vegetal and animal test systems. *Acta Scientiarum*, *23*, 593–598.

Wall, M. E., Wani, M. C., Manikumar, G., Taylor, H., Hughes, T. J., Gaetano, K., & McPhail, D. R. (1989). Plant antimutagenic agents 7.Structure and antimutagenic properties of cymobarbatol and 4-isocymbarbatol, new cymopols from green alga (Cymopolia barbata). *Journal of Natural Products*, *52*(5), 1092–1099.

Waters, M. D., Brady, A. L., Stack, H. F., & Brockman, H. E. (1990). Antimutagenicity profiles of some model compounds. *Mutation Research*, *238*, 57–85.

Wattemberg, L. W. (1985). Chemoprevention of cancer. *Cancer Res.*, *45*, 1–8.

Whelan, L. C., & Ryan, M. F. (2003). Ethanolic extracts of Euphorbia and other ethnobotanical species as inhibitors of human tumour cell growth. *Phytomedicine*, *10*(1), 53–58.

Wilpart, M., Speder, A., Ninane, P., & Roberfroid, M. (1986). Antimutagenic effects of natural and synthetic hormonal steroids. *Teratogenesis, Carcinogenesis, and Mutagenesis*, *6*(4), 265–273.

Wolf, H. T., van den Berg, T., Czygan, F. C., Mosandl, A., Winckler, T., Zündorf, I., & Dingermann, T. (1999). Identification of Melissa officinalis Subspecies by DNA Fingerprinting. *Planta Medica*, *65*(1), 83–85.

Xie, J. W., Hu, W., Zhou, Z. L., Huang, L. F., Wang, Y. L., Fang, J. J., & He, Y. B. (2010). Determination of the Volatile Constituents in Radix Flemingiae Philippinensis by GC-MS and a heuristic evolving latent projection method. *Molecules, 15*(6), 4055–4066.

Yaacob, N. S., Hamzah, N., Kamal, N. N. N. M., Abidin, S. A. Z., Lai, C. S., Navaratnam, V., & Norazmi, M. N. (2010). Anticancer activity of a sub-fraction of dichloromethane extract of Strobilanthes crispus on human breast and prostate cancer cells in vitro. *BMC Complementary and Alternative Medicine, 10*(42), 1–14.

Ying, D., Kevin, P., Weihan, Z., Xiaoqiang, Y. and Jianrong, H. (2005). US Patent, 124684 A1.

Yu, F. R., Lian, X. Z., Guo, H. Y., McGuire, P. M., Li, R. D., Wang, R., & Yu, F. H. (2005). Isolation and characterization of methyl esters and derivatives from Euphorbia kansui (Euphorbiaceae) and their inhibitory effects on the human SGC-7901 cells. *Journal of Pharmacology and Pharmaceutical Science, 8*(3), 528–535.

Zeiger, E. (2000). An interesting state of affairs in genetic toxicology. *Environmental and Molecular Mutagenesis, 35*(2), 82–85.

Zhao, L., Chen, J., Su, J., Li, L., Hu, S., Li, B., & Chen, T. (2013). In vitro antioxidant and antiproliferative activities of 5-hydroxymethylfurfural. *Journal of Agricultural and Food Chemistry, 61*(44), 10604–10611.

Zhou, J., Xie, G., & Yan, X. (2011). *Encyclopedia of Traditional Chinese Medicines Molecular Structures, Pharmacological Activities, Natural Sources and Applications*. Berlin, Heidelberg: Springer Berlin Heidelberg.

CHAPTER 2

Unraveling the mode of action of medicinal plants in delaying age-related diseases using model organisms

Mani Iyer Prasanth[a], Bhagavathi Sundaram Sivamaruthi[b], Periyanaina Kesika[b], Pulikkottil Stanes Rosmol[c], Tewin Tencomnao[a]

[a]Age-Related Inflammation and Degeneration Research Unit, Department of Clinical Chemistry, Faculty of Allied Health Sciences, Chulalongkorn University, Bangkok, Thailand

[b]Innovation Center for Holistic Health, Nutraceuticals and Cosmeceuticals, Faculty of Pharmacy, Chiang Mai University, Chiang Mai, Thailand.

[c]Mookken House, Haritha Nagar, Viyyur, Trissur, Kerala, India

Introduction

Aging is the irreversible damages that take place in cells, tissues, and organs of a living system during its lifetime. The aging process is universal, unique and inevitable that continues until death. It can bring a cascade of changes to the organism such as, weakening of the immune system, reduced elasticity of the skin, slower cell to cell communication, delayed neurological response, delayed or insufficient supply of oxygen, lower level of antioxidants, accumulation of toxins, etc., which culminates the lifespan and health span of the living system.

In simple terms, aging can be described as the total time an organism survives in a healthy manner (Tissenbaum, 2012). However, proper nutrition and diet along with physical exercise can control or slow down the rate of aging and age-related health complications. There are many living examples in and around us who look younger than their biological age. On the other hand, high-calorie intake, poor nutrition, and lack of exercise could lead to many metabolic diseases and disorders like obesity and diabetes, which can accelerate the rate of aging (López-Otín, Galluzzi, Freije, Madeo, & Kroemer, 2016). This suggests that the aging process can be taken under control by modulating the lifestyle choices and thereby can decelerate the rate of aging and improve the overall health (Martel et al., 2019).

Medicinal and Aromatic Plants: Expanding their Horizons through Omics
http://dx.doi.org/10.1016/B978-0-12-819590-1.00002-1

Plants are considered as the most convenient source for food and medicine from primitive age. Usage of plants dates back to 2600 BC in the Mesopotamian civilization and 2900 BC in the Egyptian medicine. The oldest known record of use of plants for medicinal purposes dates back to the "Ebers Papyrus" dating from 1500 BC, which has almost 700 plant derivatives used as medicine (Cragg & Newman, 2013; Borchardt, 2002). Traditional medicine is gaining more popularity and researchers in this field are coming up with new varieties of bioactive phyto-chemical compounds, which have potential even to treat most dreadful diseases including cancer (Martel et al., 2019). However, it is important to understand the mode of action elicited by these plants and its bioactives. In this regard, the present book chapter focuses on the major molecular pathways conserved in the host, which can regulate aging and the role played by different plants in mediating them.

Scenario of increased lifespan expectancy and synthetic medicine

The last century witnessed several advancements in the field of medicine, which had led to an increased life expectancy in humans. Knowledge and awareness about sanitation, nutrition, and other medical conditions have also contributed to this (Vaupel, 2010; Oeppen & Vaupel, 2002). The average life expectancy was between 45 and 50 years in the mid-1800s that gained a tremendous increase up to 80 years within a century (Martel et al., 2019; Crimmins, 2015). The average life expectancy of males and females in the United States was about 55 and 58 years, respectively in 1900 that reached up to 78 and 83 years, respectively by 2010 (Crimmins, 2015). This could be attributed to the decline in the level of deaths caused by infectious diseases. The discovery of penicillin in the early 1940s was a groundbreaking achievement in the field of medicine (Fleming, 1929), which eventually brought down bacterial infections mediated death at a significant level. The focus of medical scientists now is on reducing the level of deaths caused by cardiovascular diseases, age-related diseases and cancer (Crimmins, 2015). Thus, the development of synthetic drugs paved way in treating and curing a wide variety of diseases, which was very specific in action and increased life expectancy. Even though modern drugs can offer so many advantages including better and faster response against diseases, there are some disadvantages also. It is believed that in the United States itself, 8% of the hospital admitted patients are diagnosed with side effects of synthetic drugs (Karimi, Majlesi, & Rafieian-Kopaei, 2015). Natural and herbal medicine becomes an effective alternative at this juncture (Karimi et al., 2015). The World Health Organization (WHO) in 1985 estimated that approximately 65% of the world's population relies on plant-derived traditional medicines for their primary health care (Farnsworth, Akerele, Bingel, Soejarto, & Guo, 1985).

Majority of the plants possess antioxidant properties, which enables them to exhibit positive activity against cancer, memory deficits and neurodegenerative diseases, atherosclerosis, diabetes and other cardiovascular diseases (Karimi et al., 2015). Interestingly, the individual pure compound(s) identified from the plant extracts individually are not sufficient to exhibit the same effect as that of the plant extract (Rafieian-Kopaei, Baradaran, & Rafieian, 2013) indicating that the plant extract with all the macro and micro constituents in that particular predefined ratio is necessary to impart the observed effects in a synergistic manner.

The need for a model system

One of the early study identified that could possibly decelerate the aging process is the dietary restriction process (McCay, Crowell, & Maynard, 1989), which was conserved in many models including yeasts (*Saccharomyces cerevisiae*), worms (*Caenorhabditis elegans*), fruit flies (*Drosophila melanogaster*), mice (*Mus musculus*), and nonhuman primates (Fontana, Partridge, & Longo, 2010). The antiaging effects, which extend both lifespan and healthspan, exhibited by the dietary restriction mechanism are majorly mediated by insulin and IGF-1 signaling, mammalian target of rapamycin (mTOR), and sirtuins (Fontana et al., 2010; Kenyon, 2010). Additionally, the level of reactive oxygen species (ROS) and telomere length are some other factors that could mediate the rate of the aging process. The on or off mechanism of the pathway could reduce the possibilities of developing a disease condition. In this regard, plants or the bioactive derived from it, can regulate any of these pathways; can be considered as an effective drug against various diseases. Some of the well-known compounds are metformin, rapamycin, resveratrol, quercetin, epigallocatechin gallate, naringin, curcumin, and acetic acid (Prasanth, Sivamaruthi, Chaiyasut, & Tencomnao, 2019; Lee & Min, 2013; Fontana et al., 2010; Vaupel, 2010; Oeppen & Vaupel, 2002).

Conserved pathways involved

Plants and/or its metabolites that were able to extend lifespan of various model organisms are described in Table 2.1. Different pathways that modulate the lifespan extension effects by these bioactive compounds are explained further.

Table 2.1 Plant extracts or bioactives that improve lifespan and healthspan in different model organisms.

S. No.	Plant	Active compound	Mode of action reported	Model used	References
1	*Toxicodendronvernicifluum*	Butein	Activates sirtuins	*S. cerevisiae*	Howitz et al. (2003)
2	Apple	Phloridzin	Activates sirtuins and SOD and reduces ROS	*S. cerevisiae*	Xiang et al. (2011)
3	*Aspalathus linearis*	Aspalathin	Reduce the level of ROS and activate DAF-16/FOXO	*C. elegans*	Chen et al. (2013)
4	Coffee and Chocolate	Caffeine	Activate DAF-16/FOXO	*C. elegans*	Lublin et al. (2011)
5	Vegetables	Chlorophyll, Kaempferol	Activate DAF-16/FOXO and enhance stress resistance	*C. elegans*	Wang & Wink (2016); Grünz et al. (2012); Kampkötter et al. (2007)
6	Vegetables and tea	Myricetin	Activate DAF-16/FOXO and enhance stress resistance and reduce ROS	*C. elegans*	Büchter et al. (2013); Grünz et al. (2012)
7	Vegetables and fruits	Tannic acid	Activate DAF-16/FOXO, longevity, stress resistance and neuroprotection	*C. elegans*	Lublin et al. (2011); Saul et al. (2011); Saul et al. (2010) Saul, Pietsch, Menzel, Stürzenbaum, Steinberg (2010)
8	Blueberry fruits	Proanthocyanidins	Increase lifespan, stress resistance and thermo tolerance	*C. elegans*	Mekheimer et al. (2012); Wilson et al. (2006)
9	Grape	Polydatin	Activate DAF-16/FOXO and SOD-3	*C. elegans*	Wen, Gao, & Qin (2014)
10	*Vignaangularis*	Genistein	Activates HSPs and SOD-3 and extends lifespan	*C. elegans*	Lee et al. (2015)

11	Simaroubaceae plants	Glaucarubinone	Promotes mitochondrial metabolism,reduces body fat, and extends lifespan	*C. elegans*	Zarse et al. (2011)
12	*Epimediumbrevicornum*	Icariside II	Activate DAF-16/FOXO and HSPs and extend healthspan	*C. elegans*	Cai et al. (2011)
13	*Panaxnotoginseng*	Polysaccharides	Enhance thermal stress resistance by activating SOD	*C. elegans*	Feng et al. (2015)
14	*Rauvolfiaserpentina*	Reserpine	Stress resistance and lifespan extension	*C. elegans*	Srivastava et al. (2008)
15	Willow bark	Aspirin	Inhibits oxidant stress, reduces age-associated declines, and extends lifespan	*C. elegans*	Ayyadevara et al. (2013)
16	*Silybummarianum*-seed	Silymarin	Activate DAF-16/FOXO and SOD-3 along with reducing ROS	*C. elegans*	Srivastava et al. (2017)
17	*Curcuma aromatica*	Curcumin	Activates sirtuins and SOD mechanism	*C. elegans*, Drosophila	Shen et al. (2013); Liao et al. (2011); Lee et al. (2010)
18	Vegetables and fruits	Fisetin	Activates sirtuins and DAF-16/FOXO mechanism along with reducing ROS	*C. elegans*, Drosophila	Kampkötter et al. (2007); Wood et al. (2004)
19	Orange; *Sutherlandia frutescens*	Inositol; Pinitol	Activates FOXO along with reducing ROS	Drosophila	Hada et al. (2013)
20	Citrus fruits	Rutin	Activates FOXO and SOD along with reducing ROS	Drosophila	Chattopadhyay et al. (2017)
21	Black tea	Theaflavin	Extends lifespan and activates stress markers including SOD and CAT	Drosophila	Peng, Chan, Li, Huang, & Chen (2009)

(Continued)

Table 2.1 Plant extracts or bioactives that improve lifespan and healthspan in different model organisms. *(Cont.)*

S. No.	Plant	Active compound	Mode of action reported	Model used	References
22	Berry fruits	Urolithin A	Extends lifespan, improve autophagy and muscle functions	*C. elegans*, Mice	Ryu et al. (2016)
23	Vegetables and fruits	Lutein	Extends lifespan, enhance stress resistance, activates antioxidant mechanism and reduces skin aging	Drosophila, Mice	Kamoshita et al. (2016); Zhang et al. (2014) Zhang, Han, Wang, & Wang (2014); Astner et al. (2007)
24	Green tea, cocoa	Catechin, Epicatechin, Epigallocatechin galate	Activates stress resistance and enhances longevity by reducing ROS and IGF-1 along with increasing AMPK and SOD mechanism along with enhancing cognitive functions	*C. elegans*, Drosophila, Mice	Saul et al. (2011); Si et al. (2011); Abbas and Wink (2009); He et al. (2009); Saul et al. (2009)
25	Vegetables	Quercetin	Activates DAF-16/FOXO andsirtuins, reduce senescence and improve brain functions	*C. elegans*, Drosophila, Mice	Xu et al. (2018); Grünz et al. (2012); Spindler et al. (2012); Lu et al. (2010); Davis et al. (2009) Davis, Murphy, Carmichael, & Davis (2009); Kampkötter et al. (2007)
26	*Tripterygiumwilfordii*	Celastrol	Activates HSP-70 and reduces TNF-α	Mice	Kiaei et al. (2005)
27	*Epimediumbrevicornum*	Icariin	Activates SOD mechanism and extends healthspan	Mice	Zhang et al. (2015)

28	Garlic	Allicin	Activates SOD, memory and Nrf2, reduces ROS in Alzheimer's disease model	Mice	Li, Li, Lu, Tian, & Wei (2012); Li et al. (2010)
29	*Astragalusmembranaceus*	Astragaloside	Enhances memory and motor functions	Mice	Lei et al. (2003)
30	*Rehmanniaglutinosa*	Catalpol	Improves cholinergic function and reduces inflammatory cytokines	Mice	Zhang et al. (2013)
31	Rose flower	Gallic acid	Antioxidant activity in liver, kidney and brain	Mice	Li et al. (2005)
32	*Herbaepimedii*	Icariin	Activates sirtuins and reduces cardiac inflammation and inhibits NF-kappa B pathway	Mice	Chen et al. (2015)
33	*Larreatridentata*	NDGA	Enhances motor functions and inhibits TNF-α	Mice	Hou et al. (2018)
34	*Trolliuschinensis*	Orientin	Reduces the level of ROS and increases the level of antioxidant enzymes and brain weight	Mice	An et al. (2012)
35	Cranberries	Proanthocyanidins	Anti-aging and oxidative stress resistance	Mice	Jiao et al. (2017)
36	Fruits	Tiliroside	Activates AMPK, improves insulin sensitivity and regulates obesity	Mice	Goto et al. (2012)
37	*Curcuma aromatica*	Tetrahydrocurcumin	Reduces the level of ROS, decreases the chances of renal dysfunction and attenuates hypertension	Mice	Sangartit et al. (2016); Okada et al. (2001)
38	*Tsugachinensis*	Vitexin	Reduces the level of ROS and increases the activity of antioxidant enzymes	Mice	An et al. (2012)

(Continued)

Table 2.1 Plant extracts or bioactives that improve lifespan and healthspan in different model organisms. *(Cont.)*

S. No.	Plant	Active compound	Mode of action reported	Model used	References
39	Grape	Resveratrol	Activates DAF-16/FOXO, sirtuins and enhances stress resistance, motor function, autophagy and fertility, reduces obesity and neurodegenerative diseases	*S. cerevisiae*, *C. elegans*, Drosophila, Mice, Rat	Liu et al. (2013); Morselli et al. (2011); Khan et al. (2010); Morselli et al. 2010; Karuppagounder et al. (2009); Baur et al. (2006); Bauer et al. (2004); Wood et al. (2004); Howitz et al. (2003)
40	*Coptischinensis*	Berberine	Activates AMPK and improves glucose metabolism along with reducing obesity	Mice, Rat	Yin, Gao, Liu, Liu, & Ye (2008); Lee et al. (2006);
41	*Curcuma aromatica*	Curcumin	Reduces oxidative damage, improves insulin resistance, enhances memory and reduces amyloid-β plaques	Mice, Rat	Banji et al. (2013) Banji, Banji, Dasaroju, & Annamalai (2013); Fleenor et al. (2013); Na et al. (2011); Lim et al. (2001)
42	*Gastrodiaelata*	Gastrodin	Improves cognitive and motor functions in animals with Parkinson's disease and dementia	Mice, Rat	Li and Zhang, (2015); Chen et al. (2014); Wang et al. (2014)

43	Ginseng	Ginsenoside Rg1	Improves cognitive function and telomerase activity along with reducing ROS	Mice, Rat	Li et al. (2016); Zhu et al. (2014)
44	*Huperziaserrata*	Huperzine A	Improves chronic inflammation and cognitive function along with promoting hippocampal neurogenesis	Mice, Rat	Ma et al. (2013); Wang et al. (2010) Wang, Zhang, & Tang (2010)
45	*Andrographispanicu-lata*	Andrographolide	Reduces ROS and enhances cognitive functions	Rats	Thakur, Rai, Chatterjee, & Kumar (2016); Yu et al. (2003) Yu, Hung, Chen, & Cheng (2003)
46	*Zingiberofficinale*	Gingerol	Protects against gentamicin induced nephrotoxicity and reduces ROS	Rats	Rodrigues et al. (2014)
47	Lychee fruit	Oligonol	Reduces ROS and increases insulin sensitivity, protects from apoptosis and oxidative stress	Rats	Park et al. (2016)

Source: Adapted from Martel et al. (2019).

Insulin-like signaling (IIS) pathway

The pathway is otherwise known as IGF-1 signaling pathway or the Forkhead box O (FOXO) transcription pathway, which majorly regulates cell growth, development, protein synthesis, and energy storage (Martel et al., 2019). When glucose, amino acids, and other nutrients are available, insulin induces nutrient uptake through skeletal muscle and adipose tissue and store energy in the form of glycogen. On the other hand, when the level of glucose is low, the pituitary gland induces the secretion of IGF-1 that could eventually release energy and promote cell growth and development. In other words, the insulin receptor gene, once when activated (depending on the ligand that is bound to it), will activate the PI3 kinase, which will phosphorylate the FOXO transcription factor. When the insulin receptor is not activated, PI3 kinase will not phosphorylate the FOXO transcription factor which will get activated, move to the nucleus and activate the transcription of genes which are responsible for stress resistance and development.

Studies conducted in various model organisms suggest that the reduction of IIS signaling pathway (activation of FOXO) increases the lifespan. The nematode *C. elegans* exhibited almost 10-fold increase in median and maximum lifespan, which is so far the maximum extension of lifespan, mediated via inactivation of the insulin pathway (Ayyadevara, Alla, Thaden, & Shmookler Reis, 2008). Loss of function mutants of IGF-1 in female mice increased the lifespan by approximately 16% (Svensson et al., 2011). Depending on the level of inactivation or not the availability of IGF-1, there was an extension of lifespan observed in mammals also (Fontana et al., 2010; Bartke, 2005). Candidate plants such as *Aspalathus linearis*, *Epimedium brevicornum*, *Sutherlandia frutescens* and compounds like resveratrol and quercetin are known to mediate antiaging mechanism that is dependent on IIS pathway (Chen et al., 2013; Hada et al., 2013; Cai et al., 2011).

Parallel to the above-mentioned findings, high-glucose intake in both *C. elegans* and mice were observed to increase insulin signaling and eventually reduce lifespan (Lee, Murphy, Kenyon, 2009; Mlekusch et al., 1996). Transgenic mice that could over-express the level of growth hormone and IGF-1 signaling pathway showed premature aging and reduction in lifespan (Steger, Bartke, & Cecim, 1993). In the case of humans, individuals who have a health condition of overexpressing growth hormones and IGF-1 signaling pathway expressed shorter lifespan (Vaiserman, Lushchak, & Koliada, 2016), which is similar to the findings in model organisms. Also, the high-fat diet is considered as a reason for reduced lifespan in mice and humans (Levine et al., 2014; Solon-Biet et al., 2014), which can activate the IGF-1 signaling pathway. When the IGF-1 signaling is reduced, the FOXO transcription factor will initiate the activation of various genes and proteins such as superoxide dismutase (SOD), catalase (CAT), glutathione S-transferase, metallothioneins, and other chaperones (Morris et al., 2015; Murphy, 2006; Murphy et al., 2003). Together, they initiate stress resistance by neutralizing reactive oxygen species (ROS), initiating DNA repair mechanism, detoxifying the harmful heavy metals, maintaining protein structure, which eventually mediates longevity. Interestingly, in humans, FOXO3 is associated with a longer lifespan, which suggests that the lifespan regulation mediated by FOXO is conserved in humans, also (Flachsbart et al., 2009; Willcox et al., 2008).

mTOR pathway and autophagy

The mTOR literally means Mammalian Target of Rapamycin. Rapamycin is an immunosuppressive drug used to prevent organ transplant rejection in humans (Martel et al., 2019). Interestingly, it was observed to increase lifespan and healthspan in worms and mice (Johnson, Rabinovitch, & Kaeberlin, 2013Rubinsztein, Marino, & Kroemer, 2011; Harrison et al., 2009). The activity of mTOR is stimulated by insulin, growth factors and oxidative stress (Ingram & Roth, 2015), and inhibition of mTOR reduces insulin and IGF-1 signaling (Johnson et al., 2013), which suggests the correlation between mTOR with other aging-related pathways. Additionally, inhibition of mTOR can also activate autophagy, the cellular mechanism that degrades and recycles damaged molecules and organelles, within the cells, thereby maintaining cellular integrity, which is supposed to decrease during aging (Rubinsztein et al., 2011; Cuervo et al., 2005).

The decrease in the level of autophagy could lead to various neurodegenerative diseases such as Alzheimer's, Parkinson's and Huntington's disease (Martel et al., 2019; Ravikumar et al., 2010). Autophagy plays a pivotal role in degrading the protein aggregates including amyloid-β and tau which are associated with neurodegenerative diseases (Ravikumar et al., 2010). Knockout mice in which autophagy are inactivated in the brain expressed increased neuronal degeneration apart from shorter lifespan (Komatsu et al., 2006). Compounds like resveratrol and urolithin A were observed to induce anti-aging effects by improving autophagy (Ryu et al., 2016; Morselli et al., 2011).

Sirtuins and acetyltransferases

Sirtuins act as a group of histone deacylase enzymes, which helps in removing acyl groups from histones and other proteins (Verdin, 2015) and its activity depends on the energy levels of the cells, wherein proper diet and exercise increases the activity and overnutrition will decrease the same (Mouchiroud et al., 2013). The activity of sirtuin requires the cofactor nicotinamide adenine dinucleotide (NAD^+) as an acyl group acceptor and the enzyme is inhibited by the reduced form of the compound (NADH) (Verdin, 2015; Mouchiroud et al., 2013). Sirtuins modulate lifespan and aging by regulating various cellular pathways including IGF-1 signaling pathway, as it deacetylates FOXO leading to the transcription and activation of proteins that mediate stress resistance and longevity (Brunet et al., 2004). It can also activate PGC-1α which induce fatty acid oxidation (Rodgers, Lerin, Gerhart-Hines, & Puigserver, 2008), hypoxia-inducible factor-1α (HIF-1α), which may promote cell longevity via inhibition of glycolysis (Verdin, 2015; Houtkooper, Pirinen, & Auwerx, 2012), promote DNA repair and cell survival (Jeong et al., 2007), prevent senescence, and induce cell survival (Zu et al., 2010; Langley et al., 2002).

Sirtuins are essential for longevity and anti-aging mechanism in various models including yeasts, worms, fruit flies, and mice (Rogina & Helfand, 2004; Tissenbaum & Guarente, 2001). The overexpression of *sirt-1* in the brain extended lifespan in mice (Satoh et al., 2013) apart from reducing different age-related diseases, including diabetes, cardiovascular disease, inflammation, and neurodegeneration in various model organisms (Morris, 2013). In mice, overexpression of *sirt-6* was found to extend the lifespan of male mice by 15% (Kanfi et al., 2012), whereas *sirt-6* deficiency showed premature aging along with inefficient or defective DNA repair mechanism (Mostoslavsky et al., 2006). In *C. elegans* also, *sir-2.1* was able to extend the lifespan in the presence of different extracts such as green tea extract which was dependent on FOXO transcription pathway (Rathor, Akhoon, PAndey, Srivastava, & Pandey, 2015). On the whole, activation of sirtuinis directly proportional to delayed aging and lifespan extension (Martel et al., 2019).

ROS and antioxidant equilibrium

Accumulation of ROS could lead to protein damage, organelle dysfunction, DNA damage, and aging (Mikhed et al., 2015; Harman, 1956). Antioxidants present inside the system could neutralize the level of ROS which could stop the rate of aging accelerated by the ROS inside the body (Martel et al., 2019). However, there should be equilibrium between the level of

ROS and antioxidants as ROS also has some positive effects. Immune cells such as macrophages and neutrophils utilize ROS to destroy many bacterial pathogens. It can also act as a secondary messenger, which could activate cells responsible for DNA repair and stress resistance (Tiganis, 2011). In short, increase in the level of ROS could accelerate the aging process; however, complete neutralization of ROS via antioxidant mechanism can also have a negative impact on various cellular processes including stress resistance and so equilibrium should be maintained between the level of ROS and antioxidants inside the system (Martel et al., 2019). Candidate plants such as *Aspalathus linearis*, *Silybum marianum*, *Trollius chinensis*, and *Tsuga chinensis* were observed to impart antioxidant activity by reducing the level of ROS inside the host (An, Yang, Tian, & Wang, 2012).

Telomerase activity

As the cell division progresses with time, the length of the telomere will be shortened because of increase in oxidation, weakening or improper recognition of DNA repair enzymes, or due to the incomplete activity of DNA polymerase (Martel et al., 2019). Telomerase enzymes can protect telomeres by adding sequence-specific DNA at chromosomal ends. However, most of the somatic cells have low or no telomerase activity, to prevent telomere erosion (Harley et al., 1990; Harley, Futcher, & Greider, 1990) which could lead to senescence, wherein, cells stop replicating and become resistant to apoptosis (Kirkland & Tchkonia, 2017) and eventually accelerate aging. In this regard, when the telomerase genes were reintroduced to human primary cell lines, they became immortal (Bodnar et al., 1998). Mice which don't have telomerase enzymes end up having shorter telomeres and also show signs of premature tissue degeneration in bone marrow and skin (Flores, Cayuela, & Blasco, 2005; Herrera et al., 1999). Length of telomere was also observed to have a role in various age-related diseases in humans, including heart disease and various infections (Cawthon, Smith, O'Brien, Sivatchenko, & Kerber, 2003). Ginseng was observed to have positive effects in extending anti-aging effects by modulating telomerase activity in mice and rats (Li et al., 2016; Zhu et al., 2014).

Gut microbiome

The different microbes that reside in the gut region are collectively known as the gut microbiome, which apart from the digestion of food, aids in many physiological processes including longevity (Heintz & Mair, 2014; Lin et al., 2014). Some of the specific bacteria when inside the gut of *C.*

elegans were observed to protect the host against age-related tumor formation or amyloid-β deposition, or modulate the unfolded-protein response (Han et al., 2017). In mice, probiotics were observed to extend lifespan by inducing anti-inflammatory effects and improving colonic mucosal function (Matsumoto, Kurihara, Kibe, Ashida, & Benno, 2011). Another study showed that a mixture of prebiotics and probiotics could reduce hepatic cell proliferation and muscle wasting along with extended survival in leukemic mice (Bindels et al., 2016). The gut microbiome is also supposed to play a role in extending lifespan which is mediated by different phytochemicals such as metformin (Lee & Ko, 2014) and rapamycin (Jung et al., 2016).

Bioactives activating multi targets

Some bioactives such as resveratrol, quercetin and EGCG were able to act on multiple targets to invoke anti-aging mechanism along with stress resistance and neuroprotection. These compounds were observed to mediate IGF-1/FOXO pathway, enhance stress resistance, reduce the level of ROS, activate sirtuins, autophagy and other motor functions along with preventing from neurodegenerative diseases (Prasanth et al., 2019; Xu et al., 2018; Liu et al., 2013; Grünz et al., 2012; Splindler, Li, Dhahbi, Yamakawa, & Sauer, 2012; Morselli et al., 2011; Saul, Pietsch, Stürzenbaum, Menzel, Steinberg, 2011; Si et al., 2011; Khan et al., 2010; Lu et al., 2010; Morselli et al., 2010; Abbas and Wink, 2009; Davis et al., 2009; He et al., 2009; Karuppagounder et al., 2009; Saul, Pietsch, Stürzenbaum, Menzel, Steinberg, 2009; Kampkötter et al., 2007; Baur et al., 2006; Bauer, Goupil, Garber, & Helfand, 2004; Wood et al., 2004; Howitz et al., 2003). Finding out similar compounds with such properties and evaluating the same in humans will have immense benefits in healthcare and medicine as these compounds can be consumed easily and with negligible side effects which will be of greater significance.

Challenges

In humans, there is no clear idea about the optimal dosage, absorption, bioavailability, and efficacy of plant extracts or their bioactives, which were observed to have positive effects in model organisms (Table 2.1). These plant extracts and bioactives should be tested clinically on humans to have a better idea about safety and efficacy. Currently, many of the available natural compounds do not possess sufficient information regarding the purity, efficacy and the optimal dose for usage, which has to be strictly followed by the producers of these dietary supplements (Martel et al., 2017). Even though natural medicine has not have reported any toxic side effects, it should be properly addressed. Conventionally, if a drug is found to be effective, it could have some side effects also (Karimi et al., 2015). For example, the hormone melatonin found in various fruits and vegetables (Anisimov et al., 2001) and lignan compound nordihydroguaiaretic acid (Spindler et al., 2015) were observed to increase the lifespan in mice which in turn increased tumor formation, reduced the fertility, and increased peritoneal hemorrhage. Therefore, it is important to make sure that the identified plant or their bioactive compound does not cause any damages or side effects in healthy subjects.

Natural medication is a holistic approach, which includes emotional, mental, and spiritual levels of an individual (Karimi et al., 2015). So deeper understanding of the mode of action of the plant compounds in all the systems has to be done to make sure that the compound will not elicit any unwanted effects. Even though several plants, which were observed to extend lifespan by reducing the IGF-1 levels (Table 2.1), still it is, not clear that where exactly the plant compounds will initiate the response. Also, much deeper studies have to be done to see whether lifespan extension can be done without compromising on the overall health in humans by reducing IGF-1 levels, because it is necessary for the development and remodeling of neurons which is essential for learning and memory in mammals (Sonntag et al., 2013). Apart from the anti-aging efficacies, some studies have pointed out the pro-aging effects of sirtuins (Martel et al., 2019). Deficiency of *sirt-2* in non-dividing yeast cells enhanced the lifespan and extended the longevity-enhancing effects of dietary restriction mechanism (Fabrizio et al., 2005). The orthologs of the same gene in mice, *sirt-1*, when absent, expressed reduced signs of oxidative stress in the brain, even though these animals had a reduced lifespan under a normal or controlled diet (Li et al., 2008).

Increase in the level of antioxidants may not end up in lifespan extension always. Overexpression of antioxidant enzymes could increase the lifespan in *Drosophila* (Sun & Tower, 1999), but the same mechanism could not be observed in mice (Pérez et al., 2009; Huang et al., 2000). In the case of humans also, some studies suggest that increased intake of antioxidants intake could aid in reduced mortality (Zhao et al., 2017; Stepaniak et al., 2016), whereas some other studies suggest the reverse (Bjelakovic, Nikolova, Gluud, Simonetti, & Gluud, 2012; Bjelakovic, Nikolova, Gluud, Simonetti, & Gluud, 2007).

Conclusion

With the available information, it is important to understand that till now no method has been identified to halt or stop the aging process, whereas the rate of the aging process could be mediated. Studies using model organisms will give a better picture on the advantages and disadvantages of various medicinal plants and it will help to narrow down on fewer plants to be analyzed in humans. In other words, understanding the mechanism and mode of action of medicinal plants and its bioactives in model organisms will allow researchers to have a better picture of the possible outcomes in human trials.

References

Abbas, S., & Wink, M. (2009). Epigallocatechin gallate from green tea (*Camellia sinensis*) increases lifespan and stress resistance in *Caenorhabditis elegans*. *Planta Medica*, *75*(3), 216–221.

An, F., Yang, G., Tian, J., & Wang, S. (2012). Antioxidant effects of the orientin and vitexin in *Trollius chinensis* Bunge in D-galactose-aged mice. *Neural Regeneration Research*, 7(33), 2565–2575.

Anisimov, V. N., Zavarzina, N.Y., Zabezhinski, M.A., Popovich, I. G., Zimina, O.A., Shtylick, A.V., et al. (2001). Melatonin increases both life span and tumor incidence in female CBA mice. *Journals of Gerontology Series A: Biological Sciences and Medical Sciences*, *56*(7), B311–323.

Astner, S., Wu, A., Chen, J., Philips, N., Rius-Diaz, F., Parrado, C., et al. (2007). Dietary lutein/zeaxanthin partially reduces photoaging and photocarcinogenesis in chronically UVB-irradiated Skh-1 hairless mice. *Skin Pharmacology and Physiology*, *20*(6), 283–291.

Ayyadevara, S., Alla, R., Thaden, J. J., & Shmookler Reis, R. J. (2008). Remarkable longevity and stress resistance of nematode PI3K-null mutants. *Aging Cell*, 7(1), 13–22.

Ayyadevara, S., Bharill, P., Dandapat, A., Hu, C., Khaidakov, M., Mitra, S., et al. (2013). Aspirin inhibits oxidant stress, reduces age-associated functional declines, and extends lifespan of *Caenorhabditis elegans*. *Antioxidants & Redox Signaling*, *18*(5), 481–490.

Banji, D., Banji, O. J., Dasaroju, S., & Annamalai, A. R. (2013). Piperine and curcumin exhibit synergism in attenuating D-galactose induced senescence in rats. *European Journal of Pharmacology*, *703*(1–3), 91–99.

Bartke, A. (2005). Minireview: role of the growth hormone/insulin-like growth factor system in mammalian aging. *Endocrinology*, *146*(9), 3718–3723.

Bauer, J. H., Goupil, S., Garber, G. B., & Helfand, S. L. (2004). An accelerated assay for the identification of lifespan-extending interventions in *Drosophila melanogaster*. *Proceedings of the National Academy of Sciences of the United States of America*, *101*(35), 12980–12985.

Baur, J. A., Pearson, K. J., Price, N. L., Jamieson, H. A., Lerin, C., Kalra, A., et al. (2006). Resveratrol improves health and survival of mice on a high-calorie diet. *Nature*, *444*(7117), 337–342.

Bindels, L. B., Neyrinck, A. M., Claus, S. P., Le Roy, C. I., Grangette, C., Pot, B., et al. (2016). Synbiotic approach restores intestinal homeostasis and prolongs survival in leukaemic mice with cachexia. *The ISME Journal*, *10*(6), 1456–1470.

Bjelakovic, G., Nikolova, D., Gluud, L. L., Simonetti, R. G., & Gluud, C. (2012). Antioxidant supplements for prevention of mortality in healthy participants and patients with various diseases. *Cochrane Database of Systematic Reviews*, *3*, CD007176.

Bjelakovic, G., Nikolova, D., Gluud, L. L., Simonetti, R. G., & Gluud, C. (2007). Mortality in randomized trials of antioxidant supplements for primary and secondary prevention: systematic review and meta-analysis. *The Journal of the American Medical Association, 297*(8), 842–857.

Bodnar, A. G., Ouellette, M., Frolkis, M., Holt, S. E., Chiu, C. P., Morin, G. B., et al. (1988). Extension of life-span by introduction of telomerase into normal human cells. *Science, 279*(5349), 349–352.

Borchardt, J. K. (2002). The beginnings of drug therapy: ancient mesopotamian medicine. *Drug News Perspectives, 15*(3), 187–192.

Brunet, A., Sweeney, L. B., Sturgill, J. F., Chua, K. F., Greer, P. L., Lin, Y., et al. (2004). Stress-dependent regulation of FOXO transcription factors by the SIRT1 deacetylase. *Science, 303*(5666), 2011–2015.

Büchter, C., Ackermann, D., Havermann, S., Honnen, S., Chovolou, Y., Fritz, G., et al. (2013). Myricetin-mediated lifespan extension in *Caenorhabditis elegans* is modulated by DAF-16. *International Journal of Molecular Sciences, 14*(6), 11895–11914.

Cai, W. J., Huang, J. H., Zhang, S. Q., Wu, B., Kapahi, P., Zhang, X. M., et al. (2011). Icariin and its derivative icariside II extend healthspan via insulin/IGF-1 pathway in *C. elegans*. *PLOS One, 6*(12), e28835.

Cawthon, R. M., Smith, K. R., O'Brien, E., Sivatchenko, A., & Kerber, R. A. (2003). Association between telomere length in blood and mortality in people aged 60 years or older. *The Lancet, 361*(9355), 393–395.

Chattopadhyay, D., Chitnis, A., Talekar, A., Mulay, P., Makkar, M., James, J., et al. (2017). Hormetic efficacy of rutin to promote longevity in *Drosophila melanogaster*. *Biogerontology, 18*(3), 397–411.

Chen, P. Z., Jiang, H. H., Wen, B., Ren, S. C., Chen, Y., Ji, W. G., et al. (2014). Gastrodin suppresses the amyloid (-induced increase of spontaneous discharge in the entorhinal cortex of rats. *Neural Plasticity, 2014*, 320937.

Chen, W., Sudji, I. R., Wang, E., Joubert, E., van Wyk, B. E., & Wink, M. (2013). Ameliorative effect of aspalathin from rooibos (*Aspalathus linearis*) on acute oxidative stress in *Caenorhabditis elegans*. *Phytomedicine, 20*(3–4), 380–386.

Chen, Y., Sun, T., Wu, J., Kalionis, B., Zhang, C., Yuan, D., et al. (2015). Icariin intervenes in cardiac inflammaging through upregulation of SIRT6 enzyme activity and inhibition of the NF-kappa B pathway. *BioMed Research International, 2015*, 895976.

Cragg, G. M., & Newman, D. J. (2013). Natural products: a continuing source of novel drug leads. *Biochimica et Biophysica Acta, 1830*(6), 3670–3695.

Crimmins, E. M. (2015). Lifespan and Healthspan: Past, Present, and Promise. *Gerontologist, 55*(6), 901–911.

Cuervo, A. M., Bergamini, E., Brunk, U. T., Dröge, W., Ffrench, M., & Terman, A. (2005). Autophagy and aging: the importance of maintaining "clean" cells. *Autophagy, 1*(3), 131–140.

Davis, J. M., Murphy, E. A., Carmichael, M. D., & Davis, B. (2009). Quercetin increases brain and muscle mitochondrial biogenesis and exercise tolerance. *American Journal of Physiology-Regulatory, Integrative and Comparative Physiology, 296*(4), R1071–1077.

de Magalhães, J. P. (2014). The scientific quest for lasting youth: prospects for curing aging. *Rejuvenation Research, 17*(5), 458–467.

Deusing, D. J., Winter, S., Kler, A., Kriesl, E., Bonnlander, B., Wenzel, U., et al. (2015). A catechin-enriched green tea extract prevents glucose-induced survival reduction in *Caenorhabditis elegans* through sir-2.1 and uba-1 dependent hormesis. *Fitoterapia, 102*, 163–170.

Fabrizio, P., Gattazzo, C., Battistella, L., Wei, M., Cheng, C., McGrew, K., et al. (2005). Sir2 blocks extreme life-span extension. *Cell, 123*(4), 655–667.

Farnsworth, N. R., Akerele, O., Bingel, A. S., Soejarto, D. D., & Guo, Z. (1985). Medicinal plants in therapy. *Bulletin of World Health Organization, 63*(6), 965–981.

Feng, S., Cheng, H., Xu, Z., Shen, S., Yuan, M., Liu, J., et al. (2015). Thermal stress resistance and aging effects of *Panax notoginseng* polysaccharides on *Caenorhabditis elegans*. *International Journal of Biological Macromolecules, 81*, 188–194.

Flachsbart, F., Caliebe, A., Kleindorp, R., Blanché, H., von Eller-Eberstein, H., Nikolaus, S., et al. (2009). Association of FOXO3A variation with human longevity confirmed in German centenarians. *Proceedings of the National Academy of Sciences of the United States of America, 106*(8), 2700–2705.

Fleenor, B. S., Sindler, A. L., Marvi, N. K., Howell, K. L., Zigler, M. L., Yoshizawa, M., et al. (2013). Curcumin ameliorates arterial dysfunction and oxidative stress with aging. *Experimental Gerontology, 48*(2), 269–276.

Fleming, A. (1929). On the antibacterial action of cultures of a penicillium, with special reference to their use in the isolation of B. influenzae. *Bulletin of World Health Organization, 79*(8), 780–790.

Flores, I., Cayuela, M. L., & Blasco, M. A. (2005). Effects of telomerase and telomere length on epidermal stem cell behavior. *Science, 309*(5738), 1253–1256.

Fontana, L., Partridge, L., & Longo, V. D. (2010). Extending healthy life span--from yeast to humans. *Science, 328*(5976), 321–326.

Goto, T., Teraminami, A., Lee, J. Y., Ohyama, K., Funakoshi, K., Kim, Y. I., et al. (2012). Tiliroside, a glycosidic flavonoid, ameliorates obesity-induced metabolic disorders via activation of adiponectin signaling followed by enhancement of fatty acid oxidation in liver and skeletal muscle in obese-diabetic mice. *Journal of Nutritional Biochemistry, 23*(7), 768–776.

Grünz, G., Haas, K., Soukup, S., Klingenspor, M., Kulling, S. E., Daniel, H., et al. (2012). Structural features and bioavailability of four flavonoids and their implications for lifespan-extending and antioxidant actions in *C. elegans*. *Mechanisms of Ageing and Development, 133*(1), 1–10.

Hada, B., Yoo, M. R., Seong, K. M., Jin, Y. W., Myeong, H. K., & Min, K. J. (2013). D-chiro-inositol and pinitol extend the life span of *Drosophila melanogaster*. *Journals of Gerontology Series A: Biological Sciences and Medical Sciences, 68*(3), 226–234.

Han, B., Sivaramakrishnan, P., Lin, C. J., Neve, I. A. A., He, J., Tay, L. W. R., et al. (2017). Microbial genetic composition tunes host longevity. *Cell, 169*(7), 1249-1262.e13.

Harley, C. B., Futcher, A. B., & Greider, C. W. (1990). Telomeres shorten during ageing of human fibroblasts. *Nature, 345*(6274), 458–460.

Harman, D. (1956). Aging: a theory based on free radical and radiation chemistry. *Journal of Gerontology, 11*(3), 298–300.

Harrison, D. E., Strong, R., Sharp, Z. D., Nelson, J. F., Astle, C. M., Flurkey, K., et al. (2009). Rapamycin fed late in life extends lifespan in genetically heterogeneous mice. *Nature, 460*(7253), 392–395.

He, M., Zhao, L., Wei, M. J., Yao, W. F., Zhao, H. S., & Chen, F. J. (2009). Neuroprotective effects of (-)-epigallocatechin-3-gallate on aging mice induced by D-galactose. *Biological and Pharmaceutical Bulletin, 32*(1), 55–60.

Heintz, C., & Mair, W. (2014). You are what you host: microbiome modulation of the aging process. *Cell, 156*(3), 408–411.

Herrera, E., Samper, E., Martín-Caballero, J., Flores, J. M., Lee, H. W., & Blasco, M. A. (1999). Disease states associated with telomerase deficiency appear earlier in mice with short telomeres. *The EMBO Journal, 18*(11), 2950–2960.

Hou, Y., Lautrup, S., Cordonnier, S., Wang, Y., Croteau, D. L., Zavala, E., et al. (2018). NAD+ supplementation normalizes key Alzheimer's features and DNA damage responses in a new AD mouse model with introduced DNA repair deficiency. *Proceedings of the National Academy of Sciences of the United States of America, 115*(8), E1876–E1885.

Houtkooper, R. H., Pirinen, E., & Auwerx, J. (2012). Sirtuins as regulators of metabolism and healthspan. *Nature Reviews Molecular Cell Biology, 13*(4), 225–238.

Howitz, K. T., Bitterman, K. J., Cohen, H. Y., Lamming, D. W., Lavu, S., Wood, J. G., et al. (2003). Small molecule activators of sirtuins extend *Saccharomyces cerevisiae* lifespan. *Nature, 425*(6954), 191–196.

Huang, T. T., Carlson, E. J., Gillespie, A. M., Shi, Y., & Epstein, C. J. (2000). Ubiquitous overexpression of CuZn superoxide dismutase does not extend life span in mice. *Journals of Gerontology Series A: Biological Sciences and Medical Sciences, 55*(1), B5–9.

Ingram, D. K., & Roth, G. S. (2015). Calorie restriction mimetics: can you have your cake and eat it, too? *Ageing Research Reviews, 20*, 46–62.

Jeong, J., Juhn, K., Lee, H., Kim, S. H., Min, B. H., Lee, K. M., et al. (2007). SIRT1 promotes DNA repair activity and deacetylation of Ku70. *Experimental & Molecular Medicine, 39*(1), 8–13.

Jiao, J., Wei, Y., Chen, J., Chen, X., & Zhang, Y. (2017). Anti-aging and redox state regulation effects of A-type proanthocyanidins-rich cranberry concentrate and its comparison with grape seed extract in mice. *Journal of Functional Foods, 30*, 63–73.

Johnson, S. C., Rabinovitch, P. S., & Kaeberlein, M. (2013). mTOR is a key modulator of ageing and age-related disease. *Nature, 493*(7432), 338–345.

Jung, M. J., Lee, J., Shin, N. R., Kim, M. S., Hyun, D. W., Yun, J. H., et al. (2016). Chronic Repression of mTOR Complex 2 Induces Changes in the Gut Microbiota of Diet-induced Obese Mice. *Scientific Reports, 6*, 30887.

Kamoshita, M., Toda, E., Osada, H., Narimatsu, T., Kobayashi, S., Tsubota, K., et al. (2016). Lutein acts via multiple antioxidant pathways in the photo-stressed retina. *Scientific Reports, 6*, 30226.

Kampkötter, A., Gombitang Nkwonkam, C., Zurawski, R. F., Timpel, C., Chovolou, Y., Wätjen, W., et al. (2007). Effects of the flavonoids kaempferol and fisetin on thermotolerance, oxidative stress and FoxO transcription factor DAF-16 in the model organism *Caenorhabditis elegans*. *Archives of Toxicology, 81*(12), 849–858.

Kanfi, Y., Naiman, S., Amir, G., Peshti, V., Zinman, G., Nahum, L., et al. (2012). The sirtuin SIRT6 regulates lifespan in male mice. *Nature, 483*(7388), 218–221.

Karimi, A., Majlesi, M., & Rafieian-Kopaei, M. (2015). Herbal versus synthetic drugs; beliefs and facts. *Journal of Nephropharmacology, 4*(1), 27–30.

Karuppagounder, S. S., Pinto, J. T., Xu, H., Chen, H. L., Beal, M. F., & Gibson, G. E. (2009). Dietary supplementation with resveratrol reduces plaque pathology in a transgenic model of Alzheimer's disease. *Neurochemistry International, 54*(2), 111–118.

Kenyon, C. J. (2010). The genetics of ageing. *Nature, 464*, 504–512.

Khan, M. M., Ahmad, A., Ishrat, T., Khan, M. B., Hoda, M. N., Khuwaja, G., et al. (2010). Resveratrol attenuates 6-hydroxydopamine-induced oxidative damage and dopamine depletion in rat model of Parkinson's disease. *Brain Research, 1328*, 139–151.

Kiaei, M., Kipiani, K., Petri, S., Chen, J., Calingasan, N. Y., & Beal, M. F. (2005). Celastrol blocks neuronal cell death and extends life in transgenic mouse model of amyotrophic lateral sclerosis. *Neurodegenerative Diseases, 2*(5), 246–254.

Kirkland, J. L., & Tchkonia, T. (2017). Cellular Senescence: A Translational Perspective. *EBioMedicine, 21*, 21–28.

Komatsu, M., Waguri, S., Chiba, T., Murata, S., Iwata, J., Tanida, I., et al. (2006). Loss of autophagy in the central nervous system causes neurodegeneration in mice. *Nature, 441*(7095), 880–884.

Langley, E., Pearson, M., Faretta, M., Bauer, U. M., Frye, R. A., Minucci, S., et al. (2002). Human SIR2 deacetylates p53 and antagonizes PML/p53-induced cellular senescence. *The EMBO Journal, 21*(10), 2383–2396.

Lee, E. B., Ahn, D., Kim, B. J., Lee, S.Y., Seo, H. W., Cha, Y. S., et al. (2015). Genistein from *Vigna angularis* Extends Lifespan in *Caenorhabditis elegans*. *Biomolecules & Therapeutics, 23*(1), 77–83.

Lee, H., & Ko, G. (2014). Effect of metformin on metabolic improvement and gut microbiota. *Applied and Environmental Microbiology, 80*(19), 5935–5943.

Lee, K. S., Lee, B. S., Semnani, S., Avanesian, A., Um, C.Y., Jeon, H. J., et al. (2010). Curcumin extends life span, improves health span, and modulates the expression of age-associated aging genes in *Drosophila melanogaster*. *Rejuvenation Research*, *13*(5), 561–570.

Lee, S. H., & Min, K. J. (2013). Caloric restriction and its mimetics. *BMB Reports*, *46*(4), 181–187.

Lee, S. J., Murphy, C. T., & Kenyon, C. (2009). Glucose shortens the life span of *C. elegans* by downregulating DAF-16/FOXO activity and aquaporin gene expression. *Cell Metabolism*, *10*(5), 379–391.

Lee, Y. S., Kim, W. S., Kim, K. H., Yoon, M. J., Cho, H. J., Shen, Y., et al. (2006). Berberine, a natural plant product, activates AMP-activated protein kinase with beneficial metabolic effects in diabetic and insulin-resistant states. *Diabetes*, *55*(8), 2256–2264.

Lei, H., Wang, B., Li, W. P., Yang, Y., Zhou, A. W., & Chen, M. Z. (2003). Anti-aging effect of astragalosides and its mechanism of action. *Acta Pharmacologica Sinica*, *24*(3), 230–234.

Levine, M. E., Suarez, J. A., Brandhorst, S., Balasubramanian, P., Cheng, C. W., Madia, F., et al. (2014). Low protein intake is associated with a major reduction in IGF-1, cancer, and overall mortality in the 65 and younger but not older population. *Cell Metabolism*, *19*(3), 407–417.

Li, J., Cai, D., Yao, X., Zhang, Y., Chen, L., Jing, P., et al. (2016). Protective Effect of Ginsenoside Rg1 on Hematopoietic Stem/Progenitor Cells through Attenuating Oxidative Stress and the Wnt/β-Catenin Signaling Pathway in a Mouse Model of d-Galactose-induced Aging. *International Journal of Molecular Sciences*, *17*(6), pii: E849.

Li, L., Ng, T. B., Gao, W., Li, W., Fu, M., Niu, S. M., et al. (2005). Antioxidant activity of gallic acid from rose flowers in senescence accelerated mice. *Life Sciences*, 77(2), 230–240.

Li, X. H., Li, C. Y., Lu, J. M., Tian, R. B., & Wei, J. (2012). Allicin ameliorates cognitive deficits ageing-induced learning and memory deficits through enhancing of Nrf2 antioxidant signaling pathways. *Neuroscience Letters*, *514*(1), 46–50.

Li, X. H., Li, C. Y., Xiang, Z. G., Zhong, F., Chen, Z. Y., & Lu, J. M. (2010). Allicin can reduce neuronal death and ameliorate the spatial memory impairment in Alzheimer's disease models. *Neurosciences (Riyadh)*, *15*(4), 237–243.

Li, Y., Xu, W., McBurney, M. W., & Longo, V. D. (2008). SirT1 inhibition reduces IGF-I/IRS-2/Ras/ERK1/2 signaling and protects neurons. *Cell Metabolism*, *8*(1), 38–48.

Li, Y., & Zhang, Z. (2015). Gastrodin improves cognitive dysfunction and decreases oxidative stress in vascular dementia rats induced by chronic ischemia. *International Journal of Clinical and Experimental Pathology*, *8*(11), 14099–14109.

Liao, V. H., Yu, C. W., Chu, Y. J., Li, W. H., Hsieh, Y. C., & Wang, T. T. (2011). Curcumin-mediated lifespan extension in *Caenorhabditis elegans*. *Mechanisms of Ageing and Development*, *132*(10), 480–487.

Lim, G. P., Chu, T., Yang, F., Beech, W., Frautschy, S. A., & Cole, G. M. (2001). The curry spice curcumin reduces oxidative damage and amyloid pathology in an Alzheimer transgenic mouse. *The Journal of Neuroscience*, *21*(21), 8370–8377.

Lin, C. S., Chang, C. J., Lu, C. C., Martel, J., Ojcius, D. M., Ko, Y. F., et al. (2014). Impact of the gut microbiota, prebiotics, and probiotics on human health and disease. *Biomedical Journal*, *37*(5), 259–268.

Liu, M., Yin, Y., Ye, X., Zeng, M., Zhao, Q., Keefe, D. L., et al. (2013). Resveratrol protects against age-associated infertility in mice. *Human Reproduction*, *28*(3), 707–717.

López-Otín, C., Galluzzi, L., Freije, J. M. P., Madeo, F., & Kroemer, G. (2016). Metabolic control of longevity. *Cell*, *166*(4), 802–821.

Lu, J., Wu, D. M., Zheng, Y. L., Hu, B., Zhang, Z. F., Shan, Q., et al. (2010). Quercetin activates AMP-activated protein kinase by reducing PP2C expression protecting old mouse brain against high cholesterol-induced neurotoxicity. *The Journal of Pathology*, *222*(2), 199–212.

Lublin, A., Isoda, F., Patel, H., Yen, K., Nguyen, L., Hajje, D., et al. (2011). FDA-approved drugs that protect mammalian neurons from glucose toxicity slow aging dependent on cbp and protect against proteotoxicity. *PLOS One*, *6*(11), e27762.

Ma, T., Gong, K., Yan, Y., Zhang, L., Tang, P., Zhang, X., et al. (2013). Huperzine A promotes hippocampal neurogenesis in vitro and in vivo. *Brain Research*, *1506*, 35–43.

Martel, J., Ko, Y. F., Liau, J. C., Lee, C. S., Ojcius, D. M., Lai, H. C., et al. (2017). Myths and Realities Surrounding the Mysterious Caterpillar Fungus. *Trends in Biotechnology*, *35*(11), 1017–1021.

Martel, J., Ojcius, D. M., Ko, Y. F., Chang, C. J., & Young, J. D. (2019). Antiaging effects of bioactive molecules isolated from plants and fungi. *Medicinal Research Reviews,* doi: 10.1002/med.21559.

Matsumoto, M., Kurihara, S., Kibe, R., Ashida, H., & Benno, Y. (2011). Longevity in mice is promoted by probiotic-induced suppression of colonic senescence dependent on upregulation of gut bacterial polyamine production. *PLOS One*, *6*(8), e23652.

McCay, C. M., Crowell, M. F., & Maynard, L. A. (1989). The effect of retarded growth upon the length of life span and upon the ultimate body size. 1935. *Nutrition*, *5*(3), 155–171.

Mekheimer, R. A., Sayed, A. A., & Ahmed, E. A. (2012). Novel 1,2,4-triazolo[1,5-a]pyridines and their fused ring systems attenuate oxidative stress and prolong lifespan of *Caenorhabiditis elegans*. *Journal of Medicinal Chemistry*, *55*(9), 4169–4177.

Mikhed, Y., Daiber, A., & Steven, S. (2015). Mitochondrial Oxidative Stress, Mitochondrial DNA Damage and Their Role in Age-Related Vascular Dysfunction. *International Journal of Molecular Sciences*, *16*(7), 15918–15953.

Miller, R. A., Harrison, D. E., Astle, C. M., Baur, J. A., Boyd, A. R., de Cabo, R., et al. (2011). Rapamycin, but not resveratrol or simvastatin, extends life span of genetically heterogeneous mice. *Journals of Gerontology Series A: Biological Sciences and Medical Sciences*, *66*(2), 191–201.

Mlekusch, W., Lamprecht, M., Ottl, K., Tillian, M., & Reibnegger, G. (1996). A glucose-rich diet shortens longevity of mice. *Mechanisms of Ageing and Development*, *92*(1), 43–51.

Morris, B. J., Willcox, D. C., Donlon, T. A., & Willcox, B. J. (2015). FOXO3: A Major Gene for Human Longevity--A Mini-Review. *Gerontology*, *61*(6), 515–525.

Morris, B. J. (2013). Seven sirtuins for seven deadly diseases of aging. *Free Radical Biology and Medicine*, *56*, 133–171.

Morselli, E., Maiuri, M. C., Markaki, M., Megalou, E., Pasparaki, A., Palikaras, K., et al. (2010). Caloric restriction and resveratrol promote longevity through the Sirtuin-1-dependent induction of autophagy. *Cell Death & Disease*, *1*, e10.

Morselli, E., Mariño, G., Bennetzen, M. V., Eisenberg, T., Megalou, E., Schroeder, S., et al. (2011). Spermidine and resveratrol induce autophagy by distinct pathways converging on the acetylproteome. *Journal of Cell Biology*, *192*(4), 615–629.

Mostoslavsky, R., Chua, K. F., Lombard, D. B., Pang, W. W., Fischer, M. R., Gellon, L., et al. (2006). Genomic instability and aging-like phenotype in the absence of mammalian SIRT6. *Cell*, *124*(2), 315–329.

Mouchiroud, L., Houtkooper, R. H., & Auwerx, J. (2013). NAD^+ metabolism: a therapeutic target for age-related metabolic disease. *Critical Reviews in Biochemistry and Molecular Biology*, *48*(4), 397–408.

Murphy, C. T., McCarroll, S. A., Bargmann, C. I., Fraser, A., Kamath, R. S., Ahringer, J., et al. (2003). Genes that act downstream of DAF-16 to influence the lifespan of *Caenorhabditis elegans*. *Nature*, *424*(6946), 277–283.

Murphy, C. T. (2006). The search for DAF-16/FOXO transcriptional targets: approaches and discoveries. *Experimental Gerontology*, *41*(10), 910–921.

Na, L. X., Zhang, Y. L., Li, Y., Liu, L. Y., Li, R., Kong, T., et al. (2011). Curcumin improves insulin resistance in skeletal muscle of rats. *Nutrition, Metabolism & Cardiovascular Diseases*, *21*(7), 526–533.

Oeppen, J., & Vaupel, J. W. (2002). Demography. Broken limits to life expectancy. *Science, 296*(5570), 1029–1031.

Okada, K., Wangpoengtrakul, C., Tanaka, T., Toyokuni, S., Uchida, K., & Osawa, T. (2001). Curcumin and especially tetrahydrocurcumin ameliorate oxidative stress-induced renal injury in mice. *Journal of Nutrition, 131*(8), 2090–2095.

Park, C. H., Lee, J. Y., Kim, M. Y., Shin, S. H., Roh, S. S., Choi, J. S., et al. (2016). Oligonol, a low-molecular-weight polyphenol derived from lychee fruit, protects the pancreas from apoptosis and proliferation via oxidative stress in streptozotocin-induced diabetic rats. *Food & Function*, 7(7), 3056–3063.

Pendergrass, W. R., Li, Y., Jiang, D., & Wolf, N. S. (1993). Decrease in cellular replicative potential in "giant" mice transfected with the bovine growth hormone gene correlates to shortened life span. *Journal of Cellular Physiology, 156*(1), 96–103.

Peng, C., Chan, H. Y., Li, Y. M., Huang, Y., & Chen, Z. Y. (2009). Black tea theaflavins extend the lifespan of fruit flies. *Experimental Gerontology, 44*(12), 773–783.

Pérez, V. I., Van Remmen, H., Bokov, A., Epstein, C. J., Vijg, J., & Richardson, A. (2009). The overexpression of major antioxidant enzymes does not extend the lifespan of mice. *Aging Cell, 8*(1), 73–75.

Prasanth, M. I., Sivamaruthi, B. S., Chaiyasut, C., & Tencomnao, T. (2019). A Review of the role of green tea (*Camellia sinensis*) in antiphotoaging, stress resistance, neuroprotection, and autophagy. *Nutrients, 11*(2), pii: E474.

Rafieian-Kopaei, M., Baradaran, A., & Rafieian, M. (2013). Oxidative stress and the paradoxical effects of antioxidants. *Journal of Research in Medical Sciences, 18*(7), 629.

Rathor, L., Akhoon, B. A., Pandey, S., Srivastava, S., & Pandey, R. (2015). Folic acid supplementation at lower doses increases oxidative stress resistance and longevity in Caenorhabditis elegans. *Age (Dordr), 37*(6), 113.

Ravikumar, B., Sarkar, S., Davies, J. E., Futter, M., Garcia-Arencibia, M., Green-Thompson, Z. W., et al. (2010). Regulation of mammalian autophagy in physiology and pathophysiology. *Physiological Reviews, 90*(4), 1383–1435.

Rodgers, J. T., Lerin, C., Gerhart-Hines, Z., & Puigserver, P. (2008). Metabolic adaptations through the PGC-1 alpha and SIRT1 pathways. *FEBS Letters, 582*(1), 46–53.

Rodrigues, F. A., Prata, M. M., Oliveira, I. C., Alves, N. T., Freitas, R. E., Monteiro, H. S., et al. (2014). Gingerol fraction from *Zingiber officinale* protects against gentamicin-induced nephrotoxicity. *Antimicrobial Agents and Chemotherapy, 58*(4), 1872–1878.

Rogina, B., & Helfand, S. L. (2004). Sir2 mediates longevity in the fly through a pathway related to calorie restriction. *Proceedings of the National Academy of Sciences of the United States of America, 101*(45), 15998–16003.

Rubinsztein, D. C., Mariño, G., & Kroemer, G. (2011). Autophagy and aging. *Cell, 146*(5), 682–695.

Ryu, D., Mouchiroud, L., Andreux, P. A., Katsyuba, E., Moullan, N., Nicolet-Dit-Félix, A. A., et al. (2016,). Urolithin A induces mitophagy and prolongs lifespan in *C. elegans* and increases muscle function in rodents. *Nature Medicine, 22*(8), 879–888.

Sangartit, W., Pakdeechote, P., Kukongviriyapan, V., Donpunha, W., Shibahara, S., & Kukongviriyapan, U. (2016). Tetrahydrocurcumin in combination with deferiprone attenuates hypertension, vascular dysfunction, baroreflex dysfunction, and oxidative stress in iron-overloaded mice. *Vascular Pharmacology, 87*, 199–208.

Satoh, A., Brace, C. S., Rensing, N., Cliften, P., Wozniak, D. F., Herzog, E. D., et al. (2013). Sirt1 extends life span and delays aging in mice through the regulation of Nk2 homeobox 1 in the DMH and LH. *Cell Metabolism, 18*(3), 416–430.

Saul, N., Pietsch, K., Menzel, R., Stürzenbaum, S. R., & Steinberg, C. E. (2009). Catechin induced longevity in *C. elegans*: from key regulator genes to disposable soma. *Mechanisms of Ageing and Development, 130*(8), 477–486.

Saul, N., Pietsch, K., Menzel, R., Stürzenbaum, S. R., & Steinberg, C. E. (2010). The longevity effect of tannic acid in *Caenorhabditis elegans*: Disposable Soma meets hormesis. *Journals of Gerontology Series A: Biological Sciences and Medical Sciences*, *65*(6), 626–635.

Saul, N., Pietsch, K., Stürzenbaum, S. R., Menzel, R., & Steinberg, C. E. (2011). Diversity of polyphenol action in *Caenorhabditis elegans*: between toxicity and longevity. *Journal of Natural Products*, *74*(8), 1713–1720.

Shen, L. R., Xiao, F., Yuan, P., Chen, Y., Gao, Q. K., Parnell, L. D., et al. (2013). Curcumin-supplemented diets increase superoxide dismutase activity and mean lifespan in Drosophila. *Age (Dordr)*, *35*(4), 1133–1142.

Si, H., Fu, Z., Babu, P. V., Zhen, W., Leroith, T., Meaney, M. P., et al. (2011). Dietary epicatechin promotes survival of obese diabetic mice and *Drosophila melanogaster*. *Journal of Nutrition*, *141*(6), 1095–1100.

Solon-Biet, S. M., McMahon, A. C., Ballard, J. W., Ruohonen, K., Wu, L. E., Cogger, V. C., et al. (2014). The ratio of macronutrients, not caloric intake, dictates cardiometabolic health, aging, and longevity in ad libitum-fed mice. *Cell Metabolism*, *19*(3), 418–430.

Sonntag, W. E., Deak, F., Ashpole, N., Toth, P., Csiszar, A., Freeman, W., et al. (2013). Insulin-like growth factor-1 in CNS and cerebrovascular aging. *Frontiers in Aging Neuroscience*, *5*, 27.

Spindler, S. R., Li, R., Dhahbi, J. M., Yamakawa, A., & Sauer, F. (2012). Novel protein kinase signaling systems regulating lifespan identified by small molecule library screening using Drosophila. *PLOS One*, 7(2), e29782.

Spindler, S. R., Mote, P. L., Lublin, A. L., Flegal, J. M., Dhahbi, J. M., & Li, R. (2015). Nordihydroguaiaretic Acid Extends the Lifespan of Drosophila and Mice, Increases Mortality-Related Tumors and Hemorrhagic Diathesis, and Alters Energy Homeostasis in Mice. *Journals of Gerontology Series A: Biological Sciences and Medical Sciences*, *70*(12), 1479–1489.

Srivastava, D., Arya, U., SoundaraRajan, T., Dwivedi, H., Kumar, S., & Subramaniam, J. R. (2008). Reserpine can confer stress tolerance and lifespan extension in the nematode *C. elegans*. *Biogerontology*, *9*(5), 309–316.

Srivastava, S., Sammi, S. R., Laxman, T. S., Pant, A., Nagar, A., Trivedi, S., et al. (2017). Silymarin promotes longevity and alleviates Parkinson's associated pathologies in *Caenorhabditis elegans*. *Journal of Functional Foods*, *31*, 32–43.

Steger, R. W., Bartke, A., & Cecim, M. (1993). Premature ageing in transgenic mice expressing different growth hormone genes. *Journal of Reproduction and Fertility. Supplement*, *46*, 61–75.

Stepaniak, U., Micek, A., Grosso, G., Stefler, D., Topor-Madry, R., Kubinova, R., et al. (2016). Antioxidant vitamin intake and mortality in three Central and Eastern European urban populations: the HAPIEE study. *European Journal of Nutrition*, *55*(2), 547–560.

Sun, J., & Tower, J. (1999). FLP recombinase-mediated induction of Cu/Zn-superoxide dismutase transgene expression can extend the life span of adult *Drosophila melanogaster* flies. *Molecular and Cellular Biology*, *19*(1), 216–228.

Svensson, J., Sjögren, K., Fäldt, J., Andersson, N., Isaksson, O., Jansson, J. O., et al. (2011). Liver-derived IGF-I regulates mean life span in mice. *PLOS One*, *6*(7), e22640.

Thakur, A. K., Rai, G., Chatterjee, S. S., & Kumar, V. (2016). Beneficial effects of an *Andrographis paniculata* extract and andrographolide on cognitive functions in streptozotocin-induced diabetic rats. *Pharmaceutical Biology*, *54*(9), 1528–1538.

Tiganis, T. (2011). Reactive oxygen species and insulin resistance: the good, the bad and the ugly. *Trends in Pharmacological Sciences*, *32*(2), 82–89.

Tissenbaum, H. A. (2012). Genetics, life span, health span, and the aging process in *Caenorhabditis elegans*. *Journals of Gerontology Series A: Biological Sciences and Medical Sciences*, *67*(5), 503–510.

Tissenbaum, H. A., & Guarente, L. (2001). Increased dosage of a *sir-2* gene extends lifespan in *Caenorhabditis elegans*. *Nature*, *410*(6825), 227–230.

Vaiserman, A. M., Lushchak, O.V., & Koliada, A. K. (2016). Anti-aging pharmacology: promises and pitfalls. *Ageing Research Reviews*, *31*, 9–35.
Vaupel, J. W. (2010). Biodemography of human ageing. *Nature*, *464*(7288), 536–542.
Verdin, E. (2015). NAD$^+$ in aging, metabolism, and neurodegeneration. *Science*, *350*(6265), 1208–1213.
Wang, E., & Wink, M. (2016). Chlorophyll enhances oxidative stress tolerance in Caenorhabditis elegans and extends its lifespan. *PeerJ*, *4*, e1879.
Wang, J., Zhang, H. Y., & Tang, X. C. (2010). Huperzine A improves chronic inflammation and cognitive decline in rats with cerebral hypoperfusion. *Journal of Neuroscience Research*, *88*(4), 807–815.
Wang, X. L., Xing, G. H., Hong, B., Li, X. M., Zou, Y., Zhang, X. J., et al. (2014). Gastrodin prevents motor deficits and oxidative stress in the MPTP mouse model of Parkinson's disease: involvement of ERK1/2-Nrf2 signaling pathway. *Life Sciences*, *114*(2), 77–85.
Wen, H., Gao, X., & Qin, J. (2014). Probing the anti-aging role of polydatin in *Caenorhabditis elegans* on a chip. *Integrative Biology*, *6*(1), 35–43.
Willcox, B. J., Donlon, T. A., He, Q., Chen, R., Grove, J. S., Yano, K., et al. (2008). FOXO3A genotype is strongly associated with human longevity. *Proceedings of the National Academy of Sciences of the United States of America*, *105*(37), 13987–13992.
Wilson, M. A., Shukitt-Hale, B., Kalt, W., Ingram, D. K., Joseph, J. A., & Wolkow, C. A. (2006). Blueberry polyphenols increase lifespan and thermotolerance in *Caenorhabditis elegans*. *Aging Cell*, *5*(1), 59–68.
Wood, J. G., Rogina, B., Lavu, S., Howitz, K., Helfand, S. L., Tatar, M., et al. (2004). Sirtuin activators mimic caloric restriction and delay ageing in metazoans. *Nature*, *430*(7000), 686–689.
Xiang, L., Sun, K., Lu, J., Weng, Y., Taoka, A., Sakagami, Y., et al. (2011). Anti-aging effects of phloridzin, an apple polyphenol, on yeast via the SOD and Sir2 genes. *Bioscience, Biotechnology, and Biochemistry*, *75*(5), 854–858.
Xu, M., Pirtskhalava, T., Farr, J. N., Weigand, B. M., Palmer, A. K., Weivoda, M. M., et al. (2018). Senolytics improve physical function and increase lifespan in old age. *Nature Medicine*, *24*(8), 1246–1256.
Yin, J., Gao, Z., Liu, D., Liu, Z., & Ye, J. (2008). Berberine improves glucose metabolism through induction of glycolysis. *American Journal of Physiology-Endocrinology and Metabolism*, *294*(1), E148–E156.
Yu, B. C., Hung, C. R., Chen, W. C., & Cheng, J. T. (2003). Antihyperglycemic effect of andrographolide in streptozotocin-induced diabetic rats. *Planta Medica*, *69*(12), 1075–1079.
Zarse, K., Bossecker, A., Müller-Kuhrt, L., Siems, K., Hernandez, M. A., Berendsohn, W. G., et al. (2011). The phytochemical glaucarubinone promotes mitochondrial metabolism, reduces body fat, and extends lifespan of *Caenorhabditis elegans*. *Hormone and Metabolic Research*, *43*(4), 241–243.
Zhang, S. Q., Cai, W. J., Huang, J. H., Wu, B., Xia, S. J., Chen, X. L., et al. (2015). Icariin, a natural flavonol glycoside, extends healthspan in mice. *Experimental Gerontology*, *69*, 226–235.
Zhang, X., Jin, C., Li, Y., Guan, S., Han, F., & Zhang, S. (2013). Catalpol improves cholinergic function and reduces inflammatory cytokines in the senescent mice induced by D-galactose. *Food and Chemical Toxicology*, *58*, 50–55.
Zhang, Z., Han, S., Wang, H., & Wang, T. (2014). Lutein extends the lifespan of *Drosophila melanogaster*. *Archives of Gerontology and Geriatrics*, *58*(1), 153–159.
Zhao, L. G., Shu, X. O., Li, H. L., Zhang, W., Gao, J., Sun, J. W., et al. (2017). Dietary antioxidant vitamins intake and mortality: a report from two cohort studies of Chinese adults in Shanghai. *Journal of Epidemiology*, *27*(3), 89–97.
Zhu, J., Mu, X., Zeng, J., Xu, C., Liu, J., Zhang, M., et al. (2014). Ginsenoside Rg1 prevents cognitive impairment and hippocampus senescence in a rat model of D-galactose-induced aging. *PLOS One*, *9*(6), e101291.
Zu, Y., Liu, L., Lee, M. Y., Xu, C., Liang, Y., Man, R. Y., et al. (2010). SIRT1 promotes proliferation and prevents senescence through targeting LKB1 in primary porcine aortic endothelial cells. *Circulation Research*, *106*(8), 1384–1393.

CHAPTER 3

Metabolomics and fluxomics studies in the medicinal plant *Catharanthus roseus*

Hebert Jair Barrales-Cureño[a], Jorge Montiel-Montoya[b], José Espinoza-Pérez[c], Juan Antonio Cortez-Ruiz[d], Gonzalo Guillermo Lucho-Constantino[e], Fabiola Zaragoza-Martínez[f], Jesús Antonio Salazar-Magallón[g], César Reyes[h], José Lorenzo-Laureano[h], Luis Germán López-Valdez[i]

[a]Programa Institucional de Maestría en Ciencias Biológicas. Universidad Michoacana de San Nicolas Hidalgo. Calle de Santiago Tapia 403, Centro, Morelia, Michoacán, México

[b]Centro Interdisciplinario de Investigación para el Desarrollo Integral Regional. Unidad Sinaloa, Instituto Politécnico Nacional, Guasave, Sinaloa, Mexico

[c]El Colegio de la Frontera Sur-Unidad San Cristóbal de las Casas. Departamento de Agricultura, Sociedad y Ambiente. Carretera Panamericana y Periférico Sur s/n, Barrio de María Auxiliadora. San Cristóbal de Las Casas, Chiapas – México

[d]Instituto Tecnológico de Mazatlán Instituto Tecnológico de Mazatlán. Ingeniería Bioquímica. México Corsario I 203, Urías, Mazatlán, Sin

[e]Universidad Tecnológica de Gutiérrez Zamora,Veracruz. Prolongación Dr. Miguel Patiño s/n, Centro, 93556 Gutiérrez Zamora, Veracruz

[f]Centro de Investigación y de Estudios Avanzados del Instituto Politécnico Nacional, Av. Instituto Politécnico Nacional 2508, 07360, Mexico City, Mexico

[g]Unit in Development for Research and Technology Transfer in Biological Control, Postgraduate College, San Pedro Cholula, Puebla, México

[h]División de Ciencias Naturales, Universidad Intercultural del Estado de Puebla. Calle Principal a Lipuntahuaca S/N; Lipuntahuaca, Huehuetla, Puebla, Mexico

[i]Universidad Autónoma Chapingo. Carretera Federal México-Texcoco Km 38.5, Universidad Autonoma de Chapingo, Texcoco, México

List of Abbreviations

AFLP Amplified Fragment Length Polymorphism
cDNA Complementary DNA
FBA Flux Balance Analysis
^{1}H NMR Hydrogen-1 Nuclear Magnetic Resonance
ISSR Intersimple Sequence Repeat
MFA Metabolic Flux Analysis
RAPD Random Amplified Polymorphic DNA
SSR Simple Sequence Repeat
STMS Sequence-Tagged Microsatellites Sites
TIA Terpenoid Indole Alkaloid
UHPLC-MS Ultra-Performance Liquid Chromatography-Mass Spectrometry

Medicinal and Aromatic Plants: Expanding their Horizons through Omics
http://dx.doi.org/10.1016/B978-0-12-819590-1.00003-3

Introduction

Medicinal plant *Catharanthus roseus* belongs to Apocynaceae family. *C. roseus* biosynthesize a broad range of terpenoid indole alkaloid such as ajmalicine, which is used as an antihypertensive, serpentine as a sedative, and vinblastine and vincristine as an antileukemic agents. *C. roseus* synthesize more than 150 terpenoid indole alkaloids and produces a broad spectrum of phenolic compounds such as flavonoids, caffeic acid, cinnamic acid, and derivatives, besides the terpenoid indole alkaloid (Ferreres et al., 2008; Filippini et al., 2003). Unfortunately, the production of alkaloids in vivo has been relatively low; until now has not been completely elucidated the biosynthesis pathway of alkaloids in *C. roseus*. The prices of these alkaloids terpenoid indole alkaloid are too high as a result of the low level of plant production, industriousness extraction or extraction and nonfeasible chemical synthesis because of its complex structures. However, the high yield of secondary metabolites in submerged cultures in lab scale is a prerequisite to perform plant cell cultures a great scale by transferring callus tissues (a group of friable and hyaline cells) to liquid medium. The estrectosidine is a central precursor metabolite which consists of more than 150 terpenoid indole alkaloid (Verma, Mathur, Srivastava, & Mathur, 2012). Foliar extracts of *C. roseus* are the unique source of vindoline and catharanthine, both are monomeric precursors of the commercial production of terpenoid indole alkaloid (O'Keefe, Mahady, Gills, Beecher, & Schilling, 1997).

With respect to metabolomics, there are two basic techniques used to determine metabolites through metabolomics analysis—Nuclear magnetic resonance and mass spectroscopy. The use of metabolomics in plant biotechnology is observed in recent applications reported in several plant species, for example, in *Solanum lycopersicum* the accumulation of flavonoids was analyzed and their nutritional value through the mutation of the HP1/LeDDBI gene with the use of LC-ESI-MS/MS technology (Calvenzani et al., 2010). In *Arabidopsis thaliana* metabolomics was used in order to differentiate transgenic plants to the wild types with RMN technology (Ren, Wang, Peng, Xia, & Qu, 2009; Nikiforova, Kopka, & Tolstikov, 2005). In *Oryza sativa* the modulation of saline tolerance was analyzed by reduction of OsSUTI expression (sucrose transporter 1 in *O. sativa*) by using GC-TOF-MS technology (Siahpoosh et al., 2012). In *Panicum virgatum* (switchgrass) the increase of phenolic acids and analogous of monolignol which is associated with the deconstruction of cell wall was determined through GC-MS technology (Tschaplinski et al., 2012). In *Solanum tuberosum* the drought tolerance was increased by the expression of trehalose-6-phosphate synthase 1 with GC-MS technology (Kondrak, Marincs, Antal, Juhasz, & Banfalvi, 2012). In

C. roseus the productivity of anticancer indole alkaloids was increased through the overexpression of ORCA3 and G10H in plants with the use of RMN technology (Pan, Wang, & Yuan, 2012; O'keefe et al., 1997).

On the other hand, several markers systems RAPD, AFLP, ^{1}H-NMR, ISSR, phytochemical/morphochemical, phytochemical/enzymatic, isoenzymes/phytochemical, SSR/STMS) have been used in order to identify several sections of genetic diversity which are present in *C. roseus*. Actually, it is important to know the studies reported with the aid of metabolomics and fluxomics with the purpose of identifying all biosynthesis pathways of terpenoid indole alkaloids. Therefore, the objective of the current chapter is to show the spectrum of metabolomics and fluxomics studies in *C. roseus* from a point of view in the in vitro analysis directly in plants and tissue and elicitation cultures in suspended cells of *C. roseus* and the metabolic changes provoked by elicitation.

Biological activity of *C. roseus*

There are various medicinal effects of *C. roseus* which include acetylcholinesterase and cholinergic antagonism inhibition (*in vitro*: microplate assay; *in vivo*: male Wistar rats) (Pereira et al., 2009; Pereira et al., 2010); Alzheimerś syndrome (human clinical trial) (Singh, Pandey, & Verma, 2011); antihelminthic activity (*Pheretima posthuma*) (Agarwal, Jacob, & Chettri, 2011); antiandrogenic activity in mice (Murugavel and Akbarsha, 1991); antiangiogenesis activity in chicken eggs (Wang et al., 2004); antibacterial (antiseptic) activity (Virmani, Srivaatava, & Singh, 1978; Goyal, Khanna, Chauhan, Chauhan, & Kaushik, 2008; Ramya, Govindaraji, Navaneetha, & Jayakumararaj, 2008; Verma and Singh, 2010); antidysenteric activity in Wistar rats (Hassan, Brenda, Patrick, & Patrick, 2011); antifertility effect in male rat (Prajapati, Tripathi, Jain, Sharma, & Khanuja, 1998); antifungal activity (*Trichophyton rubrum* and *Hendersonula toruloidea*) (Chile and Vyas, 1984; Barde and Singh, 1983); antihyperglycemic activity antidiabetic in mice, Wistar albino rat, rabbit (Vega-Avila et al., 2012; Benjamin et al., 1994; Singh et al., 2001); antihypercholesterolemic activity type antihyperlipidemic in rabbit and rats (Chauhan, Sharma, Rohatgi, & Chauhan, 2011); antiinflammatory activity in rat (Chattopadhyay, Banerjee, Sarkar, Ganguly, & Basu, 1992); antimutagenic activity in micronucleated erythrocytes (Lim-Sylianco and Blanco, 1981); antineoplastic activity in mice, rat (clinical use) (Nobili, Lippi, & Witort, 2009; Dong, Bornmann, Nakanishi, & Berova, 1995El-Merzabani, El-Aaser, Attia, El-Duweini, & Ghazal, 1979El-Sayed, Handy, & Cordell, 1983Johnson, Wright, Svoboda, & Vlantis, 1960; Mukherjee, Basu, Sarkar, & Ghosh, 2001; Noble, 1990); antioxidant activity in rat (Jaleel, Gopi, Alagu Lakshmanan, &

Panneerselvam, 2006; Chauhan et al., 2011; Zheng and Wang, 2001); antiplasmodial activity in human erythrocytes (Gathirwa, Rukunga, & Njagi, 2007; Ponarulselvam et al., 2012); antiproliferative activity in human cells (Mans, da Rocha, & Schwartsmann, 2000; Ueda et al., 2002); antispermatogenic in male rat and mice (Gupta and Sharma, 2006; Joshi and Ambaye, 1968); blood cleanser (Moerman, 2009); cytochrome P450 inhibition (CYP2D6) (Usia, Watabe, Kadota, & Tezuka, 2005); cytotoxic activity in human cell line (Hostettmann, Marston, Ndjoko, & Wolfender, 2000; Siddiqui, Ismail, Aisha, & Abdul Majid, 2010); enhances kidney and liver functions in Wistar rat (Iweala and Okeke, 2005; Adekomi, 2010); epididymal dysfunction in Wistar rat (Averal, Stanley, Murugaian, Palanisamy, & Akbarsha, 1996); generate giant spermatogonial cells in albino rat (Stanley and Akbarsha, 1992); hypotensive activity in rat (Narayana and Dimri, 1990); larvicidal activity in *Anopheles stephensi* (malaria vector) and *Aedes aegypti* (Kuppusamy, Murugan, Arul, & Yasodha, 2009, Remia and Logaswamy, 2010); regression of accessory reproductive organs in male Wistar rats (Akbarsha, Stanley, & Averal, 1995); regression of entire reproductive system in male rat (Stanley, Averal, & Akbarsha, 1993); stomachic; tonic; tranquilizing and sedative action; wound healing in rat. Flower extract of *C. roseus* is used as an acid-base natural indicator.

Phytochemicals of *C. roseus*

Chemical compounds of major importance biosynthesized by *C. roseus* are alkaloids and phenolic compounds, the presence of several chemical groups such as polyphenols, alkaloids, steroids, flavonoid glycosides, anthocyanins, and iridoid glycosides is normally found in several structures of the plant. However, there are some evidences biosynthesis of similar chemical compounds in leaves and stems of the plant, but not in seed and petals.

Main alkaloids of *C. roseus*

Alkaloids are part of a group of secondary metabolites, many of which possess biological functions that are vital not only for own plant survival but also for the human health (De Luca, 2011). *C. roseus* biosynthesize various types of active alkaloids, which possess an indole molecule. More than 150 alkaloids are biosynthesized in the plant and some of them have several medicinal properties. The amount of alkaloids is maximum at the flowering stage. The location of the main alkaloids synthesized in several parts of *C. roseus* is shown in Fig. 3.1.

Mustafa and Verpoorte (2007) have listed the most important phenolic compounds biosynthetized in *C. roseus* (Fig. 3.2).

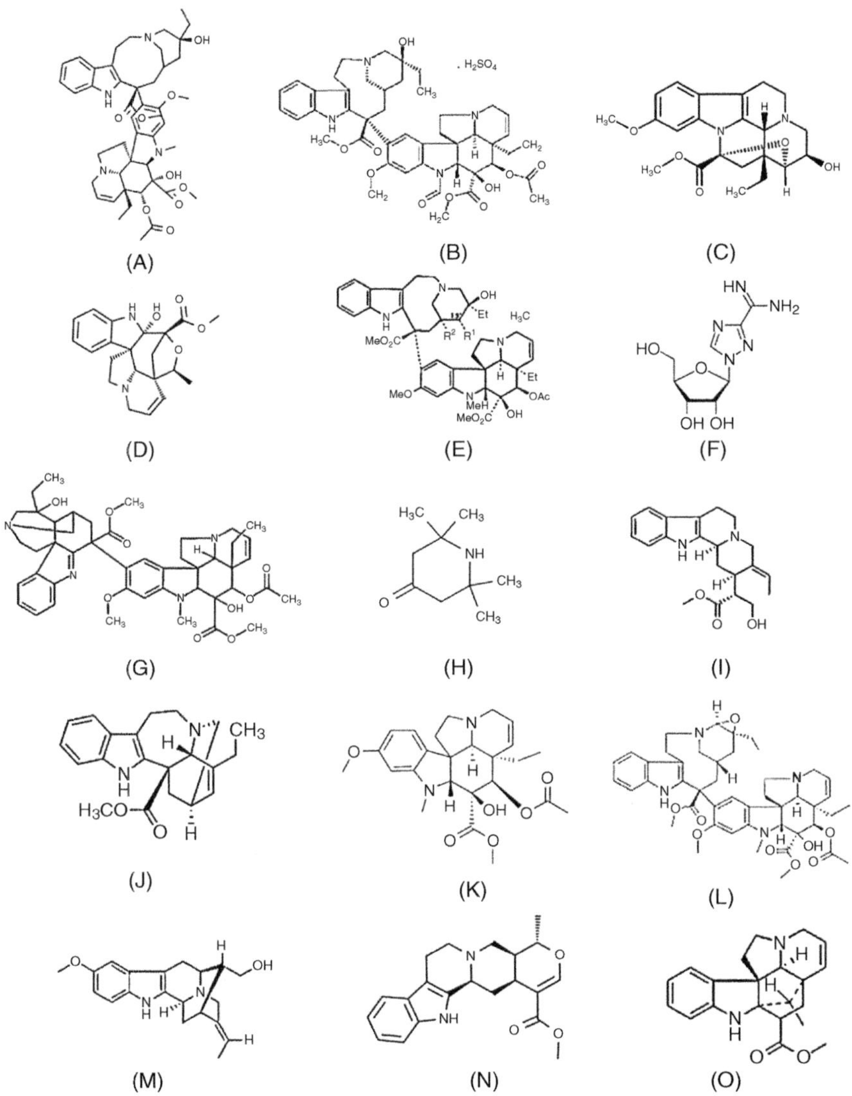

Figure 3.1 Principal active alkaloids located in the aerial parts of the plant (not floral). (A) Vincaleukoblastine (810.98 g/mol, $C_{46}H_{58}N_4O_9$); (B) (leurocristine, vincaleurocristine); (C) Vincarodine (398.45 g/mol, $C_{22}H_{26}N_2O_5$); (D) Vincoline (368.42 g/mol, $C_{21}H_{24}N_2O_4$); (E) Leurocolombine; (F) Viramidine (243.22 g/mol, $C_8H_{13}N_5O_4$); (G) Vincathicine (808.97 g/mol, $C_{46}H_{56}N_4O_9$); (H) Vincubine (156.24 g/mol, $C_9H_{18}NO$); (I) Isositsirikine (354.45 g/mol, $C_{21}H_{26}N_2O_3$); (J) Catharanthine (336.435 g/mol, $C_{21}H_{24}N_2O_2$); (K) Vindoline (456.539 g/mol, $C_{25}H_{32}N_2O_6$); (L) Leurosine (808.96 g/mol, $C_{46}H_{56}N_4O_9$);(Mm) Lochnerine (324.424 g/mol, $C_{20}H_{24}N_2O_2$), (N) Tetrahydroalstonine (352.434 g/mol, $C_{21}H_{24}N_2O_3$) and (O) Vindolinine (336.435 g/mol, $C_{21}H_{24}N_2O_2$).

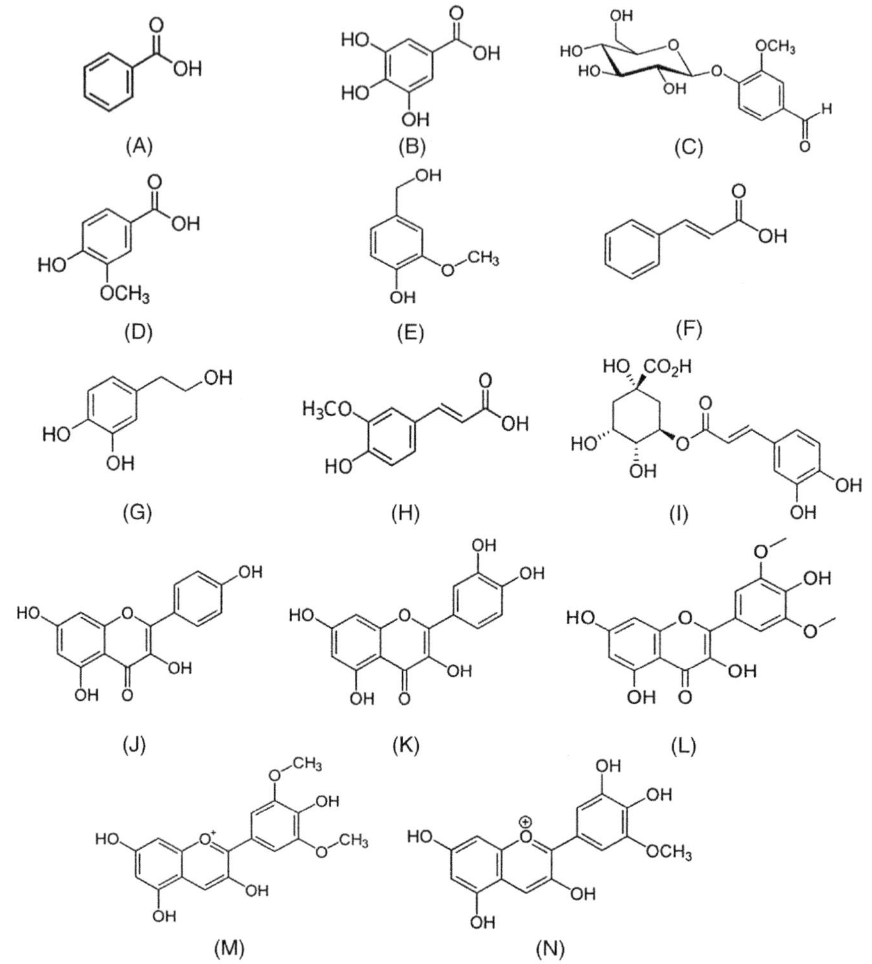

Figure 3.2 (A) Benzoic acid (122.12 g/mol, $C_7H_6O_2$); (B) Gallic acid (170.12 g/mol, $C_7H_6O_5$); (C) Glucovanillin (314.29 g/mol, $C_{14}H_{18}O_8$); (D) Vanillic acid (168.14 g/mol, $C_8H_8O_4$); (E) Vanillyl alcohol (154.17 g/mol, $C_8H_{10}O_3$); (F) *Trans*-cinnamic acid (148.15 g/mol, $C_9H_8O_2$); (G) Hydroxytyrosol (154.16 g/mol, $C_8H_{10}O_3$); (H) Ferulic acid (194.18 g/mol, $C_{10}H_{10}O_4$); (I) Chlorogenic acid (354.31 g/mol, $C_{16}H_{18}O_9$); (J) Kaempferol (286.23 g/mol, $C_{15}H_{10}O_6$); (K) Quercetin (302.23 g/mol, $C_{15}H_{10}O_7$); (L) Syringetin glycosides (346.29 g/mol, $C_{17}H_{14}O_8$); (M) Malvidin (331.29 g/mol, $C_{17}H_{15}O_7^+$); and (N) Petunidin (317.27 g/mol; $C_{16}H_{13}O_7^+$).

Compartmentalization of terpenoid indole alkaloid in *C. roseus*

Specific type cellular location of metabolic pathways of terpenoid indole alkaloid was first inferred by RNA hybridization assays in situ and the location of enzymes involved in the biosynthesis pathways was identified by immunocytochemistry techniques (St-Pierre, Vazquez, & De Luca, 1999).

Anabolism of terpenoid indole alkaloid involves more than 20 enzymatic steps, and every biocatalyst is biosynthesized in different cell types, starting in parenchymatose cells associated to the inner phloem through the epidermal cells both in idioblastic cells (specialized parenchymatose cells) and laticiferous cells (epidermic cells) where hypothetically vindoline and other terpenoid indole alkaloid are accumulated. The metabolic pathway of vinblastine from geranyl pyrophosphate has been elucidated as 31 steps. Caputi et al. (2018) reported the genes that encode absent enzymes, which complete the whole pathway of the vinblastine. Two redox enzymes convert stemmadenine acetate 7 in an unstable molecule, presumably dihydroprecondylocarpine acetate 11, which is desacetoxylated by one or two hydrolases in order to generated through cyclizations type Diels-Alder, tabersonine 2 or catharanthine 3 which finally dimerized into vinblastine and vincristine.

Metabolomics

Various medicinal plants contain secondary metabolites which are traditionally used in disease management. Traditional medicine helps promote an early disease interruption, combined therapy and personalized medicine (Wang et al., 2011a,b). The global market of the traditional medicine is expected to reach 155 billion dollars for the year 2020, for example, the net sales of Chinese herbal medicine achieved 83 billion dollars in 2008 (WHO, 2013). The main case studies performed in vivo and in vitro carried out in the medicinal plant *C. roseus* from a point of view of the metabolomics and fluxomics of the terpenoid indole alkaloid are as below.

Elicitation in cell suspended cultures of *C. roseus*

Namdeo, Patil, & Fulzele (2002) reported an elicitation with fragments of fungal cell walls of *Aspergillus niger*, *Fusarium moniliforme* and *Trichoderma viride* in suspended cultures of *C. roseus*. These authors analyzed the effects of cell walls amount, exposition time and age of the cultivation in the accumulation of ajmalicine. The production of this secondary metabolite was increased three times with the application of fungal elicitors. The maximum production of ajmalicine was around 75 $\mu g\ g^{-1}$ of dry weight in the cultures elicited with *T. viridae* and the maximum yield was about 166 $g\ g^{-1}$ of ajmalicine which was synthesized after about 20 days of cultivation. Authors argued that during the incubation time, the cell cultures added with elicitors affect the ajmalicine synthesis.

Metabolomic analysis in plant tissues of *C. roseus*

Pan et al. (2014) reported that stems and roots can allow the identification of quimomarkers related to the final expression of flower color trough analysis of metabolic profile of leaves. This expression of metabolic profile in leaves, stems, roots, and flowers of *C. roseus* and the possible correlation with all flower colors were analyzed through ^{1}H-NMR, with analysis of multivariate data. The results showed that flower color is characterized by a special metabolite pattern that include anthocyanins, flavonoids, organic acids, and sugar (orange flowers stem and leaves= kaempferol; pink and purple leaves, stems, and flowers= Sucrose, glucose, 2,3-dihydroxybenzoic acid; red flowers and leaves= malic acid, fumaric acid and asparagine). There is a correlation between the metabolites specifically associated with the expression of different flower colors and the metabolite profile of other organs of the plant and therefore, is possible to predict the flower color through specific metabolite detection in leaves, stems and roots. This fact could be an interesting application in the breeding industry. Yamamoto et al. (2016) demonstrated that catharanthine, ajmalicine, serpentine and estrictosidine are the alkaloids commonly accumulated in laticiferous and idioblastic cells; this phenomenon was corroborated through MS/MS analysis of MS spectrum of single cell from stem tissue of *C. roseus*. Liu et al., (2016) studied the combined effects of ethylene and methyl jastomate over the profiles of phenolic compounds and genetic expression in *C. roseus*. These authors, whom used a nonspecific metabolomics method, identified 34 types of phenolic compounds in the leaves; 7 of which were C_6C_1 (salicylic acid, benzoic acid), 11 were C_6C_3 (cinnamic acid, sinapic acid) and 16 were $C_6C_3C_6$.

Metabolomic analysis in suspended cell cultures of *C. roseus*

Rianika, Kyong, Hae, & Verpoorte (2009) analyzed the effect of salicylic acid in the metabolomics profile of *C. roseus* cells through nuclear magnetic resonance and multivariate analysis. The results of these authors demonstrated that elicitation with salicylic acid increased sugar level (glucose and sucrose) after 40 hours of acid addition, besides a dynamic change in the amount of amino acids, fenilpropanoides and tryptamine was observed. In these experiments, when salicylic acid was added, 2,5-dihydroxybenzoic-5-O-glucoside was detected. The biosynthesis of terpenoid indole alkaloid and phytosterols based on metabolomics and nuclear magnetic resonance analysis in cell cultures of *C. roseus* was investigated. Rischer et al. (2006) generated profiles of genome transcription through amplified fragment length polymorphism of cDNA and related to the metabolomics profile of cell cultures of *C. roseus*. It is interesting to note that such authors generated a collection of transcription labels known and not previously described before and metabolites associated with terpenoid indole alkaloids.

Fluxomics

In recent years, omics technologies have been demonstrating their usefulness in a large number of scientific disciplines, including health, bioinformatics, pharmacology and biotechnology fields. Most of the omics technologies known today are genomics, proteomics, and metabolomics, which seek to obtain a complete description of cell changes at genes expression level, protein synthesis and metabolite production that occur by certain disturbances in the living beings, for example, disease or stress conditions (Barabasi and Oltvai, 2004). Unlike other omics technologies, fluxomics has become a tool for the prediction of genes involved in the regulation of flux (flow of raw material) through the metabolic networks (Wiechert, Siefke, de Graaf, & Marx, 1997; Zamboni and Sauer, 2009). Every flow indicates the function of specific metabolic pathway during metabolism and it can define the phenotype of any organism (Ratcliffe and Shachar-Hill, 2005).

Fluxomics and metabolomics are recent omics applications (Winter and Kromer, 2013). Metabolomics analyzes all metabolites (sugars, amino acids, fatty acids) which are present in every cell, organ, and organism; while fluxomics is responsible for determining the metabolic flow. The later is considered as a biological top technique and their results represent the gene expression, protein production and its kinetics as well as the pathway metabolite regulation and synthesis (Winter and Kromer, 2013). The nature of metabolic pathways and its fundamental role in the living beings has been subject of research since the beginning of the last century (Kohler, 1977). At first, research was focused on the structural elucidation of metabolic pathways by using isotopic tracers. An isotopic tracer is a molecule in which an atom has been replaced by a different isotope in a radioactive state (^{14}C, ^{3}H) or steady state (^{13}C, $^{2}H^{18}O$, ^{15}N) (Klein y Heinzel, 2012). In the 1930ś, the isotopic tracers began to be used in the cell metabolism study (Kohler, 1977). Radioisotopes (^{3}H, ^{14}C) were used in order to catalogue many types of metabolic pathways (Wolfe, 1984). Until the 1980s, the use of radioactive tracers prevailed, later the stable isotopes become in a viable alternative because of technical advances of nuclear magnetic resonance (NMR), mass spectroscopy (MS), and computational sciences (Bier, 1987).

Empirical methods used for data analysis that were able to associate NMR and MS data by using carbon fluxes were the most common method used between the 80ś and 90ś, prior to the appearance of ^{13}C-MFA. These models were developed specifically to perform studies that involved radioisotopes; although the models showed several limitations and often were based on simplifiers with the purpose of calculations were more manageable. These studies were simple when compared with the current approaches of the Analysis of Metabolic Flow (^{13}C-MFA), but it gave origin to the quantification of substrate consumption, or fluxes of TAC cycle and determination of anaplerotic flow by using a variety of tracers such as pyruvate, acetate, lactate, glucose and fatty acids (Comte, Vicent, Bouchard, & Des Rosiers, 1997a;Comte et al., 1997b; Jones, Sherry, Jeffrey, Storey, & Malloy, 1993, Malloy, Sherry, & Jeffrey, 1990).

The methodology used by the fluxomics has its origins from the beginning of the 80ś. Current approaches of ^{13}C-MFA are based on equilibrium techniques of isotopomers which were established in the mid-1990s (Schmidit, Carlsen, Nielsen, & Villadse, 1997; Zupke and Stephanopoulos 1994). In the last decade, the quantitative evaluation of metabolic flows by ^{13}C-MFA has been applied in several research fields that include metabolic engineering, systems biology, microbiology and biomedical research; Moxley et al., (2009). The research based on ^{13}C-MFA has been enhanced by computational developments as well friendly computer programs such as ^{13}C flux, mentra (Yoo, Antoniewicz, Stephanopoulos, & Kelleher, 2008) and open flux (Quek, Wittmann, Nielsen, & Kromer, 2009), among others. With respect to the Flux Balance Analysis (FBA), the first studies were conducted by Papoutsakis in 1984, who reported the use for the first time of linear programming in order to estimate the maximum theoretical yields by bacteria during the production of butyric acid (Papoutsakis, 1984). Two years later, Fell and Small used linear programming for the study of lipogenesis (Fell and Small, 1986). In 1992, Sanivell and Palsson reported a detail analysis and development of the theory of the flux balance analysis (Savinell and Palsson, 1992).

Fluxomics or to be more precise, metabolic flux analysis-^{13}C (^{13}C-MFA), is a well-established discipline that integrates measurements in vivo of metabolic flux in a stoichiometry manner in order to determine the absolute flow through the central carbon metabolism network (Winter and Kromer 2013). MFA consists in the use of stable substrates, marked with isotope ^{13}C, this analysis has as objective the quantification of small molecules through the metabolic networks, besides to provide access to the biochemical activity in vivo as well as the elucidation of metabolic pathways in alive and intact cells (Winter and Kromer 2013). Thus, the fluxomics integrates the different components (proteins, transcripts, and metabolites) in a biological network, therefore, play a key role in systems biology and metabolic engineering (Kohlstedt, Becker, & Wittman, 2010). Metabolic flow consists in a material balance approach, which can be defined by the conversion rate of a specific metabolic precursor into a product. This can be determined by the rate of disappearance of substrate and by the accumulation of product through measurements of flow tracers (^{14}C or ^{13}C) obtained from the marked substrate; this method can be used in dynamic studies through changes in substrate concentration respect to the time (Stephanopoulos and Sinskey, 1993).

Research performed through fluxomics has made important progress in a wide range of applications, for example in the phenotype recognition of certain diseases (e.g., diabetes and cancer); development of noninvasive methods with the purpose of analysis of metabolism through blood and urine tests; reengineering of cell phenotypes in order to produce high value-added metabolites, especially when is used renewable raw material as substrate and the application of metabolic engineering. In systematic disorders such as diabetes mellitus, it is well known that a whole flux network is altered drastically. Metabolic remodeling is produced at genome, transcriptome and proteome level, including posttranslational modifications; at the end, the enzymatic activities and metabolite profiles reflect all these changes. Metabolite profile can be translated as a cluster flows which becomes in the metabolome (Cortassa, Vaceres, Bell, O'Rourke, & Paolocci, 2015). Several studies have been conducted to analyze the effects of medicines such as the use of ^{13}C tracers in the cell line earpiece murine (HL-1) with the purpose of making known the effects of verapamil, a calcium channel blocker that is used to treat hypertension, the results showed that the drug reduces significantly the flux of carbon on the glycolysis pathway, while affect in minor degree the flow in the TCA cycle, suggesting a possible explanation of its anticancer potential of the drug (Strigun et al., 2011).

In the medical field, MFA has been used to analyze the enzymes with the ability to reduce lipid accumulation in adipocytes in the context of obesity (Si, Yoon, & Lee, 2009). The particular case of macrophages, a defensive cell of the immune system, little is known about their metabolism, therefore MFA has been used in order to know if the metabolic state has a preponderant influence in their activation. Have been performed several studies by mean of isotope tracers in primary murine macrophages cultures and was observed that the activation of macrophages exhibits a similar behavior to those cells found in tumors which favored the glycolysis pathway over the oxidative metabolism (Rodríguez-Prados et al., 2010). MFA analysis can also be used to determine the flow of metabolites in plants, mainly in suspended cells, transformed hairy roots, stems, and leaf cultures (Allen et al., 2009). A particular case of the use of MFA is in studies of shear stress, which relate to flow intensity with the cell phenotype in response to the shear rate, environmental conditions or plant physiology state (Ma, Jazmin, Young, & Allen, 2014).

Fluxomics in plants

From all living organisms, plants have the most complex metabolic networks, due to the great variation of the yields in the metabolism as well as to the environmental conditions that are exposed (Stitt, Sulpice, & Keurentjes, 2010). Due to the growth and survival, plants are strongly associated with the metabolism, it is interesting to understand and to predict the behavior of the metabolism and relate it with the genotype of plants in order to design new cultures with high yields in special those species of pharmacological interest. The measurement of metabolic flow is a well-established technique in the metabolism analysis. In plants it is used for the characterization of net flow through of specific metabolic pathways, for example, $[U^{13}C_6]$-glucose was used in *Arabidopsis* for the analysis of the phosphate pentose pathway (Masakapalli et al., 2010). The flow measurements can be performed through the determination of substrate and product concentrations or it can be determined through tagged precursor tracers (ap Rees and Hill 1994).

Fluxomics has the purpose of determining the flows (interchange rates or molecules transfer) in a determined biological system, from a single cell to a whole plant. The flow measurement represents the same process that occurs in a single cell, when is compared with a tissue, organ, or whole plant. Since the study of the flow at cell level only consists of intracellular traffic (through molecular transporters) which are associated a metabolic network and only can be used to determine the efficiency use of nutrients, interchange rates between organs, xylem, and phloem or inside of the photosynthesis process (Salon et al., 2017). The methods have been developed to decipher the metabolic pathways such as metabolic flux analysis (MFA) and the flux balance analysis (FBA). The MFA is used in order to estimate metabolic flows, therefore an isotope labeled substrate is supplied, sometimes radioactive but stable with the purpose of analyze metabolic intermediaries through the distribution of C labeled (Alonso, Val, & Schachar, 2011). Tracer provides a quantitative description of cellular phenotype (it means a map of flow) based on stoichiometry balance. This type of analysis has been used to study heterogeneous tissues which were found in a metabolic steady state (Buescher et al., 2015). The nonsteady state isotope AFM method has been developed to analyze the flux dynamics with respect to the time.

Plant tissues have been studied by using different AFM focuses, including submerged cell cultures of *Arabidopsis thaliana* (Baxter, Liu, Fernie, & Sweetlove, 2007a, Baxter et al., 2007b; Kruger, Le, Brown, & Ratcliffe, 2007; Williams et al., 2008), *Lycopersicon esculetum cerastiforme* (Rontein, Dieuaide-Noubhani, Mufourc, Raymond, & Rolin, 2002) and *Oryza sativa* (Matsuda, Wakasa, Miyagawa, 2007). Also, several studies have been reported through the development of seeds of *Zea mays* (Glawisching, Gierl, Tomas, Bacher, & Eisenreich, 2002; Ettenhuber et al., 2005), *Helianthus annuus* (Alonso, Goffman, Ohlrogge, & Shachar-Hill, 2007); *Brassica napus* (Junker, Lonien, Heady, Rogers, & Schwender, 2007); *Glycine max* (Iyer et al., 2008; Allen et al., 2009); *Hordeum vulgare* (Grafahrend-Belau, Schreiber, Koschutzki, & Junker, 2009). As well as in stems of *Z. mays* (Uys, Botha, Hofmeyr, & Rohwer, 2007), leaves (McNeil, Nuccio, Rhodes, Schachar-Hill, & Hanson, 2000) and flowers (Boatright, Shmind, & Tramper, 1997; Orlova et al., 2006), besides in transformed hairy roots cultures (Siriam, Fulton, & Shank, 2007).

Also, it has been possible through AFM to establish if ribulose-1,5-bisphosphate carboxylase/oxygenase can work without the Calvin cycle in order to recycle carbon which has been lost in the form of CO_2 through the lipid synthesis in oilseeds (Schwender, Goffman, Ohlrogge, & Schahar-Hill, 2004; Allen et al., 2009). On the other hand, MFA analysis has also revealed different flow modes in the tricarboxylic acid cycle (TCA) (Sweetlove, Beard, Nunes-Nesi, Fernie, & Ractcliffe, 2010) as well as the clear demonstration of the stability of the central carbon metabolism in the face of environmental stress (Williams et al., 2008) and relationships between fluxes, metabolism levels, and enzymatic activities (Junker et al., 2007; Kruger & Ratcliffe 2009). It is important to mention that the labeling of substrate with ^{13}C does not disrupt the flux through the biochemical reactions in a metabolic network (Kruger et al., 2007). Besides, the MFA analysis can be used to probe the impact of the environmental disturbances in the working of the central metabolic network such as the TCA cycle (William et al., 2008). Also, it has been observed by means of MFA that anabolic flows (starch, polysaccharides and organic acids biosynthesis) and constitute the flexible portion of metabolism of plant cells, such metabolism fluctuations in relation to the rate growth of the cells (Rontein et al., 2002).

The response of the plants to the oxidative stress has been widely reported, actually, the dynamic of metabolic changes in *Arabidopsis thaliana* cells have been broadly characterized by MFA during the oxidative stress generated by the presence of menadiones and [U-^{13}C-]- glucose. In these studies, it was evident that oxidative stress has a profound effect over the central metabolic pathways and the metabolic inhibition ofTCA cycle occurs in large sections of amino acids metabolism, besides it was evident a metabolic change on anabolic to catabolic pathways. On the other hand, it was found that the regulatory pathways and stress signaling of the plants and microorganisms could share common elements (Baxter et al., 2007b).

The modeling by flux balance analysis (FBA) is an approach based on restrictions, where flows are predicted and involves algorithm optimization within a space linked experimentally. When metabolic processes of plant cells are analyzed by FBA is necessary take into account the metabolic functions of the plants which are based on interactions between different subcellular compartments of the cell, tissues, and organ types, therefore, the reconstruction of specific models of organs and plant cells is a prior requirement to achieve metabolic information by FBA (Grafahrend-Belau et al., 2013). Most models made by FMA analysis for plant metabolism study are restricted to a specific tissues or organs (Pilalis, Chatziioannou, Thomasset, & Kolisis, 2011 ; Mintz-Oron et al., 2012). In leaves of *Zea mays*, *Saccharum officinarum* and *Sorghum bicolor* has been reported a model that takes into account the interaction between two types of cells in the leaf sheaths during the photosynthesis (Dal'Molin, Quek, Palfreyman, Brumbley, & Nielsen, 2010).

FBA analysis has also been applied in *Arabidopsis* by the analysis of ATP consumption provoked by stimulation or production of amino acids, lipids, starch, cellulose (Poolman, Miguet, Sweetlove, & Fell, 2009) and vitamin E (Mintz-Oron et al., 2012). This technique been performed in the endosperm of barley seeds (*Hordeum vulgare*) to study the metabolic balance flow when the seeds were exposed to oxygen starvation and the study demonstrated that inorganic pyrophosphate keeps active in the seed metabolism (Grafahrend-Belau et al., 2009, 2013). Several models have been developed in *Brassica napus* that allow the simulation of some vital physiological conditions during oil production (Hay and Schwender 2011; Pilalis et al., 2011). Besides, in *Oriza sativa* cells, it has been analyzed by metabolic balance flow in order to determine the ideal conditions that are able to produce biomass precursors; the results showed that redox reactions and transporters between the chloroplasts, cytosol, and mitochondria play an important role in low

Table 3.1 Studies of Flux Balance Analysis in some plant species.

Plant	Tissue/culture	Substrate	Metabolic pathway	References
Arabidopsis thaliana	Suspension cell cultures	U-^{13}C-Glucose	TCA cycle, glycolysis, phosphate pentose pathway	Baxter et al. (2007b)
Arabidopsis thaliana	Suspension cell cultures	1-^{13}C- Glucose, 2-^{13}C- Glucose, U-^{13}C-Glucose	TCA cycle	William et al. (2008)
Oryza sativa	Suspension cell cultures	1-^{13}C-Serine	Tryptophan biosynthesis	Matsuda et al. (2007)
Zea mays	Seeds	U-^{13}C-Glucose, U-$^{13}C_{12}$-Sacarose	Starch biosynthesis	Glawisching et al. (2002)
Helianthus annuus	Seeds	1-^{13}C-Glucose, 2-^{13}C-Glucose, U-$^{13}C_5$-Glutamine	Phosphate pentose pathway	Alonso et al. (2007)

intensity levels of light conditions, besides photorespiration is activated in order to dissipate the light excess.

In FBA analysis, the results can be affected by the cell biomass presence and by the model used; but the models only work if steady state conditions are considered, FBA has the advantage of not requiring the understanding of kinetic parameters, therefore, this analysis can be applied for modeling detailed systems at great scale, Table 3.1

Catharanthus roseus and fluxomics

The tropical plant *Catharanthus roseus* (L) G. Don is one of the most studied plants because it is considered as a medicinal plant. *C. roseus* synthesize terpene indole alkaloids (TIA) with high medicinal and economical value such as ajmalicine, catharanthine, vindoline, vinblastine and vincristine, the last are considered anti-tumor agents (Verpoorte, Van der Heijden, Schripsema, Hoge, & Ten Hoopen, 1993). *C. roseus* produces secondary metabolites with high commercial value; therefore, intensive studies have been performed to optimize culture media, supplied of biosynthetic precursors which are currently obtained from shikimate and mevalonate pathways as well as by mean of the use of genetic engineering with the purpose of increase the production *in vivo* of alkaloids (Whitmer, Canel, Hallard, Goncalves, & Verpoorte, 1998).

Numerous studies have been reported in *C. roseus* where several precursors were added into the culture media. Several effects have been reported in the construction of indole and terpenoids building blocks, but in an inconsistent way. For example, the tryptophan addition into the culture media increases the production of tryptamine but without adverse effects on TIA production (Mérillon, Doireau, Guillot, Chenieux, & Rideau, 1986; Facchini and Di Cosmo, 1991). On the other hand, other authors reported the tryptophan and tryptamine addition have a negative effect on alkaloid accumulation (Dòller, Alfermann, & Reinhard, 1976). While exogenous secologanin does not show an apparent effect over alkaloids production (Facchini and Di Cosmo, 1991; Moreno, Van der Heijden, & Verpoorte, 1993). Due to the contradictory results, other strategies have been used which consider that the plant metabolism is a complex network due the pathways are correlated with each other and their function affects to the others. The secondary metabolic pathways in plant cells are strictly regulated and in the case of additional stimuli (e.g., stress or elicitation), the metabolic flows change as a response of natural plant self-defense process.

In *C. roseus*, experiments carried out with an isotope tracers ^{13}C have been conducted in submerged cell and in hairy roots cultures, where have been achieved flow measurements of several pathways. However, due to the complexity of the metabolic network; there are no reports of alkaloid biosynthesis on whole and intact plant (Whitmer et al., 1998). When a culture medium was fed with [1-^{13}C] glucose in cell cultures of *C. roseus,* and analyzed with spectroscopy ^{13}C NMR, the elucidation of secologanin biosynthesis was possible and it was documented that its molecule is synthesized from the triose phosphate pathway (Contin, van der Heijden, Lefeber, & Verpoorte, 1998). The quantification of central carbon flow metabolism on hairy roots of *C. roseus* has been analyzed by means of ^{13}C isotope labeling, and the quantification of interferences caused by the activation of the two pathways of isoprenoids biosynthesis, also has been reported in submerged cultures of *C. roseus* (Siriam et al., 2007; Schuhr et al., 2003).

On the other hand, fluxomics has been used for elicitation studies with the purpose to increase the yields of AIT when an elicitor is used. The elicitors such as jastomate and salicylic acid activates the plant defenses among which the increase of production of secondary metabolites (El-Sayed and Verpoorte, 2007; Lackman et al., 2011). Antonio et al., (2013) reported the use of suspended cell cultures of *C. roseus* in order to study the flow changes after an elicitation with methyl jastomate through the addition of pyruvate-$^{13}C_3$ which was determined by gas chromatography-mass spectrometry (GC-MS) and ultra-performance liquid chromatography-mass spectrometry (UHPLC-MS). It was found that elicitation got achieved disturbances in several metabolic pathways which were deduced by the differences in ^{13}C incorporation. This phenomenon was increased mainly by the accumulation of luganic acid, a precursor of loganin, and its molecule has analgesic activity and stimulates the central nervous system.

The studies performed in intact plants of *C. roseus* are limited because it is a metabolic system more complex to those normally found in suspended cells or hairy roots cultures, these systems only contain few types of different cells. Although it should be noted that some values TIA (vindoline, vinblastine, and vincristine) only are produced in the leaf and not in the submerged cell culture conditions (Sweetlove and Fernie 2005, Hughes, Hong, Gibson, & shanks, 2004). For this reason, a fed with $[_1C^{13}]$-glucose through the root of the plant has been performed and the authors reported that its molecule was incorporated into the roots with the aid of some amino acids (alanine, threonine, arginine, glutamine, glutamate, asparagine, aspartate, and malate) which were identified by $[_1C^{13}]$-glucose tracers in roots, stems and under the leaves. On the other hand, chlorogenic acid was found only in the leaves (Pan, Mustafa, Verpoorte, & Tang, 2016). This study opens up possibilities to make studies by MFA analysis in whole plants in vitro. Computational alternatives known as "bondomers," have been developed which become as an alternative to the isopotomers currently used in MFA analysis because the quantification of flows should be efficient and accurate. For example, in hairy roots cultures of *C. roseus* a study was reported by using U-^{13}C and bondomers; in that study parallel ways in the phosphate pentose pathway at cytosol and plastid level were identified, besides anapletoric flows between phosphoenolpyruvate and oxaloacetate at the cytosol and between pyruvate and mitochondria was reported (Siriam et al., 2007).

Conclusions

The medicinal plant *C. roseus is* an example of biological model in which several studies of metabolomics and fluxomics has been performed. Although, today it is possible to know the exact location where the alkaloids are produced into the cellular organelles, both techniques, metabolomics and fluxomics offer clues in order to know the details of terpenoid indole alkaloids. Moreover, it is also possible now to known other metabolic pathways and through metabolomics and fluxomics, it has been demonstrated that metabolite biosynthesis inside of the pathways are linked to the hormone regulations. Finally, it is possible to identify a select number of genes and metabolomics and fluxomics can contribute to the discovery of new genes and unknown metabolites in the pathway that possibly participate in the biosynthesis of terpenoid indole alkaloids.

References

Adekomi, D. A. (2010). Madagascar periwinkle (*Catharanthus roseus*) enhances kidney and liver functions in Wistar rats. *European Journal of Anatomy, 14*(3), 111–119.

Agarwal, S., Jacob, S., Chettri, N., et al. (2011). Evaluation of *invitro* antIhelminthic activity of *Catharanthus roseus* extract. *The International Journal of Pharmaceutical Sciences and Drug Research, 3*(3), 211–213.

Akbarsha, M. A., Stanley, A., & Averal, H. I. (1995). Effect of vincristine on Leydig cell and accessory reproductive organs. *Current Science, 68*(10), 1053–1057.

Allen, D. K., Ohlrogge, J. B., & Shachar-Hill, Y. (2009). The role of light in soybean seed filing metabolism. *Plant Journal, 58*, 220–234.

Alonso, A. P., Val, D. L., & Schachar-Hill, Y. (2011). Central metabolic fluxes in the endosperm of developing maize seeds and their implications for metabolic engineering. *Metabolic Engineering, 13*, 96–107.

Alonso, A. P., Goffman, F. D., Ohlrogge, J. B., & Shachar-Hill, Y. (2007). Carbon conversion efficiency and central metabolic fluxes in developing sunflower (*Helianthus annuus* L.) embryos. *The Plant Journal, 52*, 296–308.

Antonio, C., Mustafa, N. R., Osorio, S., Tohge, T., Giavalisco, P., Willmitzer, L., Rischer, H., Oksman-Caldentey, K. M., & Fernie, A. R. (2013). Analysis of the interface between primary and secondary metabolism in *Catharanthus roseus* cell culture using 13C-stable isotope feeding. *Molecular Plant, 6*, 581–584.

Averal, H. I., Stanley, A., Murugaian, P., Palanisamy, M., & Akbarsha, M. A. (1996). Specific effect of vincristine on epididymis. *Indian Journal of Experimental Biology, 34*(1), 53–56.

Barabasi, A. L., & Oltavi, Z. N. (2004). Network biology: understanding the cellś functional organization. *Natural Reviews Genetics, 5*, 101–113.

Barde, A. K., & Singh, S. M. (1983). Activity of plant extracts against *Scytalidium anamorph* of *Hendersonula toruloidea* causing skin and nail diseases in man. *Indian Drugs, 20*, 362–364.

Baxter, C. J., Liu, J. L., Fernie, A. R., & Sweetlove, L. J. (2007a). Determination of metabolic fluxes in a non-steady-state system. *Phytochemistry, 68*, 2313–2319.

Baxter, C. J., Redestig, H., Schauer, N., Repsilber, D., Patil, K. R., Nielsen, J., Selbig, J., Liu, J., Fernie, A. R., & Sweetlove, L. J. (2007b). The metabolic response of heterotrophic *Arabidopsis* cells to oxidative stress. *Plant Physiology, 143*, 312–325.

Benjamin, B. D., Kelkar, S. M., Pote, M. S., Kaklij, G. S., Sipahimalani, A. T., & Heble, M. R. (1994). *Catharanthus roseus* cell cultures: growth, alkaloid synthesis and antidiabetic activity. *Phytotherapy Research*, *8*(3), 185–186.

Bier, D. M. (1987). The use of stable isotopes in metabolic investigation. *Baillierés clinical endocrinology and metabolism*, *1*, 817–836.

Boatright, H. J., Shmind, G., & Tramper, J. (1997). Flux analysis of underdetermined metabolic networks: the quest for the missing constraints. *Trends in Biotechnology*, *15*, 308–314.

Buescher, J. M., Antoniewicz, M. R., Boros, L. G., Burgess, S. C., Brunengraber, H., Clish, C. B., et al. (2015). A roadmap for interpreting 13C metabolite labeling patterns from cells. *Current Opinion Biotechnology*, *34*, 189–201.

Calvenzani, V., Martinelli, M., Lazzeri, V., et al. (2010). Response of wild-type and high pigment-1 tomato fruit to UV-B depletion: flavonoid profiling and gene expression. *Planta*, *231*(3), 755–765.

Caputi, L., Franke, J., Farrow, S. C., Chung, K., Payne, R. M. E., Nguyen, T. D., et al. (2018). Missing enzymes in the biosynthesis of the anticancer drug vinblastine in Madagascar periwinkle. *Science*, *360*, 1235–1239.

Chattopadhyay, R. R., Banerjee, R. N., Sarkar, S. K., Ganguly, S., & Basu, T. K. (1992). Antiinflammatory and acute toxicity studies with the leaves of *Vinca rosea* linn. in experimental animals. *Indian Journal of Physiology and Pharmacology*, *36*(4), 291–292.

Chauhan, K., Sharma, S., Rohatgi, K., & Chauhan, B. (2011). Antihyperlipidemic and antioxidative efficacy of *Catharanthus roseus* Linn [Sadabahar] in streptozotocin induced diabetic rats. *Asian Journal of Pharmaceutical and Health Sciences*, *2*(1), 235–243.

Chile, S. K., & Vyas, K. M. (1984). Efficacy of *Vinca rosea* extracts against protease from human pathogenic strains of *Trichophyton rubrum* Sab. *Hindustan Antibiotics Bulletin*, *26*(3–4), 114–116.

Comte, B., Vicent, G., Bouchard, B., & Des Rosiers, C. D. (1997a). Probing the oridin of acetyl-CoA and oxaloacetate entering the citric acid cycle from the 13C labeling of citrate released by perfused rat hearts. *Journal of Biological Chemistry*, *272*, 26117–26124.

Comte, B., Vicent, G., Bouchard, B., Jette, M., Cordeaus, S., & Rosiers, C. D. (1997b). A 13C mass isotopomer study of anaplerotic pyruvate carboxylation in perfused hearts. *Journal of Biological Chemistry*, *272*, 26125–26131.

Contin, A., van der Heijden, R., Lefeber, A. M., & Verpoorte, R. (1998). The iridoid glucoside secologanin is derived from the novel triose phosphate/piruvate pathway in a *Catharanthus roseus* cell culture. *FEBS Letter*, *434*, 413–416.

Cortassa, S., Caceres, V., Bell, L. N., ÓRourke, B., & Paolocci, M. A. (2015). From metabolomics to fluxomics: a computational procedure to translate metabolite profiles into metabolic fluxes. *Biophysical Journal*, *108*, 163–172.

Cragg, G. M., & Newman, D. J. (2005). Plants as a source of anticáncer agents. *Journal of Ethnopharmacology*, *100*(1–2), 72–79.

Crown, S. B., & Antoniewicz, M. R. (2013). Parallel labeling experiments and metabolic flux analysis: past, present and future methodologies. *Metabolic Engineering*, *16*, 21–32.

Dal'Molin, C. G., Quek, L. E., Palfreyman, R. W., Brumbley, S. M., & Nielsen, L. K. (2010). C4GEM, a genome-scale metabolic model to study C4 plant metabolism. *Plant Physiology*, *154*, 1871–1885.

De Luca, V. (2011). Monoterpenoid indole alkaloid biosynthesis. In H. Ashihara, A. Grozier, & A. Komamine (Eds.), *Plant Metabolism and Biotechnology* (pp. 263–291). New York: Wiley.

Dòller, G., Alfermann, A. W., & Reinhard, E. (1976). Producktion von indol alkaloid in callus-kulturen von *Catharanthus roseus*. *Planta Medica*, *30*, 14–20.

Dong, J. G., Bornmann, W., Nakanishi, K., & Berova, N. (1995). Structural studies of vinblastine alkaloids by exciton coupled circular dichroism. *Phytochemistry*, *40*(6), 1821–1824.

El-Merzabani, M. M., El-Aaser, A. A., Attia, M. A., El-Duweini, A. K., & Ghazal, A. M. (1979). Screening system for Egyptian plants with potential anti-tumour activity. *Planta Medica*, *36*(2), 150–155.

El-Sayed, M., & Verpoorte, R. (2007). *Catharanthus* terpenoids indole alkaloids: biosynthesis and regulation. *Phytochemistry Reviews*, *6*, 277–305.

El-Sayed, A., & Cordell, G. A. (1981). Catharanthus alkaloids. Catharanthamine, a new antitumor bisindole alkaloid from *Catharanthus roseus*. *Journal of Natural Products*, *44*(3), 289–293.

El-Sayed, A., Handy, G. A., & Cordell, G. A. (1983). Catharanthus alkaloids, Confirming structural evidence and antineoplastic activity of the bisindole alkaloids Leurosine-N -oxide (pleurosine), roseadine and vindolicine from *Catharanthus roseus*. *Journal of Natural Products*, *46*(4), 517–527.

Ettenhuber, C., Spielbauer, G., Margl, L., Hannah, L. C., Gierl, A., Bacher, A., Genschel, U., & Eisenreich, W. (2005). Changes in flux pattern of the central carbohydrate metabolism during kernel development in maize. *Phytochemistry*, *66*, 2632–2642.

Facchini, P. J., & DiCosmo, F. (1991). Secondary metabolite biosynthesis in cultured cells Catharanthus roseus (L.) G. Don immobilized by adhesion to glass fibers. *Applied Microbiology and Biotechnology*, *35*, 382–392.

Fell, D. A., & Small, J. A. (1986). Fat synthesis in adipose tissue. An examination of stoichiometric constraints. *Journal of Bioengineering*, *283*, 738–781.

Fernier, A. R., & Schauer, N. (2009). Metabolomics-assisted breeding a variable option for crop improvement? *Trends Genetic*, *25*, 39–48.

Ferreres, F., Pereira, D. M., Valentão, P., Andrade, P. B., Seabra, R. M., & Sottomayor, M. (2008). New phenolic compounds and antioxidant potential of *Catharanthus roseus*. *Journal of Agricultural and Food Chemistry*, *56*, 9967–9974.

Filippini, R., Caniato, R., Piovan, A., & Cappelletti, E. M. (2003). Production of anthocyanins by Catharanthus roseus. *Fitoterapia*, *74*, 62–67.

Gathirwa, J. W., Rukunga, G. M., Njagi, N. M., et al. (2007). In vitro anti-plasmodial and *in vivo* anti-malarial activity of some plants traditionally used for the treatment of malaria by the Meru community in Kenya. *Journal of Natural Medicines*, *61*(3), 261–268.

Glawisching, E., Gierl, A., Tomas, A., Bacher, A., & Eisenreich, W. (2002). Starch biosynthesis and intermediary metabolism in maize kernels. Quantitative analysis of metabolite flux by nuclear magnetic resonance. *Plant Physiology*, *130*, 1717–1727.

Goyal, P., Khanna, A., Chauhan, A., Chauhan, G., & Kaushik, P. (2008). In vitro evaluation of crude extracts of *Catharanthus roseus* for potential antibacterial activity. *International Journal of Green Pharmacy*, *2*, 176–181.

Gupta, R. S., & Sharma, R. (2006). A review on medicinal plants exhibiting antifertility activity in males. *Natural Product Radiance*, *5*, 389–410.

Grafahrend-Belau, E., Junker, A., Eschenrôder, A., Muller, J., Schreiber, F., & Junker, B. J. (2013). Multiscale metabolic modeling: dynamic flux balance analysis on whole-plant scale. *Plant Physiology*, *163*, 637–647.

Grafahrend-Belau, E., Schreiber, F., Koschutzki, D., & Junker, B. H. (2009). Flux balance analysis of barley seeds: a computational approach to study systemic properties of central metabolism. *Plant Physiology*, *149*, 585–598.

Hassan, K. A., Brenda, A. T., Patrick, V., & Patrick, O. E. (2011). *In vivo* antidiarrheal activity of the ethanolic leaf extract of *Catharanthus roseus* linn. (Apocyanaceae) in wistar rats. *African Journal of Pharmacy and Pharmacology v*, *5*(15), 1797–1800.

Hay, J., & Schwender, J. (2011). Metabolic network reconstruction and flux variability analysis of storage synthesis in developing oilseed rape (*Brassica napus* L.) embryos. *Plant Journal*, *67*, 513–525.

Hostettmann, K., Marston, A., Ndjoko, K., & Wolfender, J. L. (2000). The potential of African plants as a source of drugs. *Current Organic Chemistry*, *4*(10), 973–1010.

Hughes, E. H., Hong, S. B., Gibson, S. I., & Shanks, J. V. (2004). Metabolic engineering of the indole pathway in *Catharanthus roseus* hair root leads to increase in tryptamine and serpentine accumulation. *Metabolic Engineering*, *6*, 268–276.

Iweala, E. J., & Okeke, C. U. (2005). Comparative study of the hypoglycemic and biochemical effects of *Catharanthus roseus* (Linn) G.Apocynaceae (Madagascar periwinkle) and chlorpropamide (diabenese) on alloxan-induced diabetic rats. *Biokemistri*, *17*(2), 149–156.

Iyer, V.V., Sriram, G., Fulton, D. B., Zhou, R., Wesgate, M. E., & Shanks, J.V. (2008). Metabolic flux maps comparing the effect of temperature on protein an oil biosynthesis in developing soybean cotyledons. *Plant, Cell & Environment*, *31*, 506–5017.

Jaleel, C. A., Gopi, R., Alagu Lakshmanan, G. M., & Panneerselvam, R. (2006). Triadimefon induced changes in the antioxidant metabolism and ajmalicine production in *Catharanthus roseus* (L.) G. Don. *Plant Science*, *171*(2), 271–276.

Johnson, I. S., Wright, H. F., Svoboda, G. H., & Vlantis, J. (1960). Antitumor principles derived from *Vinca rosea* Linn. I. Vincaleukoblastine and leurosine. *Cancer Research*, *20*, 1016–1022.

Jones, J. G., Sherry, A. D., Jeffrey, F. M., Storey, C. J., & Malloy, C. R. (1993). Spurces of acetyl-CoA entering the tricarboxylic acid cycle as determined by analysis of 13 C isotopomers. *Biochemistry*, *32*, 12240–12244.

Joshi, M. S., & Ambaye, R.Y. (1968). Effect of alkaloids from *Vinca rosea* L. on spermatogenesis in male rats. *Indian Journal of Experimental Biology*, *6*(4), 256–257.

Junker, B. H., Lonien, J., Heady, L. E., Rogers, A., & Schwender, J. (2007). Parallel determination of enzyme activities and in vivo fluxes in *Brassica napus* embryos grown on organic or inorganic nitrogen source. *Phytochemistry*, *68*, 2231–2242.

O'Keefe, B. R., Mahady, G. B., Gills, J. J., Beecher, C.W., & Schilling, A. B. (1997). Stable vindoline production in transformed cell cultures of Catharanthus roseus. *Journal of Natural Products*, *60*(3), 261–264.

Klein, S., & Heinzle, E. (2012). Isotope labeling experiments in metabolomics and fluxomics. Wiley Interdisciplinary Reviews. *Systems Biology and Medicine*, *4*, 261–272.

Kohler, R. E. (1977). Rudolf Schoenheimer, isotopic tracers and biochemistry in the 1930ś. *Historical Studies in the Physical Sciences*, *8*, 257–298.

Kohlstedt, M., Becker, J., & Wittmann, C. (2010). Metabolic fluxes and beyond-systems biology understanding and engineering of microbial metabolism. *Applied Microbiology and Biotechnology*, *88*, 1065–1075.

Kondrak, M., Marincs, F., Antal, F., Juhasz, Z., & Banfalvi, Z. (2012). Effects of yeast trehalose-6-phosphate synthase 1 on gene expression and carbohydrate contents of potato leaves under drought stress conditions. *BMC Plant Biology*, *12*(74), 1–12.

Kruger, N. J., Le, P., Brown, N. D., & Ratcliffe, R. G. (2007). Network flux analysis: impact of C-13-substrates on metabolism in *Arabidopsis thaliana* cell suspension cultures. *Phytochemistry*, *68*, 2176–2188.

Kruger, N. J., & Ratcliffe, R. G. (2009). Insights into plant metabolic networks from steady-state metabolic flux analysis. *Biochimie*, *91*, 697–702.

Kuppusamy, C., Murugan, K., Arul, N., & Yasodha, P. (2009). Larvicidal and insect growth regulator effect of α-amyrin acetate from *Catharanthus roseus* Linn against the malaria vector *Anopheles stephensi* Liston (Diptera: Culicidae). *Entomological Research*, *39*(1), 78–83.

Lackman, P., Gonzáles-Guzmán, M., Tilleman, S., Carqueijeiro, I., Pérez, A. C., Moses, T., Seo, M., Kanno, Y., Hakkinen, S. T., Van Montagu, M. C., et al. (2011). Jasmonate signalinginvolves the abscisic acid receptor PYL4 to regulate metabolic reprograming in *Arabidopsis* and tabacco. *Proceedings of the National Academy of Sciences of the United States of America*, *108*, 5891–5896.

Lim-Sylianco, C.Y., & Blanco, F. (1981). Antimutagenic effects of some anti-cancer agents. *Bulletin of the Philippine Society for Biochemistry and Molecular Biology*, *4*, 1–7.

Liu, J., Liu, Y., Wang, Y., Zhang, Z. H., Zu, Y. G., Efferth, T., et al. (2016). The combined effects of ethylene and meja on metabolic profiling of phenolic compounds in Catharanthus roseus revealed by metabolomics analysis. *Frontiers in Physiology*, 7, 217.

Mans, D. R. A., da Rocha, A. B., & Schwartsmann, G. (2000). Anticancer drug discovery and development in Brazil: targeted plant collection as a rational strategy to acquire candidate anti-cancer compounds. *Oncologist, 5*(3), 185–198.

Ma, F., Jazmin, L. J., Young, J. D., & Allen, D. K. (2014). Isotopically nonstationary 13C flux analysis of changes in Arabidopsis thaliana leaf metabolism due to high light acclimation. *Proceeding of the National Academy of Sciences of the United States of America, 111*, 16967–16972.

Malloy, C. R., Sherry, A. D., & Jeffrey, F. M. (1990). Analysis of trtricarboxylic acid cycle of the heart using 13C isotope isomers. *American Journal of Physiology, 259*, 987–995.

Mathur, R., & Chaudan, S. (1985). Antifertility efficacy of *Catharanthus roseus* Linn: a biochemical and histological study. *Acta Europaea Fertilitatis, 16*(3), 203–205.

Matsuda, F., Wakasa, K., & Miyagawa, H. (2007). Metabolic flux analysis in plans using dynamic labeling technique: application to tryptophan biosynthesis in cultured rice cells. *Phytochemistry, 68*, 2290–2301.

McNeil, S. D., Nuccio, M. L., Rhodes, D., Schachar-Hill, Y., & Hanson, A. D. (2000). Radioatracer and computer modeling evidence that phospho-base methylation is the main route of choline synthesis in tobacco. *Plant Physiology, 123*, 371–380.

Mérillon, J. M., Doireau, P., Guillot, A., Chénieux, J. C., & Rideau, M. (1986). Indole alkaloid accumulation and tryptophan descarboxylase activity in *Catharanthus roseus* cells cultured in three diferent media. *Plant Cell Reports, 5*, 23–26.

Mintz-Oron, S., Meir, S., Malistsky, S., Ruppin, E., Aharoni, A., & Shlomi, T. (2012). Reconstriction of Arabidopsis metabolic network models accounting for subcellular compartmentalization and tissue-specificity. *Proceedings of the National Academy of Sciences of the United States of America, 109*, 339–344.

Moerman, D. E. (2009). *Native American Medicinal Plants. An Ethnobotanical Dictionary*. Portland, Ore, USA: Timber Press.

Moreno, P. R. H., Van der Heijden, R., & Verpoorte, R. (1993). Effect of terpenoid precursor feeding and elicitation of formation of indole alkaloids in cell suspension cultures of *Catharanthus roseus*. *Plant Cell Reports, 12*, 702–705.

Moxley, J. F., Jewett, M. C., Antoniewicz, M. R., Villas-Boas, S. G., Alper, H., Wheeler, R. T., Tong, L., Hinnebusch, A. G., Ideker, T., Nielsen, J., & Stephanopoulos, G. (2009). Linking high-resolution metabolic flux phenotypes and transcriptional regulation in yest modulated by thr global regulator Gcn4p. *Proceeding National Academic of Science of United Stated of America, 106*, 6477–6482.

Masakapalli, S. K., Le, Lay. P., Huddleston, J. E., Pollock, N. L., Kruger, N. J., & Ratcliffe, R. G. (2010). Subcellular flux analysis of central metabolism in a heterotrophic Arabidopsis cell suspension using steady-state stable isotope labeling. *Plant Phisiology, 152*, 602–619.

Mukherjee, A. K., Basu, S., Sarkar, N., & Ghosh, A. C. (2001). Advances in cancer therapy with plant based natural products. *Current Medicinal Chemistry, 8*(12), 1467–1486.

Murugavel, T., & Akbarsha, M. A. (1991). Anti-spermatogenic effect of Vinca rosea Linn. *Indian Journal of Experimental Biology, 29*(9), 810–812.

Mustafa, N. R., & Verpoorte, R. (2007). Phenolic compounds in Catharanthus roseus. *Phytochemistry Reviews, 6*(2-3), 243–258.

Narayana, M. R., & Dimri, B. P. (1990). *Periwinkle and its Cultivation in India*. Lucknow, India: CIMAP.

Namdeo, A., Patil, S., Fulzele, D., 2008. Influence of fungal elicitors on production of ajmalicine by cell cultures of *Catharanthus roseus.* Biotechnology Progress 18 (1), 159-162.

Nikiforova, V. J., Kopka, J., & Tolstikov, V. (2005). Systems rebalancing of metabolism in response to sulfur deprivation, as revealed by metabolome analysis of Arabidopsis plants. *Plant Physiology, 138*(1), 304–318.

Nobili, S., Lippi, D., & Witort, E. (2009). Natural compounds for cancer treatment and prevention. *Pharmacological Research, 59*(6), 365–378.

Noble, R. L. (1990). The discovery of the vinca alkaloids-chemotherapeutic agents against cáncer. *Biochemistry and Cell Biology*, *68*(12), 1344–1351.

Orlova, I., Marshall-Colon, A., Schnepp, J., et al. (2006). Reduction of benzenoid synthesis in petunia flowers reveals multiple pathways to benzoic acid and enhancement in auxin transport. *The Plant Cell*, *18*, 3458–3475.

Pan, Q., Wang, Q., & Yuan, F. (2012). Overexpression of ORCA3 and G10H in Catharanthus roseus plants regulated alkaloid biosynthesis andmetabolism revealed byNMR-metabolomics. *PLOS ONE*, 7(8), 430–438.

Pan, Q., Mustafa, N. R., Verpoorte, R., & Tang, K. (2016). 13C-isotope-labeling experiments to study metabolism in *Catharanthus roseus.* In: Metabolomics fundamentals and Applications. *Jeevan Prasain.*, *5*, 703–766.

Pan, Z., Li, Y., Deng, X., & Xiao, S. (2014). Non-targeted metabolomic analysis of orange (*Citrus sinensis* (L.) Osbeck) wild type and bud mutant fruits by direct analysis in real-time and HPLC-electrospray mass spectrometry. *Metabolomics*, *10*, 508–523.

Patil, P. J., & Ghosh, J. S. (2010). Antimicrobial activity of *Catharanthus roseus*—a detailed study. *British Journal of Pharmacology and Toxicology*, *1*, 40–44.

Papoutsakis, E. T. (1984). Equations and calculations for fermentations of butyric acid bacteria. *Biotechnology and Bioengineering*, *26*, 174–187.

Pereira, D. M., Ferreres, F. J., Oliveira, P., Valentao, P., Andrade, B., & Sottomayor, M. (2009). Targeted metabolite analysis of *Catharanthus roseus* and its biological potential. *Food and Chemical Toxicology*, *47*(6), 1349–1354.

Pereira, D. M., Ferreres, F., Oliveira, J. A., et al. (2010). Pharmacological effects of *Catharanthus roseus* root alkaloids in acetylcholinesterase inhibition and cholinergic neurotransmission. *Phytomedicine*, *17*(8–9), 646–652.

Pierre, B., Vazquez, F. A., & De Luca, V. (1999). Multicellular compartmentation of *Catharanthus roseus* alkaloid biosynthesis predicts intercellular translocation of a pathway intermediate. *Plant Cell*, *11*(5), 887–900.

Pilalis, E., Chatziioannou, A., Thomasset, B., & Kolisis, F. (2011). A in silico compartmentalized metabolic model of *Brassipa napus* enables the systematic study of regulatory aspects of plant central metabolism. *Biotechnology and Bioengineer*, *108*, 1570–1581.

Ponarulselvam, S., Panneerselvam, C., Murugan, K., Aarthi, N., Kalimuthu, K., & Thangamani, S. (2012). Synthesis of silver nanoparticles using leaves of *Catharanthus roseus* Linn. G. Don and their antiplasmodial activities. *Asian Pacific Journal of Tropical Biomedicine*, *2*(7), 574–580.

Poolman, M. G., Assmus, H. E., & Fell, D. A. (2004). Modelling of photosynthesis and its control. *Journal of Experimental Botany*, *51*, 319–328.

Poolman, M. G., Miguet, L., Sweetlove, L. J., & Fell, D. A. (2009). A genome scale metabolic model of Arabidopsis and some of its properties. *Plant Physiology*, *151*, 1570–1581.

Prajapati, V., Tripathi, A. K., Jain, D. C., Sharma, S., & Khanuja, P. S. (1998). Sensitivity of *Spilarctia obliqua* to root extracts of *Catharanthus roseus*. *Phytotherapy Research*, *12*(4), 270–274.

Quek, L. E., Wittmann, C., Nielsen, L. K., & Kromer, J. O. (2009). OpenFLUX: efficient modelling software for 13C-based metabolic flux analysis. *Microbial Cell Factories*, *8*(5), 1–15.

Ramya, S., Govindaraji, V., Navaneetha, K., & Jayakumararaj, R. (2008). *In vitro* evaluation of antibacterial activity using crude extracts of *Catharanthus roseus* L. (G.) Don. *Ethnobotanical Leaflets*, *12*, 1067–1072.

Ratcliffe, R. G., & Shachar-Hill, Y. (2005). Revealing metabolic phenotypes in plants: inputs from NMR analysis. *Biological Reviews*, *80*, 27–43.

Remia, K. M., & Logaswamy, S. (2010). Larvicidal efficacy of leaf extract of two botanicals against the mosquito vector *Aedes aegypti* (Diptera: Culicidae). *Indian Journal of Natural Products and Resources*, *1*(2), 208–212.

Ren, Y., Wang, T., Peng, Y., Xia, B., & Qu, L. J. (2009). Distinguishing transgenic from non-transgenic *Arabidopsis* plants by 1HNMR-based metabolic fingerprinting. *Journal of Genetics and Genomics, 36*(10), 621–628.

Rees, T., & Hill, S. A. (1994). Metabolic control analysis of plant metabolism. *Plant Cell Environmental, 17*, 587–599.

Rianika, N., Kyong, H., Hae, Y., & VerpoorteF R. (2009). Metabolic changes of salicylic acid-elicited Catharanthus roseus cell suspension cultures monitored by NMR-based metabolomics. *Biotechnology Letters, 31*, 1967–1974.

Rischer, H., Orešič, M., Seppänen-Laakso, T., Katajamaa, M., Lammertyn, F., Ardiles-Diaz, W., et al. (2006). Gene-to-metabolite networks for terpenoid indole alkaloid biosynthesis in *Catharanthus roseus* cells. *Proceedings of the National Academy of Sciences, 103*, 5614–5619.

Rodríguez-Prados, J. C., Traves, P. G., Cuenca, J., Rico, D., Aragones, J., Martin-Sanz, P., Cascante, M., & Bosca, L. (2010). Substrate fate in activated macrophages: a comparison betwwen innate, classic and alternative activation. *Journal of Immunology, 185*, 605–614.

Rohwer, J. M., & Botha, F. C. (2001). Analysis of sucrose accumulation in the sugar cane culm on the basis of in vitro kinetic data. *Biochemical Journal, 358*, 437–445.

Rontein, D., Dieuaide-Noubhani, M., Mufourc, E. J., Raymond, P., & Rolin, D. (2009). The metabolic architecture of plant cells-stability of central metabolism and flexibility of anabolic pathways during the growth cycle of tomato cells. *Journal of Biological Chemistry, 277*, 43948–43960.

Salon, C., Avice, J. C., Colombié, M., Gallardo, K., Jeudy, C., Ourry, A., Prudent, M., Voisin, A. S., & Rolin, D. (2017). Fluxomic links cellular functional analyses to whole-plant phenotyping. *Journal of Experimental Botany, 68*(9), 2083–2098.

Savinell, J. M., & Palsson, B. O. (1992). Optimal selection of metabolic fluxes for in vivo measurement. I. Development of mathematical methods. *Journal of Theoretical Biology, 155*, 201–214.

Schmidit, K., Carlsen, M., Nielsen, J., & Villadse, J. (1997). Modeling isotopomer distributions in biochemical networks using isostopomer mapping matrices. *Biotechnology and Bioengineering, 55*, 831–840.

Schuhr, C. A., Radykewicz, T., Sagner, S., Latzel, C., Zenk, M. H., et al. (2003). Quantitative assessment of crosstalk between the two isoprenoid biosynthesis pathways in plants by NMR spectroscopy. *Phytochemistry Reviews, 2*, 3–16.

Schwender, J., Goffman, F., Ohlrogge, J. B., & Schachar-Hill, Y. (2004). Rubisco without the Calvin Cycle improves the carbon efficiency of developing green seeds. *Nature, 432*, 779–782.

Si, Y., Yoon, J., & Lee, K. (2009). Flux profile and modularity analysis of time-dependent metabolic changes of de novo adiposyte formarion. *American Journal of Physiology and Endocrinology Metabolism, 291*, 1637–1646.

Siahpoosh, M. R., Sanchez, D. H., Schlereth, A., et al. (2012). Modification of *OsSUT1* gene expression modulates the salt response of rice *Oryza sativa* cv. Taipei 309. *Plant Science, 182*(1), 101–111.

Siddiqui, M. J., Ismail, Z., Aisha, A. F. A., & Abdul Majid, A. M. S. (2010). Cytotoxic activity of *Catharanthus roseus* (Apocynaceae) crude extracts and pure compounds against human colorectal carcinoma cell line. *International Journal of Pharmacology, 6*(1), 43–47.

Singh, N., Pandey, B. R., & Verma, P. (2011). An overview of phytotherapeutic approach in prevention and treatment of Alzheimer's Syndrome & Dementia. *International Journal of Pharmaceutical Sciences and Drug Research, 3*(3), 162–172.

Singh, S. N., Vats, P., Suri, S., et al. (2001). Effect of an antidiabetic extract of *Catharanthus roseus* on enzymic activities in streptozotocin induced diabetic rats. *Journal of Ethnopharmacology, 76*(3), 269–277.

Siriam, G., Fulton, D. B., & Shank, J.V. (2007). Flux quantification in central carbon metabolism of *Catharanthus roseus* hair roots by C-13 labeling and comprehensive bondomer balancing. *Phytochemistry, 68*, 2243–2257.

Stanley, A., & Akbarsha, M. A. (1992). Giant spermatogonial cells generated by vincristine and their uses. *Current Science, 63*(3), 144–147.

Stanley, A., Averal, H. A., & Akbarsha, M. A. (1993). Reproductive toxicity of vincristine inmale rats. *Indian Journal of Experimental Biology, 31*(4), 380–382.

Stephanopoulos, G., & Sinskey, A. J. (1993). Metabolic engineering- methodologies and future prospects. *Trends Biotechnology, 11*(9), 392–396.

Stitt, M., Sulpice, R., & Keurentjes, J. (2010). Metabolic networks in the regulation of metabolism and growth. *Plant Physisology, 152*, 428–444.

Strigun, A., Noor, F., Pironti, A., Niklas, J., Yang, T. H., & Heinzel, E. (2011). Metabolic flux analysis gives an insight on verapamil induced changes in central metabolism of HL-1cells. *Journal Biotechnology, 155*, 299–307.

Sweetlove, L. J., Beard, K. F., Nunes-Nesi, A., Fernie, A. R., & Ractcliffe, R. G. (2010). Not just circle: flux modes in the plant TCA cycle. *Trends Plant Science, 15*, 462–470.

Sweetlove, L. J., & Fernie, A. R. (2005). Regulation of metabolic networks: understanding metabolic complexity in the system biology era. *New Phytologist, 168*, 9–23.

Tschaplinski, T. J., Standaert, R. F., Engle, N. L., et al. (2012). Downregulation of the caffeic acid O-methyltransferase gene in switchgrass reveals a novel monolignol analog. *Biotechnology for Biofuels, 5*(71), 1–15.

Ueda, J. Y., Tezuka, Y., Banskota, A. H., et al. (2002). Antiproliferative activity of Vietnamese medicinal plants. *Biological and Pharmaceutical Bulletin, 25*(6), 753–760.

Usia, T., Watabe, T., Kadota, S., & Tezuka, Y. (2005). Cytochrome P4502D6 (CYP2D6) inhibitory constituents of *Catharanthus roseus. Biological and Pharmaceutical Bulletin, 28*(6), 1021–1024.

Uys, L., Botha, F. C., Hofmeyr, J. S., & Rohwer, J. M. (2007). Kinetic model of sucrose accumulation in maturing sugarcane culm tissue. *Phytochemistry, 68*, 2375–2392.

Vega, E., Cano, J. L., Alarcón, F. J., Fajardo, D. C., Almanza, J. C., & Román, R. (2012). Hypoglycemic activity of aqueous extracts from *Catharanthus roseus. Evidence-Based Complementary and Alternative Medicine.*

Verma, A. K., & Singh, R. R. (2010). Induced dwarfmutant in *Catharanthus roseus* with enhanced antibacterial activity. *Indian Journal of Pharmaceutical Sciences, 72*(5), 655–657.

Verma, P., Mathur, A. K., Srivastava, A., & Mathur, A. (2012). Emerging trends in research on spatial and temporal organization of terpenoid indole alkaloid pathway in *Catharanthus roseus*: a literature update. *Protoplasma, 249*(2), 255–268.

Verpoorte, R., Van der Heijden, R., Schripsema, J., Hoge, J. C., & Ten Hoopen, H. G. (1993). Plant cell biotechnology for the production of alkaloids: present status and prospects. *Journal Natural Products, 56*, 186–207.

Virmani, O. P., Srivastava, G. N., & Singh, P. (1978). *Catharanthus roseus* the tropical periwinkle. *Indian Drugs, 15*, 231–252.

Wang, Z., Klipfell, E., Bennett, B. J., Koeth, R., Levison, B. S., Dugar, B., et al. (2011a). Gut flora metabolism of phosphatidylcholine promotes cardiovascular disease. *Nature, 472*, 57–63.

Wang, X., Sun, H., Zhang, A., Sun, W., Wang, P., & Wang, Z. (2011b). Potential role of metabolomics apporoaches in the area of traditional Chinese medicine: as pillars of the bridge between Chinese and Western medicine. *Journal of Pharmaceutical and Biomedical Analysis, 55*, 859–868.

Wang, S., Zheng, Z., Weng, Y., et al. (2004). Angiogenesis and antiangiogenesis activity of Chinesemedicinal herbal extracts. *Life Sciences, 74*(20), 2467–2478.

WHO Traditional Medicine Strategy: 2014-2023 (2013) WHO.. (Online)Available from: http://www.who.int/medicines/publications/traditional/trm_strategy14_23/en/..

Whitmer, S., Canel, C., Hallard, D., Goncalves, C., & Verpoorte, R. (1998). Influence of precursor availability of alkaloid accumulation by transgenic cell line of *Catharanthus roseus*. *Plant Physiology*, *116*, 853–857.

Wiechert, W., Siefke, C., de Graaf, A. A., & Marx, A. (1997). Biodirectional reaction steps in metabolic networks: II. Flux estimation and statistical analysis. *Biotechnology and Bioengineering*, *55*, 118–135.

Williams, T. R., Miguet, L., Masakapalli, S. K., Kruger, N. J., Sweetlove, L. J., & Ratcliffe, R. G. (2008). Metabolic network fluxes in heterotrophic *Arabidopsis* cells-: stability on the flux distribution under different oxygenation conditions. *Plant Physiology*, *148*, 704–718.

Winter, G., & Kromer, J. O. (2013). Fluxomics- connecting omics analysis and phenotypes. *Environmental Microbiology*, *15*(7), 1426–2920.

Wolfe, R. R. (1984). Tracers in metabolic research: radioisotope and stable isotope/mass spectrometry methods. *Laboratory and research methods in biology and medicine*, *9*, 1–287.

Yang, S. O., Kim, S. H., Kim, Y., Kim, H. S., Chun, Y. J., & Choi, H. K. (2009). Metabolic discrimination of *Catharanthus roseus* calli according to their relative locations using (1) H-NMR and principal component analysis. *Bioscience, Biotechnology, and Biochemistry*, *73*(9), 2032–2036.

Yamamoto, K., Takahashi, K., Mizuno, H., Anegawa, A., Ishizaki, K., Fukaki, H., et al. (2016). Cell-specific localization of alkaloids in *Catharanthus roseus* stem tissue measured with imaging MS and single-cell MS. *Proceedings of the National Academy of Sciences*, *113*, 3891–3896.

Yoo, T. H., Antoniewicz, M. R., Stephanopoulos, G., & Kelleher, J. K. (2008). Quantifying reductive carboxylation flux of glutamine to lipid a brown adipocyte cell line. *Journal of Biological Chemistry*, *283*, 20621–20627.

Kotaro, Y., Katsutoshi T., Hajime M., Aya A., Kimitsune, I., Hidehiro, F., et al., Cell-specific localization of alkaloids in *Catharanthus roseus* stem tissue measured with Imaging MS and Single-cell MS. Proceedings of the National Academy of Sciences 113 (14) 3891-3896.

Zamboni, N., & Sauer, U. (2009). Novel biological insights through metabolomics and 13C-flux analysis. *Current Opinion in Microbiology*, *12*(5), 553–558.

Zheng, W., & Wang, S. Y. (2001). Antioxidant activity and phenolic compounds in selected herbs. *Journal of Agricultural and Food Chemistry*, *49*(11), 5165–5170.

Zupke, C., & Stephanopoulos, G. (1994). Modeling of isotope distributions and intracellular fluxes in metabolic networks using atom mapping matrices. *Biotechnology Progress*, *10*, 489–498.

CHAPTER 4

Multivariate analysis of herbal drugs with diverse pharmacological activities: metabolomics study

Lubna Azmi[a], Ashish Srivastava[b], Ila Shukla[a], Arti Gautam[a]
[a]Pharmacognosy and Ethnopharmacology Division, National Botanical Research Institute (CSIR), Lucknow, Uttar Pradesh, India
[b]Pharmaceutics and Pharmacokinetics Division, CSIR- Central Drug Research Institute, Lucknow, Uttar Pradesh, India

Introduction

Nowadays Multivariate analysis is the most efficient and widely used analytical technique. It is the new determination of traditional statistical studies of experiments data of several measurements that are made on each and every experimental unit and for representing this interaction between multivariate data and their relative structure are very important to properly understand the experimental model (Markowitz, 2018). Multivariate analysis requires much computational effort, Due to the level of complexity and size of any data which need more computational work (Fig. 4.1). This analytical technique plays a very important role in the analysis of data of several synthetic and herbal drugs (Beckett, Eriksson, Johansson, & Wikström, 2017).

Quantitative and qualitative analysis of the target set of metabolites or all metabolites which are non-targeted in the specified cells, tissue, and organism were analyzed by metabolomics studies (Härdle & Simar, 2013). In metabolomics studies nontargeted approach, a chemometric approach which involves spectral data, raw peaks which are listed and spectral profile determination with the help of multivariate analysis to identification of spectral properties in different sample sets. Noise and false peaks in a spectrum of the profile during handling is a major challenge in compound identification (Tabachnick & Fidell, 2012). Targeted metabolic profiling generally requires identification of compound and quantification of such procedures which are achieved by comparing samples with a reference or standard spectral library. The traditional method of separation, purification

Medicinal and Aromatic Plants: Expanding their Horizons through Omics
http://dx.doi.org/10.1016/B978-0-12-819590-1.00004-5

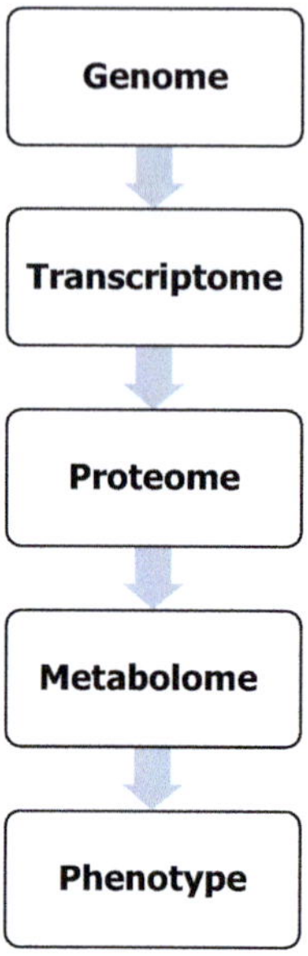

Figure 4.1 *Flow chart for genome to phenotype.*

and structure determination of bioactive secondary metabolites of plant extracts is very cumbersome and time taking for identification of chemical standards. Lack of chemical standards library is a big hurdle in herbal plant metabolomics studies (Anderson, 2017).

Medicinal plants produce a broad range metabolite, near about 200,000–1,000,000 worldwide plant species. Plant develops these metabolites for their defense mechanism, overcome from environmental stress and natural enemies. According to the World health organization (WHO) report, near about 80% of the total world population depends on herbal drugs as their primary source of treatment (Akerele, 1993; United Nations Office on Drugs and Crime, 2017). Recently herbal drugs and their secondary metabolites have shown a broad range of pharmacological activity, and they act as therapeutic agents and lead for new drug discovery (Fig. 4.2) (Parveen, Parveen, Parveen, & Ahmad, 2015).

Metabolomics approach in pharmaceutical industries

Metabolomics is the latest technology which is basically concerned with the identification and quantification of bioactive compounds and small molecules present in the cell, this technology is one of the latest breakthroughs in analytical techniques for identification of new lead. As we know

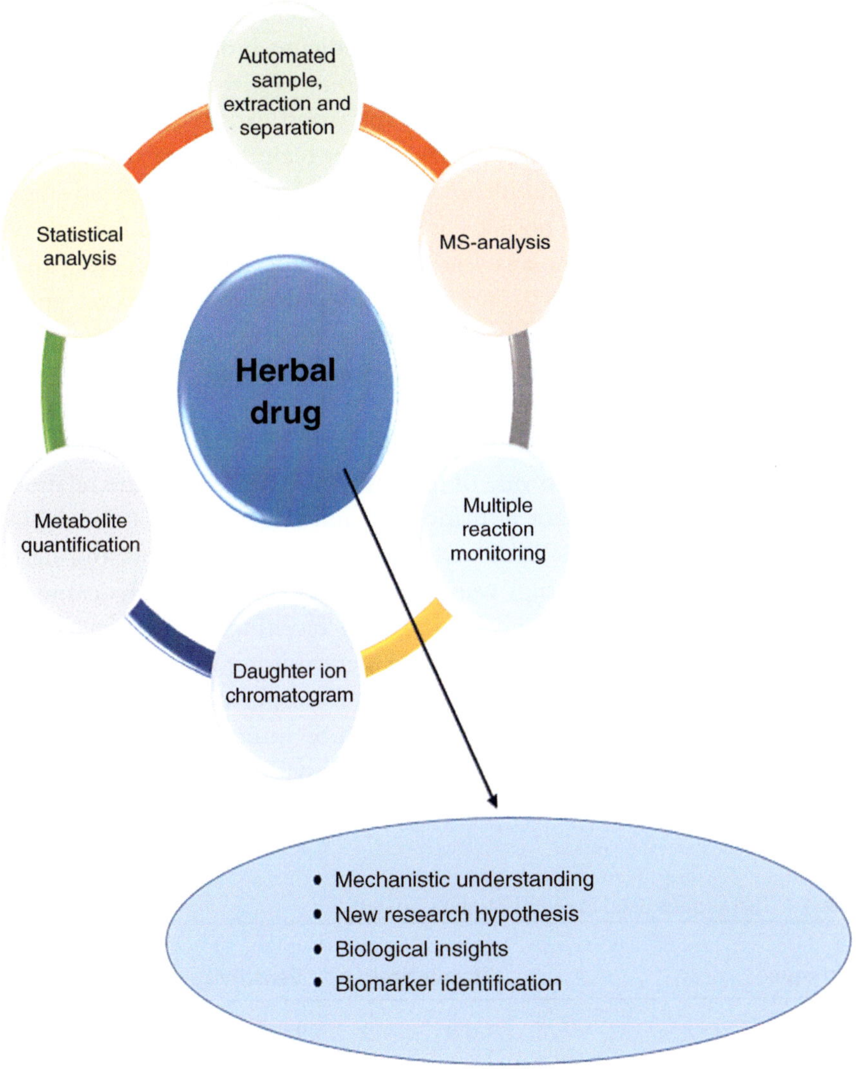

Figure 4.2 ***Metabolomics workflow.***

that the latest drug development and discovery of new leads have become very costly but when we see the report on a new drug which is developed their number is very low (Lindon, Holmes, & Nicholson, 2007). Complex diseases which are not treated with single, targeted drug therapy, like Parkinson's diseases, obesity, cancer, and Alzheimer's disease, these diseases need a new concept of new drug discovery with a new technology, which can target on different aspects of disease (Nicholson & Lindon, 2008; Kordalewska & Markuszewski, 2015). Herbal medicine, broad source of bioactive compounds targeting on particular site, recognized as multi-content drugs and these drugs were determined and investigated by latest technology of metabolomics with support of different technology of spectroscopy and chromatography for multidrug targeting screening. Recent advancements in metabolomics technology have improved the system biology, diagnostic platforms, and gene analysis (Fillet & Frédérich, 2015). New challenges in drug discovery needs latest scientific platform and the concept of metabolomics is such an effective tool for the determination of secondary metabolites for better diagnosis and treatment of diseases.

Advantage and disadvantage of metabolomics

Herbal drug metabolomics is one of the bases for determining the relationship between content present in plant drug and targeted site of disease, for drug-pharmacology effects. Plant extracts obtained from plants and herbs which are used in herbal drugs were analyzed by different analytical technologies (GC-MS, LC-MS, and NMR, etc.) resulting in plant metabolic profiling (Table 4.1) (Wishart, 2016). For determining the pharmacological properties of herbal drugs data obtained from plant metabolic profiles, animal studies, and bioactivity can be assessed by multivariate data analysis. Some latest examples are interpretation and visualization of data with the help of NMR and their correlation with the bioactivity of the compound

Table 4.1 Techniques used in metabolomics analysis.

Technique	Throughput	Sensitivity	Comprehensiveness
Infrared	Low	High	Low
Nuclear magnetic resonance (NMR)	Low	Low-high	Low-high
Gas chromatography (MS)-Liquid chromatography (LC)	High	High	High
LC-NMR	Low	Low	High
LC-MS	Medium	High	High
LC-Electrochemistry-MS	High	High	High
Capillary electrophoresis (CE)-MS	High	Medium	High
LC-Ultraviolet	Medium-high	High	Very low

(Xia, Broadhurst, Wilson, & Wishart, 2013). Metabolomic profiling database represents a single component and all type of extracts obtained from medicinal plant source, which were commercialized (Lei, Huhman, & Sumner, 2011). In case of various known compounds metabolomics help to determine the new pharmacological activity of that particular compound. There is some disadvantage of this approach, like requirement of a large number of samples and difficulty to determine bioassay of fractions (van den Berg, Hoefsloot, Westerhuis, Smilde, & van der Werf, 2006). However, by administrating medicinal plants extracts of various composition using metabolomics studies, it will be easy to determine which compounds and their selective analogs were associated with the highest pharmacological activity (Markley et al., 2017).

Application of metabolomics in pharmacological studies

Pharmacological properties of herbal drugs depend not only on secondary metabolites present in medicinal drugs but also depend on some minor constituents which are present in plants. They have a broad spectrum of biological activity which was used for the treatment of a different type of complicated disease (Kaddurah-Daouk, Kristal, & Weinshilboum, 2008). That is why, new lead discovery form secondary metabolites, quality control determination and their efficacy have been considered very complicated and challenging task, in the advancement of herbal drugs (Raterink, Lindenburg, Vreeken, Ramautar, & Hankemeier, 2014). Due to the reductionistic type of approach, identification of least abundant phytochemicals from herbal drugs is hardly approachable and determine synergistic effect for the various constituents from single plant herb or from the different herbal formulation is very tough (Allwood, Ellis, & Goodacre, 2008). Chromatographic analysis with help of LC-MS, GC-MS, and LC-NMR represents chemical property and pharmacological activity of plant constituents and are a very important approach for identification and standardization of herbal drugs and hold great value in drug discovery (Trushina & Mielke, 2014). LC–ESI-TOF-MS a new latest technology which was used to determine chemical content of PHY906, an herbal formulation prepared from a combination of four meditational plants, for prevention of gastrointestinal toxicity, caused by adjuvant cancer chemotherapy (Lam et al., 2010; Saif et al., 2010). LC-MS analysis was reported to determine Phytomics QC data platform for fingerprinting, chemical characterization, and gene expression for in vivo studies which can allow quality control determination of herbal products (Nagrath, Caneba, Karedath, & Bellance, 2011). Ephedra and ginseng plant products were an example of NMR-based metabolomics study (Kim, Choi, Erkelens, Lefeber, & Verpoorte, 2005; Yang et al., 2006). Recently the concept of comparative metabolomics was more popular. GC-MS, supercritical fluid technology, and data mining were used to determine Echinacea plant species *E. pallida, E. purpurea,* and *E. angustifolia,* which are commercialized for Echinacea products in the market (Wanigesekara, Wijeratne, Arnold, & Gunatilaka, 2019).

The traditional herbal formulation has strong pharmacological activity, due not because of single bioactive compound but due to the synergy of active components present in particular herbal drug, which were broadly analyzed by metabolomics. Antimicrobial activity of berberine an alkaloid is nearly 90% increased by 5′ methoxyhydnocarpin (5′ MHC) which proved by LC-NMR (Kumar et al., 2015). This is a bioactive compound which is present in the same herbal plant having alkaloid but having no antimicrobial activity of its own (Stermitz, Lorenz, Tawara, Zenewicz, & Lewis, 2000). Ginkgo plant extracts with antioxidant, antiestrogenic, stress controlling, vasoregulatory, and gene regulation properties and their anticlastogenic, antioxidant, vasoregulatory, cognition-enhancing, stress alleviating and gene-regulatory effects with help of combined property of bioactive compound present in it were determined by GC-MS (Qian et al., 2017; Wang, Zhang, Ren, & Dong, 2016). Tetrahydrocannabinol a major bioactive constituent of cannabis extract show synergistic effect with all other compounds present in the extract, However from resenting report it was found that tetrahydrocannabinol with the combination of cannabidiol and some remaining constituents is very much effective as a herbal drug when compared to a single component (Berman et al., 2018; K.M. et al., 2015). Peel and seed of pomegranate extract showed strong antiproliferative activity (Nuñez-Sánchez et al., 2014). Solanine and chaconine show strong antifungal activity and their antifungal activity is attributed to glycoalkaloid ratios determined by computational analysis (Weltring, Wessels, & Geyer, 1997). Single bioassay was used to increase the effectiveness of multiple herbal mixtures. Multivariate analysis approach in herbal drugs was used to understand phytomedicine, as can be understood with some examples. Acetylsalicylate is one of the best and effective examples of this technology, it was synthesized near about 100 years ago. This compound was traditionally obtained from the bark of Salix which is used to treat mainly headaches and pain. The expected activity was based on the use of bark of Salix to treat pain and headaches (Oltean et al., 2014). One very interesting thing about salicylate, that it is not found as a chemical constituent of bark of Salix, it is a glucoside which converts salicylate to saligenin after hydrolytically oxidized in the gut and thus salicylic acid is released (Oltean et al., 2014). This concept of medicinal plant and strong pharmacological activity is fully explained by metabolomics. The conventional screening of salicylic acid was developed to determine the presence of salicylate in the herbal drugs. Herbal drugs are lifesaving with no side effect. Near about 15 currently used drugs in the life saving treatment of cancer were obtained from a plant source. Metabolomics determines the open source for the pharmacological activity of herbal drugs in for new era drug discovery.

Future perspectives

The broad spectrum application of metabolomics in herbal drug discovery and research, provide evidence for plant-based pharmaceuticals and determine a new paradigms for the discovery of such complex plants and plant-based phytocompounds resulting in the new system of medicine (Arumugam et al., 2012). Development of a direct link between bioactive markers and spectral profiling of herbal drugs is a very important approach for the discovery and development of herbal-based medicine for the treatment of human disease (Escandar, Damiani, Goicoechea, & Olivieri, 2006). Latest advancements in metabolomics are number of bioactive components which were detected with the method of conventional analysis, compound identification, the throughput of the method, and quantification of compounds with a wide range of samples (Haneef et al., 2013). That is why, it is very important to improve the spectral data, for the development of a new database for drug analysis and determining the accuracy of various analytical signals for metabolomics profiling of the scientific compound, and for integration, data transformation, and data normalization. Metabolomics is a latest and very effective approach for determining active phytochemicals for medicinal plants (Hamill et al., 2003). Phyto-formulations, which are very difficult to evaluate qualitatively and quantitatively with bioactivity of formulation of medicinal plants for the treatment of human diseases, can be analyzed by integrating data of metabolomics and related information to determine the pharmacological activity of herbal drugs will lead to the concept of metabolome for new phytomedicine, for the human healthcare (Hamill et al., 2003).

Conclusion

A recent development in proteomics, genomics, bioinformatics, and metabolomics is a modern and efficient scientific approach of herbal drugs with the help of the integrative approach of a biological system. Moreover, these latest technologies show the determination of herbal drugs as compared to the previous one. Metabolomics has proved very important tool in herbal drug research with help of these deliverables, development of biomarkers at the particular endpoint, efficacy study with herbal preparation, fingerprinting of herbal products, validation of new target, and new leads. Very important next step is to understand the interaction of the herbal drug with human physiology. Finally, the full pharmacological activity of herbal drugs in terms of interplay with human health can be studied, which is a strong base for studying modern personalized drug for health care.

References

Akerele, O. (1993). Summary of WHO guidelines for the assessment of herbal medicines. *Herbal Gram, 28*, 13–19.

Allwood, J. W., Ellis, D. I., & Goodacre, R. (2008). Metabolomic technologies and their application to the study of plants and plant-host interactions. *Physiologia Plantarum, 132*(2), 117–135. https://doi.org/10.1111/j.1399-3054.2007.01001.x.

Anderson, M.J. (2017). Permutational Multivariate Analysis of Variance (PERMANOVA). In *Wiley StatsRef: Statistics Reference Online,* 1-15. https://doi.org/10.1002/9781118445112.stat07841.

Arumugam, R., Ragupathi Raja Kannan, R., Jayalakshmi, J., Manivannan, K., Karthikai Devi, G., & Anantharaman, P. (2012). Determination of element contents in herbal drugs: chemometric approach. *Food Chemistry, 135*, 2372–2377. https://doi.org/10.1016/j.foodchem.2012.07.040.

Beckett, C., Eriksson, L., Johansson, E., & Wikström, C. (2017). Multivariate Data Analysis (MVDA). In *Pharmaceutical quality by design: A practical approach* (3rd revised ed.). *Umetries Academy,* pp. 201–225. https://doi.org/10.1002/9781118895238.ch8.

Berman, P., Futoran, K., Lewitus, G. M., Mukha, D., Benami, M., Shlomi, T., & Meiri, D. (2018). A new ESI-LC/MS approach for comprehensive metabolic profiling of phytocannabinoids in Cannabis. *Scientific Reports, 8*(14280), 1–15. https://doi.org/10.1038/s41598-018-32651-4.

Escandar, G. M., Damiani, P. C., Goicoechea, H. C., & Olivieri, A. C. (2006). A review of multivariate calibration methods applied to biomedical analysis. *Microchemical Journal, 82*, 29–42 https://doi.org/10.1016/j.microc.2005.07.001.

Fillet, M., & Frédérich, M. (2015). The emergence of metabolomics as a key discipline in the drug discovery process. *Drug Discovery Today: Technologies, 13*, 19–24. https://doi.org/10.1016/j.ddtec.2015.01.006.

Hamill, F. A., Apio, S., Mubiru, N. K., Bukenya-Ziraba, R., Mosango, M., Maganyi, O. W., & Soejarto, D. D. (2003). Traditional herbal drugs of Southern Uganda, II: literature analysis and antimicrobial assays. *Journal of Ethnopharmacology, 84*, 57–78. https://doi.org/10.1016/S0378-8741(02)00289-1.

Haneef, J., Shaharyar, M., Husain, A., Rashid, M., Mishra, R., Siddique, N. A., & Pal, M. (2013). Analytical methods for the detection of undeclared synthetic drugs in traditional herbal medicines as adulterants. *Drug Testing and Analysis, 5*, 607–613. https://doi.org/10.1002/dta.1482.

Härdle, W. K., & Simar, L. (2013). *Applied multivariate statistical analysis, 1*, 183–192. https://doi.org/10.1007/978-3-642-17229-8.

K.M., K.-S., H.-J., B., M., B., W., V., L., K., T., H., & M.M., S. (2015). Epigenetic upregulation of the o-linked beta-n-acetylglucosamine transferase (OGT) in response to dronabinol results in antileukemic efficacy in vivo. *Blood, 126,* 2437–2437.

Kaddurah-Daouk, R., Kristal, B. S., & Weinshilboum, R. M. (2008). Metabolomics: a global biochemical approach to drug response and disease. *Annual Review of Pharmacology and Toxicology, 48,* 653–683. https://doi.org/10.1146/annurev.pharmtox.48.113006.094715.

Kim, H. K., Choi, Y. H., Erkelens, C., Lefeber, A. W. M., & Verpoorte, R. (2005). Metabolic fingerprinting of ephedra species using 1H-NMR spectroscopy and principal component analysis. *Chemical & Pharmaceutical Bulletin, 53,* 105–109. https://doi.org/10.1248/cpb.53.105.

Kordalewska, M., & Markuszewski, M. J. (2015). Metabolomics in cardiovascular diseases. *Journal of Pharmaceutical and Biomedical Analysis, 72,* 629–641. https://doi.org/10.1016/j.jpba.2015.04.021.

Kumar, A., Ekavali, Chopra, K., Mukherjee, M., Pottabathini, R., & Dhull, D. K. (2015). Current knowledge and pharmacological profile of berberine: an update. *European Journal of Pharmacology, 15,* 288–297. https://doi.org/10.1016/j.ejphar.2015.05.068.

Lam, W., Bussom, S., Guan, F., Jiang, Z., Zhang, W., Gullen, E. A., & Cheng, Y. C. (2010). Chemotherapy: the four-herb Chinese medicine PHY906 reduces chemotherapy-induced gastrointestinal toxicity. *Science Translational Medicine, 18,* 45–59. https://doi.org/10.1126/scitranslmed.3001270.

Lei, Z., Huhman, D. V., & Sumner, L. W. (2011). Mass spectrometry strategies in metabolomics. *Journal of Biological Chemistry, 22,* 25435–25442. https://doi.org/10.1074/jbc.R111.238691.

Lindon, J. C., Holmes, E., & Nicholson, J. K. (2007). Metabonomics in pharmaceutical R&D. *FEBS Journal, 274,* 1140–1151. https://doi.org/10.1111/j.1742-4658.2007.05673.x.

Markley, J. L., Brüschweiler, R., Edison, A. S., Eghbalnia, H. R., Powers, R., Raftery, D., & Wishart, D. S. (2017). The future of NMR-based metabolomics. *Current Opinion in Biotechnology, 43,* 34–40. https://doi.org/10.1016/j.copbio.2016.08.001.

Markowitz, J.S. (2018). Multivariate analysis. In *SpringerBriefs in public health 286,* 25435–25442. https://doi.org/10.1007/978-3-319-77203-5_8.

Nagrath, D., Caneba, C., Karedath, T., & Bellance, N. (2011). Metabolomics for mitochondrial and cancer studies. *Biochimica et Biophysica Acta - Bioenergetics, 1807,* 650–663. https://doi.org/10.1016/j.bbabio.2011.03.006.

Nicholson, J. K., & Lindon, J. C. (2008). Systems biology: metabonomics. *Nature, 27,* 1054–1056. https://doi.org/10.1038/4551054a.

Nuñez-Sánchez, M. A., García-Villalba, R., Monedero-Saiz, T., García-Talavera, N. V., Gómez-Sánchez, M. B., Sánchez-Álvarez, C., & Espín, J. C. (2014). Targeted metabolic profiling of pomegranate polyphenols and urolithins in plasma, urine and colon tissues from colorectal cancer patients. *Molecular Nutrition and Food Research, 58,* 1191–1211. https://doi.org/10.1002/mnfr.201300931.

Oltean, H., Robbins, C., van Tulder, M. W., Berman, B. M., Bombardier, C., & Gagnier, J. J. (2014). Herbal medicine for low-back pain. *Cochrane Database of Systematic Reviews, 23,* 12. https://doi.org/10.1002/14651858.CD004504.pub4.

Parveen, A., Parveen, B., Parveen, R., & Ahmad, S. (2015). Challenges and guidelines for clinical trial of herbal drugs. *Journal of Pharmacy and Bioallied Sciences, 7,* 329–333. https://doi.org/10.4103/0975-7406.168035.

Qian, Y., Peng, Y., Shang, E., Zhao, M., Yan, L., Zhu, Z., & Duan, J. ao. (2017). Metabolic profiling of the hepatotoxicity and nephrotoxicity of Ginkgolic acids in rats using ultra-performance liquid chromatography-high-definition mass spectrometry. *Chemico-Biological Interactions, 273,* 11–17. https://doi.org/10.1016/j.cbi.2017.05.020.

Raterink, R. J., Lindenburg, P. W., Vreeken, R. J., Ramautar, R., & Hankemeier, T. (2014). Recent developments in sample-pretreatment techniques for mass spectrometry-based metabolomics. *TrAC - Trends in Analytical Chemistry*, *85*, 7762–7768. https://doi.org/10.1016/j.trac.2014.06.003.

Saif, M. W., Lansigan, F., Ruta, S., Lamb, L., Mezes, M., Elligers, K., & Cheng, Y. C. (2010). Phase I study of the botanical formulation PHY906 with capecitabine in advanced pancreatic and other gastrointestinal malignancies. *Phytomedicine*, *17*, 161–169. https://doi.org/10.1016/j.phymed.2009.12.016.

Stermitz, F. R., Lorenz, P., Tawara, J. N., Zenewicz, L. A., & Lewis, K. (2000). Synergy in a medicinal plant: antimicrobial action of berberine potentiated by 5'-methoxyhydnocarpin, a multidrug pump inhibitor. *Proceedings of the National Academy of Sciences*, *15*, 1433–1437. https://doi.org/10.1073/pnas.030540597.

Tabachnick, B. G., & Fidell, L. S. (2012). *Using multivariate statistics* (7th ed.). New York: Harper and Row, Pearson. https://doi.org/10.1037/022267.

Trushina, E., & Mielke, M. M. (2014). Recent advances in the application of metabolomics to Alzheimer's Disease. *Biochimica et Biophysica Acta - Molecular Basis of Disease*, *1842*, 1232–1239. https://doi.org/10.1016/j.bbadis.2013.06.014.

United Nations Office on Drugs and Crime. (2017). World Drug Report 2017: Executive Summary Conclusions and Policy Implications. *World Drug Report*, 1–34.

van den Berg, R. A., Hoefsloot, H. C. J., Westerhuis, J. A., Smilde, A. K., & van der Werf, M. J. (2006). Centering, scaling, and transformations: improving the biological information content of metabolomics data. *BMC Genomics*, 1272–1281 https://doi.org/10.1186/1471-2164-7-142.

Wang, Z., Zhang, J., Ren, T., & Dong, Z. (2016). Targeted metabolomic profiling of cardioprotective effect of Ginkgo biloba L. extract on myocardial ischemia in rats. *Phytomedicine*, *1*, 621–631. https://doi.org/10.1016/j.phymed.2016.03.005.

Wanigesekara, W. M. A. P., Wijeratne, E. M. K., Arnold, A. E., & Gunatilaka, A. A. L. (2019). 10′-Deoxy-10′ α-hydroxyascochlorin, a New Cell Migration Inhibitor and Other Metabolites from Acremonium sp., a Fungal Endophyte in Ephedra trifurca. *Natural Product Communications*, *61*, 7727–7737. https://doi.org/10.1177/1934578x1300800515.

Weltring, K. M., Wessels, J., & Geyer, R. (1997). Metabolism of the potato saponins α-chaconine and α-solanine by Gibberella pilicaris. *Phytochemistry*, *46*, 1005–1009. https://doi.org/10.1016/S0031-9422(97)00388-9.

Wishart, D. S. (2016). Emerging applications of metabolomics in drug discovery and precision medicine. *Nature Reviews Drug Discovery*, *15*, 473–484. https://doi.org/10.1038/nrd.2016.32.

Xia, J., Broadhurst, D. I., Wilson, M., & Wishart, D. S. (2013). Translational biomarker discovery in clinical metabolomics: an introductory tutorial. *Metabolomics*, *9*, 280–299. https://doi.org/10.1007/s11306-012-0482-9.

Yang, S.Y., Kim, H. K., Lefeber, A. W. M., Erkelens, C., Angelova, N., Choi, Y. H., & Verpoorte, R. (2006). Application of two-dimensional nuclear magnetic resonance spectroscopy to quality control of ginseng commercial products. *Planta Medica*, *72*, 364–369. https://doi.org/10.1055/s-2005-916240.

CHAPTER 5

Metabolomics: a recent advanced omics technology in herbal medicine research

Siva Nageswara Rao Gajula, Satheeshkumar Nanjappan
Drug Metabolism and Interactions Research Lab, Department of Pharmaceutical Analysis, National Institute of Pharmaceutical Education and Research (NIPER), Hyderabad, Telangana, India

Introduction

Plants produce a broad range of secondary metabolites, to protect from natural enemies, to overcome the hostile environmental stress and to serve as medical attractants (Benderoth et al., 2006). These metabolites have some pharmacological action related to their biochemical structure and can act as disease treating agents in humans. Almost 80% of people are using herbal medicines to treat various diseases. The enormous potential role of these secondary metabolites of plants to serve in the public health and lead compound identification in the drug discovery process leads to more interest in the herbal medicine research (Shyur & Yang, 2008). The application of system biology technologies and omics approaches like proteomics, genomics and metabolomics to the phytomedical science assisted the herbal medicines development. However, metabolomics has shown to be fast and better to any other technique for pattern identification of biological samples.

Metabolomics is a rapidly growing and promising omics approach for the relative comprehensive quantitative and qualitative analysis of all small metabolites in an organism, which can clearly represent the metabolic picture of that living system. The two main approaches in the metabolomics are the targeted and non-targeted (global) analysis of metabolites. The global metabolic analysis is an unbiased way of analysis and used for the unknown metabolites profiling. However, targeted metabolite analysis targets a set of targeted known metabolites for profiling (Shyur & Yang, 2008). Due to its wide range of applications in the basic and applied science, metabolomics became a powerful technique in system biology research. There is increasing knowledge in the understanding of some complex diseases and disorders like cancer, Alzheimer's disease and Parkinson's diseases are not present

Medicinal and Aromatic Plants: Expanding their Horizons through Omics
http://dx.doi.org/10.1016/B978-0-12-819590-1.00005-7

with simple or single targets. Hence, multi-compound drugs are required to threat those complex diseases. Herbal medicines are recognized as the multi-compound drugs and can be investigated by using metabolomics approach for multi-compound screening. For the profiling of plant secondary metabolites and also for the assessment of quality control of herbal medicines the metabolomics strategy is the superior approach (Lee, Jeon, Lee, Lee, & Choi, 2017).

Significant improvements in the analytical instruments like mass spectrometry (MS), high-field nuclear magnetic resonance spectroscopy (NMR) and their coupling with chemometrics offers a comprehensive analysis of metabolites in the matrices. Both MS and NMR are the most widely used analytical techniques in metabolomics for the identification of metabolites in the matrices. Hyphenated techniques like LC-MS, GC-MS are effective tools in the profiling of metabolite and also for the investigation of their quality control. Additionally, these techniques provide the toxicological proof of some herbal medicines after their metabolism in humans (Weckwerth, 2018). GC-MS is one of the most popular analytical platforms in metabolomics for the highly sensitive and selective profiling of the volatile, derivatized primary and secondary metabolites of plants in the plant crude extracts (Fiehn et al., 2001). Metabolome analysis with the help of LC-MS is a powerful unique technique for the profiling of plant secondary metabolites like alkaloids, glycosides, tannins, flavonoids, and carotenoids (Moco, Schneider, & Vervoort, 2009; Hou et al., 2010).

Hyphenated NMR is also a powerful platform for the chemical composition elucidation of a complex mixture of the natural extract. The LC-SPE-NMR offers a high sensitive NMR analysis and can characterize both low and high abundance of metabolites in the crude plant extracts (Mutlib, Strupczewski, & Chesson, 1995; Hostettmann, Wolfender, & Rodriguez, 1997).

Plant metabolomics

Working Group of experts from Max Planck Institute investigated the applications of metabolomics techniques in plant science for the first time. By applying the GC/MS technique Weckwerth and coworkers obtained a set of potato tubers metabolome data (Weckwerth, Loureiro, Wenzel, & Fiehn, 2004). Metabolite levels are evaluated by applying pair-wise comparison method and this revealed that the existence of numerous pairs of metabolites. The levels of these metabolites showed high-correlation coefficients in spite of variations in biological samples. Researchers revealed that the correlation network structure can be altered in a mutant with a silent phenotype, in which metabolic changes could be observed. The correlation between metabolic levels depends on some factors such as coordinated regulation of gene expression and neighboring metabolites having equilibrium with other metabolites (Camacho, De La Fuente, & Mendes, 2005). Their work along with the application of statistical analysis illustrated that the data on the plant metabolic system and the unusual relationship among metabolites of a metabolome dataset. Investigators also specified that the metabolome indicate the definitive phenotype of cells. Metabolomics approach has several technical advantages. Hence, it has a superior role in the decoding of functional genes and understanding of cellular system biology (Saito & Matsuda, 2010).

Environmental and developmental conditions of a plant are the two important factors, which regulate the transcriptional changes. This in turn controls the metabolic state of the plant. The mechanism behind the expression of genes and resulting metabolic phenotype is still mysterious; therefore a comprehensive study of the behavior of plant metabolic systems exists as the greatest challenge for plant system biologists. A global survey related to the expression of genes and metabolites accumulation is a propitious strategy to guess the mechanisms. This strategy is applied in the examination of the expression of gene reprogramming and also in the study of metabolism stimulated by nutritional stress, such as sulfur starvation. The metabolism of the plant against stress induced by the environment changes the regulation of gene-to-metabolite network. These changes and also application of rules leading the relationship of metabolite accumulation with the gene expression are the two major advancements happened by using the above-mentioned strategy (Hirai et al., 2004; Nikiforova et al., 2005). This strategy has also been applied to other species of plants. Identification of *A. thaliana* genome sequence shows that the production of numerous phytochemicals is through the several metabolic genes of that plant. On the other hand, diversity of phytochemicals in plant species with basic for metabolic systems is not understood. Phytochemicals produced in *A. thaliana* during development was investigated by studying the samples of different stages in the growth and organs. Using dataset, signals of 1589 metabolite signals were identified and 167 metabolites were elucidated in those metabolites. Various types of phytochemicals produced by Arabidopsis were observed by metabolite data set and it revealed that the production of these principle compounds in a tissue-specific manner (Schmid et al., 2005).

Need for the detection and identification of different phytochemicals in plants has encouraged plant scientist to develop new technologies. Plants have developed different phytochemicals toward self-defense, interaction with organisms and environmental adaptation. As we use different phytochemicals for the production of pharmaceuticals and other purposes, the study of the genetic background behind the variety of secondary metabolites of plants helps in extensive application of these phytochemicals. But, the current edge in metabolome analysis of plant metabolites is the marginal note of the metabolite signals (Iijima et al., 2008; Böttcher et al., 2008; Matsuda et al., 2011). For the metabolite structures elucidation through the database, 3 categories of data and infrastructures are needed. Those are (1) Tandem mass spectra data for metabolites structural elucidation. (2) Comprehensive phytochemicals mass spectral database. (3) A suitable method to determine the false discovery. Further development in these methodologies and databases is challenging in the metabolomics but must need and can therefore be used for exploring the diversity of secondary plant metabolites (Matsuda et al., 2011; Horai et al., 2010; Matsuda et al., 2009b).

Advantages of metabolomics in herbal medicine research

Measurement of metabolites is crucial and they are important as components of a biochemical pathway, plays the potential role as biomarkers meant for an extensive scope of biological conditions and provide significance in the diet (Hall, 2006). Improvements in the characterization of metabolic changes in unconditional pathways provides the understanding of cell prioritize and partitioning of important nutrients in different conditions. Some studies are implicating metabolites in mediating gene expression and signifying their importance in determining responses to supposed stress and developmental cues (Lancien & Roberts, 2006). The investigation of entities at the metabolite level may directly impact the biological function. Hence, it can be promisingly applied in the herbal metabolomics to establish the disease-modifying action of the herbal medicinal products.

Commonly used metabolomics approaches in the herbal medicine research

NMR-based metabolomics approach is the principle and uniform detection for the clear identification of unknown metabolites. But, this platform has less sensitivity than the mass spectrometry (MS). At present, the most widely used platform in plant metabolomics is based on MS (Hall, 2006). The first approach applied in plant metabolomics is the GC coupled to electron impact and time-of-flight mass spectrometry (GC-TOF-MS) at a large-scale level (Fiehn et al., 2000). GC-TOF-MS can provide high reproducibility in the separation and fragmentation patterns of non-volatile metabolites of primary metabolism as well as the lipophilic compounds (Schauer et al., 2005). CE-MS provides the best separation efficiency and detection of several polar primary metabolites. However, it is rarely used techniques in comparison with GC-TOF-MS (Sato, Soga, Nishioka, & Tomita, 2004). For the identification and elucidation of semi-polar metabolites, LC-MS with atmospheric ionization technique such as atmospheric pressure chemical ionization and electrospray ionization is the best-suited technique. Usually, secondary metabolites of plants such as saponins, alkaloids, flavonoids, polyamines, phenolic acids, and derivates can be detected by LC-MS (Tolstikov, Lommen, Nakanishi, Tanaka, & Fiehn, 2003; Rischer et al., 2006). These metabolites are generally extracted with hydroalcoholic solutions and thereby can analyzed directly without any derivatization (Tolstikov & Fiehn, 2002). High-resolution mass spectrometers like Time of flight mass spectrometry and Fourier transform-ion cyclotron resonance mass spectrometry allows the estimation of the accurate mass of the detected metabolites. Direct flow injection mass spectrometry (DFI-MS) can develop a direct comparison of metabolic fingerprints without prior separation. But, in this direct injection approach, may lead to face the ion suppression and significant adduct formation problems. And also DFI-MS cannot differentiate when there are many molecular isomers. Hence, MS-based platform must require a separation technique in the plant metabolomics studies.

Generally, isomeric compounds are abundantly present in plants. LC coupled to MS offers the detection of such isomeric compounds. In addition, along with the enabling, it also provides valuable information related to the structure of the metabolites to be collect online (Huhman & Sumner, 2002). Coupling of LC with high- resolution MS results in the detection of hundreds of metabolites in a single crude plant extract. With the rapid improvement in the data acquisition and processing methods, LC-MS will provide the most valuable technique in metabolite identification. LC-NMR and LC-NMR-MS are some of the other most advanced identification tools used in the plant metabolomics (Peterman, Duczak, Kalgutkar, Lame, & Soglia, 2006; Exarchou, Godejohann, van Beek, Gerothanassis, & Vervoort, 2003; Wilson & Brinkman, 2003; Wolfender, Ndjoko, & Hostettmann, 2003).

LC-MS-based comprehensive analysis of untargeted metabolites in medicinal plants

LC-MS offers the analysis of thermolabile and non-volatile compounds that ranges from polar sugars to various lipids. LC is having universal adaptability to most of the compound and hence it was used as the most successful technique in herbal medicine research. UPLC, monolithic column LC and capillary LC were developed to improve the efficiency of separation. This approach provides high sensitivity and untargeted analysis. Hence it can be used for the analysis wide range of plant metabolites. Untargeted metabolomics examines the complete plant metabolome and does not show any bias toward metabolites. For both comprehensive qualitative and quantitative analysis of herbal medicines, untargeted metabolomics is the best approach.

A typical workflow of LC-MS based untargeted metabolomics studies

LC-MS based untargeted metabolomics studies have some fundamental steps that shown in Fig. 5.1. The first step is the experimental design

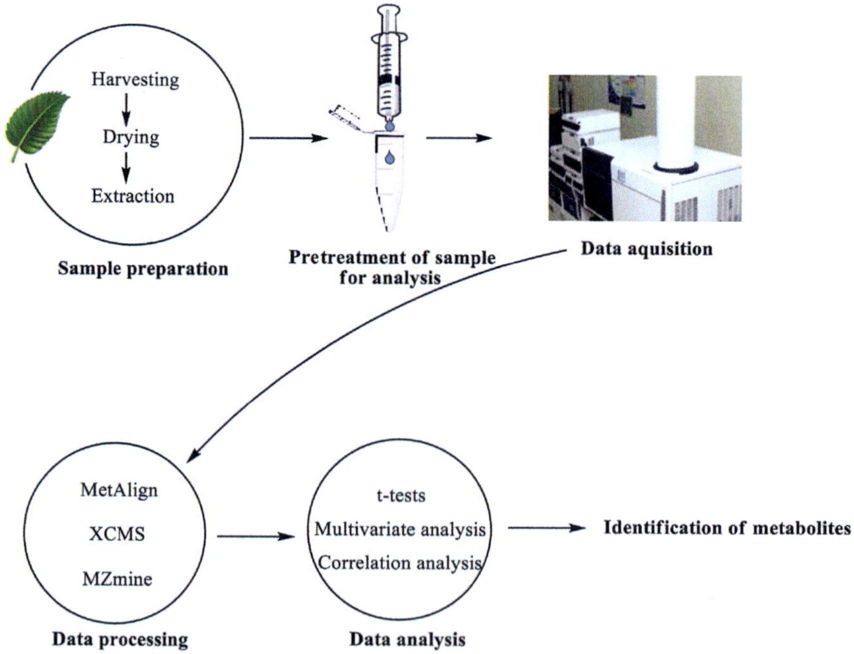

Figure 5.1 *Schematic workflow of LC-MS based untargeted metabolomics in herbal medicine research.*

followed by sample preparation and the data acquisition are mainly depends on the aim and objective of the research investigation (Cevallos-Cevallos, Reyes-De-Corcuera, Etxeberria, Danyluk, & Rodrick, 2009). Sample preparation is a very crucial step and it can depend on several factors of sample integrity (Heyman & Meyer, 2012). The yield of metabolites and the range of metabolites extracted depend on the solvents and method used for the extraction. Generally, a single solvent system is used for the extraction of metabolites. However, a combination of solvents can also be used for the extraction of diverse metabolites from the plant sample (Kim, Choi, Erkelens, Lefeber, & Verpoorte, 2005; van der Kooy, Maltese, Choi, Kim, & Verpoorte, 2009).

The significant developments in automated procedures and data processing tools like MZmine, Metalign, and XCMS helps in the processing of challenging untargeted metabolomics data and extraction of appropriate information from crude chromatographic data. This allows the processing of large data of untargeted metabolomics and makes it a practical reality (Theodoridis, Gika, Want, & Wilson, 2012; Eliasson et al., 2012).

Once data processing is done, the obtained data has to analyze according to the aim and objective of the study by a suitable approach. The basic aim of the untargeted metabolomics is the identification and quantification of whole plant metabolome (both known and unknown metabolites). The use of univariate statistical analysis and other classical statistical methods are not much feasible with untargeted metabolomics. Hence multivariate statistical methods are used in this approach for analysis of large data sets of unknown metabolites. The synergistic effects, which cannot be detected at the individual level, can also be detected in untargeted metabolomics. To solve the trouble being investigated, projection-based multivariate statistical methods can be used. In this method, the examined variables can be combined with the latent variable. To the investigation of exploratory data, principle component analysis (PCA) is commonly used. On the other hand, for the extraction of information hidden that classifies and explains the system behavior, orthogonal projection to latent structures discriminant analysis (OPLS-DA) and projection to latent structures discriminant analysis (PLS-DA) are the best-preferred statistical method (Trygg & Wold, 2003).

In plants, glycosylation and esterification are the main pathways for the transformation of secondary metabolites. These diverse secondary metabolites produce a pharmacological effect. But, these are very challenging to identify the metabolites with accuracy in untargeted metabolomics. However, several metabolite databases like METLIN, Campus Chemical

Instrument Center, MS2T, MASSBANK, Birmingham Metabolite Library and HMDB are available but the metabolite coverage by these databases is inadequate due to a wide diversity of plant metabolites (Matsuda et al., 2009a; Smith et al., 2005). Determination of exact mass and fragmentation pattern of plant metabolites by tandem mass spectrometry permits the structural elucidation of metabolites de novo.

NMR-based comprehensive analysis of untargeted metabolites in medicinal plants

The recent development in analytical chemistry toward identification and characterization of small molecular compounds in high-field NMR together with modern multivariate statistical analysis allows the high efficiency systems for comprehensive analysis of the data obtained in metabolomics experiments. NMR methods are rapid and robust and can produce reproducible results. It has some benefits more than mass spectrometry-based methods that includes it is a universal detection method for all NMR active compounds and it can allow the direct comparison of concentrations of all compounds by the relation of signal intensity of all protons is proportional to the concentration of metabolites. Coupling of LC with NMR (HPLC-NMR) and also in combination with this with solid phase extraction technique (HPLC-SPE-NMR) had rapidly grown in the field of plant metabolomics.

Workflow of NMR-based untargeted metabolomics studies

The fundamental steps in the NMR based untargeted metabolomics studies are the same as steps involved in the LC-MS based studies. The workflow of NMR based metabolomics shown in Fig. 5.2. The sample collection and extraction step are crucial in these studies. The solvent used for the extraction is generally deuterated solvents. For the extraction of the complete metabolome, D_2O and CD_3OD or KH_2PO_4 combinations are used for the NMR analysis (Kim et al., 2005). Data processing tools such as SIMCA and ACD Labs have been designed to extract the useful and required information from the NMR spectroscopic data and thereby it will helpful for the untargeted metabolomics studies through quick processing of the multiple variables. The data acquired from NMR will use multivariate statistical methods in untargeted metabolomics approach for analysis of a wide range of unknown plant metabolites. The multivariate statistical methods used in the LC-MS based metabolomics can be used in this platform also (Commisso, Strazzer, Toffali, Stocchero, & Guzzo, 2013).

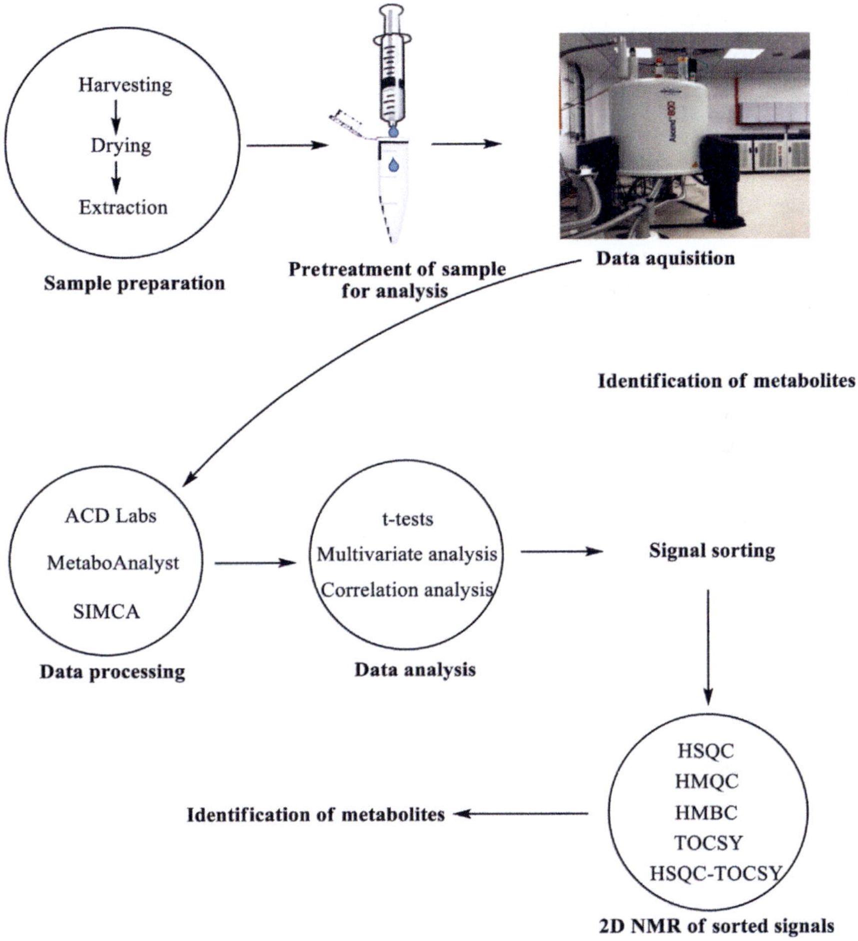

Figure 5.2 ***Schematic workflow of NMR-based untargeted metabolomics in herbal medicine research.***

NMR parameters can be sensitive to several factors such as ionic condition, temperature, pH and concentration (Weljie, Newton, Mercier, Carlson, & Slupsky, 2006). Hence, the identification of metabolites is most challenging. For the better identification of the plant metabolites, the NMR spectra should be acquired under the same as acquired conditions in the database. A series of multidimensional NMR experiments such as HMQC, TOCSY, HSQC, HMBC or the combination of either HMQC or HSQC and TOCSY (HMQC/HSQC -TOCSY) should be conducted to verify the identified metabolites in a set of similar samples (Carvalho, Jeffrey, Sherry, & Malloy, 1998). Additionally, derivatizing agents (chemically selective probes) that can be target specific functional groups that include ketones, aldehydes, carboxylates and thiols can be used for the determination of functional groups. By the introduction of ^{15}N or ^{13}C at a specific position in the chemically selective probe, the desired metabolites with functional groups of interest can be detected by utilizing the isotope-editing feature of NMR (Bousamra et al., 2014; Gori et al., 2014).

Quality control of herbal medicine by the metabolomics approach

Quality control of herbal medicines and their standardization is extremely important and necessary in the maintenance of its integrity for pharmaceutical quality (Heyman & Meyer, 2012). Assessment of quality control of herbal medicines is by its pharmacological phytoconstituents evaluations. So far, the normal assessment of quality control is based on the detection of one or two kinds of indicator markers of pharmacologically active constituents. But, herbal medicine usually contains more chemical constituents and hence assessment of only those few biomarkers does not provide a holistic view of quality control of the product. Hence, the investigation of all the phytochemical constituents in the herbal products is required for a better understanding of the biological activities of the product (Lee et al., 2017). Nowadays, metabolomics is uniquely suited and extensively used in the evaluation of all the phytochemical constituents and thereby it is showing as a leading approach in the assessment of quality control of herbal medicines. Good quality control investigations help for the validating the repeatability and reproducibility of the pharmacological and clinical research and also play a role in the identification of possible side effects of active phytoconstituents. Currently, chromatographic and spectroscopic techniques are used for the sufficient assessment of quality control of herbal medicines. With the help of chromatographic and spectroscopic techniques, it is possible to provide a standardized metabolite fingerprint of the entire herbal product and thereby help in the analysis of full herbal medicine. Metabolite profiling studies can be integrated to identify the phytoconstituents that comprise the metabolite fingerprint (Bailey, Sampson, Hylands, Nicholson, & Holmes, 2002; Valentão, Andrade, Areias, Ferreres, & Seabra, 1999). The two main chromatographic techniques used for the metabolite fingerprinting are high-performance liquid chromatography and gas chromatography. Generally, for targeted metabolomics, the mass spectrometry coupled to a chromatographic technique is employed and for untargeted metabolomics

NMR and MS are used without chromatographic separation. NMR and MS are the most widely used analytical platforms for the quality control assessment through metabolite fingerprinting approach. Recent developments in chemometrics further strengthen the quality control studies when combined with the advanced chromatographic and spectroscopic techniques through the establishing of more accurate data and thus enhancing the integrity of the herbal medicine. The strength of the chemometrics due to the integration of data in different dimensions and explain the similarities and differences of the data in an easy to use graphical manner (Valentão et al., 1999; Peishan, 2001).

Chemical profile and bioactivity linkage of herbal medicines by metabolomics approach

The reductionist approach will mainly focus on the single active compound. However, in some cases like St John's Wort extract where a group of principle active compounds or no single compound is found to be accountable for its pharmacological activity, then there may be the chance of synergism which is evident. Hence, for the valuable assurance of therapeutic efficiency can be obtained from a holistic in vivo approach (Verpoorte, Choi, & Kim, 2005). Generally, animal studies and clinical trials are the two traditional approaches followed to conduct in vivo studies. However, the main drawback of these methods is that it is a very tedious process and not very cost effective. But, if we use this approaches for extracts with uncertain composition, the results we get may be variable or inconsistent.

By considering the numerous chemical constituents rather than focusing only on the single and specific active constituent of a natural product provides a better view of the mechanism of action of herbal medicine and also gives us reliable information related to its pharmacological action over a different sets of batches. Comparison of changes by the sample on transcriptome, proteome, or metabolome pattern to the changes obtained after treating with known drugs can be investigated in these approaches. Study on genomics, proteomics, and metabolomics is important as they help in the identification of compounds, which may relate to the pharmacological activity of complex extract (Ulrich-Merzenich et al., 2007). Due to this, there may be a chance of changing the concept of improvement and application of complex herbal compound mixtures in modern medicine (Shyur & Yang, 2008). However, we can correlate the result (measured after performing in vitro and in vivo tests of the complex chemical extract) to the activity. Bioavailability, efficacy, and toxicity parameters of various natural products can be evaluated by using the metabolomic approach.

Identification of mode of action and efficiency of herbal medicines by metabolomics approach

With the use of proteomic and transcriptome platforms, in vitro effects at the cellular level of cannabis from diverse cultivars and preparations were analyzed. With the help of ^{1}H-NMR and multivariate data analysis method, metabolites of *Ginkgo* in rat urine are measured and effect of three traditional Chinese medicines were studied for the assessment responsive in plasma lipids to dietary treatment in insulin-resistant APOE*3-Leiden transgenic mice. Three TCM preparations were analyzed using LC-MS with multivariate data analysis for the screening of the most effective with low toxicity preparation (Wang et al., 2005).

Recently metabolomic profiling has also been applied to the study of the mode of action of many natural products. With the use of HPLC/DAD/ESI-MS and principle component analysis, it was discovered that dihydro-cucurbitacin F-25-*O*-acetate is responsible for the antimicrobial activity in *Hemsleya pengxianensi*. The metabolite profiles are compared between the commercial antibacterials and *S. aureus* treated by the plant extract with a known mode of action for the determination of the possible mode of antibacterial activity of *Staphylococcus aureus* (Biao-Yi, Yu, & Zeng-Liang, 2008). Some papers reported that the assessment of the efficiency of herbal medicine by the metabolomics approach. A popular Chinese herbal medicine named as *Herba Epimedii* (TFE) contains many flavones used for the treatment of some age-related diseases. Total ten ageing marker metabolites such as lactate, dimethylamine, succinate, creatinine, acetate, trimethylamine-*N*-oxide, alanine, methylamine, allantoin, and acetone were identified in at urine by using ^{1}H NMR and Principle component analysis. This study showed that the 24-month-old rats aging features and markers were shifted toward that of 18-month-old rats with the intervention of TFE (Wu et al., 2008).

Epimedium brevicornum Maxim is an herbal medicine, widely used in China for strengthening bones and toning the kidney. However, the principle active chemical constituents responsible for its pharmacological activity are unknown. UPLC-MS and principle component analysis was used for the identification of plasma metabolites in post hydrocortisone treatment, prehydrocortisone treatment, and *Epimedium brevicornum* Maxim-treated groups. This study identified that the four compounds were identified in the plasma such as icariin, 2″-O-rhamnosoylicaride, epimedin C and icariside II and among these four metabolites icariin and epimedin C were the active ones (Li, Lu, Liu, Liu, & Xiong, 2007).

A Chinese herbal medicine formulation named as Xindi soft capsules and it contains sea buckthorn berry oil sea buckthorn flavonoids. It is generally used for the treatment of nasal congestion, blood stasis, local ischemia, hemorrhage, and thrombosis. Rat urine metabolites were identified in acute blood stasis model group, healthy control group, high-dose group of Xindi soft capsule, middle dose group of Xindi soft capsule and low dose group of Xindi soft capsule by UPLC coupled with Q-TOF mass spectrometry. This was followed by data analysis with principle component analysis and partial least squares data analysis methods. The obtained results showed that the treated animal group is located between the healthy control group and acute blood stasis model group. Potential biomarkers, such as phenylalanine, cholic acid, and kynurenic acid were also identified (Zhao et al., 2008).

Assessment of bioavailability and fate of the herbal medicine by metabolomics approach

Metabolomics approach is successfully applied in the assessment of bioavailability and fate of the herbal medicines. In China, a decoction (ODD) of huangqi (*Radix astragali*)-danggui (*Radix angelicae sinensis*) is used as a tonic. It is a very complex mixture containing hundreds of constituents. Among all of them, only 70 were isolated and identified till the date and few were reported for its pharmacological activity. For the assessment of in vivo bioavailability of these metabolites of ODD, metabolite fingerprinting by LC-MS was applied. The metabolite finger printing of ODD treated rabbit plasma with fingerprint of ODD formula was compared to assess the bioavailability. The detected metabolites were confirmed by comparing the UV and MS spectra with the reference spectra obtained from the literature. Some ODD active compounds such as phthalides, flavonoids, and triterpene saponins were well absorbed into the body (Wang et al., 2007).

Safety and toxicity assessment of herbal medicines by metabolomics approach

Safety and toxicity are the two major issues related to the quality of the herbal medicine. Metabolomics approach is applied to solve these two issues and was reported for various herbal medicines. For the toxicity assessment of aristolochic acid, the rat urine samples were collected after the intraperitoneal injection of aristolochic acid and some toxins with known mode of action. It was concluded with the help of principle component analysis (PCA) and biochemical parameter analysis that it would induce the renal toxicity accomplished by slight hepatic lesion (Zhang et al., 2006). Likewise, ^{1}H NMR spectral metabolite profiling was used in the evaluation of *Hei-shun-pian* toxicity in rats. This herb has analgesic, antirheumatoid arthritis, antipyretic and anti-inflammatory activities. The active principles of this herb are the C-19 diterpenoid alkaloid aconitine and a series of its derivatives. Some of these active principles have narrow therapeutic windows and highly toxic. Processing of the obtained ^{1}H NMR data by principle least squares regression (PLS) and PCA showed that the difference between the plasma of control and treated group was the decrease in the concentration of some markers such as *O*- and *N*-acetyl glycoproteins and phosphatidylcholine and increase in the concentration of lipids. Whereas, other markers that includes 2-oxoglutarate, taurine, dimethylamine, creatinine, acetate, and hippurate levels in urine contributed to the difference between high, medium and control groups (Li et al., 2008).

Detection of active principle in herbal product and quantitative prediction of its bioactivity by metabolomics approach

^{1}H NMR in combination with chemometrics was used to discriminate analysis of *Artemisia annua* L. samples obtained from various sources. PCA showed that artemisinin is the main discriminant factor for clustering. For the quantitative prediction of antiplasmodial activity and toxicity values of different Artemisia annua extracts, PLS was applied (Bailey et al., 2004). This approach was also applied to reveal the correlation between the spectral pattern and opioid receptor binding properties of St. John's Wort extracts prepared with different solvents (Roos, Röseler, Büter, & Simmen, 2004). Six samples of Galphimia glauca were differentiated by the application of ^{1}H NMR and PCA analysis. From the data, it was found that 1,3,4,5-tetra-Ogalloylquinic acid contents and Galphimines are the major discriminating factors for PCA. Additional PLS-DA analysis confirmed that both the compounds were correlated with anxiolytic and sedative effects in mice (Cardoso-Taketa, Pereda-Miranda, Choi, Verpoorte, & Villarreal, 2008). To distinguish F2 hybrids (thrips-resistant and susceptible) of *Senecio aquaticus and Senecio jacobaea,* the same technique was applied. It was observed that some active principle components such as kaempfero, pyrrolizidine, jacaranone were high in resistant plants (Leiss, Choi, Abdel-Farid, Verpoorte, & Klinkhamer, 2009). The anti-inflammatory activity of different sources of ginger and other zingiber species and metabolite profiling data obtained from GC-MS showed that the anti-inflammatory activity was shown by other species is similar to *Z. officinale* but there is no proper correlation between their gingerol content. Hence, it was concluded that there are other active principle compounds unidentified to date but are have therapeutic activity. These have to be detected and quantified in order to promise the bioactivity (Jiang et al., 2006).

Challenges in untargeted metabolomics studies and promising way to overcome them

Untargeted metabolomics platforms such as NMR and LC-MS offers a high-throughput, fast, automated analysis of crude plant extracts, quantification of different metabolites and also provides information related to structure that including the stereochemical details (Kim, Choi, & Verpoorte, 2010; Seger & Sturm, 2007). NMR provides the metabolites structural information but it has low chemical resolution and low resolution (Schultz et al., 2013). The sensitivity and intensity of the NMR signals depend on the magnetic field strength. Hence, more powerful magnets are required for improving the performance of the NMR technique. In recent, complementary metal oxide semiconductors that would consist of a series of high-sensitivity micro-coils integrated with interfacing radio-frequency circuits on the same chip are being used in the advanced high-throughput NMR spectrometer to enhance sensitivity and throughput (Pourmodheji, Ghafar-Zadeh, & Magierowski, 2016). Additionally, the quantification and identification of metabolites could be obstructed due to extensive resonance overlap in the evaluation of plant extraction by NMR spectroscopy. This can be overcome by a number of NMR studies using enrichment or fraction step (Simmler, Napolitano, McAlpine, Chen, & Pauli, 2014).

LC-MS platform provides more sensitivity and selectivity than the NMR. It also allows the identification and quantification of very low concentration of metabolites (Sumner, Mendes, & Dixon, 2003). The basic aim of the metabolomics is to identify the as many metabolites as possible. That includes several altered metabolites which may be known and unknown, responsive to a given set of conditions can also be identified (Fan & Lane, 2016). Plants have wide diverse coverage of metabolites, due to this most database do not provide the complete coverage of plant metabolome. If the accurate structure of the metabolites is not available in the database, then it is very difficult to identify those metabolites. High-resolution mass spectrometry provides high mass accuracy in the identification of metabolites (Kamleh et al., 2008). Although the potential molecular formulas will gets decrease by the high mass accuracy corresponding to an individual metabolic peak. Issues arise with this approach as there may be several potential isomers for each given molecular formula and several molecular formulas appropriate for the accurate mass data (Schultz et al., 2013). So, for providing some structural information, there is a need of fragmentation behavior. Compared to accurate mass data, the matching of accurate mass data and fragmentation pattern with standard spectra available in the database may give promising metabolite identification. LC-MS is widely used for the identification of semi-polar phytochemicals. Multidimensional and heteronuclear NMR methods like ^{31}P and ^{15}N could be important in the identification of unknown spectral features in NMR-based studies. The combination of chemical shift values along with heteronuclear scalar coupling can provide sufficient information in the identification of metabolites with a comparison of the database. Sometimes it is quite unclear in the identification of correct functional groups by chemical shift values. Hence, additional experiments such as pH titrations are needed to confirm the presence of carboxylic and amine functional groups (Fan & Lane, 2016).

Conclusion

The increased demand and growth of the global market for herbal medicines proposed that modern research technologies have to be utilized to study and understanding the lead finding in the herbal medicines. Usage of different analytical platforms, in combination with different approaches and online bioassays are helpful for the estimation of efficiency. Till the date, the reductionist approach is mostly used approach for the identification of interesting compounds in the drug discovery. However, metabolomics approach in combination with multivariate data analysis tools provides an excellent means of identification of active compounds. Recent advancements in the NMR and MS offers a wide range of compounds estimation and support the data obtained for new compounds with the open access natural products library. Without any doubt, metabolomics is the best tool for the quality control of herbal medicines. This approach captures the differences in all of the factors that significantly affect the active principle components. A quick analysis of a wide range of metabolites is possible with the metabolomics approach. Hence, metabolomics is increasing in the field of personalized medicine. This approach offers the metabolite profiling of individual patient linked to a specific response to a specific herbal drug in terms of toxicity, efficiency and clearance. It is clearly evident that the metabolomics approach is the powerful facilitator in the identification of novel leads and the discovery of novel targets. Mode of action and toxicity effects can be obtained from the metabolic studies.

Acknowledgments

The authors would like to thank Dr. Shashi Bala Singh, Director, NIPER-Hyderabad for her motivation and support. The authors sincerely appreciate the financial support from the Science and Engineering Research Board (SERB), DST, New Delhi, India, (EMR/2016/002098).

Competing interests

The authors declare no conflict of interest.

References

Baier, M., Hemmann, G., Holman, R., Corke, F., Card, R., Smith, C., Rook, F., & Bevan, M. W. (2004). Characterization of mutants in Arabidopsis showing increased sugar-specific gene expression, growth, and developmental responses. *Plant Physiology*, *134*(1), 81–91.

Bailey, N. J., Sampson, J., Hylands, P. J., Nicholson, J. K., & Holmes, E. (2002). Multi-component metabolic classification of commercial feverfew preparations via high-field 1H-NMR spectroscopy and chemometrics. *Planta Medica*, *68*(08), 734–738.

Bailey, N. J., Wang, Y., Sampson, J., Davis, W., Whitcombe, I., Hylands, P. J., Croft, S. L., & Holmes, E. (2004). Prediction of anti-plasmodial activity of Artemisia annua extracts: application of 1H NMR spectroscopy and chemometrics. *Journal of Pharmaceutical and Biomedical Analysis*, *35*(1), 117–126.

Benderoth, M., Textor, S., Windsor, A. J., Mitchell-Olds, T., Gershenzon, J., & Kroymann, J. (2006). Positive selection driving diversification in plant secondary metabolism. *Proceedings of the National Academy of Sciences*, *103*(24), 9118–9123.

Biao-Yi, Z., Yu, Y., & Zeng-Liang, Y. (2008). Investigation of antimicrobial model of Hemsleya pengxianensis WJ Chang and its main active component by metabolomics technique. *Journal of Ethnopharmacology*, *116*(1), 89–95.

Böttcher, C., von Roepenack-Lahaye, E., Schmidt, J., Schmotz, C., Neumann, S., Scheel, D., & Clemens, S. (2008). Metabolome analysis of biosynthetic mutants reveals a diversity of metabolic changes and allows identification of a large number of new compounds in Arabidopsis. *Plant Physiology*, *147*(4), 2107–2120.

Bousamra, M., II, Schumer, E., Li, M., Knipp, R. J., Nantz, M. H., Van Berkel, V., & Fu, X. -A. (2014). Quantitative analysis of exhaled carbonyl compounds distinguishes benign from malignant pulmonary disease. *The Journal of Thoracic and Cardiovascular Surgery*, *148*(3), 1074–1081.

Camacho, D., De La Fuente, A., & Mendes, P. (2005). The origin of correlations in metabolomics data. *Metabolomics*, *1*(1), 53–63.

Cardoso-Taketa, A. T., Pereda-Miranda, R., Choi, Y. H., Verpoorte, R., & Villarreal, M. L. (2008). Metabolic profiling of the Mexican anxiolytic and sedative plant *Galphimia glauca* using nuclear magnetic resonance spectroscopy and multivariate data analysis. *Planta Medica*, *74*(10), 1295–1301.

Carvalho, R. A., Jeffrey, F. M. H., Sherry, A. D., & Malloy, C. R. (1998). 13C isotopomer analysis of glutamate by heteronuclear multiple quantum coherence-total correlation spectroscopy (HMQC-TOCSY). *FEBS Letters*, *440*(3), 382–386.

Cevallos-Cevallos, J. M., Reyes-De-Corcuera, J. I., Etxeberria, E., Danyluk, M. D., & Rodrick, G. E. (2009). Metabolomic analysis in food science: a review. *Trends in Food Science & Technology*, *20*(11–12), 557–566.

Commisso, M., Strazzer, P., Toffali, K., Stocchero, M., & Guzzo, F. (2013). Untargeted metabolomics: an emerging approach to determine the composition of herbal products. *Computational and Structural Biotechnology Journal*, *4*(5), e201301007.

Eliasson, M., Rännar, S., Madsen, R., Donten, M. A., Marsden-Edwards, E., Moritz, T., Shockcor, J. P., Johansson, E., & Trygg, J. (2012). Strategy for optimizing LC-MS data processing in metabolomics: a design of experiments approach. *Analytical Chemistry*, *84*(15), 6869–6876.

Exarchou, V., Godejohann, M., van Beek, T. A., Gerothanassis, I. P., & Vervoort, J. (2003). LC-UV-solid-phase extraction-NMR-MS combined with a cryogenic flow probe and its application to the identification of compounds present in Greek oregano. *Analytical Chemistry*, *75*(22), 6288–6294.

Fan, T. W. -M., & Lane, A. N. (2016). Applications of NMR spectroscopy to systems biochemistry. *Progress in Nuclear Magnetic Resonance Spectroscopy*, *92*, 18–53.

Fiehn, O., Kopka, J., Dörmann, P., Altmann, T., Trethewey, R. N., & Willmitzer, L. (2000). Metabolite profiling for plant functional genomics. *Nature Biotechnology*, *18*(11), 1157.

Fiehn, O., Kopka, J., Dörmann, P., Altmann, T., Trethewey, R., & Willmitzer, L. (2001). Erratum: metabolite profiling for plant functional genomics. *Nature Biotechnology*, *19*(2), 173.

Gori, S. S., Lorkiewicz, P., Ehringer, D. S., Belshoff, A. C., Higashi, R. M., Fan, T. W. -M., & Nantz, M. H. (2014). Profiling thiol metabolites and quantification of cellular glutathione using FT-ICR-MS spectrometry. *Analytical and Bioanalytical Chemistry*, *406*(18), 4371–4379.

Hall, R. D. (2006). Plant metabolomics: from holistic hope, to hype, to hot topic. *New Phytologist*, *169*(3), 453–468.

Heyman, H. M., & Meyer, J. J. M. (2012). NMR-based metabolomics as a quality control tool for herbal products. *South African Journal of Botany*, *82*, 21–32.

Hirai, M. Y., Yano, M., Goodenowe, D. B., Kanaya, S., Kimura, T., Awazuhara, M., Arita, M., Fujiwara, T., & Saito, K. (2004). Integration of transcriptomics and metabolomics for understanding of global responses to nutritional stresses in *Arabidopsis thaliana*. *Proceedings of the National Academy of Sciences*, *101*(27), 10205–10210.

Horai, H., Arita, M., Kanaya, S., Nihei, Y., Ikeda, T., Suwa, K., Ojima, Y., Tanaka, K., Tanaka, S., & Aoshima, K. (2010). MassBank: a public repository for sharing mass spectral data for life sciences. *Journal of Mass Spectrometry*, *45*(7), 703–714.

Hostettmann, K., Wolfender, J. -L., & Rodriguez, S. (1997). Rapid detection and subsequent isolation of bioactive constituents of crude plant extracts. *Planta Medica*, *63*(01), 2–10.

Hou, C. -C., Chen, C. -H., Yang, N. -S., Chen, Y. -P., Lo, C. -P., Wang, S. -Y., Tien, Y. -J., Tsai, P. -W., & Shyur, L. -F. (2010). Comparative metabolomics approach coupled with cell-and gene-based assays for species classification and anti-inflammatory bioactivity validation of Echinacea plants. *The Journal of Nutritional Biochemistry*, *21*(11), 1045–1059.

Huhman, D.V., & Sumner, L.W. (2002). Metabolic profiling of saponins in *Medicago sativa* and *Medicago truncatula* using HPLC coupled to an electrospray ion-trap mass spectrometer. *Phytochemistry*, *59*(3), 347–360.

Iijima, Y., Nakamura, Y., Ogata, Y., Tanaka, K. i., Sakurai, N., Suda, K., Suzuki, T., Suzuki, H., Okazaki, K., & Kitayama, M. (2008). Metabolite annotations based on the integration of mass spectral information. *The Plant Journal, 54*(5), 949–962.

Jiang, H., Xie, Z., Koo, H. J., McLaughlin, S. P., Timmermann, B. N., & Gang, D. R. (2006). Metabolic profiling and phylogenetic analysis of medicinal Zingiber species: Tools for authentication of ginger (*Zingiber officinale Rosc.*). *Phytochemistry, 67*(15), 1673–1685.

Kamleh, A., Barrett, M., Wildridge, D., Burchmore, R., Scheltema, R., & Watson, D. (2008). Metabolomic profiling using Orbitrap Fourier transform mass spectrometry with hydrophilic interaction chromatography: a method with wide applicability to analysis of biomolecules. *Rapid Communications in Mass Spectrometry: An International Journal Devoted to the Rapid Dissemination of Up-to-the-Minute Research in Mass Spectrometry, 22*(12), 1912–1918.

Kim, H. K., Choi, Y. H., Erkelens, C., Lefeber, A. W., & Verpoorte, R. (2005). Metabolic fingerprinting of Ephedra species using 1H-NMR spectroscopy and principal component analysis. *Chemical and Pharmaceutical Bulletin, 53*(1), 105–109.

Kim, H. K., Choi, Y. H., & Verpoorte, R. (2010). NMR-based metabolomic analysis of plants. *Nature Protocols, 5*(3), 536.

Lancien, M., & Roberts, M. R. (2006). Regulation of *Arabidopsis thaliana* 14-3-3 gene expression by γ-aminobutyric acid. *Plant, Cell & Environment, 29*(7), 1430–1436.

Lee, K. -M., Jeon, J. -Y., Lee, B. -J., Lee, H., & Choi, H. -K. (2017). Application of metabolomics to quality control of natural product derived medicines. *Biomolecules & Therapeutics, 25*(6), 559.

Leiss, K. A., Choi, Y. H., Abdel-Farid, I. B., Verpoorte, R., & Klinkhamer, P. G. (2009). NMR metabolomics of thrips (*Frankliniella occidentalis*) resistance in Senecio hybrids. *Journal of Chemical Ecology, 35*(2), 219–229.

Li, F., Lu, X., Liu, H., Liu, M., & Xiong, Z. (2007). A pharmaco-metabonomic study on the therapeutic basis and metabolic effects of *Epimedium brevicornum* Maxim. on hydrocortisone-induced rat using UPLC-MS. *Biomedical Chromatography, 21*(4), 397–405.

Li, L., Sun, B., Zhang, Q., Fang, J., Ma, K., Li, Y., Chen, H., Dong, F., Gao, Y., & Li, F. (2008). Metabonomic study on the toxicity of Hei-Shun-Pian, the processed lateral root of *Aconitum carmichaelii Debx.* (Ranunculaceae). *Journal of Ethnopharmacology, 116*(3), 561–568.

Matsuda, F., Yonekura-Sakakibara, K., Niida, R., Kuromori, T., Shinozaki, K., & Saito, K. (2009a). MS/MS spectral tag-based annotation of non-targeted profile of plant secondary metabolites. *The Plant Journal, 57*(3), 555–577.

Matsuda, F., Shinbo, Y., Oikawa, A., Hirai, M. Y., Fiehn, O., Kanaya, S., & Saito, K. (2009b). Assessment of metabolome annotation quality: a method for evaluating the false discovery rate of elemental composition searches. *Plos One, 4*(10), e7490.

Matsuda, F., Nakabayashi, R., Sawada, Y., Suzuki, M., Hirai, M. Y., Kanaya, S., & Saito, K. (2011). Mass spectra-based framework for automated structural elucidation of metabolome data to explore phytochemical diversity. *Frontiers in Plant Science, 2*, 40.

Moco, S., Bino, R. J., Vorst, O., Verhoeven, H. A., de Groot, J., van Beek, T. A., Vervoort, J., & De Vos, C. R. (2006). A liquid chromatography-mass spectrometry-based metabolome database for tomato. *Plant Physiology, 141*(4), 1205–1218.

Moco, S., Schneider, B., & Vervoort, J. (2009). Plant micrometabolomics: the analysis of endogenous metabolites present in a plant cell or tissue. *Journal of Proteome Research, 8*(4), 1694–1703.

Mutlib, A., Strupczewski, J., & Chesson, S. (1995). Application of hyphenated LC/NMR and LC/MS techniques in rapid identification of in vitro and in vivo metabolites of iloperidone. *Drug Metabolism and Disposition, 23*(9), 951–964.

Nikiforova, V. J., Kopka, J., Tolstikov, V., Fiehn, O., Hopkins, L., Hawkesford, M. J., Hesse, H., & Hoefgen, R. (2005). Systems rebalancing of metabolism in response to sulfur deprivation, as revealed by metabolome analysis of Arabidopsis plants. *Plant Physiology, 138*(1), 304–318.

Peishan, X. (2001). A feasible strategy for applying chromatography fingerprint to assess quality of Chinese herbal medicine [j]. *Traditional Chinese Drug Research & Clinical Pharmacology*, *12*(3), 141–169.

Peterman, S. M., Duczak, N., Kalgutkar, A. S., Lame, M. E., & Soglia, J. R. (2006). Application of a linear ion trap/orbitrap mass spectrometer in metabolite characterization studies: examination of the human liver microsomal metabolism of the non-tricyclic anti-depressant nefazodone using data-dependent accurate mass measurements. *Journal of the American Society for Mass Spectrometry*, *17*(3), 363–375.

Pourmodheji, H., Ghafar-Zadeh, E., & Magierowski, S. (2016). A multidisciplinary approach to high throughput nuclear magnetic resonance spectroscopy. *Sensors*, *16*(6), 850.

Rischer, H., Orešič, M., Seppänen-Laakso, T., Katajamaa, M., Lammertyn, F., Ardiles-Diaz, W., Van Montagu, M. C., Inzé, D., Oksman-Caldentey, K. -M., & Goossens, A. (2006). Gene-to-metabolite networks for terpenoid indole alkaloid biosynthesis in *Catharanthus roseus* cells. *Proceedings of the National Academy of Sciences*, *103*(14), 5614–5619.

Roos, G., Röseler, C., Büter, K. B., & Simmen, U. (2004). Classification and correlation of St. John's wort extracts by nuclear magnetic resonance spectroscopy, multivariate data analysis and pharmacological activity. *Planta Medica*, *70*(08), 771–777.

Saito, K., & Matsuda, F. (2010). Metabolomics for functional genomics, systems biology, and biotechnology. *Annual Review of Plant Biology*, *61*, 463–489.

Sato, S., Soga, T., Nishioka, T., & Tomita, M. (2004). Simultaneous determination of the main metabolites in rice leaves using capillary electrophoresis mass spectrometry and capillary electrophoresis diode array detection. *The Plant Journal*, *40*(1), 151–163.

Schauer, N., Steinhauser, D., Strelkov, S., Schomburg, D., Allison, G., Moritz, T., Lundgren, K., Roessner-Tunali, U., Forbes, M. G., & Willmitzer, L. (2005). GC–MS libraries for the rapid identification of metabolites in complex biological samples. *FEBS Letters*, *579*(6), 1332–1337.

Schmid, M., Davison, T. S., Henz, S. R., Pape, U. J., Demar, M., Vingron, M., Schölkopf, B., Weigel, D., & Lohmann, J. U. (2005). A gene expression map of *Arabidopsis thaliana* development. *Nature Genetics*, *37*(5), 501.

Schultz, A. W., Wang, J., Zhu, Z. -J., Johnson, C. H., Patti, G. J., & Siuzdak, G. (2013). Liquid chromatography quadrupole time-of-flight characterization of metabolites guided by the METLIN database. *Nature Protocols*, *8*(3), 451.

Seger, C., & Sturm, S. (2007). Analytical aspects of plant metabolite profiling platforms: current standings and future aims. *Journal of Proteome Research*, *6*(2), 480–497.

Shyur, L. -F., & Yang, N. -S. (2008). Metabolomics for phytomedicine research and drug development. *Current Opinion in Chemical Biology*, *12*(1), 66–71.

Simmler, C., Napolitano, J. G., McAlpine, J. B., Chen, S. -N., & Pauli, G. F. (2014). Universal quantitative NMR analysis of complex natural samples. *Current Opinion in Biotechnology*, *25*, 51–59.

Smith, C. A., O'Maille, G., Want, E. J., Qin, C., Trauger, S. A., Brandon, T. R., Custodio, D. E., Abagyan, R., & Siuzdak, G. (2005). METLIN: a metabolite mass spectral database. *Therapeutic Drug Monitoring*, *27*(6), 747–751.

Sumner, L. W., Mendes, P., & Dixon, R. A. (2003). Plant metabolomics: large-scale phytochemistry in the functional genomics era. *Phytochemistry*, *62*(6), 817–836.

Theodoridis, G. A., Gika, H. G., Want, E. J., & Wilson, I. D. (2012). Liquid chromatography–mass spectrometry based global metabolite profiling: a review. *Analytica Chimica Acta*, *711*, 7–16.

Tolstikov, V. V., & Fiehn, O. (2002). Analysis of highly polar compounds of plant origin: combination of hydrophilic interaction chromatography and electrospray ion trap mass spectrometry. *Analytical Biochemistry*, *301*(2), 298–307.

Tolstikov, V. V., Lommen, A., Nakanishi, K., Tanaka, N., & Fiehn, O. (2003). Monolithic silica-based capillary reversed-phase liquid chromatography/electrospray mass spectrometry for plant metabolomics. *Analytical Chemistry*, *75*(23), 6737–6740.

Trygg, J., & Wold, S. (2003). O2-PLS, a two-block (X–Y) latent variable regression (LVR) method with an integral OSC filter. *Journal of Chemometrics*, *17*(1), 53–64.

Tyler, V. E. (1999). Phytomedicines: back to the future. *Journal of Natural Products*, *62*(11), 1589–1592.

Ulrich-Merzenich, G., Zeitler, H., Jobst, D., Panek, D., Vetter, H., & Wagner, H. (2007). Application of the "-Omic-" technologies in phytomedicine. *Phytomedicine*, *14*(1), 70–82.

Valentão, P., Andrade, P. B., Areias, F., Ferreres, F., & Seabra, R. M. (1999). Analysis of vervain flavonoids by HPLC/diode array detector method, its application to quality control. *Journal of Agricultural and Food Chemistry*, *47*(11), 4579–4582.

van der Kooy, F., Maltese, F., Choi, Y. H., Kim, H. K., & Verpoorte, R. (2009). Quality control of herbal material and phytopharmaceuticals with MS and NMR based metabolic fingerprinting. *Planta Medica*, *75*(07), 763–775.

Verpoorte, R., Choi, Y. H., & Kim, H. K. (2005). Ethnopharmacology and systems biology: a perfect holistic match. *Journal of Ethnopharmacology*, *100*(1–2), 53–56.

von Roepenack-Lahaye, E., Degenkolb, T., Zerjeski, M., Franz, M., Roth, U., Wessjohann, L., Schmidt, J., Scheel, D., & Clemens, S. (2004). Profiling of Arabidopsis secondary metabolites by capillary liquid chromatography coupled to electrospray ionization quadrupole time-of-flight mass spectrometry. *Plant Physiology*, *134*(2), 548–559.

Wang, M., Lamers, R. J. A., Korthout, H. A., van Nesselrooij, J. H., Witkamp, R. F., van der Heijden, R., Voshol, P. J., Havekes, L. M., Verpoorte, R., & van der Greef, J. (2005). Metabolomics in the context of systems biology: bridging traditional Chinese medicine and molecular pharmacology. *Phytotherapy Research: An International Journal Devoted to Pharmacological and Toxicological Evaluation of Natural Product Derivatives*, *19*(3), 173–182.

Wang, P., Liang, Y., Zhou, N., Chen, B., Yi, L., Yu, Y., & Yi, Z. (2007). Screening and analysis of the multiple absorbed bioactive components and metabolites of Dangguibuxue decoction by the metabolic fingerprinting technique and liquid chromatography/diode-array detection mass spectrometry. *Rapid Communications in Mass Spectrometry: An International Journal Devoted to the Rapid Dissemination of Up-to-the-Minute Research in Mass Spectrometry*, *21*(2), 99–106.

Weckwerth, W. (2018). Metabolomics: integrating the metabolome and the proteome for systems biology. *Annual Plant Reviews Online*, *35*, 258–289.

Weckwerth, W., Loureiro, M. E., Wenzel, K., & Fiehn, O. (2004). Differential metabolic networks unravel the effects of silent plant phenotypes. *Proceedings of the National Academy of Sciences*, *101*(20), 7809–7814.

Weljie, A. M., Newton, J., Mercier, P., Carlson, E., & Slupsky, C. M. (2006). Targeted profiling: quantitative analysis of 1H NMR metabolomics data. *Analytical Chemistry*, *78*(13), 4430–4442.

Wilson, I., & Brinkman, U. T. (2003). Hyphenation and hypernation: the practice and prospects of multiple hyphenation. *Journal of Chromatography A*, *1000*(1–2), 325–356.

Wolfender, J. -L., Ndjoko, K., & Hostettmann, K. (2003). Liquid chromatography with ultraviolet absorbance–mass spectrometric detection and with nuclear magnetic resonance spectrometry: a powerful combination for the on-line structural investigation of plant metabolites. *Journal of chromatography A*, *1000*(1–2), 437–455.

Wu, B., Yan, S., Lin, Z., Wang, Q., Yang, Y., Yang, G., Shen, Z., & Zhang, W. (2008). Metabonomic study on ageing: NMR-based investigation into rat urinary metabolites and the effect of the total flavone of Epimedium. *Molecular BioSystems*, *4*(8), 855–861.

Zhang, X., Wu, H., Liao, P., Li, X., Ni, J., & Pei, F. (2006). NMR-based metabonomic study on the subacute toxicity of aristolochic acid in rats. *Food and Chemical Toxicology*, *44*(7), 1006–1014.

Zhao, X., Zhang, Y., Meng, X., Yin, P., Deng, C., Chen, J., Wang, Z., & Xu, G. (2008). Effect of a traditional Chinese medicine preparation Xindi soft capsule on rat model of acute blood stasis: a urinary metabonomics study based on liquid chromatography–mass spectrometry. *Journal of Chromatography B*, *873*(2), 151–158.

CHAPTER 6

Genome editing: applications for medicinal and aromatic plants

Summia Rehman, Ishfaq Ul Rehman, Bushra Jan, Irfan Rashid, Zafar Ah Reshi, Aijaz H Ganie
Department of Botany, University of Kashmir, Srinagar, Jammu and Kashmir, India

Introduction

With the advances in recombinant DNA technology, genetic engineering has achieved enormous success. Knowledge of many molecular and genetic mechanisms and phenomena studied in detail has led researchers to carry out experiments in vitro. The techniques in molecular genetics of bacteria and viruses have led to development of DNA manipulation through various vector systems with their delivery into the cell. These achievements not only allow creation of transgenic microorganisms but also genetically modified higher organisms like various plant and crop species. These novel techniques used in genetic engineering, has received significant importance however conventional genetic engineering strategy has several limitations, like complexity associated with the manipulation of large genomes of higher plants and public acceptance which itself is a serious concern preventing commercialization of transgenic crops. Currently, several tools help to solve the problems of precise genome editing of plants. In the recent past, genome editing by engineered nucleases such as "zinc fingers" coupled with *Fok*I endonuclease domains act as site-specific nucleases [zinc-finger nucleases (ZFNs)], transcription activator-like effector nucleases (TALENs) and Clustered Regularly Interspaced Short Palindromic Repeats/CRISPR-associated 9 (CRISPR/Cas9) (Cong et al., 2013) has proved a viable tool to alter the targeted site in the genome and has been widely used in several agricultural crops. ZFNs cleave the DNA in vitro in strictly defined regions. Such chimeric macromolecule incorporates a standard structure, because each of the "zinc finger" domains recognizes one triplet of nucleotides. This method in 1996 for the first time became the basis for the editing of cultured cells, including model and non-model plants. The designing TALENs need reengineering of a brand new macromolecule for every target. However,

Medicinal and Aromatic Plants: Expanding their Horizons through Omics
http://dx.doi.org/10.1016/B978-0-12-819590-1.00006-9

the planning method has been efficient recently by creating the modules of repeat combinations obtainable that basically reduces the biological research needed for the planning. On the opposite hand, designing and use of CRISPR are simple. Both TALEN and CRISPR systems have been shown to work in human cells, animals, and plants. Such piece of writing systems once used for economical manipulation of genomes might solve advanced issues together with the creation of mutant and transgenic plants. Moreover, chimeric proteins containing zinc finger and activation domains of different proteins and those based on the TALE DNA-binding domain and Cas9 nuclease were employed in experiments for regulation of gene transcription, study of epigenomes, and the behavior of chromosome loci in cell cycle.

Asia is one among the most diverse region containing the richest plant resources in the world. It includes but not limited to numerous medicinal and aromatic plant species, well documented traditional knowledge, a long-standing practice of traditional medicine, and the potential for social and economic development of medicinal and aromatic plants (MAPs). Six of the world's 18 biodiversity hot spots, namely eastern Himalaya, North Borneo, Peninsular Malaysia, Sri Lanka, Philippines, and the Western Ghats of South India, lie in Asia. Sustainable industrial exploitation of such a valuable bioresource, through use of appropriate technologies, can substantially contribute to the socioeconomic growth of Asian countries.

In this review, we briefly described the mechanisms of different genome-editing systems and their use for medical and aromatic plants and also highlighted the multiple advantages and applications of engineered nucleases as well as biosafety and regulatory aspects of plants generated using engineered nuclease based technologies.

Tools

The past decade has been one among speedy innovation in genome-editing technology. The opportunity now exists for investigators to edit the genome using artificial nucleases has the potential to accelerate basic research as well as plant breeding by providing the means to modify genomes rapidly in an exceedingly precise and certain manner. Here we describe the clustered regularly interspaced short palindromic repeat (CRISPR)/ CRISPR-associated protein 9 (Cas9) system, a recently developed tool for the introduction of site-specific double-stranded DNA breaks, ZFNs- are fusion proteins comprising associate array of site-specific DNA-binding domains—adapted from zinc finger–containing transcription factors—connected to the domain of the bacterial FokI restriction endonuclease,

TALENs a class of proteins called transcription activator-like effector nucleases (TALENs), exclusive to a group of plant pathogens, has led to the characterization of a novel DNA-binding domain, termed TALE repeats.

The CRISPR/Cas9 system

The discovery of bacterial adaptive immune systems known as CRISPR and CRISPR-associated (Cas) systems has led to the newest set of genome-editing tools. CRISPR-Cas systems target specific DNA sequences for cleavage using combination of proteins and short RNAs. The bacterial vectors collect "protospacers" from foreign DNA sequences (e.g., from bacteriophages), integrate them into their genomes, and employ them to express short guide RNAs, which can then be used by a CRISPR-Cas system to destroy any DNA sequences matching the protospacers.

Heterologous expression of a CRISPR-Cas system from *Streptococcus pyogenes* in mammalian cells was demonstrated by four teams in early 2017. It comprises of Cas9 protein along with guide RNA(s) (either two separate RNAs, as found in bacteria, or a single chimeric RNA), which results in DSBs at target sites having (1) a 20-bp sequence which matches the protospacer of the guide RNA and (2) an adjacent downstream NGG nucleotide sequence [termed the protospacer-adjacent motif (PAM)] (Cong et al., 2013; Mali et al., 2013; Jinek et al., 2012; Cho, Kim, Kim, & Kim, 2013). This happens with the formation of a ternary complex comprises of Cas9 binding to the PAM in the DNA, and then binding to the nonprotospacer portion of the guide RNA, upon which the protospacer of the guide RNA hybridizes with one strand of the genomic DNA. Cas9 then catalyzes the DSB in the DNA at a position 3 bp upstream of the PAM

CRISPR-Cas9 system can easily target any genomic sequence thereby changing the 20-bp protospacer of the guide RNA, which can be accomplished by subcloning this nucleotide sequence into the guide RNA plasmid backbone. The Cas9 protein component remains unchanged. This easy use for CRISPR-Cas9 may be a vital advantage over ZFNs and TALENs, particularly in generating a large set of vectors to target numerous sites (Cong et al., 2013) or perhaps genome-wide libraries (Wang, Wei, Sabatini, & Lander, 2014; Shalem et al., 2014; Koike-Yusa, Li, Tan, Velasco-Herrera Mdel, & Yusa, 2014; Zhou et al., 2014) .Another potential advantage of CRISPR-Cas9 is that the ability to multiplex, that is, to use multiple guide RNAs in parallel to focus on multiple sites at the same time within the same cell (Mali et al., 2013; Cong et al., 2013). This makes it simple to mutate multiple genes at once or to engineer precise deletions in a genomic region, though it ought to be noted that coinciding use of multiple ZFN or TALEN pairs are able to do the same outcomes (Park et al., 2014).

The *S. pyogenes* CRISPR-Cas system (most flexible one) limits the site selection to 23-bp sequences on either strand that ends in an NGG motif (the PAM for S. pyogenes Cas9), and this sequence occurs on average once every 8 bps (Cong et al., 2013). Most CRISPR-Cas9 systems express the guide RNA from a plasmid using a RNA polymerase III promoter like the U6 promoter, which requires a G in the first position, or the T7 promoter, which needs Gs within the initial two positions; but the G or Gs can simply be added to the 5′ end of the 20-nucleotide protospacer in the guide RNA as needed and thus do not affect site selection in the genome. CRISPR-Cas systems from alternative species are setting out to use in mammalian cells (Cong et al., 2013; Hou et al. 2013; Esvelt et al., 2013), and their versions of Cas9 have completely different PAM needs, that permits for targeting of sites in the genome for which the S. pyogenes system is not optimal. For example, the canonical *Neisseria meningitid* is Cas9 PAM that has been reported to be NNNNGATT, although it appears to be more tolerant of PAM variation than the *S. pyogenes* Cas9 (Hou et al. 2013; Esvelt et al., 2013).

One of the disadvantages of CRISPR-Cas9 is the size of the Cas9 protein itself. The size of the cDNA encoding *S. pyogenes* Cas9 is approximately 4.2 kb which makes it larger than a TALEN monomer and much more larger than a ZFN monomer (although both TALENs and ZFNs require dimerization, making their effective sizes larger). This large size makes Cas9 challenging to deliver via viral vectors (which would additionally require a promoter and a polyadenylation sequence) or as an RNA molecule. Another biggest concern regarding CRISPR-Cas9 is the issue of off-target effects. Since binding and specificity of *S. pyogenes* Cas9 is contributed by every nucleotide within the 20 nucleotide protospacer, single nucleotide mismatches are often well tolerated and depending on the locations in the protospacer multiple nucleotide mismatches can also be tolerated (Fu et al., 2013; Hsu et al., 2013; Mali et al., 2013; Pattanayak et al., 2013). The systematic analysis of the effect of mismatches in the protospacer shows an increasing tolerance for mismatches with increasing distance from the PAM.

Zinc-finger nucleases (ZFNs)

Zinc-finger nucleases are widely used in research for the generation of animal models to human medical therapies (Urnov, Rebar, Holmes, Zhang, & Gregory, 2010). These are fusion proteins comprising multiple site-specific

DNA-binding domains — adapted from zinc finger which contains transcription factors. These are intern attached to the endonuclease domain of the bacterial FokI restriction enzyme. Each zinc-finger domain identifies 3–4 bp DNA sequence and tandem domains can potentially bind to an extended nucleotide sequence (typically with a length that is a multiple of 3, usually 9–18 bp) that is distinctive at intervals within a cell's genome.

In order to cut a specific site within the genome, ZFNs act as a pair which identifies two flanking sequence sites, one on the forward strand and the other on the reverse strand. When ZFNs binds on either side of the sequence site, the pair of FokI domains dimerizes and cut the DNA at the particular site, generating a double-strand break (DSB) with 5′ overhangs (Urnov et al., 2010). The cells then repair DSBs using either (1) nonhomologous end joining (NHEJ), which can occur during any phase of the cell cycle, but occasionally results in erroneous repair, or (2) homology-directed repair (HDR), which typically happens during late S phase or G2 phase when a sister chromatid is available to serve as a repair template.

The NHEJ can be exploited to introduce frameshift alterations into the coding sequence of a gene, potentially helping in gene knock out by a combination of two mechanisms—premature truncation of the protein and nonsense-mediated decay of the mRNA transcript in which the latter is not always that much efficient. On the other hand, HDR mechanism can be utilized to insert a specific mutation by the introduction of a repair template containing the desired mutation flanked by homology arms. In response to a DSB in DNA, HDR uses another closely matching DNA sequence to repair the break.

Although there are many advantages of using ZFNs for genome editing, it has several potential disadvantages also. It is not that simple to assemble zinc-finger domains to bind an extended stretch of nucleotides with high affinity (Ramirez et al., 2008). This makes it difficult for non-specialists to use ZFNs. To overcome this problem, an academic consortium has developed an open-source library of zinc finger components and protocols to perform screens to identify ZFNs that bind with high affinity to a desired sequence (Maeder et al., 2008; Maeder, Thibodeau-Beganny, Sander, Voytas, & Joung, 2009), however, it can still take many months for nonspecialists to obtain optimized ZFNs. Another potential disadvantage is limitation of target site selection, that is, only an open-source ZFN components can only be used to target binding sites after every 200 bps in random DNA sequence. While this may not be an issue if an researcher seeks to knock out a gene by a frameshift introduction anywhere in the early coding sequence of the gene which in turn can produce the desired result (Sander et al., 2011; Gupta et al., 2012; Gaj, Guo, Kato, Sirk, & Barbas, 2012). Finally, an important concern about ZFNs designed to introduce DSBs into the genome is that they will do so not only at the desired site but also at off-target random sites. In one study, ZFNs were used for genome editing in human pluripotent stem cells but investigators identified ten possible off target genomic sites based on high-sequence similarity to the on-target site and found a single off-target mutation in the 184 clones assessed (Hockemeyer et al., 2009). Thus, investigators should take into consideration the possibility that ZFNs designed for a particular purpose may incur undesired off target events at a low rate. One strategy to reduce off-target events is to make use of a pair of ZFNs that have different FokI domains that are obligate heterodimers (Miller et al., 2007; Szczepek et al., 2007; Doyon et al., 2011). This can prevent a single ZFN from binding to two adjacent off-target sites and generating a DSB; rather, the only way an off-target event could occur is if both ZFNs in a pair bind adjacently and thus allow the FokI dimer to form. Another way to reduce off-target events is the introduction of purified form of ZFN proteins into cells (Gaj et al., 2012).

Transcription activator-like effector nucleases (TALENs)

The another class of proteins called transcription activator-like effectors (TALEs) is exclusive to a group of plant pathogens that has led to the characterization of a novel DNA-binding domain, termed TALE repeats. The naturally occurring TALE repeats consists of tandem arrays with 10 to 30 repeats that bind and recognize extensive DNA sequences (Bogdanove & Voytas, 2011). Each TALE repeat is 33 to 35 amino acids long with two adjacent amino acids [termed the repeat-variable di-residue (RVD)] which contribute specificity for one of the four DNA base pairs (Moscou & Bogdanove, 2009; Boch et al., 2009; Morbitzer, Romer, Boch, & Lahaye, 2010; Streubel, Blucher, Landgraf, & Boch, 2012; Cong, Zhou, Kuo, Cunniff, & Zhang, 2012). The elucidation of the RVD code has led to creation of a new type of engineered site-specific nuclease which fuses domain of TALE repeats to the FokI endonuclease domain, termed TAL effector nucleases (TALENs) (Christian et al., 2010; Miller et al., 2011; Li et al. 2011). TALENs work similar to ZFNs in generating DSBs at a desired target site in the genome and so can be used to knock out genes or knock in mutations in the same way.In comparison with ZFNs, TALENs are easier to design. The RVD code has been used to create many TALE repeat arrays that bind with high affinity to desired genomic DNA sequences; it appears that de novo–engineered TALE repeat arrays will bind to desired DNA sequences with high affinity at rates as high as 96% (Miller et al., 2011; Reyon et al., 2012). TALENs can be designed and constructed in two days that too hundreds in number at a time Reyon et al., 2012; Cermak et al. 2011); indeed, a library with TALENs targeting all of the genes in the genome has been constructed (Kim et al., 2013).

The advantage of TALE repeat array over ZFNs is that these can be easily extended to any desired length. Whereas engineered ZFNs typically can bind 9–18 bp sequences only. TALENs often bind 18 bp sequences or even longer, though recent studies suggests that the use of larger TALENs may result in less specificity (Guilinger et al., 2014). Another advantage of TALENs over ZFNs is that there appear to be fewer constraints on site selection as there are multiple possible TALEN pairs available for each bp of a random DNA sequence (Reyon et al., 2012). A research study in which TALENs were used for genome editing in human pluripotent stem cells found low but countable rates of mutagenesis at some of 19 possible off-target sites based on sequence similarity to the on-target sites. Although comparative data are limited, one study found that for TALENs and ZFNs targeting the same site in the CCR5 gene, the TALENs produced fewer off target mutations than the ZFNs at a highly similar site in the CCR2 gene (Mussolino et al., 2011). Further, the ZFNs produced greater cell toxicity (i.e., inhibited their growth) when introduced into target cells compared with the TALENs. So, TALENs with obligate heterodimer FokI domains are usually used to minimize the possibility of off-target events. Recently, whole-genome sequencing studies of human pluripotent stem cell clones edited with TALENs showed that the overall off-target event rate is very low (Smith et al. 2014; Suzuki et al., 2014). A major disadvantage ofTALENs is that, these are significantly larger in size when compared with ZFNs. The typical size for a TALEN encoding cDNA is approximately 3 kb, whereas a cDNA encoding a ZFN is only approximately 1 kb. This huge size of TALENs makes it harder to deliver and express into target cells compared with ZFNs, and thus size complexity makes them less useful for therapeutic applications in which they must be delivered in viral vectors with limited cargo size [such as adeno associated virus (AAV), with less than 5 kb] or as RNA molecules. Furthermore, the highly repetitive nature of the TALENs weaken their ability to be packaged and delivered by some viral vectors (Holkers et al., 2013), though this can apparently be overcome by diversifying the coding sequences of the TALE repeats (Yang et al., 2013).

Agrobacterium rhizogenes-mediated genetic transformations: a powerful tool for the production of metabolites

Hairy root is associated with an infectious disease caused by the soil bacteria *Agrobacterium rhizogenes* which occurs is a naturally in the soil. The mechanism and principles of the infection led many scientists to imitate the phenomenon. Nowadays, genetic transformations are been carried out using the "natural genetic engineer" *Agrobacterium rhizogenes* for induction of the so-called "hairy root" plant in vitro systems. Such systems are important for the production of economically important biologically-active substances like various secondary metabolites. Furthermore the expression and integration of foreign genes into plant cells via *Agrobacterium rhizogenes* plasmids have allowed the mass production of desirable phytochemicals, medicinally important enzymes and foreign proteins either in in vitro conditions (bioreactors) or through plant regeneration (in greenhouses or in the field).

The *Agrobacterium* has broad host range and can transfer DNA to numerous dicot and monocot angiosperm species (Gelvin, 2003) and gymnosperms (McAfee, White, Pelcher, & Lapp, 1993). In addition, *Agrobacterium* can also transform several types of eukaryotic cells, including but not limited to phytopathogenic fungi (Rolland, Jobic, Fevre, & Bruel, 2003), yeasts, ascomycetes, basidiomycetes (Tzfira & Citovsky, 2003) as well as animal cells (Pelczar, Kalck, Gomez, & Hohn, 2004). Recently, it has been reported that *Agrobacterium* can transfer DNA to human cells as well. The molecular mechanism of hairy root induction is still implicit (Sevón and Oksman-Caldentey 2002; Veena & Taylor, 2007). However, the process of hairy root formation involves five basic steps:

1. Attachment of Agrobacterium to host plant cell and activation by chemo attractants released by the host plant,
2. Processing of integrated DNA in the bacterial cell,
3. Transport of T-complex part of bacterial DNA to plant cell,
4. Integration of DNA and expression,
5. Subsequently induction of hairy-root disease.

The transferred DNA (T-DNA) is a segment of the *Agrobacterium* Ri-plasmid, which is flanked by 25-bp imperfect direct border repeats present on both left and right borders. The T-DNA is transferred as a single-stranded intermediate (T-strand) from the *Agrobacterium* into the host plant cells, where it is integrated into the nuclear genome of the host (Bako, Umeda, Tiburcio, Schell, & Koncz, 2003). The mechanism followed in the T-DNA transfer is similar for *A. tumefaciens* and *A. rhizogenes* (Altamura, 2004; Gelvin, 2003; Nilsson & Olsson, 1997; Tzfira & Citovsky, 2003). It has also been found that the genes on the T-DNA and flanking border are responsible for the transfer and integration and thus any fragment DNA, placed between the left and right borders, could be transferred to the host cell (Valentine, 2003). This phenomenon makes it possible to clone the gene of interest between the border sequences and for it to be transferred to the host cells (Valentine, 2003). Furthermore, *Agrobacterium* mediated transformations are considered preferable to genetic transformation relying on artificial approaches because of the straightforward, low cost procedure and the relatively low complexity of intact transgenes integrated into the plant genome (Gelvin 2003). The transformation begins with the bacteria–plant attachment (Tzfira & Citovsky, 2006). However, little is known about the assembly of the transfer apparatus or the molecular details of its operation. It is known that the apparatus has a pilus and may form a transmembrane channel for cell-to cell trafficking of the T complex by a type IV secretion system (Tzfira & Citovsky, 2003). For successful transformation the following must be present: the major loci (virA, virB, virC, virD, virE, virG, and virH) of the vir region of the Ti- or Ri-plasmids, T-DNA with left and right borders and some virulent chv genes, localized in the bacterial chromosome (Tzfira & Citovsky, 2000). The products of the virD2 and virE2 genes, that is,e. VirD2 and VirE2 proteins, play a key role in the processes of transfer, nuclear localization and integration into the host plant chromosome (Bako et al., 2003; Tzfira & Citovsky, 2000; Ward & Zambryski, 2001). Experiments have shown that the integration of T-DNA into the plant genome is performed in unspecified regions, but tends to occur in regions of plant DNA that are A-T-rich (Brunaud et al., 2002). Most of the *Agrobacterium* strains transfer a single fragment, however there are some (the agropine type of *A. rhizogenes* for example), which transfer two independent T-DNAs, denoted TL-DNA and TR-DNA (Nilsson and Olsson 1997), these are separated by a non-integrated region (Altamura, 20040. TRDNA is highly homologous to the T-DNA of the Ti-plasmid of *A. tumefaciens*, while TLDNA is completely different and possesses similarities to the T-DNA of the Ri-plasmid of mannopine *A. rhizogenes* strains (Nilsson & Olsson, 1997). Both fragments (TL-DNA and TR-DNA) are known to be independently transferred and integrated to the host plant genome, however, the transfer of TL-DNA is both necessary and sufficient to induce hairy root syndrome (Altamura, 2004; Nilsson & Olsson, 1997; Sevón & Oksman-Caldentey, 2002). Homologues of both the auxin biosynthesis genes (iaaM and iaaH) and the genes encoding synthases for the opines mannopine (mas1' and mas2') and agropine (ags) have been found in the TR-DNA genes (Altamura 2004; Nilsson &

Olsson, 1997). The auxin gene homologues in *A. rhizogenes* are known as aux1 and aux2; they are considered to play an accessory role in hairy root induction. The sequence analysis of TL-DNA identified 18 open reading frames (ORF), of which four are essential for hairy root induction. These loci were denoted root locus A, B, C and D (rolA, rolB, rolC and rolD) and assigned to ORF10, ORF11, ORF12 and ORF15, respectively (Nilsson & Olsson, 1997). The molecular mechanism of root formation after infection with *A. rhizogenes* has still not been fully explained; in the case of *A. tumefaciens,* the formation of tumor growth is due to changes in the phytohormonal balance in the plant cells. The neoplastic root formation in plants infected with *A. rhizogenes* is not, however, due to changes in phytohormonal balance in plant cells and is probably because the cells become more sensitive to auxins (Altamura 2004; Nilsson & Olsson, 1997). The products of the four rol (A, B, C, and D) genes play key roles in the process of hairy root formation. The rolA gene contains an ORF of about 300 bp, encoding a 100-amino acid protein (Nilsson & Olsson, 1997) and is believed to be involved in the generation of a functional imbalance in phytohormone levels [Dehio, Grossmann, Schell, & Schmulling, 1993]. The rolB gene contains an ORF of about 777 bp, encoding a 259 amino acid protein, and appears to be the most important in hairy root induction, since mutation in this locus renders the plasmid avirulent (Altamura 2004; Nilsson and Olsson 1997). In fact, once rolB is introduced into the host plant genome as a single gene, it is capable of formation hairy root disease (Altamura, 2004). It has been elucidated that the role of rolA and rolB effects are antagonistic to each other (Veena & Taylor, 2007). The rolC gene containing an ORF of 540 bp which encodes a cytokinin-β-glucosidase (Nilsson & Olsson, 1997; Veena and Taylor 2007) plays an important role in formation of shoot. Also, rolD gene contains an ORF of 1032 bp which encodes a protein made of 344 amino acids (Nilsson & Olsson, 1997). The RolD protein bears sequence homology with ornithine cyclodeaminase and thus acts as a stress-related osmoprotectant and it is also involved in the production of proline at the flowering stage (Altamura, 2004).

Applications

Genetic and metabolic engineering techniques have enabled manipulation of the production of specific plant secondary metabolites of interest by modifying the genes that play a key role in the biosynthetic pathway. Any perturbation in a given biosynthetic pathway is likely to cause a series of alterations in the transcription of the whole system. Those alterations may involve the plant's regulatory system which is designed to tightly control secondary metabolite production. Frequently the mechanism for this regulation is poorly understood. Overexpressing the rate-limiting enzyme not only led to the transcriptional change of closely related secondary metabolite pathway genes, but also to the broader transcriptional changes ranging from primary to other secondary metabolite pathways. Applications of the genome editing in some plants for increased secondary metabolite production are summarized as.

Rhazya stricta

Rhazya stricta Decne is an evergreen dwarf shrub belonging to the Apocynaceae family and is widely distributed in the Middle East and Indian sub-continent. The family consists of the plants well known for their exclusive array of terpenoid indole alkaloids (TIAs) with a diverse range of biological activities. *Rhazya stricta* has a long history in folk medicine and has been used to treat several diseases e.g. fever and chronic rheumatism (Atta-ur-Rahman, Qureshi, Zaman, Malik, & Ali, 1989). The phytochemical, pharmacological and ethnobotanical studies of R. stricta have been extensively studied (Atta-ur-Rahman et al., 1989; Ali, Al-Qarawi, Bashir, & Tanira, 2000; Gilani, Kikuchi, Shinwari, Khattak, & Watanabe, 2007).

A total of 75 alkaloids of *R. stricta* were classified into 17 sub-groups until late 1980s. After that *R. stricta* extracts were fractionated by TLC, medium performance liquid chromatography (MPLC) and HPLC and consequently subjected to GC–MS for identification of compounds (Atta-ur-Rahman et al., 1989; Kostenyuk, Lyubarets, Endress, Gleba, & Stockigt, 1995; Aimi et al., 1996; Sheludko, Gerasimenko, Unger, Kostenyuk, & Stöckigt, 1999). More than 100 alkaloids have been isolated, structurally characterized and identified from intact plants (Gilani et al., 2007), mostly from leaves (Ali et al., 2000). Many of the *Rhazya* alkaloids have been isolated from other Apocynaceae plants too, and are thus Apocynaceae-specific. However, some alkaloids like aspidospermiose, aspidospermidose, leepacine, strictanine, strictibine and strictanol (Atta-ur-Rahman et al., 1989) are species-specific TIAs and have only been reported from *R. stricta*. Alkaloids in *R. stricta*, first detected by Hooper (1906), TIAs, as secondary metabolites, usually present in small quantities in plants and their total chemical synthesis is often difficult and uneconomical because of their very complex stereochemical structures (Rischer et al., 2013)

The pharmacological properties of *R. stricta* extracts and the isolated compounds have been evaluated in numerous in vitro and animal models.

At least 15 TIAs from *R. stricta* were subjected to pharmacological and toxicological assays and indicated predominant anticancer and antibacterial properties of this plant (Gilani et al., 2007). The anticancer compound, rhazinilam, is unique due to its activity on tubulin. This compound induces spinalization of tubulin similar to vinblastine and inhibits the cold-induced disassembly of microtubules in a manner similar to paclitaxel (David et al., 1994). Sewarine, vallesiachotamine and tetrahydrosecamine exhibit cytotoxic activity against Eagle's KB carcinoma of the nasopharynxin a cell culture model. The onomericindole alkaloid, 16-epi-Z-isositsirikine, from the leaves of *R. stricta* displays antineoplastic activity both in eagle's KB carcinoma of the nasopharynx invitro and in P-388 lymphocytic leukaemia in vivo test systems (Mukhopadhyay, El-Sayed, Handy, & Cordell, 1983). Stemmadenine exhibits antimicrobial activity against *Pseudomonas aeruginosa, Escherichia coli, Staphylococcus aureus and Candida albicans* (Mariee, Khalil, Nasser, Al-Hiti, & Ali, 1988).

In order to increase the production of secondary metabolites *Agrobacterium* mediated genome editing technique has been utilized. Induction of transgenic hairy roots by *Agrobacterium rhizogenes*, a natural plant genetic engineer, is an effective, low cost and simple method to achieve stable transgene expression in plants. Transgenic hairy roots have become a core research tool for the production of secondary metabolites and engineering of metabolic pathways, and are used as a model system for functional genomics analysis (Ono & Tain, 2011; Miralpeix et al., 2013). Hairy root cultures were previously established from important medicinal plants in the Apocynaceae an easy and convenient method for establishing transgenic hairy roots of *R. stricta* through infection with wild type and gus (betaglucuronidase) gene harbouring *A. rhizogenes* LBA 9402 was developed by Akhgari et al. (2015). When comparing various explants only leaves were susceptible to Agrobacterium infection and subsequent root induction. The transformation efficiency of *R. stricta* was 74 and 83% for wild-type and GUS clones, respectively. Akhgari (2015) also reported the transformation of hairy roots with single heterologous genes including geraniol synthase (ges), geraniol 8-oxidase (g8o) and strictosidine synthase (str) using leaf explants.

Hairy roots produce a spectrum of secondary metabolites that are not detectable in the parent plant (Chandra, 2012). Nine compounds, including pleiocarpamine, fluorocarpamine, vincamine, ajmalicine, yohimbine isomers, serpentine isomers and tetrahydrosecodinol, were identified for the first time in *R. stricta* (Akhgari et al., 2015b; Akhgari et al., 2015c). However, ajmalicine and yohimbine isomers had previously been isolated from *R.*

serpentina and *R. stricta* hybrid cell cultures. These two alkaloids are derived from the *R. serpentina* parent since they are considered as typical *R. rauvolfia* alkaloids (Sheludko et al., 1999; Gerasimenko, Sheludko, Unger, & Stöckigt, 2001).

A total of 31 alkaloids, including eight isomers, were identified in hairy root cultures of *R. stricta.* Using a fully validated GC–MS method for the identification and quantification of low molecular weight, thermo-stable, non-polar alkaloids in a single run the presence of 20 monomeric TIAs in *R. stricta* wild type hairy roots with molecular weights ranging from 278 to 354 were reported (Akhgari et al., 2015b). GC–MS analysis of hairy root extracts indicated the occurrence of a total of 20 indole alkaloids. HPLC was used for the separation of alkaloids from *R. stricta* hairy root clones (Akhgari et al., 2015a). HPLC separation revealed the presence of five abundant alkaloids including, vincanine, leepacine isomers, serpentine and strictosidine lactam. On the basis of the UV and MS spectra, 17 alkaloids from different sub-groups were identified including 10 alkaloids, which could not have been analysed by GC–MS. Eburenine and vincanine were the main alkaloids followed by polar glucoalkaloids, strictosidine lactam and strictosidine. Secodine-type alkaloids, tetrahydrosecodinol, tetrahydro- and dihydrosecodine were detected.

Metabolic engineering of TIA pathway in hairy roots

Preliminary attempts for pathway engineering in *R. stricta* hairy roots were reported by Akhgari (2015a). Three heterologous key genes from the upstream TIA biosynthesis pathway including geraniol synthase (ges, *Valeriana officinalis*) and geraniol-8oxidase (g8o, *Catharanthus roseus*) from the terpenoid biosynthesis pathway and strictosidine synthase (str, *C. roseus*), as an intermediate gene between the upstream and downstream TIA biosynthesis pathways, were introduced to *R. stricta* hairy root. Hairy roots were analysed by HPLC and their alkaloid contents were compared to those of wild type hairy root clones. The concentrations of vincanine and serpentine were significantly increased in the hairy root clones overexpressing str compared to wild types. However, in comparison to control cultures, the accumulation of leepacine isomers remained the same and stricitosidine lactam decreased in str clones. The overexpression of ges gene in hairy roots did not statistically change the contents of vincanine and strictosidine lactam compared to wild type clones. However, the amounts of leepacine isomers and serpentine decreased. Introduction of the g8o gene into hairy root clones decreased the production of all studied alkaloids. It must be taken into account that in that study a real-time quantitative reverse transcription PCR (q-RT-PCR) was not performed to investigate the expression levels of selected genes, as it is possible that a gene is integrated into the genome but not expressed. Furthermore, intergeneric somatic hybrid cell cultures for the biosynthesis of novel and known indole alkaloid represent an alternative approach to secondary pathway engineering in *R. stricta*. Hybrid cell lines of *R. serpentina* and *R. stricta* showed enhanced alkaloid metabolism resulting in a greater diversity of alkaloids than in the parental species (Kostenyuk et al., 1991; Aimi et al., 1996). A new monoterpenoid indole alkaloid, 3-oxo-rhazinilam, was isolated from intergeneric somatic hybrid cells.

Papaver bracteatum

Papaver bracteatum (Iranian poppy) is naturally distributed in Iran and has some important medicinal alkaloids such as thebaine, codeine, morphine, noscapine and papaverine, however, thebaine is the major morphinan alkaloids in this plant (Tisserat & Berhow, 2009). Thebaine is used as an intermediate for the semi-synthesis of other pentacyclic morphinan based drugs such as oxycodone, naltrexone and buprenorphine, codeine is a cough suppressant, morphine is a narcotic drug, noscapineis an antitumor agent, sanguinarine and berberineare an antimicrobial drug and papaverine is used as a vasodilator for the treatment of vasospasms (Pienky, Brandt, Schmidt, Karmell, & Zigler, 2009) These important drugs are produced in low quantities. In order to increase the drug production Agrobacterium mediated genome editing of drug producing genes is done.

Agrobacterium rhizogenes mediated transformation is a rapid and convenient transformation system for functional characterization and manipulation of target genes to study plant secondary metabolism. Transgenic hairy root not only used as an efficient expression system in functional genomics but also as a source for developing transgenic plants (via direct or indirect plant regeneration) (Chattopadhyay, Roy, Mitra, & Maiti, 2011). Hairy-root production mediated with *Agrobacterium rhizogenes* creates a rapid and simple means to integrate and express foreign genes in plant cells, which are capable of synthesizing specific secondary metabolites (Ono & Tain, 2011). It was found that SalAT gene is responsible for the production of morphanian alkaloid. The acetyl coenzyme A-dependent enzyme salutaridinol 7-o-acetyltransferase (SalAT, EC 2.3.1.150) converts the salutaridinol to salutaridinol-7-0-acetate (Lenz & Zenk, 1994). Grothe, Lenz, & Kutchan (2001) reported molecular characterization of SalAT and indicated that the presence of SalAT transcript predicts the ability to accumulate morphinan alkaloids. Allen et al. (2008) demonstrated both over expression and suppression of SalAT in *P. somniferum* transgenic plants. Application of new biotechnological methods for improving of morphinane alkaloids compound has been reviewed by Dehghan, Hosseini, Naghdi Badi, & Shariari Ahmadi (2010). *Agrobacterium rhizogenes* can carry a binary vector containing a gene of interest to be transferred into the plant genome. The bacterial T-DNA from the Ri plasmid and the T-DNA from binary vector integration (co-transformation) into a single explant cell induces a hairy root expressing the integrated gene (Cao et al., 2011). As *A. rhizogenes* is able to transfer T-DNA of binary vectors, it is enabling to produce transformed root including other foreign genes on a second plasmid. Presently focus is on over expression of SalAT in transgenic hairy root by *A. rhizogenes* mediated co-transformation. The rapidly growing transformed hairy root cultures could serve as a simple system to study molecular regulation of SalAT gene encoding salutaridinol 7-o-acetyltransferase enzyme to evaluate its potential use in metabolic engineering of *P. bracteatum*.

Excised shoots (i.e., with the roots removed) from 3 weeks *P. bracteatum* seedlings were used as explants material for co-cultivation with *A. rhizogenes*. The excised shoots were randomly wounded using a sterile needle, dipped in *A. rhizogenes* inoculation medium for 10 min, blotted dry on sterile filter paper, and were cultured for 3 days on the cocultivation medium, which was the same as inoculation medium but solidified with 7 g 1-1 agar at 25°C in dark. After 2 days of cocultivation, the explants were transferred to selection medium (hormone free MS medium containing 20 mg l-1 kanamaycin and 400 mg l-1 cefotaxime). Within 3 weeks numerous hairy root emerged from the wound sites. Two weeks later the hairy roots were excised on selection medium at 25°C in the dark. Fast growing hairy root cultures were obtained after repeated transfer to fresh selection medium. Wild-type root culture was produced by inoculating liquid MS medium with excised roots from *P. bracteatum* seedling grown in vitro.

The amounts of codeine and morphine in transgenic hairy root lines of *P. bracteatum* increased by 160% and 86%, respectively, by over expression of codeinone reductase gene. This resulted in increasing codeine of up to 0.04% DW and morphine of up to 0.28% DW (Sharafi et al., 2013). Transgenic SalAT over expressing line with the highest alkaloid content showed on average 40% greater total alkaloids content compared to the control in three independent trials over 3 years. Novel accumulation of the alkaloid salutaridine at up to 23% of total alkaloid was detected by

DNA-encoded hairpin RNA-mediated suppression of SalAT, which was not detectable in the parental genotype. Salutaridine is not a substrate of SalAT but the substrate of the precursor enzyme in the pathway, salutaridine reductase. Detection of salotaridine, but not salutaridinol revealed the existence of a metabolic complex involving salutaridinol reductase (SalR) and SalAT in Papaver (Kempe, Higashi, Frick, Sabarna, & Kutchan, 2009). Correlation in transcript levels of SalR, SalAT and contents of salotaridine and thebaine supported the co-regulation of genes encoding enzymes operating in this part of the pathway (Desgagne-Penix, Farrow, Cram, Nowak, & Facchini, 2012). These results are in agreement with the cooperative analysis of ESTs from Papaver species with different alkaloid profiles (Ziegler et al., 2006). Our results confirmed that over expression SalAT gene using *A. rhizogenes* mediated transformation could provide a fast, simple and reliable model system to investigate the molecular regulation of morphinan alkaloids biosynthesis in *P. bracteatum*. Moreover, similar strategy could be used to consider other candidate genes involving in benzylisoquinoline alkaloid pathway in *Papaver* species.

Rauwolfia serpentine

In Apocynaceae family, the members produce more than 100 medicinally-important terpenoid indole alkaloids (TIAs). In this family, the genus *Rauwolfia,* mainly comprised of small trees and shrubs, is found in tropical regions of Africa, Asia, and Latin America. The genus is characterized by the presence of some important TIAs such as reserpine, ajmalicine, ajmaline, serpentine, vomiline, yohimbine, etc. These alkaloids are mostly concentrated in the roots which contain about 85%–90 % of the total alkaloid content (Pathania, Randhawa, & Bagler, 2013). Synthesis and accumulation of these molecules, particularly TIAs, amounts to the medicinal properties of roots. The extract of roots is used as an important ingredient in drugs for the treatment of hypertension, high-blood pressure, mental illness, and problems related to central nervous system. Root extracts having sedative, aphrodisiac, and antispasmodic properties, also possesses hypoglycemic and hypolipidemic activities against animal models (Pathania et al., 2013). Moreover, the root extract of many other species of this genus like *R. vomitoria*, the African tree species has antidiabetic activity (Campbell-Tofte, 2006). Such reports specify the medicinal value of the roots of different species of this genus. The over exploitation of *R. serpentina* for medicinal uses has resulted in its gradual decline in the wild and, consequently, it has been given an endangered status by International Union for the Conservation of Nature and Natural Resources (IUCN) and enlisted in CITES (Convention on International Trade in Endangered Species). It has been placed among 32 plants identified and prioritized for cultivation and conservation through conventional and unconventional methods by the National Medicinal Plant Board, India (NMPB; www.nmpb.nic.in). The continuous increase in commercial demand and the restricted supply of TIAs from natural resources has promoted the use of *A. rhizogenes* mediated hairy root cultures of *R. serpentin* (Goel, Goel, Banerjee, Shanker, & Kukreja, 2010; Mehrotra, Goel, Rahman, & Kukreja, 2013a). Hairy root cultures of various species of *Rauwolfia* produce obvious amounts of pharmaceutically-important TIAs (Falkenhagen, Stockigt, Kuzovkina, Alterman, & Kolshorn, 1993; Sudha, Reddy Obul, Ravishankar, & Seeni, 2003; Goel et al., 2010; Liu et al., 2012). Hairy root cultures of *R. serpentina* have broadly been studied for TIA production. This review provides a comprehensive account of the hairy root cultures of *R. serpentina*, their biosynthetic potential and various biotechnological methods attempted for metabolite production using these roots.

Agrobacterium rhizogenes-mediated hairy root cultures

Benjamin, Roja, & Heble (1993) in his experiment reported hairy roots from the leaf explants of *R. serpentina* through the infection of *A. rhizogenes* strain ATCC 15834. Falkenhagen et al. (1993) also reported hairy root stimulation in leaf explants infected with *A. rhizogenes* strain A4. The appreciable amount of alkaloid content and different alkaloid profile of these A4 generated hairy root cultures justified attention. A novel compound, 3-epi-a-yohimbine, was also isolated from these hairy roots. Transformation efficiency (interns of number of hairy root lines produced per explant) of different *A. rhizogenes* strains, such as SV4, LBA9402 and SV2, have also been examined where SV2 was found to be more efficient in inducing the hairy roots in *R. serpentina* leaf explants (Sarma, Kukreja, & Baruah, 1997). However, in majority of reported hairy root induction experiments, irrespective of bacterial strain, the emergence of roots at infection sites of leaves was observed within 15 days of infection. These reports indicate the high susceptibility of *R. serpentina* for different strains of *A. rhizogenes*. *A. rhizogenes*-mediated hairy roots of *R. serpentina* signify a rich repository of a range of terpenoid indole alkaloids (Sheludko, Gerasimenko, Kolshorn, & Stockigt, 2002; Madhusudanan, Banerjee, Khanuja, & Chattopadhyay, 2008). The exploitation of *R. serpentina* hairy roots is presently focused on reserpine biosynthesis (Goel et al., 2010; Mehrotra et al., 2013a). *A. rhizogenes* *A4*-mediated hairy roots show reserpine contents, varying from 0.0064%

to 0.088% dry weight (DW). In comparison to field-grown plants (variety cim-sheel; 0.03%–0.034% DW reserpine) harvested after 18–24 months, the hairy roots had a much higher reserpine content (0.086% DW) in 10–12 week cultures (Goel et al., 2010; Mehrotra et al., 2013a). Hairy root clones showed a varied morphology; however, no relationship between root morphology and reserpine content has been established. The variation in root morphology can be better explained by studying the number of T-DNA copies inserted in host cell during transformation.

Scale-up of *R. serpentina* hairy root culture

With the improvement in the techniques for hairy root-based secondary metabolite production at an industrial level, large scale or bioreactor cultivation is the ultimate step (Kim, Wyslouzil, & Weathers, 2002; Srivastava & Srivastava, 2007). *R. serpentina* hairy root cultures can be a hopeful source for the production of reserpine and other appreciated alkaloids. In laboratory, it has been observed that the feasibility of a scale-up protocol to grow *R. serpentina* hairy roots in a bench top and mechanically-agitated bioreactor. Many experiments used the bioreactor that were with a modified air-lift and were fitted with all necessary probes. Same bioreactors had been previously used to scale-up hairy root cultures of *Glycyrrhiza glabra* (Mehrotra, Kukreja, Khanuja, & Mishra, 2008). To avoid submergence of inoculated tissue and its damage from the impeller an autoclavable nylon mesh were fabricated that divided the vessel into upper and lower halves. The mesh also provided anchorage to the growing tissue in the upper half of the vessel. A fast growing hairy-root clone is usually being selected for the cultivation in bioreactor. In order to compare the growth performance of hairy roots in reactor vessel and culture flasks, the same clone is used to be inoculated in the shake-flasks (150 mL). The initial inoculum has been kept constant (2 g/L) in both culture vessels.

TIA biosynthetic pathway engineering

Pathway engineering demonstrates that the introduction and expression of foreign genes into the host genome for alteration in cell metabolism leading to an increased flux of target compounds in a biosynthetic pathway (Mehrotra, Rahman, & Kukreja, 2010; Zhou, Zhu, Shao, Tang, & Wu, 2011). This target compound could be the metabolite of interest or any conversion product, the conversion of which is the rate limiting step within the pathway. The revelation of TIA biosynthetic pathway and cloning of various key genes have promoted pathway engineering in *Rauwolfia* hairy roots. The pathway of begins with the combination of tryptamine and monoterpenoid glycoside secologanin. For operation of the pathway a continuous supply of tryptamine is necessary which comes from the conversion of tryptophan through the activity of tryptophan decarboxylase (EC 4.1.1.28). This step is rate-limiting step as the overproduction/accumulation of tryptophan results in its feedback inhibition of anthranilate synthase activity. The TIA pathway engineering in *R. serpentina* hairy roots through heterologous expression of tryptophan decarboxylase (tdc) has been attempted (Mehrotra, Srivastava, Rahman, & Kukreja, 2013c). Over-expression of *Catharanthus* tryptophan decarboxylase (Crtdc) in six transgenic hairy root lines of *R. serpentina* increased the content of reserpine and ajmalicine. Almost a two-fold increase in total alkaloid content was observed in transgenic hairy root lines. The presence of the transgene in hairy root lines was confirmed by Southern blot analysis. The study promoted the viability of pathway engineering and offered striking opportunities for productivity enhancement in Rauwolfia hairy roots. Liu et al. (2012) revealed in his experiments a regular expression level of *R. verticillata* tdc with the accumulation of ajmalicine in different plant parts as well as in hairy roots. Thus tdc expression is a crucial event for the accumulation of pathway products in different TIA harboring or non-harboring plants (Goddijn, Lohman, de Kam, Schilperoort, & Hoge, 1994; Geerlings et al., 1999; Hughes, Hong, Gibson, Shanks, & San, 2004). However, in *R. verticillata,* abscisic acid (ABA)-treated hairy root cultures were examined for the expression of the five genes of the MEP (methylerythritol phosphate) pathway and three genes of the ajmalicine (TIA) pathway including tdc. It has been seen that all the MEP pathway genes showed transcriptional up-regulation while all the ajmalicine pathway genes (tdc, str and sgd) did not. The study concluded in increased ajmalicine biosynthesis (Chang et al., 2014). The results revealed that there is a role of the MEP pathway in ajmalicine biosynthesis in *R. verticillata* hairy roots and also concluded that alkaloid biosynthesis in *Rauwolfia* requires multichannel pathway operation. Therefore, a base of more conceptual inputs and a thorough knowledge is necessary for overexpression studies in *Rauwolfia* hairy roots for improved alkaloid flux. The role of various physical, chemical and biological elicitors in affecting the pathway genes directly or by influencing the regulatory factors needs attention. However, the expression studies of tdc against elicitor challenge and its correlation with ORCA3 transcriptional factor have demonstrated the role of elicitor-responsive gene regulation in biosynthetic pathways (Ouwerkerk & Memelink, 1999). The role of *C. roseus* WRKY transcription factor (CrWRKY1) in TIA biosynthesis has also been reported and suggests an association of a complex transcriptional control network (Suttipanta et al., 2011). These reports clearly indicate the intricacy of TIA pathway regulation at transcriptional level. A rational analysis of these points is thus worth reviewing for describing the practical viewpoint of this comprehensive study.

Conclusion

Plants have been an imperative source of medicine for thousands of years. Medicines that are common in use are derived from plants, and new transgenic varieties could be created as efficient green production lines for other pharmaceuticals as well as vaccines and anticancer drugs. Tissue culture is worthwhile for multiplying and conserving the species, which are difficult to regenerate by conventional methods and save them from extinction. The production of secondary metabolites can be improved using bioreactors. Bioreactors offer a pronounced hope for the large-scale synthesis of therapeutically active compounds in medicinal plants. Since the biosynthetic efficiency of populations varies, a high yielding variety is recommended as a starting material. On the other hand genome editing may provide increased and efficient system for in vitro production of secondary metabolites. In this book chapter we reviewed three medicinal plants and their elevation in secondary metabolites by genome editing. The plants are *Rhazya stricta, Papaver bracteatum, and Rauwolfia serpentine.* The hasty development and improvement of genome-editing tools provide investigators with three well-characterized options for experiments as diverse as forward genetic screens to correction of pathogenic mutations in iPSC-derived human cells. CRISPRs, ZFNs and TALENs can all generate site-specific DSBs with varying degrees of specificity and efficiency. The earlier uses of these systems have proved remarkable new possibilities and allowed for the creation of model systems in a wide variety of organisms. With each iterations the technology has improved, and the prospects for the study and treatment of human disease with genome editing have never been better.

Acknowledgments

We acknowledge the facilities provided by the Department of Botany, University of Kashmir, Srinagar, J&K, India, to carry out this work. Also thankful to MOEF & CC for providing grant through National Mission on Himalayan Studies (NMHS) and also to all those who made us available their literature which helped us to compile the catalogue.

References

Aimi, N., Kitajima, M., Oya, N., Nitta, W., Takayama, H., Sakai, S., et al. (1996). Isolation of alkaloids from cultured hybrid cells of *Rauwolfia serpentine* and *Rhazya stricta*. *Chemical and Pharmaceutical Bulletin, 44*, 1637–1639.

Akhgari, A., Laakso, I., Seppänen-Laakso, T., Yrjönen, T., Vuorela, H., Oksman-Caldentey, K. M., & Rischer, H. (2015b). Determination of terpenoid indole alkaloids in hairy roots of *Rhazya stricta* (Apocynaceae) by gas chromatography-mass spectrometry. *Phytochemical Analysis, 26*, 331–338.

Akhgari, A., Laakso, I., Seppänen-Laakso, T., Yrjönen, T., Vuorela, H., Oksman-Caldentey, K. M., & Rischer, H. (2015c). Analysis of indole alkaloids from *Rhazya stricta* hairy roots by ultraperformance liquid chromatography-mass spectrometry. *Molecules, 20*, 22621–22634.

Akhgari, A., Yrjönen, T., Laakso, I., Vuorela, H., Oksman-Caldentey, K. M., & Rischer, H. (2015a). Establishment of transgenic *Rhazya stricta* hairy roots to modulate terpenoid indole alkaloid production. *Plant Cell Reports, 34*, 1939–1952.

Ali, B. H., Al-Qarawi, A. A., Bashir, A. K., & Tanira, M. O. (2000). Phytochemistry, pharmacology and toxicity of *Rhazya stricta* Decne: a review. *Phytotherapy Research, 14*, 229–234.

Allen, R. S., Miller, J. A. C., Chitty, J. A., Fist, A., Gerlach, W. L., & Larkin, P. J. (2008). Metabolite engineering of morphinan alkaloids by over expression and RNAi suppression of Salutaridinol 7-O-acetyltransferase in opium poppy. *Plant Biotechnology Journal, 6*, 22–30.

Altamura, M. M. (2004). Agrobacterium rhizogenes rolB and rolD genes: regulation and involvement in plant development. *Plant Cell, Tissue and Organ Culture*, 77, 89101.

Atta-ur-Rahman, Qureshi, M. M., Zaman, K., Malik, S., & Ali, S. S. (1989). The alkaloids of *Rhazya stricta* and *R. orientalis. A Review Fitoterapia., 60*, 291–322.

Bako, L., Umeda, M., Tiburcio, A. F., Schell, J., & Koncz, C. (2003). The VirD2 pilot protein of Agrobacterium-transferred DNA interacts with the TATA box-binding protein and a nuclear protein kinase in plants. *Proceedings of the National Academy of Sciences of the USA, 100*, 10108–10113.

Benjamin, B. D., Roja, G., & Heble, M. R. (1993). *Agrobacterium rhizogens* mediated transformation of *Rauvolfia serpentina*: Regeneration and alkaloid synthesis. *Plant cell, Tissue and Organ Culture, 35*(3), 253–257.

Boch, J., et al. (2009). Breaking the code of DNA binding specificity of TAL-type III effectors. *Science, 326*(5959), 1509–1512.

Bogdanove, A. J., & Voytas, D. F. (2011). TAL effectors: customizable proteins for DNA targeting. *Science, 333*(6051), 1843–1846.

Brunaud, V., Balzergue, S., Dubreucq, B., Aubourg, S., Samson, F., Chauvin, S., et al. (2002). T-DNA integration into the Arabidopsis genome depends on sequences of pre-insertion sites. *EMBO Reports, 12*, 1152–1157.

Campbell-Tofte J Anti-diabetic extract isolated from *Rauvolfia vomitoria* and *Citrus aurantium*, and methods of using same. US patent,7 49395 B2.

Cao, D., Hou, W., Liu, W., Yao, W., Wu, C., Liu, X., & Han, T. (2011). Overexpression of TaNHX2 enhances salt tolerance of composite and whole transgenic soybean plants. *Plant Cell, Tissue and Organ Culture, 107*, 541–552.

Cermak, T., et al. (2011). Efficient design and assembly of custom TALEN and other TAL effector-based constructs for DNA targeting. *Nucleic Acids Research, 39*(12), e82.

Chandra, S. (2012). Natural plant genetic engineer *Agrobacterium rhizogenes*: role of T-DNA in plant secondary metabolism. *Biotechnology Letters, 34*, 407–415.

Chang K, Chen M, Zeng L, Lan X, Wang Q, & Liao Z. Abscisic acid enhanced ajmalicine biosynthesis in hairy roots of Rauvolfia verticillata by upregulating expression of the MEP pathway genes. *Russian Journal of Plant Physiology* 61:136-140.

Chattopadhyay, T., Roy, S., Mitra, A., & Maiti, M. K. (2011). Development of a transgenic hairy root system in jute (*Corchorus capsularis* L.) with gusA reporter gene through Agrobacterium rhizogenes mediated co-transformation. *Plant Cell Reports, 30*, 485–493.

Cho, S.W., Kim, S., Kim, J. M., & Kim, J. S. (2013). Targeted genome engineering in human cells with the Cas9 RNA-guided endonuclease. *Nature Biotechnology*, *31*(3), 230–232.

Christian, M., et al. (2010). Targeting DNA double-strand breaks with TAL effector nucleases. *Genetics*, *186*(2), 757–761.

Cong, L., et al. (2013). Multiplex genome engineering using CRISPR/Cas systems. *Science*, *339*(6121), 819–823.

Cong, L., Zhou, R., Kuo, Y. C., Cunniff, M., & Zhang, F. (2012). Comprehensive interrogation of natural TALE DNA-binding modules and transcriptional repressor domains. *Nature Community*, *3*, 968.

David, B., Sevenet, T., Morgat, M., Guenard, G., Moisand, A., Tollon, Y., Thoison, O., & Wright, M. (1994). Rhazinilam mimics the cellular effects of taxol by different mechanisms of action. *Cell Motility and Cytoskeleton*, *28*, 317–326.

Dehghan, E., Hosseini, B., Naghdi Badi, H., & Shariari Ahmadi, F. (2010). Application of conventional and new biotechnological approaches for improving of morphinane alkaloids production. *Journal of Medicinal Plants*, *9*, 33–50.

Dehio, C., Grossmann, K., Schell, J., & Schmulling, T. (1993). Phenotype and hormonal status of transgenic tobacco plants overexpressing the rolA gene of *Agrobacterium rhizogenes* TDNA. *Plant Molecular Biology*, *23*, 1199–1210.

Desgagne-Penix, I., Farrow, S. C., Cram, D., Nowak, J., & Facchini, P. J. (2012). Integration of deep transcript and targeted metabolite profiles for eight cultivars of opium poppy. *Plant Molecular Biology*, *79*, 295–313.

Doyon, Y., et al. (2011). Enhancing zinc-finger-nuclease activity with improved obligate heterodimeric architectures. *Nature Methods*, *8*(1), 74–79.

Esvelt, K. M., Mali, P., Braff, J. L., Moosburner, M., Yaung, S. J., & Church, G. M. (2013). Orthogonal Cas9 proteins for RNA-guided gene regulation and editing. *Nature Methods*, *10*(11), 1116–1121.

Falkenhagen, H., Stockigt, J., Kuzovkina, I. N., Alterman, I. E., & Kolshorn, H. (1993). Indole alkaloids from the hairy roots of Rauvolfia serpentina. *Canadian Journal of Chemistry*, *71*, 2201–2203.

Fu, Y., et al. (2013). High-frequency off-target mutagenesis induced by CRISPR-Cas nucleases in human cells. *Nature Biotechnology*, *31*(9), 822–826.

Gaj, T., Guo, J., Kato, Y., Sirk, S. J., & Barbas, C. F. (2012). Targeted gene knockout by direct delivery of zinc-finger nuclease proteins. *Nature Methods*, *9*(8), 805–807.

Geerlings, A., Hallard, D., Caballero, A. M., Cardoso, I. L., van der Heijden, R., & Verpoorte, R. (1999). Alkaloid production by a *Cinchona officinalis* 'Ledergriana' hairy root culture containing constitutive expression constructs of tryptophan decarboxylase and strictosidine synthase cDNAs from *Catharanthus roseus*. *Plant Cell Reports*, *19*, 191–196.

Gelvin, S. B. (2003). *Agrobacterium*-mediated plant transformation: The biology behind the "Gene-Jockeying" tool. *Microbiology and Molecular Biology Reviews*, *67*, 16–37.

Gelvin, S. B. (2003). Improving plant genetic engineering by manipulating the host. *Trends in Biotechnology*, *21*, 95–98.

Gerasimenko, I., Sheludko, Y., Unger, M., & Stöckigt, J. (2001). Development of an efficient system for the separation of indole alkaloids by high performance liquid chromatography and its applications. *Phytochemical Analysis*, *12*, 96–103.

Gilani, S. A., Kikuchi, A., Shinwari, Z. K., Khattak, Z. I., & Watanabe, K. N. (2007). Phytochemical, pharmacological and ethnobotanical studies of *Rhazya stricta* Decne. *Phytotherapy Reseacrh*, *21*, 301–307.

Goddijn, O. J., Lohman, F. P., de Kam, R. J., Schilperoort, R. A., & Hoge, J. H. C. (1994). Nucleotide sequence of the tryptophan decarboxylase gene of *Catharanthus roseus* and expression of tdc-gus A gene fusions in *Nicotiana tabacum*. *Molecular Genes and Genetics*, *242*, 217–222.

Goel, M. K., Goel, S., Banerjee, S., Shanker, K., & Kukreja, K. (2010). *Agrobacterium rhizogenes*-mediated transformed roots of *Rauwolfia serpentina* for reserpine biosynthesis. *Medicinal and Aromatic Plant Science Biotechnology*, *4*, 8–18.

Grothe, T., Lenz, R., & Kutchan, T. M. (2001). Molecular characterization of the salutaridinol 7-O-acetyltransferase involved in morphine biosynthesis in opium poppy (*Papaver somniferum*). *Journal of Biological Chemistry*, *276*, 30717–30723.

Guilinger, J. P., et al. (2014). Broad specificity profiling of TALENs results in engineered nucleases with improved DNA-cleavage specificity. *Nature Methods*, *11*(4), 429–435.

Gupta, A., Christensen, R. G., Rayla, A. L., Lakshmanan, A., Stormo, G. D., & Wolfe, S. A. (2012). An optimized two finger archive for ZFN-mediated gene targeting. *Nature Methods*, *9*(6), 588–590.

Hockemeyer, D., et al. (2009). Efficient targeting of expressed and silent genes in human ESCs and iPSCs using zinc-finger nucleases. *Nature Biotechnology*, *27*(9), 851–857.

Holkers, M., et al. (2013). Differential integrity of TALE nuclease genes following adenoviral and lentiviral vector gene transfer into human cells. *Nucleic Acids Research*, *41*(5), e63.

Hooper, D. (1906). Notes on Indian drugs. *Pharmaceutical, J* 77, 258–260.

Hou, Z., et al. (2013). Efficient genome engineering in human pluripotent stem cells using Cas9 from *Neisseria meningitidis*. *Proceedings of National Academy of Science U S A.*, *110*(39), 15644–15649.

Hsu, P. D., et al. (2013). DNA targeting specificity of RNA-guided Cas9 nucleases. *Nature Biotechnology*, *31*(9), 827–832.

Hughes, E. H., Hong, S. B., Gibson, S. I., Shanks, J.V., & San, K.Y. (2004). Metabolic engineering of the indole pathway in *Catharanthus roseus* hairy roots and increased accumulation of tryptamine and serpentine. *Metabolic Engineering*, *6*, 268–276.

Jinek, M., Chylinski, K., Fonfara, I., Hauer, M., Doudna, J. A., & Charpentier, E. (2012). A programmable dual-RNA-guided DNA endonuclease in adaptive bacterial immunity. *Science*, *337*(6096), 816–821.

Kempe, K., Higashi, Y., Frick, S., Sabarna, K., & Kutchan, T. M. (2009). RNAi suppression of the morphinan biosynthetic gene SalAT and evidence of association of pathway enzymes. *Phytochemistry*, *70*, 579–589.

Kim, Y., et al. (2013). A library of TAL effector nucleases spanning the human genome. *Nature Biotechnology*, *31*(3), 251–258.

Kim, Y. J., Wyslouzil, B. E., & Weathers, P. J. (2002). Secondary metabolism of hairy root cultures in bioreactors. *In vitro Cellular and Developmental Biology—Plant*, *38*, 1–10.

Koike-Yusa, H., Li, Y., Tan, E. P., Velasco-Herrera Mdel, C., & Yusa, K. (2014). Genome-wide recessive genetic screening in mammalian cells with a lentiviral CRISPR-guide RNA library. *Nature Biotechnology*, *32*(3), 267–273.

Kostenyuk, I., Lubaretz, O., Borisyuk, N., Voronin, V., Stockigt, J., & Gleba, Y. (1991). Isolation and characterization of intergeneric somatic hybrids in the Apocynaceae family. *Theory of Applied Genetics*, *82*, 713–716.

Kostenyuk, I. A., Lyubarets, O. F., Endress, S., Gleba, Y.Y., & Stockigt, J. (1995). Alkaloids isolated from somatic hybrid cell cultures of the species combination *Rauwolfia serpentina* and *Rhazya stricta*. *Nature Product Letters*, *5*, 303–307.

Lenz, R., & Zenk, M. H. (1994). Closure of the oxide bridge in morphine biosynthesis. *Tetrahedron Letters*, *35*, 3897–3900.

Li, T., et al. (2011). TAL nucleases (TALNs): hybrid proteins composed of TAL effectors and FokI DNA-cleavage domain. *Nucleic Acids Research*, *39*(1), 359–372.

Liu, W., Chen, R., Chen, M., Zhang, H., Peng, M., Yang, C., Ming, X., Lan, X., & Liao, Z. (2012). Tryptophan decarboxylase plays an important role in ajmalicine biosynthesis in *Rauvolfia verticillata*. *Planta*, *236*, 239–250.

Madhusudanan, K. P., Banerjee, S., Khanuja, S. P., & Chattopadhyay, S. K. (2008). Analysis of hairy root culture of *Rauvolfia serpentina* using direct analysis in real time mass spectrometric technique. *Biomedical Chromatography*, *22*, 596–600.

Maeder, M. L., et al. (2008). Rapid "open-source" engineering of customized zinc-finger nucleases for highly efficient gene modification. *Molecular Cell*, *31*(2), 294–301.

Maeder, M. L., Thibodeau-Beganny, S., Sander, J. D., Voytas, D. F., & Joung, J. K. (2009). Oligomerized pool engineering (OPEN): an 'open-source' protocol for making customized zinc-finger arrays. *Nature Protocol*, *4*(10), 1471–1501.

Mali, P., et al. (2013). CAS9 transcriptional activators for target specificity screening and paired nickases for cooperative genome engineering. *Nature Biotechnology*, *31*(9), 833–838.

Mali, P., et al. (2013). RNA-guided human genome engineering via Cas9. *Science*, *339*(6121), 823–826.

Mariee, N. K., Khalil, A. A., Nasser, A. A., Al-Hiti, M. M., & Ali, W. M. (1988). Isolation of the antimicrobial alkaloid stemmadenine from Iraqi *Rhazya stricta*. *Journal of Natural Products*, *51*, 186–187.

McAfee, B. J., White, E. E., Pelcher, L. E., & Lapp, M. S. (1993). Root induction in pine (*Pinus*) and larch (*Larix*) spp. using *Agrobacterium rhizogenes*. *Plant Cell, Tissue and Organ Culture*, *34*, 53–62.

Mehrotra, S., Goel, M. K., Rahman, L. U., & Kukreja, A. K. (2013a). Molecular and chemical characterization of plants regenerated from Ri mediated hairy root cultures of *Rauwolfia serpentina*. *Plant Cell, Tissue, and Organ Culture*, *114*, 31–38.

Mehrotra, S., Kukreja, A. K., Khanuja, S. P. S., & Mishra, B. N. (2008). Genetic transformation studies and scale-up of hairy root culture of *Glycyrrhiza glabra* in bioreactor. *Electronic Journal of Biotechnology*, *11*, 69–75.

Mehrotra, S., Rahman, L. U., & Kukreja, A. K. (2010). An extensive case study of hairy-root cultures for enhanced secondary metabolite production through metabolic-pathway engineering. *Biotechnology and Applied Biochemistry*, *56*, 161–172.

Mehrotra, S., Srivastava, V., Rahman, L. U., & Kukreja, A. K. (2013c). Overexpression of a *Catharanthus* tryptophan decarboxylase (tdc) gene leads to enhanced terpenoid indole alkaloid (TIA) production in transgenic hairy root lines of *Rauwolfia serpentina*. *Plant Cell, Tissue and Organ Culture*, *115*, 377–384.

Miller, J. C., et al. (2011). A TALE nuclease architecture for efficient genome editing. *Nature Biotechnology*, *29*(2), 143–148.

Miller, J. C., et al. (2007). An improved zinc-finger nuclease architecture for highly specific genome editing. *Nature Biotechnology*, *25*(7), 778–785.

Miralpeix, B., Rischer, H., Ha kkinen, S. T., Ritala, A., Seppa ńenLaakso, T., Oksman-Caldentey, K. M., Capell, T., & Christou, P. (2013). Metabolic engineering of plant secondary products: which way forward? *Current Pharmaceutic Design*, *19*, 5622–5639.

Morbitzer, R., Romer, P., Boch, J., & Lahaye, T. (2010). Regulation of selected genome loci using de novo engineered transcription activator-like effector (TALE)-type transcription factors. *Proceedings of National Academy of Science U S A.*, *107*(50), 21617–21622.

Moscou, M. J., & Bogdanove, A. J. (2009). A simple cipher governs DNA recognition by TAL effectors. *Science*, *326*(5959), 1501.

Mukhopadhyay, S., El-Sayed, A., Handy, G. A., & Cordell, G. A. (1983). Catharanthus alkaloids. XXXVII. 16-Epi-Z-isositsirikine, a monomeric indole alkaloid with antineoplastic activity from Catharanthus roseus and Rhazya stricta. *Journal of Natural Products*, *46*, 409–413.

Mussolino, C., Morbitzer, R., Lutge, F., Dannemann, N., Lahaye, T., & Cathomen, T. (2011). A novel TALE nuclease scaffold enables high genome editing activity in combination with low toxicity. *Nucleic Acids Research*, *39*(21), 9283–9293.

Nilsson, O., & Olsson, O. (1997). Getting to the root: The role of the *Agrobacterium rhizogenes* rol-genes in the formation of hairy roots. *Physiologia Plantarum*, *100*, 463–473.

Ono, N. N., & Tain, L. (2011). The multiplicity of hairy root cultures: prolific possibilities. *Plant Science*, *180*, 439–446.

Ouwerkerk, P. B. F., & Memelink, J. (1999). Elicitor responsive promoter regions in the tryptophan decarboxylase gene from *Catharanthus roseus*. *Plant Molecular Biology*, *39*, 129–136.

Park, C.Y., et al. (2014). Targeted inversion and reversion of the blood coagulation factor 8 gene in human iPS cells using TALENs. *Proceedings of National Academy of Science U S A.*, *111*(25), 9253–9258.

Pathania, S., Randhawa, V., & Bagler, G. (2013). Prospecting for novel plant-derived molecules of *Rauvolfia serpentina* as inhibitors of aldose reductase, a potent drug target for diabetes and its complications. *PLoS One*, *8*, e61327.

Pattanayak, V., Lin, S., Guilinger, J. P., Ma, E., Doudna, J. A., & Liu, D. R. (2013). High-throughput profiling of offtarget DNA cleavage reveals RNA-programmed Cas9 nuclease specificity. *Nature Biotechnology*, *31*(9), 839–843.

Pelczar, P., Kalck, V., Gomez, D., & Hohn, B. (2004). Agrobacterium proteins VirD2 and VirE2 mediate precise integration of synthetic T-DNA complexes in mammalian cells. *EMBO Reports*, *5*, 632–637.

Pienky, S., Brandt, W., Schmidt, J., Karmell, R., & Zigler, J. (2009). Functional characterization of involvement in papaverine biosynthesis in opium poppy (*Papaver somniferum*). *Plant Journal*, *60*, 56–67.

Ramirez, C. L., et al. (2008). Unexpected failure rates for modular assembly of engineered zinc fingers. *Nature Methods*, *5*(5), 374–375.

Reyon, D., Tsai, S. Q., Khayter, C., Foden, J. A., Sander, J. D., & Joung, J. K. (2012). FLASH assembly of TALENs for high-throughput genome editing. *Nature Biotechnology*, *30*(5), 460–465.

Rischer, H., Hakkinen, S.T., Ritala, A., Seppanen-Laakso, T., Miralpeix, B., Capell, T., Christou, P., & Oksman-Caldentey, K. M. (2013). Plant cells as pharmaceutical factories. *Current Pharmaceutical Design.*, *19*, 5640–5660.

Rolland, S., Jobic, C., Fevre, M., & Bruel, C. (2003). *Agrobacterium*-mediated transformation of *Botrytis cinerea*, simple purification of monokaryotic transformants and rapid conidia based identification of the transfer-DNA host genomic DNA flanking sequences. *Current Genetics*, *44*, 164–171.

Sander, J. D., et al. (2011). Selection-free zinc-finger-nuclease engineering by context-dependent assembly (CoDA). *Nature Methods*, *8*(1), 67–69.

Sarma, D., Kukreja, A. K., & Baruah, A. (1997). Transforming ability of two *Agrobacterium rhizogenes* in *Rauwolfia serpentina* (L.). *Indian Journal of Plant Physiology*, *2*, 166–168.

Sevón, N., & Oksman-Caldentey, K. -M. (2002). *Agrobacterium rhizogenes*-mediated transformation: Root cultures as a source of alkaloids. *Planta Medica*, *68*, 859–868.

Shalem, O., et al. (2014). Genome-scale CRISPR-Cas9 knockout screening in human cells. *Science*, *343*(6166), 84–87.

Sharafi, A., Hashemi Sohi, H., Mousavi, A., Azadi, P., Razavi, K., & Ntui, V. O. (2013). A reliable and efficient protocol for inducing hairy roots in *Papaver bracteatum*. *Plant Cell and Tissue Organ Culture*, *113*, 1–9.

Sheludko, Y., Gerasimenko, I., Kolshorn, H., & Stockigt, J. (2002). Isolation and structure elucidation of a new indole alkaloid from *Rauvolfia serpentina* hairy root cultures: the first naturally occurring alkaloid of the raumalicine group. *Planta Medica*, *68*, 435–443.

Sheludko, Y., Gerasimenko, I., Unger, M., Kostenyuk, I., & Stöckigt, J. (1999). Induction of alkaloid diversity in hybrid plant cell cultures. *Plant Cell Reports*, *18*, 911–918.

Smith, C., et al. (2014). Whole-genome sequencing analysis reveals high specificity of CRISPR/Cas9 and TALEN-based genome editing in human iPSCs. *Cell Stem Cell*, *15*(1), 12–13.

Srivastava, S., & Srivastava, A. K. (2007). Hairy root culture for mass production of high-value secondary metabolites. *Critical Reviews in Biotechnology*, *27*, 29–43.

Streubel, J., Blucher, C., Landgraf, A., & Boch, J. (2012). TAL effector RVD specificities and efficiencies. *Nature Biotechnology*, *30*(7), 593–595.

Sudha, C. G., Reddy Obul, B., Ravishankar, G. A., & Seeni, S. (2003). Production of ajmalicine and ajmalinein hairy root cultures of *Rauvolfia micrantha* Hook f., a rare and endemic medicinal plant. *Biotechnology Letters*, *5*, 631–636.

Suttipanta, N., Pattanaik, S., Kulshrestha, M., Patra, B., Singh Sanjay, K., & Yuan, L. (2011). The transcription factor CrWRKY1 positively regulates the terpenoid indole alkaloid biosynthesis in *Catharanthus roseus*. *Plant Physiology*, *157*, 2081–2209.

Suzuki, K., et al. (2014). Targeted gene correction minimally impacts whole-genome mutational load in human-disease-specific induced pluripotent stem cell clones. *Cell Stem Cell*, *15*(1), 31–36.

Szczepek, M., Brondani, V., Buchel, J., Serrano, L., Segal, D. J., & Cathomen, T. (2007). Structure-based redesign of the dimerization interface reduces the toxicity of zinc-finger nucleases. *Nature Biotechnology*, *25*(7), 786–793.

Tisserat, B., & Berhow, M. (2009). Production of pharmaceutical from *Papaver* cultivars in vitro. *Engineering in Life Sciences*, *3*, 190–196.

Tzfira, T., & Citovsky, V. (2006). Agrobacterium-mediated genetic transformation of plants: Biology and biotechnology. *Current Opinion in Biotechnology*, *17*, 147–154.

Tzfira, T., & Citovsky, V. (2000). From host recognition to T-DNA integration: the function of bacterial and plant genes in the *Agrobacterium*–plant cell interaction. *Molecular Plant Pathology*, *1*, 201–212.

Tzfira, T., & Citovsky, V. (2003). The *Agrobacterium*-plant cell interaction. Taking biology lessons from a bug. *Plant Physiology*, *133*, 943–947.

Urnov, F. D., Rebar, E. J., Holmes, M. C., Zhang, H. S., & Gregory, P. D. (2010). Genome editing with engineered zinc finger nucleases. *Nature Review Genetics*, *11*(9), 636–646.

Valentine, L. (2003). *Agrobacterium tumefaciens* and the plant: the David and Goliath of modern genetics. *Plant Physiology*, *133*, 948–955.

Veena, V., & Taylor, C. G. (2007). *Agrobacterium rhizogenes*: recent developments and promising applications. *In Vitro Cellular and Developmental Biology - Plant*, *43*, 383–403.

Wang, T., Wei, J. J., Sabatini, D. M., & Lander, E. S. (2014). Genetic screens in human cells using the CRISPR-Cas9 system. *Science*, *343*(6166), 80–84.

Ward, D.V., & Zambryski, P. C. (2001). The six functions of Agrobacterium VirE2. *Proceedings of the National Academy of Sciences of the USA*, *98*, 385–386.

Yang, L., et al. (2013). Optimization of scarless human stem cell genome editing. *Nucleic Acids Research*, *41*(19), 9049–9061.

Zhou, M. L., Zhu, X. M., Shao, J. R., Tang, J. R., & Wu, Y. M. (2011). Production and metabolic engineering of bioactive substances in plant hairy root culture. *Applied Microbiology and Biotechnology*, *90*, 1229–1239.

Zhou, Y., et al. (2014). High-throughput screening of a CRISPR/Cas9 library for functional genomics in human cells. *Nature*, *509*(7501), 487–491.

Ziegler, J., Voigtlander, S., Schmidt, J., Kramell, R., Miersch, O., Ammer, C., Gesell, A., & Kutchan, T. M. et al. (2006). Comparative transcript and alkaloid profiling in *Papaver* species identifies a short chain dehydrogenase/reductase involved in morphine biosynthesis. *Plant Journal*, *48*, 177–192.

CHAPTER 7

Cytogenetic and bioactive attributes of *Crocus sativus* (Saffron): a tool to unfold its medicinal mystery

Shafat A. Mir[a], Javeed I. A. Bhat[a], Rouf Ahmad Bhat[a], Bilal A. Beigh[b], Hafiz ul Islam[b], Shakeel Ahmad Dar[a], Ishrat Bashir[a], Gowhar Rashid[c]

[a]Sher-e-Kashmir University of Agricultural Sciences and Technology, Srinagar, Jammu and Kashmir, India
[b]Barkatullah University, Bhopal, Madhya Pradesh, India
[c]Department of Clinical Biochemistry, Sher-e-Kashmir Institute of Medical Sciences, Soura, Jammu and Kashmir, India

Introduction

Therapeutic plants are acknowledged as main resources of remedy. Reports point out that about 80% of the Asian inhabitants were using these plants as primary health care (Kiran Kumari, Sridevi, Chandana Lakshmi, & Manasa, 2012). As food and drug demand projected to increase in the future, related devastation of regular resources from this area will also rise (Fischer et al., 2002). Saffron (*Crocus sativus* L.) is a perennial aromatic crop, having its place in family Iridaceae (Champalal, Nilakshi, Vijay, & Abhyankar, 2011) and famous as Red Gold because of its red-orange triple stigmas in producer countries. Being the utmost costly cultured herb in the world (Saeidnia, 2012), is extensively cultivated in Azerbaijan, Egypt, France, China, Greece, Iran, Israel, Italy, India, Mexico, Morocco, Spain, and Turkey (Kumar et al., 2011). Its fresh stigmas are dried through dehydration process (Ordoudi et al., 2015), which are generally traditional, vary among the producer countries (Ordoudi et al., 2015). Saffron is used as a spice for essence improvement and food preservative (Kyriakoudi, O'Callaghan, Galvin, Tsimidou, & O'Brien, 2015). The stigmas of saffron have good proportion of vitamin B_2 (Schmidt, Betti, & Hensel, 2007), which also adds to it yellow color besides the highly water-soluble compound crocin (Tsatsaroni & Liakopoulou-Kyriakides, 1995). Crocin, crocetin, picrocrocin,

Medicinal and Aromatic Plants: Expanding their Horizons through Omics
http://dx.doi.org/10.1016/B978-0-12-819590-1.00007-0

and safranal being the main components of saffron, impart foods, distinctive color and aroma qualities (Ordoudi et al., 2015) and contains many non-volatile constituents like carotenoids, α and β- carotene, lycopene and zeaxanthin (Ordoudi & Tsimidou, 2004; Srivastava, Ahmed, Dixit, Dharamveer, & Saraf, 2010). On hydrolysis crocin yields gentiobiose and crocetin, while picrocrocin yields glucose and safranal (Evans, 1996). Saffron stigmas are characterized by a bitter taste and an iodoform or straw-like smell caused by chemicals picrocrocin and safranal (Katzer, 2006). The major volatile compound of *C. sativus* is a carboxaldehyde called safranal produced by de-glucosylation of picrocrocin (Himeno & Sano, 1987). Approximately 36,000 flowers are used to harvest just one pound of stigmas (Rechinger, 1975; Leffingwell, 2002; Wani, Hamza, & Mohiddin, 2011). Valijonovich (2018) pointed out that the saffron plant contains 34 volatile compounds. Hexane extract of *C. sativus* has shown the following composition as depicted in Fig. 7.1.

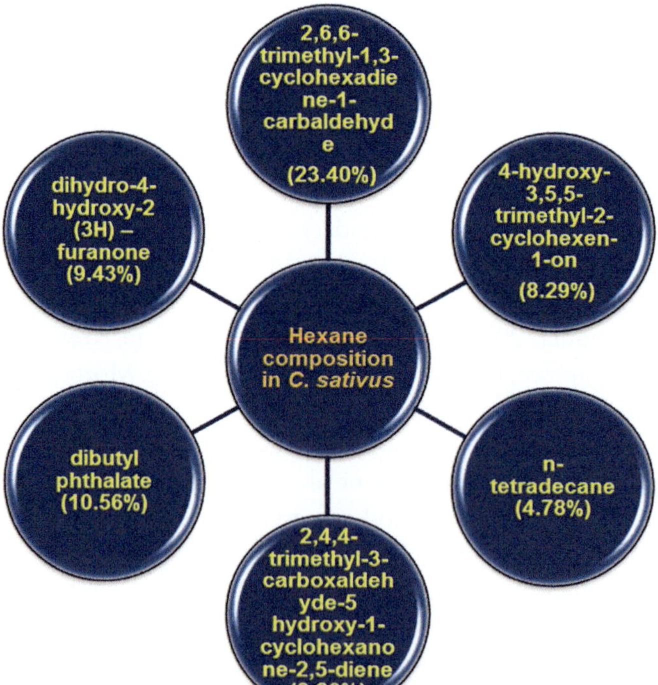

Figure 7.1 ***Composition of the hexane extracted from saffron.***

Saffron and its constituents are widely analyzed for their pharmacological activities such as treatment of memory impairment, antidepressant, anticonvulsant, and particularly for their antitumor effect (Abdullaev & Espinosa-Aguirre, 2004). It can also be used during spinal cord injury, myocardial injury, neuronal injury, menstruation distress, erectile dysfunction, lung inflammation, arthritis, and parkinsonism (Champalal et al., 2011; Wani et al., 2011; Akowuah & Htar, 2014; Malathi, Devi, & Hari, 2014; Kyriakoudi et al., 2015; Hamidpour, Hamidpour, Hamidpour, & Shahlari, 2013). Based on FDA rules (Kakhki, 2001), the food and drug administration approves saffron as a natural food preservative and flavoring agent without any constraint in culinary purposes. Therefore, the Saffron products should possess the following characteristics:

1. Saffron stigmas should be yellow and the concentration of foreign organic compounds should not exceed more than 10%
2. When the saffron is dried at 100°C the concentration of volatiles and humidity should not be more than 14%
3. Maximum total ash and the amount of soluble ash must not more than 1%

Origin and history of saffron

There is an ambiguity in the exact origin of saffron. However, in literature, there are some suggestions about its origin from east. The archeological and historical data provide evidences of its domestic cultivation around 2000–1500 years BC. According to Srivastava et al. (2010), the recent reference of the cultivation of saffron revolves around 2300 BC. The fresco paintings of saffron cultivation can be found in the history that reveals the cultivation of saffron in the Palace of Minos at Knossos in Crete. Additional records are found in the palace of Akrotiri in Thera (now Santorini, Greece, 1700–1450 BC). The crop was grown in regular patterns of rows, but there is still confusion about the type of species grown during this era (*C. sativus* or *C. cartwrightianus*). During the same period, the evidence of existence of saffron cultures can be traced at the Nile delta of Egypt. In history, the main harvesting species of saffron was *C. cartwrightianus*. Later on, a mutant of *C. cartwrightianus* was isolated, which is now known as *C. sativus*. The distinguishing feature of this species is its elongated stigma. It was "domesticated on Crete during the Late Bronze Age." From this knowledge, it can be traced out that *C. cartwrightianus* is the wild precursor of domesticated saffron (Srivastava et al., 2010).

Terminology

The term saffron originated from the Arabic word "zafaran" or "zaafar" which means yellow. There are the two subgenera of *Crocus, namely, Crocus* with extrose anthers and *Crociris* with introse anthers. The subgenus *Crociris* has only one species, *C. banaticus*, while the remaining series is included in the subgenus *Crocus*. There are further two sections of *Crocus, namely,* "*Crocus* and *Nudiscapus* and each of them is again divided into series a-f and g-o, respectively."

Other taxonomic characteristics

Saffron flowers flourish in autumn having six tepals (three inner & outer, respectively) typically violet in color with darker veins. Flowers are 2–3 in number per stem and the plant has 2–3 stems. It has numerous leaves ranging from approximately 5–30 in numbers. Bracts flaccid, usually not strictly covering the perianth-tube, membranous, white, or apparent with no marking and its anther is yellow in color, style is made up of three branches; the dense mat of papillae covers the seed (Saxena, 2002).

Cytogenetics of saffron

Genetically saffron has large genomic size roughly 30,000 Mbp and is twice, 60 and 240 times higher than *Triticum aestivum, Oriza sativa, and Arabidopsis thaliana* and shows broad cytological variation. The large genome size of Saffron has been assessed on the basis of the diploid specie *Crocus vernus* (11,000 Mbp; Chichiriccò, 1984). The different species of *Crocus* shows the wide series of chromosome numbers and a few species show variation in chromosome number interspecifically. A range of chromosome numbers of dissimilar species of *Crocus* is shown in Table 7.1.

Saffron is a triploid monocot sterile specie with a $3n = 24$ (basic chromosome number $x = 8$) number of chromosomes (Chichiriccò, 1984). The karyotype of saffron has been studied on numerous ecotypes from a number of nations and it has always resulted as $2n = 3x = 24$ without any noteworthy variations in karyology. Thus, the acknowledged karyotype consists of 8 triplets:

1. subarcocentric (1, 2)
2. metacentric (3, 4, and 8)
3. submetacentric (6, 7)

Table 7.1 A series of different species of *Crocus* with chromosome numbers.

Crocus sp.	Chromosome number	Reference
C. pallasii	2n = 12, 14, 16	Jacobsen and Orgaard (2004)
C. matheweii	2n = 16	
C. thomasii	2n = 16	
C. cartwrightianus	2n = 16	
C. sativus	2n = 3x = 24	
C. moabiticus	2n = 14	
C. oreocreticus	2n = 16	
C. asumaniae	2n = 26	
C. hadriaticus	2n = 16	
C. naqabensis	2n = 14	

Triplet 5, however, has two different chromosome subtypes:

1. metacentric 5(1)
2. subarcocentric and smaller 5(2, 3)

Biochemical composition of saffron and their medicinal attributes

Phytochemically saffron is composed of volatile, non-volatile, and aromatic compounds (Abdullaev, 2002). The key components found in the saffron are crocin, crocetin, picrocrocin, and safranal, found naturally in stigma (Finley & Gao, 2017). Among the volatile compounds, safranal forms the major component of the saffron (Finley & Gao, 2017). Crocin, crocetin, picrocrocin, flavonoids (quercetin and kaempferol), isophorones are among the non-volatile components of saffron (Finley & Gao, 2017) and low levels of other carotenoids including zeaxanthin, lycopene, and various α- and β-carotenes (Liakopoulou-Kyriakides, 2002; Finley & Gao, 2017). It also contains vitamins in traces (Rios, Recio, Giner, & Manez, 1996; Bouvier, Suire, Mutterer, & Camara, 2003; Rubio-Moraga, Castillo-Lapez, Gamez-Gamez, & Ahrazem, 2009). The main secondary metabolites of saffron consist of a chain of carotenoid-glycosyl esters of C_{20}-dicarboxylic acids. The chemical composition of saffron with percent mass percentage is shown in Table 7.2.

Some of the main and essential biochemical compounds derived from the saffron are discussed further.

Crocin and crocetin

Crocin is a comprehensive term meant for a chain of hydrophilic carotenoids that are either monoglycosyl or diglycosyl polyene esters of crocetin (Finley

Table 7.2 Chemical composition of saffron.

Component	Mass (%)
Carbohydrates	12.0 to 15.0
H_2O	19.0 to 14.0
Polypeptides	11.0 to 14.0
Cellulose	4.0-7.0
Lipids	3.0-8.0
Minerals	1.0-1.5
Miscellaneous	40.0

Source: Srivastava et al. 2010.

& Gao, 2017). Crocetin is a conjugated polyene dicarboxylic acid and is lipophilic in nature (Finley & Gao, 2017). The resultant products of the crocetin are hydrophilic when esterified with hydrophilic gentiobiose(s) or any other sugar precursors (Finley & Gao, 2017). Enzymatic or acid hydrolysis of crocin yields crocetin as an end product. The crocin and crocetin derivatives are presented in Table 7.3 (Liakopoulou-Kyriakides & Kyriakidis, 2002).

Crocin content of the saffron crop is governed by the geographical area where it is grown and the method through which the extraction of crocin is carried out. Crocin contributes about 30% to the total metabolites present in the saffron. Crocin and Crocetin commonly exist in trans form, but some minor constituents of saffron such as cis-crocetin and its glycosides. "Trans-crocetin di- β-D-gentiobiosyl ester (crocin, α-crocin or crocin I) in saffron has been reported to have the highest coloring capacity among extractions." However, lipophilic metabolites of saffron contribute comparatively less toward the red color of the extracts (Finley & Gao, 2017). The general structure of crocin and crocetin is shown in Fig. 7.2.

Table 7.3 Derivatives of *Crocus*

Compound	Sugar moieties	Chemical formula	Isomer occurrence in saffron
Crocetin	R1 = R2 = OH	$C_{20}H_{24}O_4$	cis – trans
Crocin 1	R1 = β-D-glucosyl, R2 = H	$C_{26}H_{34}O_9$	Trans
Crocin 2	R1 = β-D-gentiobiosyl, R2 = H	$C_{32}H_{44}O_{14}$	cis – trans
Crocin 2^0	R1 = R2 = β-D-glucosyl	$C_{32}H_{44}O_{14}$	cis – trans
Crocin 3	R1 = β-D-gentiobiosyl, R2 = β-D-glucosy	$C_{38}H_{54}O_{19}$	cis – trans
Crocin 4	R1 = R2 = β-D-gentiobiosyl	$C_{44}H_{64}O_2$	cis – trans
Crocin 5	RI = 3β-D-glucosyl, R2 = β-D- gentiobiosy	$C_{50}H_{24}O_{29}$	cis – trans

Source: Liakopoulou-Kyriakides & Kyriakidis, 2002; Finley & Gao, 2017.

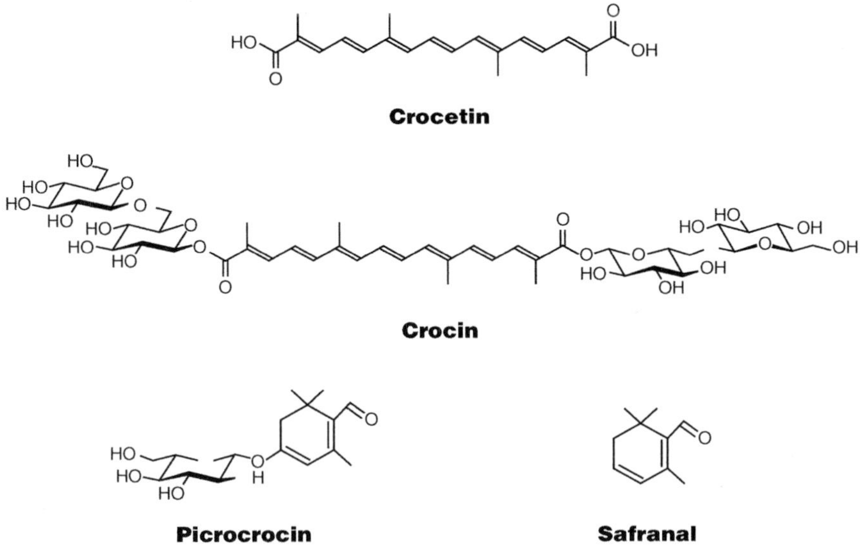

Figure 7.2 ***Different chemical components of saffron.*** *Source: Srivastava et al., 2010*

Medicinal attributes of crocin and crocetin

Crocin has been found to exert hypolipidemic effect when applied in the range of 25 mg/kg to 100 mg/kg body weight in hyper-lipidemic rats. Crocin inhibits pancreatic enzyme lipase thus decreases the absorption of fat (Papandreou et al., 2006). Crocin has a great antioxidant property as compared to alpha-tocopherol. In the neuronally distinguished pheochromocytoma cells deficient of glucose, the absence of which causes peroxidation of the cell membrane lipids and reduce the intercellular superoxide dismutase enzyme activity. These effects are reversed by crocin application which makes it an exceptional and potential antioxidant to overcome oxidative pressure in neurons (Chatterjee, Datta, Bhattacharyya, & Bandopadhyay, 2005. It also raises the levels of various enzymes such as the glutathione reductase, glutathione-transferase. Crocin derivatives from saffron increases blood flow through vasodilation to the retina and choroid, maintains retinal function and in this way helps in preventing ischeamic retinopathy and macular degeneration leading to blindness. Crocin and diglucosylcrocetin has the potential to inhibit the early tumor antigen expression of adenovirus infected cells, crocetin esters were less potent than crocin itself in this concern (Bhargava, 2011). Crocetin has the potential to act as cardiac protectant. It decreases the level of "cardiac marker," that is, lactate dehydrogenase enzyme activity and increases the mitochondrion potential in a cardiac myocyte that is treated with noradrenaline (Premkumar, Abraham,

Santhiya, Gopinath, & Ramesh, 2001). Crocetin has an antidiabetic activity. It has shown antidiabetic activity when applied to fructose induced diabetic rats (Liu, Yang, Mo, Liao, & Jin, 2005). Because of its high antioxidant and calcium antagonistic capacity, it can be a good medicine for diabetic vascular problems (Shen & Qian, 2006). Crocetin has anti-cancerous characteristics. It decreases the peroxidation of lipids, glutathione metabolizing enzymes and also regress the histopathological variations pertinent to tumor occurrence ascertaining it as a good antitumor agent (Inoue et al., 2005). Crocetin is also helpful in preventing Parkinsonism (Ahmad, 2005).

Picrocrocin

It is a monoterpene glycoside of safranal and this metabolite is responsible for the bitter taste of saffron stigmas (Caballero-Ortega et al., 2007). It accounts for 5%–15% of metabolites of saffron. Its chemical formula is $C_{16}H_{26}O_7$ (Fig. 7 2) with a molar weight of 330.37 g/mol. When β-Glucosidase acts on picrocrocin it releases the aglycone (4-hydroxy-2,6,6-trimethyl-1-cyclohexene-1-carboxaldehyde ($C_{10}H_{16}O_2$) (Lage & Cantrell, 2009), which is then transformed to safranal during the course of the drying process of plant material (Lage & Cantrell, 2009; Lozano, Delgado, Gomez, Rubio, & Iborra, 2000).

Medicinal attributes picrocrocin

Picrocrocin is responsible for antiproliferative activity in human cancer cells (Kyriakoudi et al., 2015).

Safranal

It is the volatile compound of the saffron stigma primarily responsible for the aroma of the saffron (Finley & Gao, 2017). It constitutes about 0.5% of volatile compounds. Safranal is considered a degradation product of zeaxanthin formed through a pathway where picrocrocin is an intermediate. Safranal has a molar mass of 150.21 g/mol with chemical formula $C_{10}H_{14}O$ (Fig. 7 2) and is a monoterpene glycoside.

Medicinal attributes

Safranal shows anticonvulsant activity in model animals and act as an agonist at GABAA receptors. It has good antioxidant and free radical scavenging properties with cytotoxic properties against cancer cells in vitro (Escribano, Alonso, Coca-Prados, & Fernandez, 1996). Safranal has also shown antidepressant activity in model animals and in pilot studies of human subjects (Akhondzadeh, Fallah-Pour, Afkham, Jamshidi, & Khalighi-Cigaroudi, 2004;

Hosseinzadeh, Karimi, & Niapoor, 2004). In general, the various medicinal properties and biosynthetic pathways of saffron derivatives are shown in Fig. 7.3, respectively.

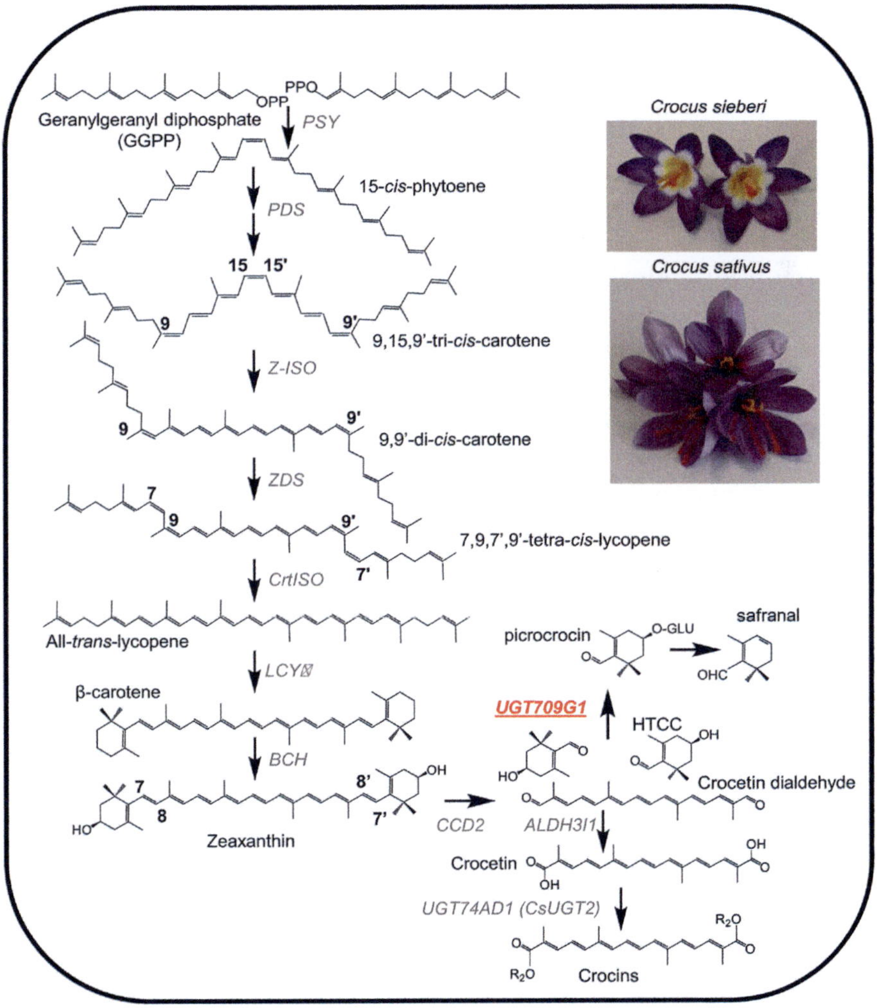

Figure 7.3 ***Schematic biosynthetic pathway for crocetin, crocins, and picrocrocin in Crocus sativus stigmas and Crocus sieberi stigmas and yellow tepal sector.*** PSY, phytoene synthase; PDS, phytoene desaturase; Z-ISO, 15-cis-f-carotene isomerase; ZDS, f-carotene desaturase; CrtISO, carotene isomerase; LCYB, lycopene-b-cyclase; BCH, b-carotene hydroxylase; CCD2, carotenoid cleavage dioxygenase 2; ALDH3I1, aldehyde dehydrogenase; UGT74AD1 (CsUGT2), crocetin glucosyltransferase; HTCC, hydroxy-2,6,6-trimethyl-1-cyclohexen-1-carboxaldehyde. R2 refers to the number of glucose molecules at the end of the crocin molecule, which can be present in the following number and combinations (1-0, 1-1, 2-0, 1-2, 2-2, 2-3) generating the different crocins molecules in saffron. (Diretto et al., 2019)

General pharmacological activities of saffron

Saffron is an important crop for pharmacological studies and is in medicinal use for the treatment of various kinds of diseases. Pharmacological and clinical research of saffron has revealed its use against cancer and depression. Rahmani, Khan, and Aldebasi (2017) has summed up the biological activity of saffron and its properties as mentioned in the Table 7.4.

Important properties of saffron crop at molecular level

Metabolomics

Metabolome is an exclusive collection of cellular functioning parts that are linked with the expression of the sequenced genomes in all living organism including plant, animal and bacteria (Ordoudi et al., 2015). Metabolomic studies assists in identifying substrates and products of enzymes with no need for going through heterologous expression systems (Beale & Sussman, 2011). Metabolomics of saffron has provided an unprejudiced, comprehensive, qualitative and quantitative overview of its metabolites such as crocetin esters, picrocrocin and safranal, elucidating their alliance with therapeutic and aesthetic properties (Ordoudi et al., 2015). Recently, saffron authenticity and quality issues have been assessed using FT-IR-based metabolomics, micro-DRIFTS method of measurement along with chemometrics (PCA) Anastasaki et al., 2009). Their study intended to differentiate authentic saffron samples originated from four countries (Iran, Spain, Italy, Greece); an objective that had been investigated via FT-NIR some years ago with the ultimate goal to offer tools for identification of mislabelling of geographical origin (Zalacain et al., 2005). Ordoudi, de los Mozos Pascual, and Tsimidou (2014) comprehensively extended the application of the FT-MIR coupled with chemometrics (PCA) towards the quality control of traded saffron and, more purposely, for the assessment of saffron freshness. The authors recommended that specific infrared bands in the sugar region (e.g., 1028 and 1175–1157 cm) associated with the existence of glucose moieties and breakage of glycosidic bonds, are valuable tools to monitor storage effects on saffron quality and, accordingly, that was the first attempt to trace back the age of saffron with FT-IR metabolomics (Zalacain et al., 2005). Further insights of the nature and fate of the glycosidic bonds during storage was then gained using a 1H-NMR-based metabolomics approach (Ordoudi et al., 2015). In that study, 1H-NMR data of 98 saffron samples of different source and harvest year, stored under different conditions for various periods of time, were subjected to

Table 7.4 Bioactive constituents and their role extracted from *Crocus sativus*.

Biological activity	Aim of study	Finding/Outcome	Reference
Antioxidant	Evaluation of antioxidant activity of stigmas extract	Saffron stigma showed antioxidant activity.	Karimi, Oskoueian, Hendra, & Jaafar, 2010
Antioxidant	Measurement of antioxidant	Antioxidant activity has been observed.	Goli, Mokhtari, & Rahimmalek, 2012
Anti-inflammatory	Effects of crocin, safranal on local inflammation	Crocin and safranal have a role in the suppression of inflammatory pain responses and decreased the number of neutrophils	Tamaddonfard, Farshid, Eghdami, Samadi, & Erfanparast, 2013
Anti-inflammatory	Anti-inflammatory effects of crocin	Crocin, a constituent showed anti-inflammatory effects and modulate inflammatory processes	Xu et al., 2009
Antibacterial	*C. sativus* extract tested against bacteria.	Strong activity of against bacteria and fungi was noted	Vahidi, Kamalinejad, & Sedaghati, 2002
Hepatoprotective	Evaluation of protective effects of extract against hepatotoxicity	Finding demonstrated that petals ameliorate acute liver injury.	Omidi, Riahinia, Montazer, & Behdani, 2014
Protective effects against nephrotoxicity	Effect of safranal against nephrotoxicity	Safranal has a protective effect against nephrotoxicity	Boroushaki, Mofidpour, & Sadeghnia, 2007
Cardioprotective	Cardio-protective effect of saffron and safranal.	Results revealed that myocardial injury preserved nearly normal tissue architecture with saffron or safranal pretreatment.	Mehdizadeh, Parizadeh, Khooei, Mehri, & Hosseinzadeh, 2013
Inhibitory action on AChE	Inhibitory action on AChE via saffron extract and its constituents	Results indicated that Saffron extract showed moderate AChE inhibitory activity.	Geromichalos et al., 2012

(*Continued*)

Table 7.4 Bioactive constituents and their role extracted from *Crocus sativus*. (*Cont.*)

Biological activity	Aim of study	Finding/Outcome	Reference
Antihyperglycemic	Saffron, crocin and safranal effects on the blood levels of fasting glucose, HbA1c and liver/kidney function tests	Results demonstrated that saffron extract, crocin and safranal significantly reduced the fasting blood glucose levels, but significantly increased the blood insulin levels in diabetic rats compared with the control diabetic rats	Kianbakht and Hajiaghaee, 2011
Antiobesity and anorectic	Effects of saffron and crocin on body weight, food intake and blood leptin levels	Results concluded that saffron has antiobesity and anorectic effects and lowered leptin levels	Kianbakht and Hashem, 2015
Antiobesity	Evaluation of antiobesity effects of saffron and crocin	Results showed that saffron extract significantly decreased food consumption in obese rats. Furthermore, crocin showed a significant decrease on the rate of body weight gain	Mashmoul et al., 2014
Aphrodisiac activities	Aphrodisiac activities of stigma and safranal and crocin	The study results revealed that aphrodisiac activity of saffron extract and its constituent crocin.	Hosseinzadeh, Ziaee, & Sadeghi, 2008
Anxiolytic properties	Evaluation of anxiolytic and hypnotic effect of saffron extract, crocin and safranal	Results confirmed safranal at higher doses demonstrated anxiolytic effects whereas crocin did not show anxiolytic properties.	Hosseinzadeh & Noraei, 2009
Immuno-stimulatory	Effects of saffron petal extract on blood parameters and immune system	Saffron petal extract use causes an increase in antibody response	Babaei et al., 2014
Anticonvulsant	Evaluation of anticonvulsant of safranal and crocin	Findings confirmed that safranal reduced the seizure duration, delayed the onset of tonic convulsions	Hosseinzadeh & Talebzadeh, 2005

Source: Rahmani et al., 2017.

OPLS-DA (Ordoudi et al., 2015). The chemometric investigation showed that crocetin esters and picrocrocin, being abundant in fresh samples, can be considered as biomarkers for saffron stored for less than four years after processing (Ordoudi et al., 2014). These results strengthened the opinion that evolution of hydrolytic degradation of crocetin esters and picrocrocin represents a key factor for saffron quality deterioration, as already observed by using FT-IR (Ordoudi et al., 2014).

Many authors have proposed quite a few analytical methods for the detection of plant adulterants in saffron, for instance UV–vis spectrophotometric measurements (Zalacain et al., 2005; Sanchez et al., 2008; Maggi et al., 2011), near infrared spectroscopy (Zalacain et al., 2005), Raman and nuclear magnetic resonance spectroscopy (NMR) (Assimiadis, Tarantilis, & Polissiou, 1998; Tarantilis & Polissiou, 2004), capillary electrophoresis (Zougagh, Simonet, Rios, & Valcarcel, 2005) and high-performance liquid chromatography without and with mass spectrometry (MS) detection (Sabatino et al., 2011; Caballero-Ortega, Pereda-Miranda, & Abdullaev, 2007; Lage and Cantrell 2009; Loskutov, Beninger, Hoseld, & Sink, 2000). Mostly, these methods are based on the analysis of target compounds used to detect adulteration of saffron and consequently have a major drawback related to the fact that they can only detect a limited number of adulterations (Babaei, Talebi, & Bahar, 2014; Torelli, Marieschi, & Bruni, 2014). Ultimately, molecular methods of encouraging results has been employed to detect DNA markers (Javanmardi, Bagheri, Moshtaghi, Sharifi, & Hemati-Kakhki, 2011; Marieschi, Torelli, & Bruni, 2012; Babaei et al., 2014; Torelli et al., 2014). Metabolomics is one of the best strategies for the characterization of complex biological samples as it allows the production of a chemical signature or fingerprint (Antignac et al., 2011). NMR and MS are the most extensive analytical techniques used in metabolomics (Antignac et al., 2011; Hu and Xu, 2013). As compared to NMR, MS is much more sensitive, enables to measure low abundance species that can offer useful information to look for new markers. "Metabolite fingerprinting obtained using ^{1}H NMR spectra and chemometrics" was reported for the validation of both Iranian and Italian saffron (Yilmaz, Nyberg, Molgaard, Asili, & Jaroszewsk, 2010; Cagliani, Culeddu, Chessa, & Consonni, 2015) and adulteration through certain plant adulterants (Petrakis, Cagliani, Polissiou, & Consonni, 2015). The potential of applying the NMR-based metabolomic move towards controlling the saffron quality deterioration has not been examined (Yilmaz et al., 2010; Cagliani et al., 2015).

Genomics

Saffron being a perennial sterile plant reproduces only vegetative by means of the corms. Even though a lot of work has been carried out via tissue culture and hybridization (Rubio-Moraga et al., 2014; Mir et al., 2015), propagation through corms offers no or little genetic variation in the form of somatic mutations, transversions, segregation distortions, etc., neither of them combining in a population nor bringing heritable variations due to its sterility (Agayev, Fernandez, & Zarifi, 2009). Saffron a triploid specie with a basic chromosome number of x = 8 (Chichiriccò, 1984) whose karyotype has been studied by quite a lot of authors and on different ecotypes from several countries (Azerbaijan, Iran, Italy, Turkey, France, and United Kingdom) and it is always resulted as 2n = 3x = 24, without any significant karyologycal variation (Chichiriccò, 1984). Nevertheless, many studies recognize variations in phenotypic and phytochemical traits due to the epigenetic changes urging the instantaneous need for developing molecular markers to identify these variations at molecular level, which can be further exploited for the enhancement of saffron (Mir et al., 2015). Moreover, 27 SSRs markers were evaluated on eight Iranian-cultivated saffron ecotypes and 29 wild alleles to calculate the molecular variability and discriminating the capability of these markers regarding their effectiveness in establishing genetic relationships in these *Crocus* ecotypes (Nemati et al., 2014). In recent times, an alternative to empiric and imprecise morphological study, the latest molecular techniques were used to distinguish authentic plant exemplars, with respect to other species (Mir et al., 2015). Specifically, the DNA barcode process employs standard nucleotide sequence analysis (maturase K, matK; ribulose 1,5-bisphosphate carboxylase/oxygenase large subunit, rbcL; intragenic spacer between tRNAHisGUG gene and photosystem II thylakoid membrane protein of Mr 32.000 gene, trnHpsbA; internal transcribed spacer of nuclear ribosomal DNA, ITS) for botanical species identification (Gugerli, Parducci, & Petit, 2005; Gismondi et al., 2012; Kress and Erickson, 2007, 2008; Seberg & Petersen, 2009;Yao et al., 2009).

In recent times, many efforts have been performed to attain a better understanding of the genomic organization of Crocus species. Seberg et al. (2009) presented the analysis of a projected barcode set of genes (Chase et al., 2007) in the genus *Crocus*. RpoC1, matK, and tmH-psbAregions were analyzed on 86 species of the *Crocus* genus and the proposing sets were further extended with several other genomic regions to achieve a final diagnostic set for 79 out ofthe 86 analyzed species. Recently, Moraga, Rambla, Ahrazem, Granell, and Gómez-Gómez (2009) investigated the

RAPD profile of 43 isolates of *C. sativus*to verifies the morphism of this species. Three different approaches, random amplified polymorphic DNA, inter simple sequence repeats (ISSR) and microsatellite analysis were used to assess the variability of saffron from several different geographic areas and concluded that *C. sativus* is a monomorphic species (Moraga et al., 2009). The rolling circle amplification of genomic templates for inverse PCR (RCA-GIP) has been proposed (Tsaftaris, Pasentzis, & Argitiou, 2009) is a new genome walking approach. By using this method, four promoter regions of flowering genes (Tsaftaris, Polidoros, Pasentsis, & Kalivas, 2007) from *C. sativus* were isolated. The RCA-GIP method authorized the isolation of genomic sequence flanking several genes; this is a vital step to understand the complicated mechanisms of gene regulation (Tsaftaris et al., 2007).

Transcriptomics

The most valued metabolites in *Crocus* are synthesized in stigma tissue, and that too in the developmental stage specific manner (Moraga et al., 2009). Hence, characterization of the transcriptome of saffron stigmas is essential for throwing light on the molecular basis of flavor, color, biogenesis, genomic organization and the biology of the gynoecium of spices and saffron in particular (Husaini et al., 2009a). Several volatile and non-volatile metabolites are also present in other *Crocus* tissues. The prototype of expression of numerous candidate genes and compare them with volatile production and organoleptic features of saffron throughout the development" (Moraga et al., 2009). The picrocrocin and crocin which were detected in the early stages, increased rapidly during the following stages of development (D'Agostino, Pizzichini, Chiusano, & Giuliano, 2007). In order to recognize candidate genes encoding enzymes involved in volatile biosynthesis, in silico screening of the stigma cDNA database previously described (D'Agostino et al., 2007) was completed and a comparison was drawn among the apocarotenoid content and the expression profiles. The results showed that throughout the development of *C. sativus* stigmas, the 1 deoxyxylulose 5 phosphate synthase (DXS), was expressed at all the stages while 3 hydroxy 3 methylglutaryl CoA reductase (HMGR) was expressed at low-levels signifying that DXS plays an important role in apocarotenoid accumulation (D'Agostino et al., 2007).

Transcriptomic and genomic studies on the saffron has received a little interest (Piqueras et al., 1999). The abundance of ESTs in a particular contig indicates the mRNA abundance of that particular gene in the stigma tissue (Piqueras et al., 1999). At present, there are 6768 saffron Expressed Sequence Tags (ESTs) since the initial set of 6603 high quality ESTs from cDNA library of a saffron stigma (D'Agostino et al., 2007). The global sequencing of a cDNA library formed from mature stigmas of *C. sativus* attached to extensive and bioinformatic analyses of sequence data approved the construction of the first saffron database (D'Agostino et al., 2007). Saffron genes are freely available resource for the research community concerned in saffron genomics (D'Agostino et al., 2007). Rubio Moraga et al.(2009) used a blend of approaches to study the accumulation of color and aroma compounds during stigma development performing an in silico screening of the stigma cDNA database earlier described (D'Agostino et al., 2007). The RNA-seq transcriptome (Wang, Gerstein, & Snyder, 2009; Flintoft, 2008) of saffron was separately sequenced by two research groups in order to achieve deeper insights into the genes involved in the biosynthesis of apocarotenoid (Jain, Srivastava, Verma, Ghangal, & Garg, 2016; Baba, Mohiuddin, Basu, Swarnkar, & Malik, 2015). Previously, the YeATS suite identified some artifacts arising from RNA-seq assembly (Chakraborty et al., 2015) and has been used to analyze the walnut transcriptome revealing the biodiversity and plant microbe interactions in twenty different tissues from walnut in California (Chakraborty, Britton, Martnez-Garca, & Dandekar, 2016).

Proteomics

The complete understanding of the biological function of proteins requires knowledge of their structure and function (Pieper et al., 2006). Proteomic investigation carried out previously identified differentially accumulated proteins in somatic embryos of saffron, which offer insights into the underlying molecular mechanisms (Sharifi, Ebrahimzadeh, Ghareyazie, Gharechahi, & Vatankhah, 2012). In addition, lack of validated structure information for a majority of plant proteins is a key obstruction to functional annotation, evolutionary analyses and building interaction networks (Pentony et al., 2012). Saffron has earned a lot of interest due to its therapeutic potential over the last few years (Naghshineh et al., 2015). In silico, molecular dynamics and docking approach have been employed to study interactions between secondary metabolites of saffron (safranal, crocetin, and dimethylcrocetin) and transport proteins such as β-lactoglobulin, may well be valuable factors in controlling their transport to biological sites (Sahihi, 2015). Reports on saffron have frequently highlighted the need for refining bioinformatics tools accessible with transcriptomic and genomic data (Husaini et al., 2009b).

Conclusion

Although saffron has visible positive impacts on human health, still the exact mechanism of this spice is not clear. The medical applicability of saffron belongs to class of carotenoids viz. crocin, crocetin, picrocrocin and safranal. Saffron needs to be studied at molecular level, so that its medicinal attributes can be exploited to its full capacity.

References

Abdullaev, F. I. (2002). Cancer chemo-preventive and tumoricidal properties of saffron (*Crocus sativus* L.). *Experimental Biology and Medicine, 227*, 20–25.

Abdullaev, F. I., & Espinosa-Aguirre, J. J. (2004). Biomedical properties of saffron and its potential use in cancer therapy and chemoprevention trials. *Cancer Detection and Prevention, 28*, 426–432.

Agayev, Y. M. O., Fernandez, J. A., & Zarifi, E. (2009). Clonal selection of saffron (*Crocus sativus* L): the first optimistic experimental results. *Euphytica, 169*, 81–99.

Ahmad, A. S. (2005). Biological properties and medicinal use of saffron. *Pharmacology and Biochemistry Behavior, 81*, 805–813.

Akhondzadeh, S., Fallah-Pour, H., Afkham, K., Jamshidi, A. H., & Khalighi-Cigaroudi, F. (2004). Comparison of *Crocus sativus* L. and imipramine in the treatment of mild to moderate depression: A pilot double-blind randomized trial. *BMC Complement. Alternative Medicine, 4*, 12.

Akowuah, G. A., & Htar, T. T. (2014). Therapeutic properties of saffron and its chemical constituents. *Journal of Natural Products*, 7, 5–13.

Anastasaki, E., Kanakis, C., Pappas, C., Maggi, L., del Campo, C. P., Carmona, M., Alonso, G. L., & Polissiou, M. G. (2009). Geographical differentiation of saffron by GC–MS/FID and chemometrics. *European Food and Research Technology, 229*, 899–905.

Antignac, J. P., Courant, F., Pinel, G., Bichon, E., Monteau, F., Elliott, C., & Le Bizec, B. (2011). Mass spectrometry-based metabolomics applied to the chemical safety of food. *Trends in Analytical Chemistry, 30*, 292–301.

Assimiadis, M. K., Tarantilis, P. A., & Polissiou, M. G. (1998). UV-Vis, FT-Raman and 1H-NMR spectroscopies of cis-trans carotenoids from saffron (Crocus sativus L.). *Applied Spectroscopy, 52*, 519–522.

Baba, S. A., Mohiuddin, T., Basu, S., Swarnkar, M. K., & Malik, A. H. (2015). Comprehensive transcriptome analysis of *crocus sativus* for discovery and expression of genes involved in apocarotenoid biosynthesis. *BMC Genomics, 16*, 1.

Babaei, S., Talebi, M., & Bahar, M. (2014). Developing an SCAR and ITS reliable multiplex PCR-based assay for safflower adulterant detection in saffron samples. *Food Control, 35*, 323–328.

Beale, M. H., & Sussman, M. R. (2011). Metabolomics of *Arabidopsis thaliana*. *Annals of Plant Reviews, 43*, 157–180.

Bhargava, V. K. (2011). Medicinal Uses and Pharmacological Properties of *Crocus sativus* L.(Saffron). *International Journal of Pharmacy and Pharmaceutical Science, 3*(3), 2011–2226.

Boroushaki, M. T., Mofidpour, H., & Sadeghnia, H. (2007). Protective effect of safranal against hexachlorobutadiene-induced nephrotoxicity in rat. *Iran Journal of Medical Science, 32*, 173–176.

Bouvier, F., Suire, C., Mutterer, J. R. M., & Camara, B. (2003). Oxidative remodeling of chromoplast carotenoids identification of the carotenoid dioxygenase CsCCD and CsZCD genes involved in crocus secondary metabolite biogenesis. *Plant Cell, 15*, 47–62.

Caballero-Ortega, H., Pereda-Miranda, R., & Abdullaev, F. I. (2007). HPLC quantification of major active components from 11 different saffron (*Crocus sativus* L.) sources. *Food Chemistry*, *100*(3), 1126–1131.

Cagliani, L. R., Culeddu, N., Chessa, M., & Consonni, R. (2015). NMR investigations for quality assessment of Italian PDO saffron (*Crocus sativus* L). *Food Control*, *50*, 342–348.

Chakraborty, S., Britton, M., Martnez-Garca, P., & Dandekar, A. M. (2016). Deep RNA-seq pro le reveals biodiversity, plant microbe interactions and a large family of NBS-LRR resistance genes in walnut (*juglans regia*) tissues. *AMB Express*, *6*, 1.

Chakraborty, S., Britton, M., Wegrzyn, J., Buttereld, T., Martinez-Garcia, P.J., 2015. YeATS-a tool suite for analyzing RNA-seq derived transcriptome identi es a highly transcribed putative extensin in heartwood/sapwood transition zone in black walnut.

Champalal, K. D., Nilakshi, N., Vijay, G. R., & Abhyankar, M. M. (2011). Detailed profile of *Crocus sativus*. *International Journal of Pharmacy and Biological Science*, *2*(1), 530–540.

Chase, M. W., Cowan, R. S., Hollingsworth, P. M., van den Berg, C., Madriñán, S., Petersen, G., Seberg, O., Jørgsensen, T., Cameron, K. M., Carine, M., Pedersen, N., Hedderson, T. A. J., Conrad, F., Salazar, G. A., Richardson, J. E., Hollingsworth, M. L., Barraclough, T. G., Kelly, L., & Wilkinson, M. (2007). A proposal for a standardized protocol to barcode all land plants. *Taxon*, *56*, 295–299.

Chatterjee, S., Datta, R. N., Bhattacharyya, D., & Bandopadhyay, S. K. (2005). Emollient and antipruritic effect of Itch cream in dermatological disorders: A randomized controlled trial. *Research Letters*, *37*(4), 253–254.

Chichiriccò, G. (1984). Karyotype and meiotic behaviour of the triploid *Crocus sativus* L. *Caryologia*, *37*, 233–239.

D'Agostino, N., Pizzichini, D., Chiusano, M. L., & Giuliano, G. (2007). An EST database from saffron stigmas. *BMC Plant Biology*, *7*, 53–61.

Diretto, G., Ahrazem, O., Rubio-Moraga, A., Fiore, A., Sevi, F., Argandona, J., Gomez-Gomez, L., 2019. UGT709G1: a novel uridine diphosphate glycosyltransferase involved in the biosynthesis of picrocrocin, the precursor of safranal in saffron (Crocus sativus). New Phytologist 224, 725–740.

Escribano, J., Alonso, G. L., Coca-Prados, M., & Fernandez, J. A. (1996). Crocin, safranal and picrocrocin from saffron (*Crocus sativus* L.) inhibit the growth of human cancer cells in vitro. *Cancer Letters*, *100*(1–2), 23–30.

Evans, W. C. (1996). *Trease and Evans-Pharmacognosy*. China: Saunders© Elsevier Limited 438.

Finley, J. W., & Gao (2017). A Perspective on *Crocus sativus* L. (Saffron) constituent crocin: a potent water-soluble antioxidant and potential therapy for Alzheimer's disease. *Journal of Agriculture and Food Chemistry*, *65*, 1005–1020.

Fischer, G., Van Velthuizen, H.T., Shah, M.M., Nachtergaele, F.O., 2002. Global Agro-ecological Assessment for Agriculture in the 21st Century: Methodology and Results. International Institute for Applied Systems Analysis, Laxenburg, Austria. 1-155.

Flintoft, L. (2008). Transcriptomics: digging deep with RNA-seq. *Nature Reviews of Genetics*, *9*, 568.

Geromichalos, G.D., Lamari, F.N., Papandreou, M.A., Trafalis, D.T., 2012. Margarity M, Papageorgiou A, et al. Saffron as a source of novel acetylcholinesterase inhibitors: molecular docking and in vitro enzymatic studies. Journal of Agriculture and Food Chemistry. 60(24), 6131-8.

Gismondi, A., Rolfo, M. F., Leonardi, D., Rickards, O., & Canini, A. (2012). Identification of ancient *Olea europaea* L. and *Cornus mas* L.seeds by DNA barcode. *Comptes Rendus Biologies*, *335*, 472–479.

Goli, S. A., Mokhtari, F., & Rahimmalek, M. (2012). Phenolic compounds and antioxidant activity from Saffron (*Crocus sativus* L.) petal. *Journal of Agricultural Science*, *4*(10), 175–181.

Gugerli, F., Parducci, L., & Petit, R. J. (2005). Ancient plant DNA: review and prospects. *New Phytolology*, *166*, 409–418.

Hamidpour, R., Hamidpour, S., Hamidpour, M., & Shahlari, M. (2013). Effect of *Crocus sativus* and its active compounds for the treatment of several diseases: A review. *International Journal of Case Reports and Images*, *4*(12), 666–670.

Hemmati Kakhki, A. (2001). Optimization of effective parameters on production of food color from Saffron petals. *Agriculture, Science and Technology*, *15*, 13–20.

Himeno, H., & Sano, K. (1987). Synthesis of crocin, picrocrocin and safranal by saffron stigma-like structures proliferated *in vitro*. *Journal of Agricultural and Biological Chemistry*, *51*, 2395–2400.

Hosseinzadeh, H., Karimi, G., & Niapoor, M. (2004). Antidepressant effect of *Crocus sativus* L. stigma extracts and their constituents, crocin and safranal, in mice. *Acta Horticulturae*, *650*, 435–445.

Hosseinzadeh, H., & Noraei, N. B. (2009). Anxiolytic and hypnotic effect of *Crocus sativus* aqueous extract and its constituents, crocin and safranal, in mice. *Phytotherapy Research*, *23*(6), 768–774.

Hosseinzadeh, H., & Talebzadeh, F. (2005). Anticonvulsant evaluation of safranal and crocin from *Crocus sativus* in mice. *Fitoterapia*, *76*(7–8), 722–724.

Hosseinzadeh, H., Ziaee, T., & Sadeghi, A. (2008). The effect of saffron, *Crocus sativus* stigma, extract and its constituents, safranal and crocin on sexual behaviors in normal male rats. *Phytomedicine*, *15*(6–7), 491–495.

Hu, C., & Xu, G. (2013). Mass-spectrometry-based metabolomics analysis for foodomics. *Trends in Analytical Chemistry*, *52*, 36–46.

Husaini, A. M., Wani, S. A., Sofi, P., Rather, A. G., & Mir, J. I. (2009a). Bioinformatics for saffron (*Crocus sativus* L.) improvement. *Communications in Biometry and Crop Science*, *4*, 1–6.

Husaini, A. M., Wani, S. A., Sofi, P., Rather, A. G., Parray, G. A., Shikari, A. B., & Mir, J. I. (2009b). Bioinformatics for saffron (*Crocus sativus* L) improvement. *Communications in Biometry and Crop Science*, *4*, 3–8.

Inoue, E., Shimizu, Y., Shoji, M., Tsuchida, H., Sano, Y., & Ito, C. (2005). Pharmacological properties of N-095, a drug containing red ginseng, polygala root, saffron, antelope horn and aloe wood". *American Journal of Chinese Medicine*, *33*(1.), 49–60.

Jacobsen, N., & Orgaard, M. (2004). *Crocus cartwrightianus* on the Attica Peninsula. *Acta Horticulturae*, *650*, 65–69.

Jain, M., Srivastava, P. L., Verma, M., Ghangal, R., & Garg, R. (2016). De novo transcriptome assembly and comprehensive expression pro ling in *crocus sativus* to gain insights into apocarotenoid biosynthesis. *Scientific Reports*, 6.

Javanmardi, N., Bagheri, A., Moshtaghi, N., Sharifi, A., & Hemati-Kakhki, A. (2011). Identification of safflower as a fraud in commercial saffron using RAPD/SCAR marker. *Journal of Cell and Molecular Research*, *3*, 31–37.

Karimi, E., Oskoueian, E., Hendra, R., & Jaafar, H. Z. (2010). Evaluation of *Crocus sativus* L. stigma phenolic and flavonoid compounds and its antioxidant activity. *Molecules*, *15*(9), 6244–6256.

Katzer, G., Saffron (*Crocus sativus* L.) Gernot Katzer's Spice Pages. [last accessed on 2006 Jan 10]. Available from: http://www.unigraz.at/~katzer/engl/croc_sat.html

Kianbakht, S., & Hajiaghaee, R. (2011). Anti-hyperglycemic effects of saffron and its active constituents, crocin and safranal, in alloxan-induced diabetic rats. *Journal of Medicinal Plants*, *3*(39), 82–89.

Kianbakht, S., & Hashem, D. F. (2015). Anti-obesity and anorectic effects of saffron and its constituent crocin in obese Wistar rat. *Journal of Medicinal Plants*, *1*(53), 25–33.

Kiran Kumari, S. P. V., Sridevi, M. V. V., Chandana Lakshmi, M., & Manasa (2012). Comparative studies and elemental analysis of fertilizer effected medicinal plant samples using Sem-Eds. Research Journal of Pharmaceutical. *Biological and Chemical Sciences*, *3*, 214–222.

Kress, W. J., & Erickson, D. L. (2007). A two-locus global DNA barcode for land plants: the coding rbcL gene complements the non-coding trnH-psbA spacer region. *PLoS ONE, 6*(508), 1–10.

Kress, W. J., & Erickson, D. L. (2008). DNA barcodes: genes, genomics, and bioinformatics. *Proceedings.*

Kumar, V., Bhat, Z. A., Kumar, D., Khan, N. A., Chashoo, I. A., & Shah, M. (2011). Pharmacological profile of *Crocus sativus* - A comprehesive review. *Pharmacology Online, 3,* 799–811.

Kyriakoudi, A., O'Callaghan, Y. C., Galvin, K., Tsimidou, M. Z., & O'Brien, N. M. (2015). Cellular Transport and Bioactivity of a Major Saffron Apocarotenoid, Picrocrocin (4-(β-d-Glucopyranosyloxy)-2,6,6-trimethyl-1-cyclohexene-1-carboxaldehyde). *Journal of Agriculture and Food Chemistry, 63,* 8662–8668.

Kyriakoudi, A., Ordoudi, S. A., Roldán-Medina, M., & Tsimidou, M. Z. (2015). Saffron, a functional spice. *Austin Journal of Nutrition and Food Science, 3*(1), 1059–1063.

Lage, M., & Cantrell, C. L. (2009). Quantification of saffron (*Crocus sativus* L.) metabolites crocins, picrocrocin and safranal for quality determination of the spice grown under different environmental Moroccan conditions. *Scientia Horticulturae, 121,* 366–373.

Leffingwell, J. C. (2002). Saffron. *Leffingwell Reports, 2,* 1–6.

Liakopoulou-Kyriakides, M., & Kyriakidis, D. (2002). *Crocus sativus*: biological active constituents. *Studies in Natural Products and Chemistry, 26,* 293–312.

Liu, N., Yang, Y., Mo, S., Liao, J., & Jin, J. (2005). Calcium antagonistic effects of Chinese crude drugs: preliminary investigation and evaluation by 45 Ca. *Applied Radiation and Isotopes, 63*(2), 151–155.

Loskutov, A.V., Beninger, C.W., Hoseld, G. L., & Sink, K. C. (2000). Development of an improved procedure for extraction and quantitation of safranal in stigmas of *Crocus sativus* L. Using high per-formance liquid chromatography. *Food Chemistry, 69,* 87–95.

Lozano, P., Delgado, D., Gomez, D., Rubio, M., & Iborra, J. L. (2000). A non-destructive method to determine the safranal content of saffron (*Crocus sativus* L.) by supercritical carbon dioxide extraction combined with high-performance liquid chromatography and gas chromatography. *Journal of Biochemistry and Biophysical Methods, 43,* 367–378.

Maggi, L., Sanchez, M. A., Carmona, M., Kanakis, D. C., Anastasaki, E., & Tarantilis, A. P. (2011). Rapid determination of safranal in the qualitycontrol of saffron spice (*Crocus sativus* L). *Food Chemistry, 127,* 369–373.

Malathi, M., Devi, D. R., & Hari, B. N.V. (2014). *Crocus sativus* Linn - A Potential Source for Diverse Therapeutic Applications. *International Journal of Pharmaceutical and Pharmacological Reviews, 26*(2), 299–305.

Marieschi, M., Torelli, A., & Bruni, R. (2012). Quality control of saffron (*Crocus sativus* L). Development of SCAR markers for the detec-tion of plant adulterants used as bulking agents. *Journal of Agriculture and Food Chemistry, 60,* 10998–11004.

Mashmoul, M., Azlan, A., Yusof, B. N., Khaza'ai, H., Mohtarrudin, N., & Boroushaki, M. T. (2014). Effects of saffron extract and crocin on anthropometrical, nutritional and lipid profile parameters of rats fed a high fat diet. *Journal of Functional Foods, 8,* 180–187.

Mehdizadeh, R., Parizadeh, M. R., Khooei, A. R., Mehri, S., & Hosseinzadeh, H. (2013). Cardioprotective effect of saffron extract and safranal in isoproterenol-induced myocardial infarction in Wistar Rats. *Iranian Journal of Basic Medical.Science, 16*(1), 56–63.

Mir, J. I., Ahmed, N., Singh, D. B., Khan, M. H., Zffer, S., & Shafi, W. (2015). Breeding and biotechnological opportunities in saffron crop improvement. *African Journal of Agriculture Research, 10,* 1970–1974.

Moraga, A. R., Rambla, J. L., Ahrazem, O., Granell, A., & Gómez-Gómez, L. (2009). Metabolite and target transcript analyses during Crocus sativus stigma development. *Phytochemistry, 70,* 1009–1016.

Naghshineh, A., Dadras, A., Ghalandari, B., Riazi, G.H., Modaresi, S.M.S., Afrasiabi, A., Aslani, M.K., 2015. Safranal as a novel anti-tubulin binding agent with potential use in cancer therapy: An in vitro study. Chemico-biological interactions. 238, 151-160. National Academy of Science USA. 105, 2761-2762.

Nemati, Z., Mardi, M., Majidian, P., Zeinalabedini, M., Pirseyedi, S.M., Bahadori, M. Saffron (*Crocus sativus* L), a monomorphic or polymorphic species? Spanish Journal of Agricultural Research 12, 753-762.

Omidi, A., Riahinia, N., Montazer, T. M., & Behdani, M. (2014). Hepatoprotective effect of *Crocus sativus* (saffron) petals extract against acetaminophen toxicity in male Wistar rats. *Avicenna Journal of Phytomedicine*, *4*(5), 330–336.

Ordoudi, S. A., Cagliani, L. R., Lalou, S., Naziri, E., Tsimidou, M. Z., & Consonni, R. (2015). H NMR-based metabolomics of saffron reveals markers for its quality deterioration. *Food Research International*, *70*, 1–6.

Ordoudi, S. A., de los Mozos Pascual, M., & Tsimidou, M. Z. (2014). On the quality control of traded saffron by means of transmission Fourier-transform mid-infrared (FT-MIR) spectroscopy and chemometrics. *Food Chemistry*, *150*, 414–421.

Ordoudi, S. A., & Tsimidou, M. Z. (2004). Production practices and quality assessment of food crops. In R. Dris, & S. M. Jain (Eds.), *Saffron Quality: Effect of agricultural practices, processing and storage* (pp. 209–260). Netherlands: Kluwer Academic Publ. Dordrecht.

Papandreou, M. A., Kanakis, C. D., Polissiou, M. G., Efthimiopoulos, S., Cordopatis, P., Margarity, M., & Lamari, F. N. (2006). Inhibitory activity on amyloid beta aggregation and antioxidant properties of *Crocus sativus* stigmas extract and its crocin constituents". *Journal of Agriculture and Food Chemistry*, *54*(23), 8762–8768.

Pentony, M. M., Winters, P., Penfold-Brown, D., Drew, K., Narechania, A., DeSalle, R., Bonneau, R., & Purugganan, M. D. (2012). The Plant Proteome folding project: structure and positive selection in plant protein families. *Genome Biology Evolution.*, *4*, 360–371.

Petrakis, E. A., Cagliani, L. R., Polissiou, M. G., & Consonni, R. (2015). Evaluation of saffron (*Crocus sativus* L.) adulteration with plant adulterants by 1H NMR metabolite fingerprinting. *Food Chemistry*, *173*, 890–896.

Pieper, U., Eswar, N., Davis, F. P., Braberg, H., Madhusudhan, M. S., Rossi, A., Marti-Renom, M., Karchin, R., Webb, B. M., Eramian, D., Shen, M. Y., Kelly, L., & Melo, F. (2006). MODBASE: a database of annotated comparative protein structure models and associated resources. *Nucleic Acid Research*, *34*, 291–295.

Piqueras, A., Han, B. H., Escribano, J., Rubio, C., Hellín, E., & Fernández, J. A. (1999). Development of cormogenic nodules and microcorms by tissue culture, a new tool for the multiplication and genetic improvement of saffron. *Agronomics*, *19*, 603–610.

Premkumar, K., Abraham, S. K., Santhiya, S. T., Gopinath, P. M., & Ramesh, A. (2001). Inhibition of genotoxicity by saffron (*Crocus sativus* L.) in reply to cyclophsophamide mice. *Drug and Chemical Toxicology*, *24*(4), 421–428.

Qadri, H., & Iqbal, A. M. (2017). Medicinal properties of saffron. *International Journal of Life Science*, *6*(1), 39–46.

Rahmani, A. H., Khan, A. A., & Aldebasi, Y. H. (2017). Saffron (*Crocus sativus*) and its active ingredients: role in the prevention and treatment of disease. *Pharmacognosy Journal*, *9*(6), 873–879.

Rechinger, K. H. (1975). Graz, Austria: Academische Druck-U-Verganstalt. Flora Iranica. *Iridaceae*, 1–79.

Rios, J. L., Recio, M. C., Giner, R. M., & Manez, S. (1996). An update review of saffron and its active constituents. *Phytotherapy Research*, *10*, 189–193.

Rubio-Moraga, A., Ahrazem, O., Pérez-Clemente, R. M., Gómez-Cadenas, A., Yoneyama, K., López-Ráez, J. A., Molina, R.V., & Gómez-Gómez, L. (2014). Apical dominance in saffron and the involvement of the branching enzymes CCD7 and CCD8 in the control of bud sprouting. *BMC Plant Biology*, *14*, 171.

Rubio-Moraga, A., Castillo-Lapez, R., Gamez-Gamez, L., & Ahrazem, O. (2009). Saffron is a monomorphic species as revealed by RAPD, ISSR and microsatellite analyses. *BMC Research Notes, 2*, 189.

Sabatino, L., Scordino, M., Gargano, M., Belligno, A., Traulo, P., & Gagliano, G. (2011). HPLC/PDA/ESI-MS evaluation of saffron(*Crocus sativus* L.) adulteration. *Natural Product Communication, 6*, 1873–1876.

Saeidnia, S. (2012). Future position of *crocus satives* as a valuable medicinal herb in phytotherapy. *Pharmacognosy Journal, 4*, 71.

Sahihi, M. (2015). In-silico study on the interaction of saffron ligands and beta-lactoglobulin by molecular dynamics and docking approach. *Journal of Macromolecular Science, 55*(1), 73–84. doi:10.1080/00222348.2015.1125066.

Sanchez, A. M., Carmona, M., Zalacain, A., Carot, T. M., Jabaloyes, J. M., & Alonso, G. L. (2008). Rapid determination of crocetin esters and picrocrocin from saffron spice (*Crocus sativus* L.) using UV visiblespectrophotometry for quality control. *Journal of Agriculture and Food Chemistry, 56*, 3167–3175.

Saxena, R.B., 2002. Series recent progress in medicinal plants.Vol. 5. USA: SCI TECH Pub; 2002. A review on cultivation of saffron (*crocus sativus* L). 295-319.

Schmidt, M., Betti, G., & Hensel, A. (2007). Saffron in phytotherapy: pharmacology and clinical uses. *Wien Med Wochenschrift, 157*, 315–319.

Sharifi, G., Ebrahimzadeh, H., Ghareyazie, B., Gharechahi, J., & Vatankhah, E. (2012). Identification of differentially accumulated proteins associated with embryogenic and non-embryogeniccalli in saffron (*Crocus sativus* L). *Proteome Science, 10*, 3–18.

Shen, X. C., & Qian, Z.Y. (2006). Effects of crocetin on antioxidant enzymatic activities in cardiac hypertrophy induced by norepinephrine in rats". *Pharmazie., 61*(4), 348–352.

Srivastava, R., Ahmed, H., Dixit, R. K., Dharamveer, & Saraf, S. A. (2010). *Crocus sativus* L.: a comprehensive review. *Pharmacognosy Reviews, 4*(8), 200–208.

Tamaddonfard, E., Farshid, A. A., Eghdami, K., Samadi, F., & Erfanparast, A. (2013). Comparison of the effects of crocin, safranal and diclofenac on local inflammation and inflammatory pain responses induced by carrageenan in rats. *Pharmacological Reports, 65*(5), 1272–1280.

Tarantilis, P. A., & Polissiou, M. G. (2004). Chemical analysis and antitumor activity of natural and semi-natural carotenoids of saffron. *Acta. Horticulturae, 650*, 447–461.

Torelli, A., Marieschi, M., & Bruni, R. (2014). Authentication of saffron (*Crocus sativus* L.) in different processed, retail products by means of SCAR markers. *Food Control, 36*, 126–131.

Tsaftaris, A., Pasentzis, K., & Argitiou, A. (2009). Rolling circle amplification of genomic templates for inverse PCR (RCA-GIP): a method for 5 #- and 3#genome walking without anchoring. *Biotechnology Letters, 32*(1), 157–161.

Tsaftaris, A., Polidoros, A., Pasentsis, K., & Kalivas, A. (2007). Cloning, structural characterization, and phylogenetic analysis of flower MADS-box genes from crocus (*Crocus sativus* L.). *Scientific World of Journal*, 7, 1047–1062.

Tsatsaroni, E., & Liakopoulou-Kyriakides, M. (1995). Effect of enzymatic treatment on the dyeing of cotton and wool fibers with natural dyes. *Dyes pigment, 29*, 203–209.

Vahidi, H., Kamalinejad, M., & Sedaghati, N. (2002). Antimicrobial Properties of *Crocus sativus* L. *Iranic Journal of Pharmaceutical Research, 1*, 33–35.

Valijonovich, M. A. (2018). Creation of plantation *Crocus sativus* L. in the conditions of Uzbekistan. *Journal of Agricultural Science and Food Research, 9*(2), 1–5.

Wang, Z., Gerstein, M., & Snyder, M. (2009). RNA-seq: a revolutionary tool for transcriptomics. *Nature Reviews of Genetics, 10*, 57–63.

Wani, B. A., Hamza, A. K. R., & Mohiddin, F. A. (2011). Saffron: A repository of medicinal properties. *Journal of Medicinal Plants Research, 5*(11), 2131–2135.

Xu, G. L., Li, G., Ma, H. P., Zhong, H., Liu, F., & Ao, G. Z. (2009). Preventive effect of crocin in inflamed animals and in LPS-challenged RAW 264.7 cells. *Journal of Agricultre and Food Chemistry*, *57*(18), 8325–8330.

Yao, H., Song, J. Y., Ma, X. Y., Liu, C., Li, Y., Xu, H. X., Han, J. P., Duan, L. S., & Chen, S. L. (2009). Identification of Dendrobium species by a candidate DNA barcode sequence: the chloroplast psbA-trnH intergenic region. *Planta Medica*, *75*, 667–669.

Yilmaz, A., Nyberg, N. T., Molgaard, P., Asili, J., & Jaroszewsk, J. W. (2010). HNMR metabolic fingerprinting of saffron extracts. *Metabolomics*, *6*, 511–517.

Zalacain, A., Ordoudi, A. S., Blazquez, I., Diaz-olaza, E. M., Carmona, M., & Tsimldou, M. Z. (2005). Screening method for the detection of artificial colours in saffron using derivative UV–vis spectrometry after precipitation of crocetin. *Food and Additional Contamination*, *22*, 607–615.

Zougagh, M., Simonet, B. M., Rios, A., & Valcarcel, M. (2005). Use ofnonaqueous capillary electrophoresis for the quality control of com-mercial saffron samples. *Journal of Chromatography A.*, *1085*, 293–298.

CHAPTER 8

Metabolic engineering for the production of plant therapeutic compounds

Mauji Ram, Himanshu Misra

Greentechnology Department, Ipca Laboratories Ltd., Sejavta, Ratlam, Madhya Pradesh, India

List of Abbreviations

IDP isopentenyl diphosphate
MEP 2-C-methyl-D-erytritol 4-phosphate
DXP 1-deoxy-D-xylulose-5-phosphate
DXS 1-deoxy-D-xylulose-5-phosphate synthase
DXR 1-deoxy-D-xylulose-5-phosphate reducto-isomerase
ABA abscisic acid
GA gibberellic acid
HDR hydroxyl-methyl-butenyl diphosphate reductase
DMADP dimethylallyl diphosphate
TXS taxadiene synthase
GDP geranyl diphosphate
FDP farnesyl diphosphate
MVA mevalonic acid
HMGR 3-hydroxy-3-methylglutaryl Coenzyme A
IPP isopentenyl pyrophosphate
DMAPP dimethylallyl pyrophosphate
MsLs *Mentha spicata* limonene synthase
CYP71AV1 cytochrome P450 monooxygenase
Dbr2 artemisinic aldehyde reductase
FaNES1 *Fragaria ananassa* nerolidol synthase1
TIAs terpenoid indole alkaloids
h6h hyoscyamine-6-hydroxylase
pmt putrescine N- methyl-transferase
PKSs polyketide synthases
SQS squalene synthase
4-HRAA 4-hydroxyphenylaldehyde
STR strictosidine synthase
STS stilbene synthase
PAL phenylalanine ammonia lyase
C4H cinnamate-4-hydroxylase
CHS chalcone synthase
4CL 4-coumarate: coenzyme A ligase.

Medicinal and Aromatic Plants: Expanding their Horizons through Omics
http://dx.doi.org/10.1016/B978-0-12-819590-1.00008-2

Introduction

Increasing the production of pharmacologically attractive natural products represents on the main targets for the genetic manipulation of medicinal plants. An increasing number of natural products are being biosynthesized in low quantities, with known examples being artemisinin, paclitaxel, podophyllotoxin, and *Vinca*-alkaloids (Fig. 8.1). The use of genetically modified plant cell cultures, such as hairy root cultures for Solanaceae (Baiza, Quiroz-Moreno, Ruiz, & Victor, 1999) or for artemisinin (Liu, Wang, Ouyang, Ye, & Li, 1998), offers a rational approach to allow the over-expression of genes encoding biosynthetic enzymes and to overcome the rate-limiting steps of the biosynthesis.

There are many speculations about how evolution diverge biosynthetic pathways (Pichersky and Gang, 2000). Metabolic engineering in plants involves the modification of endogenous pathways to increase flux toward particular desirable molecules. In some cases the aim is to enhance the production of a natural product, whereas in others it is to synthesize a novel compound or macromolecule. In additional, altering metabolic enzymes or pathways has become an important approach for investigating cell physiology (Farmer & Liao, 2001)

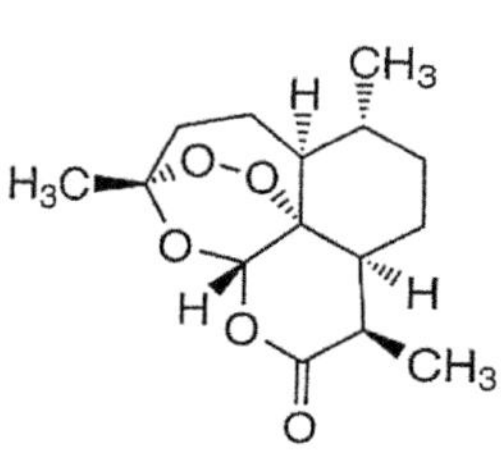

Artemisinin

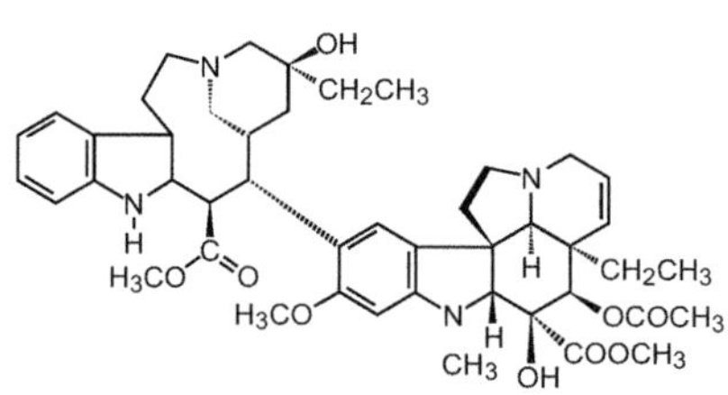

Vinblastine

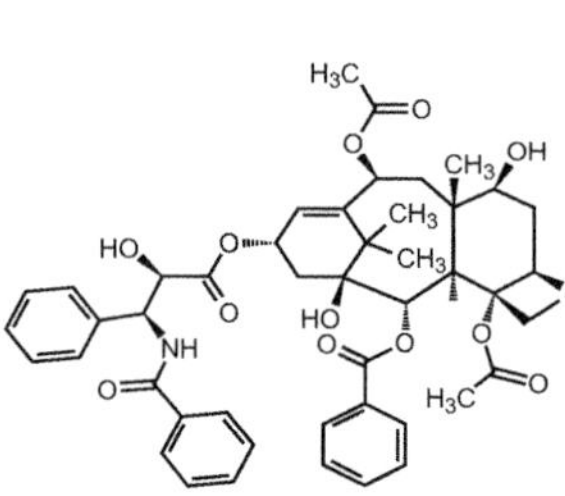

Paclitaxel

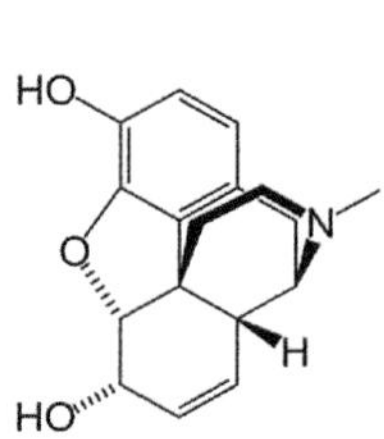

Morphine

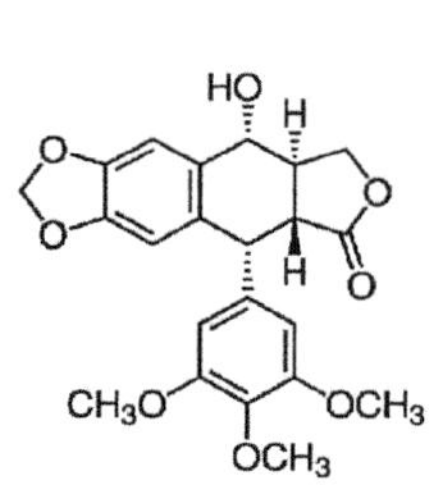

Podophyllotoxin

Figure 8.1 ***Medicinally important plant natural secondary metabolites subject to combinatorial metabolic engineering studies.***

Although 200,000 different secondary metabolites are estimated to occur in plant kingdom, they all arise from a rather limited number of chemical scaffolds divided into three major groups—terpenoids, alkaloids, and phenolic compounds, the last one being further divided into benzoates, central phenylpropanoids, and flavonoids. The enormous diversity is based on substrate and/or regio-specific enzymes that decorate the molecular backbones with different chemical modifications by, for example, hydroxylation, methylation, acylation and glucosylation (Dixon, 2001; Dixon, 2005). Genetic modification of all major classes of secondary metabolites has been described. The nitrogen-containing alkaloids (e.g., morphine, codeine, caffeine, nicotine, cocaine, and scopolamine) have been exploited as pharmaceuticals, stimulants, narcotics, and poisons. Many of the promising pharmaceuticals are still extracted from their natural plant sources. Alkaloid biosynthesis has also been the target of intensive metabolic engineering, mainly to increase the amount of the target compounds but also to conduct the whole biosynthesis in cell culture systems (Facchini, 2001). Terpenoids form the largest group of plant natural products, consisting of over 40,000 compounds. Like alkaloids, terpenoids have pharmaceutical activity and include several important drugs against human ailments such as cancer (taxanes) and malaria (artemisinin). Mass production of terpenoids by metabolic engineering is thus of great interest (Roberts, 2007). In addition, terpenes are used as flavor andcolor enhancers. Plants also use terpenoids in their communication in the environment, and metabolic engineering has been applied, for example, in *Arabidopsis* to produce a terpenoid that attracts natural enemies to herbivorous pests (Kappers et al., 2005). Some compounds in the terpenoid pathway, necessary for cellular function and maintenance, are classified as primary metabolites, that is, carotenoids, gibberellins, sterols and vitamins A and E (Roberts, 2007). Both qualitative and quantitative metabolic engineering of the carotenoid pathway has shown potential to enhance the nutritional quality of food. The "golden rice" was first developed in 2000 (Ye et al., 2000) and an improved modification with higher provitamin A content was introduced five years later (Paine et al., 2005) to solve one of the serious malnutrition problems of people in developing countries. Despite the fact that the outcome of genetic modification of carotenoid metabolism has often been unexpected it has given valuable information about the biosynthetic pathway (Sandmann, Römer, & Fraser, 2006; Galili, Galili, Lewinsohn, & Tadmor, 2002; Chen, Li, &

Liu, 2006). Indeed, a major hindrance in the engineering of secondary metabolism has been the lack of comprehensive understanding of the biosynthetic pathways. Similarly, knowledge of the regulation of the various pathways in the complex metabolic network is limited, and thus insertion of a single gene may have unpredictable effects on the metabolic balance of plant cellular systems (Dixon & Steele, 1999; Forkmann and Martens, 2001; Dixon, 2005). However, progress in molecular genetics and knowledge of plant metabolism will make targeted engineering of both primary (Carrari, Urbanczyk-Wochniak, Willmitzer, & Fernie, 2003; Zimmermann & Hurrell, 2002) and secondary (Dixon, 2005) metabolism increasingly feasible. Moreover, as more holistic approaches are taken by systems biology (i.e., integrated transcriptomics, proteomics, and metabolomics), "predictive metabolic engineering" process where models are tested to understand the regulatory network of the modified pathway —can be introduced as part of experimental design (Sweetlove, Last, & Fernie, 2003).

Recently, combinatorial metabolism emerges as a potential tool to commercial production of plant natural products. Combinatorial biosynthesis has been discussed for important classes of natural products, including alkaloids (vinblastine, vincristine), terpenoids (artemisinin and paclitaxel), and flavonoids (Julsing, Koulman, Herman, Wim, & Oliver, 2006). The main problem with combinatorial biosynthesis, however, is that most biosynthetic pathways are still poorly understood at the genetic level, with relatively few genes involved in regulation and biosynthesis in plants having been sequenced and functionally elucidated. Therefore, no complete biosynthetic pathway has been completely transferred to a heterologous host. Recent achievements with the polyketide biosynthesis from microorganisms, especially in *Streptomyces*, prove the potential of combinatorial biosynthesis (Hranueli, Cullum, Basrak, Goldstein, & Long, 2005; Moore, Kalaitzis, & Xiang, 2005; Weber, Welzel, Pelzer, Vente, & Wohlleben, 2003 & Pfeifer & Khosla, 2001).

Moreover, there are some natural products that cannot produce in vitro because of their high cost, energy requirement, and organism lacks the enzymatic machinery to perform a specific-chemical reaction. In other words, the biodiversity is endless and there are still possibilities to enlarge the diversity from a chemical point of view, by combining genes and products from different sources that in nature would never meet. This strategy will deliver compounds that are not influenced by selection pressures, by a habitat, or the biochemical limitation of an organism (such as compartmentalization or storage). These compounds can be selected for a specific pharmaceutical mode of action or an activity can be adjusted to a more specific pharmaceutical demand.

Pathway engineering in medicinal plants

The production of active phytochemical constituents is a well-established target for genetic manipulation but possess some severe challenges. In particular, the metabolic pathways by which active compounds are biosynthesized are mostly poorly understood, and relatively few genes for key enzymatic or regulatory steps have been isolated. However, there are many examples of pathway engineering leading to improvements of potential value in the breeding of medicinal plants (Ferreira and Duke, 1997; Charlwood & Pletsch, 2002). A ninefold enhancement in production of the sedative compound scopolamine in hairy-root cultures of *Hyoscyamus niger* (black henbane), has been described by simultaneously over-expressing two genes encoding the rate-limiting upstream and downstream biosynthetic enzymes (Zhang et al., 2004). Yun, Hashimoto, & Yamada (1992) increased the production of scopolamine in *A. belladona*, from the naturally occurring chemical precursor hyoscyamine, by transformation with the enzyme hyoscyamine 6 ß-hydroxylase from *Hyoscyamus*. Preliminary progress has been made toward engineering alkaloid production in *P. somniferum* (Facchini, Park, Bird, & Samanani, 2000). A threefold enhancement in production of the putative anti-malarial, anti-cancer agent artemisinin has been reported in transgenic *Artemisia* plants by over-expressing farnesyl diphosphate synthase, the enzyme immediately preceding the first committed biosynthetic step of artemisinin (St-Pierre, Vazquez-Flota, & De Luca, 1999; Chen, Ye, & Li, 2000). As an alternative to targeting an individual rate-limiting enzyme reaction, exploiting transcription factors that turn whole secondary pathways on or off shows great promise as a metabolic engineering strategy (Fig. 8.2) (Robbins et al., 2003). New genomic approaches and efficient gene isolation methods applied to difficult secondary metabolic pathways in medicinal plant metabolism will undoubtedly expand the range and precision of manipulations via transgenesis, providing potentially superior material for the breeder.

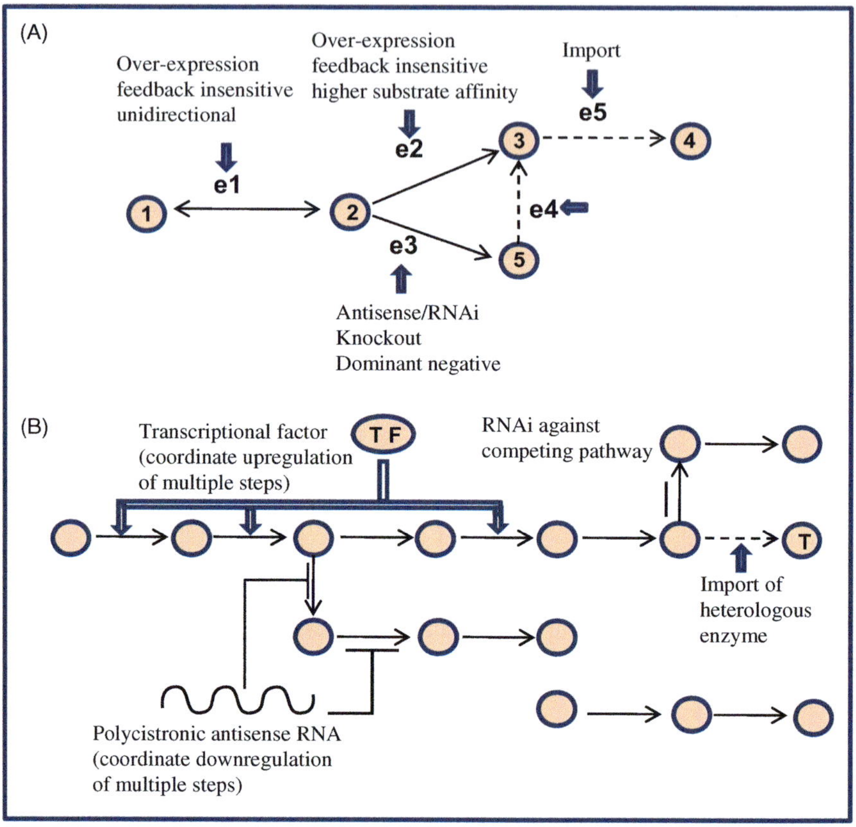

Figure 8.2 ***Approaches to plant metabolic engineering.*** (A) Strategies for single-step metabolic engineering. To increase production of a desired compound (*3*) or a novel compound (*4*) genes encoding endogenous enzymes in the biosynthetic pathway can be overexpressed (*e1, e2*) and novel enzymes can be imported (*e5*). Further increase in flux can be achieved by using feedback-insensitive enzymes or those modified to favor unidirectional catalysis in reversible reactions (*e1*). At branch points, over-expression of key enzymes or their modification to out-compete other enzymes using the same substrate (*e2*) can divert flux into appropriate pathways, while competing enzymes can also be inhibited directly (*e3*). If possible, competing pathways can be short-circulated by the import of enzyme shunts (*e4*) that redirect flux in the appropriate direction. *Solid arrows* represent endogenous metabolism and *broken arrows* represent imported (heterologous) metabolism. (B) Instead of modulating steps, the future of metabolic engineering will probably involve holistic approaches in which multiple steps are targeted simultaneously. In this complex pathway to a desired compound T, a transcription factor is being used to coordinately upregulate several enzymes, a polycistronic antisense RNA is being used to suppress several enzymes in a competing pathway, a novel enzyme is being expressed to extend the endogenous pathway beyond its normal end point, and other competing pathway are being suppressed by RNAi. *Solid arrows* represent endogenous metabolism and the *broken arrow* represents imported (heterologous) metabolism. *Part (A) Robbins et al., 2003*

Metabolic engineering of terpenoids

Engineering precursor availability

The availability of precursors is an important issue in metabolic engineering. Whether the concentration of a certain isoprenoid precursor is limiting for the production of terpenoids probably depends on the plant species, the tissue and the physiological state of the plant. Theinitial step toward the biosynthesis of IDP (isopentenyl diphosphate) through the MEP pathway is catalyzed by 1-deoxy-D-xylulose-5-phosphate (DXP) synthase (DXS). Over-expression and down-regulation of DXS in Arabidopsis affected the levels of various isoprenoids including chlorophylls, tocopherols, carotenoids, ABA, and GAs (Estevez et al., 2001 Estevez, Cantero, Reindl, & Leon, 2001). Up-regulation of DXP reductoisomerase (DXR), which converts DXP to methylerythritol phosphate, in transgenic peppermint (*Mentha piperita*) plants resulted in a 50% increase in essential oil yield (Mahmoud & Croteau, 2001). As well as DXS and DXR, hydroxymethylbutenyl diphosphate reductase (HDR), the enzyme conducting the last step in the pathway generating IDP and DMADP, has also been suggested to be a rate-limiting step in the MEP pathway in both tomato (*Lycopersicon esculentum*) and *Arabidopsis* (Botella-Pavia et al., 2004). When plants expressing taxadiene synthase (TXS) (Engineering diterpenoids and triterpenoids) under the cauliflower mosaic virus (CaMV) 35S promoter (Besumbes et al., 2004) were crossed with plants over-expressing DXS or HDR, increases of 6.5 times and 13 times, respectively, were detected in the accumulation of taxadiene compared with plants over-expressing only TXS (Botella-Pavia et al., 2004). The results of many studies reported to date suggest that, in general, the direct precursor for monoterpene biosynthesis (i.e., GDP) is largely available to introduced monoterpene synthases and, in some cases, does not seem to be limiting. For example, Lucker et al. (2004) reported that, in the flowers of transgenic tobacco (*Nicotiana tabacum*) plants expressing three different monoterpene synthases, the levels of products corresponding to the three enzymes were high but did not affect the level of the endogenous linalool production. Enhancing the precursor supply for the biosynthesis of isoprenoids in the mevalonate pathway has been attempted mainly by altering expression levels of the gene encoding 3-hydroxy-3-methylglutaryl-CoA reductase (HMGR), which is considered to be a rate-limiting step in the pathway (Chappell, Wolf, Proulx, & Saunders, 1995; Harker, Hellyer, John, Lanot, & Safford, 2003; Schaller et al., 1995). Constitutive over-expression of HMGR in tobacco resulted in a three- to tenfold increase in total-sterol levels (Chappell et al., 1995). Levels of other isoprenoids, including sesquiterpenes, were not altered, possibly because of compartmentation, channelling or the presence of other rate-limiting steps. In a study, overproduction of FDP synthase in *Artemisia annua* resulted in 2-3 fold increases in levels of the artemisinin (sesquiterpenes) in transgenic plants compared with those in wild-type plants (Chen et al., 2000).

Engineering monoterpenoids

The first reports of the feasibility of altering the monoterpene profile of plants focused on mint species (*Mentha* spp.) ((Table 8.1). Early work with the aim of over-expressing the gene from spearmint (*Mentha spicata*) encoding limonene synthase (MsLS) in peppermint (*Mentha piperita*) reported stable integration of the gene, but the terpene profile of the plants was not clearly altered (Krasnyanski, May, Loskutova, Ball, & sink, 1999). Later, transgenic peppermint and cornmint (*Mentha arvensis*) plants over-expressing MsLS were reported to have quantitative changes in levels of monoterpenes (Diemer, Caissard, Moja, Chalchat, & Jullien, 2001). Levels of several monoterpene pathway components were altered, including 1, 8-cineole and ß-ocimene (one-step enzymatic products from GDP), and monoterpene end products such as pulegone or piperitone (formed from limonene via isopiperitenone). The effects were caused not only by over-expression but probably also by cosuppression resulting from the introduction of a homologous gene. The over-expression of the gene encoding the *Clarkia breweri* S-linalool synthase (Lis), a heterologous monoterpene synthase, indeed resulted in monoterpene production in petunia (Lucker et al., 2001; Dudareva, Cseke, Blanc, & Pichersky, 1996). However, they also showed that the newly formed linalool is efficiently converted to the non-volatile S-linalyl-ß-D-glucopyranoside, probably by the action of an endogenous glucosyl transferase. In tobacco, Lucker et al. (2004) achieved substantial production of three new monoterpene products [γ-terpinene, (+)-limonene and (-)-ß-pinene] by introducing the three corresponding lemon (*Citrus limon*) monoterpene synthases into a single plant. The products were emitted by the leaves as well as the flowers of the transgenic plants. A model plant such as *Arabidopsis* would be extremely helpful for evaluating metabolic engineering strategies. Flowers of *Arabidopsis* produce many mono- and sesquiterpenes, whereas its leaves produce only trace amounts of limonene and ß-myrcene (Chen et al., 2004; Aharoni et al., 2003 & Chen et al., 2003). To evaluate the potential of Arabidopsis to produce monoterpenes and sesquiterpenes, Aharoni et al. (2004) used the strawberry (*Fragaria ananassa*)

Table 8.1 Reports of metabolic engineering of terpenoids, alkaloids, polyketids, and flavonoids.

Compounds class	Engineered species	Target	Subcellular location	Regulation	Altered profile[a]	References
Terpenoids	*A. annua*	hmgr	Cytosol	Constitutive	Artemisinin ↑	Aquil et al. (2009)
	A. annua	sqs	Cytosol	Constitutive	Artemisinin ↑	Zhang et al. (2009)
	A. annua	FDP synthase	Cytosol	Constitutive	Artemisinin ↑	Han et al. (2006)
	A. annua	ipt	Cytosol	Constitutive	Artemisinin ↑	Sa et al. (2001)
	Chicory	cyp71av1	Cytosol	Constitutive	Artemisininc acid ↑	de Kraker et al. (2003)
	Yeast	cyp71av1	Cytosol	Constitutive	Dihydroartemisinic acid↑	Chang et al. (2007)
	Mint	Limonene synthase	Plastid	Constitutive	Several products[b] ↑↓	Diemer et al. (2001)
	Mint	Menthofuran synthase	ER	Constitutive	Menthofuran ↓, pulegone ↓, menthol ↑	Mahmoud & Croteau, (2001)
	Mint	Limonene hydroxylase	ER	Constitutive	Limonene ↑, menthofuran ↓, isomenthone ↓, menthol ↓	Mahmoud et al. (2004)
	Tobacco	Limonene synthase	Cytosol	Constitutive	Limonene ↑	Lucker et al. (2004)
	Tobacco	Limonene synthase	Plastid	Constitutive	Limonene ↑	Ohara et al. (2003)
	Tobacco	Trichodiene synthase	Cytosol	Constitutive	Trichodiene ↑, oxygenated trichodiene[c] ↑	Hohn et al. (1991)
	Tobacco	Amorpha-4,11–diene synthase	Cytosol	Constitutive	Amorpha-4,11–diene ↑	Wallaart et al. (2001)
	Tobacco	Cembratrien-ol hydroxylase	ER	Constitutive	Cembratriene-diol ↓	Wang et al. (2001)
	Tobacco	Limonene hydroxylase	ER	Constitutive	Isopiperitenol and derivatives↓	Lucker et al. (2004)
	E. coli	Paclitaxel synthase	Cytosol	Constitutive	Paclitaxel ↑	Huang et al. (2001)

(*Continued*)

Table 8.1 Reports of metabolic engineering of terpenoids, alkaloids, polyketids, and flavonoids. (*Cont.*)

Compounds class	Engineered species	Target	Subcellular location	Regulation	Altered profile[a]	References
Alkaloids	*A. belladonna*	H6H	Cytosol	Constitutive	Scopolamine ↑	Yun et al. (1992)
	Datura	PMT/TRP I/ TRP II/H6H	Cytosol	Costitutive	Scopolamine ↑	Palazon et al. (2008)
	C. roseus	str/tdc	Plastid	Constitutive	Several alkaloids ↑	Whitmer (2000)
	C. roseus	DXS/G10H, ORCA3	Plastid	Constitutive	Tepenoid indole alkaloids↑	Peebles (2009)
Polyketides	*S. colicolor*	PKS	Cytosol	Constitutive	Deoxyerythromycin B ↑	Fong et al. (2007)
	S. erythraea	vhb	Cytosol	Constitutive	Erythromycin A ↑	Minas et al. (1998)
	E. coli	DEBS	Cytosol	Constitutive	Deoxyerthronolide B ↑	Pfeifer et al. (2002)
Flavonoids	*E. coli*	PAL/4CL/CHS	Cytosol	Constitutive	Several flavonones ↑	Hwang et al. (2003)
	E. coli	PAL/C4H/CHS	Cytosol	Constitutive	Naringenin ↑ pinocenbrim ↑ coumaric acid ↑	Kaneko et al. (2003)
	E. coli /Yeast	TAL	Cytosol	Constitutive	p-coumaric acid ↑	Vannelli et al. (2007)
	E. coli	MatB	Cytosol	Constitutive	flavanones ↑	Leonard et al. (2007)

HMGR, hydroxymethylglutaryl CoA reductase; *FDP*, farnesyl diphosphate; *ipt*, isopentenyl 5-phosphate transferase; *SQS*, squalene synthase; cyp71av1, cytochrome P450 monooxygenase; *H6H*, hyoscyamine-6-hydroxylase; *PMT*, putrescine N- methyltransferase; *TRP I*, tropinone reductase I; *TRP II*, tropinone II; *str*, strictosidine synthase; *tdc*, tyrosine decarboxylase; *DXS*, 1-deoxy-D-xuylulose 5-phosphate synthase; *G10H*, geraniol-10-hydroxylase; *PKS*, polyketid synthase; *vhb*, Vitreoscilla hemoglobin; *DEBS*, deoxyerythronolide B synthase ; *PAL*, phenylalanine ammonia lyase; *4CL*, 4-coumarate: coenzyme A ligase; *CHS*, chalcone synthase; *C4H*, cinnamate-4-hydroxylase; *TAL*, tyrosine ammonia lyase; *MatB*, manolyl-CoA synthetase.

[a] Compounds that have been reduced (↓) or increased (↑) in amount.

[b] Overexpression sometimes resulted in cosuppression. Therefore, in different plants, levels of certain compounds were up- or downregulated.

[c] Formation of oxygenated trichodiene was detected after induction by an elicitor.

gene FaNES1 (Nerolidol synthase 1). This gene encodes a dual-function monoterpene and sesquiterpene synthase that catalyzes the formation of both linalool from GDP and nerolidol from FDP with equal efficiency, and hence is called a linalool/nerolidol synthase. Leaves of transgenic Arabidopsis plants constitutively over-expressing FaNES1 produced free, hydroxylated and glycosylated linalool derivatives (Aharoni et al., 2003). The newly formed monoterpene alcohol was converted to various derivatives by endogenous enzymes, probably hydroxylases and glycosyl transferases. In chrysanthemum (*Chrysanthemum grandiflorum*), transformation with the same construct resulted in the emission of large amounts of unmodified linalool. The different profiles of linalool derivatives identified in all these transgenic plants illustrate the dramatic influence of the plant genetic makeup on the outcome of metabolic engineering (Aharoni et al., 2003).

Engineering sesquiterpenoids

Production of sesquiterpenes in transgenic plants is a more challenging task than generating monoterpenes due to unavailability of precursors. Tobacco plants have been transformed with a fungal trichodiene synthase (Hohn & Ohlrogge, 1991) and with the amorpha-4, 11-diene synthase of *A. annua* (Wallaart, Bouwmeester, Hille, Poppinga, & Maijers, 2001) (Table 8.1). In both cases, only low levels of the expected sesquiterpenoids could be detected. The levels of trichodiene in cell suspension cultures of the transgenic tobacco expressing trichodiene synthase could be increased tenfold by elicitation with cellulase (Zook, Johnson, Hohn, & Hammerschmidt,). The elicited transgenic suspension culture also accumulated an oxygenated trichodiene derivative. Expressing strawberry FaNES1 in *Arabidopsis* using plastidic targeting resulted in the production of small amounts of the sesquiterpene nerolidol (Aharoni et al., 2003). This was unexpected because it is generally assumed that sesquiterpenes are only produced in the cytosol but this shows that FDP is also available in the plastids. However, mitochondria are also involved in isoprenoid biosynthesis (they are the site of ubiquinone biosynthesis) and *Arabidopsis* has an FPP synthase with a mitochondrial-targeting signal, they have also targeted FaNES1 to the mitochondria (Kappers et al., 2005). The transgenic plants indeed produced 3(S)-E-nerolidol (and no linalool), confirming that FDP is also present and available in the mitochondria. The nerolidol produced was partly converted, by endogenous Arabidopsis enzymes, to the C11 homoterpene 4, 8-dimethyl-1, 3(E), 7-nonatriene [(E)-DMNT]. The enzyme catalyzing the first step in Taxol biosynthesis, TXS (Jennewein & Croteau, 2001), has been produced in

Arabidopsis (Besumbes et al., 2004). When an inducible promoter was used, taxadiene levels in Arabidopsis leaves reached ~0.6 μg g^{-1} (dry weight) (Besumbes et al., 2004), which is 10–60 times higher than the levels obtained with sesquiterpene synthases in tobacco (Wallaart et al., 2001) but 1000 times lower than obtained with monoterpenes in Arabidopsis (Aharoni et al., 2003). With regard to triterpenes, there are no examples of metabolic engineering yet, with the exception of the above-described enhancement of sterol formation caused by increasing precursor availability. In *Artemisia annua*, the artemisinin content was enhanced 2-3 fold by developing transgenic lines. In this plant, the expression of genes such as FPS and SQS were modulated (Han et al., 2006; Zhang et al., 2009). In our lab HMGR gene from *Catharanthus roseus* was cloned and over-expressed in *Artemisia annua*, 22.5% enhancement in artemisinin content was recorded (Aquil, Husaini, Abdin, & Rather, 2009). It is unclear whether no attempts were made to modify triterpene metabolism or whether these attempts were unsuccessful. Indeed, the fact that triterpenes are also cytosolic products and derived from FDP suggests that just as for sesquiterpenoids engineering is not straightforward.

Engineering alkaloid biosynthesis

Alkaloids: a large family of plant pharmaceutical compounds and is a diverse group of small, heterocyclic, nitrogen-containing molecules that are thought to be involved in defending plants against herbivores and pathogens. Several alkaloids are exploited for their pharmaceutical properties. For example, morphine and codeine are analgesics, vinblastine, vincristine and taxol are anti-cancer agents, scopolamine widely used in medicine for its anticholinergic activity, colchicine is a gout suppressant, tubocurarine is a muscle relaxant and sanguinarine is an antibiotic. Most of these alkaloids are extracted from plants or are derived either chemically or via biotransformation from natural precursors. There is an increasing demand from the pharmaceutical industry for anti-tumor agents like taxol and vinblastine, and for nicotine for the fabrication of gums and patches designed to help with tobacco weaning. Alkaloid biosynthesis pathways are often more complex than the flavonoid pathway and, at the moment, only a few structural genes from the tropane and benzylisoquinoline alkaloid pathways have been isolated (Facchini, 2001; Memerlink. Verpoorte, & Kijne, 2001). To date, the best-known alkaloid biosynthetic pathway at the gene level leads to the formation of terpenoid indole alkaloids (TIAs) in periwinkle (*Catharanthus roseus*) (Gantet & Memelink, 2002; Fig. 8.3).

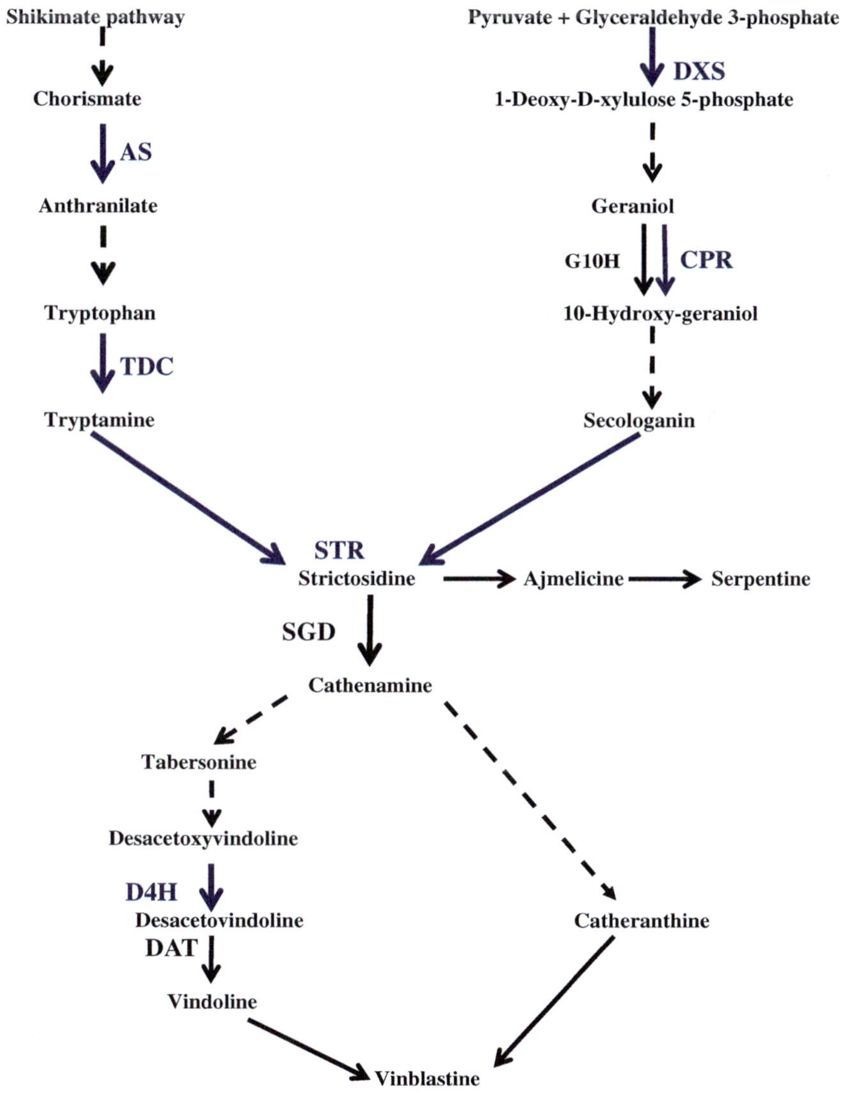

Figure 8.3 ***Regulation of terpenoid indole alkaloid biosynthesis in*** **Catharanthus roseus** ***cells by ORCA3*** . *Solid arrows* indicate a single enzymatic step and *dashed arrows* indicate multiple enzymatic steps. *Blue arrows* indicate that the gene encoding the enzyme is regulated by ORCA3. Abbreviations: *AS*, anthranilate synthase; *CPR*, cytochrome P450-reductase; *DAT*, deacetylvindoline acetyltransferase, *D4H*, desacetoxyvindoline 4-hydroxylase, *DXS*, D-1-deoxyxylulose 5-phosphate synthase; *G10H*, geraniol-10-hydroxylase; *ORCA3*, octadecanoidresponsive Catharanthus AP2-domain protein 3; *SGD*, strictosidine β-D-glucosidase; *STR*, strictosidine synthase; *TDC*, tryptophan decarboxylase. *(Gantet & Memelink, 2002)*

This plant species produces the monomeric alkaloids serpentine and ajmalicine, which are used as a tranquilizer and to reduce hypertension, respectively. Dimeric alkaloids from periwinkle, vincristine and vinblastine, and their hemisynthetic derivatives, including vinorelbine and vinflunine, are used extensively in the treatment of many cancers. Dimeric alkaloids are present at very low levels in the plant and are restricted to specific leaf cell types (St-Pierre et al., 1999). For this reason the more abundant monomeric precursors, catharanthine and vindoline, are isolated and coupled chemically to form dimeric vinblastine. But even monomeric alkaloid levels are low in plants. Palazon, Ocana,Vazquez, & Mirjalili (2008) found that the scopolamine content of *A. belladonna* was relatively high in the seedling and vegetative stages but progressively decreased toward the flowering stage and in the same time, the total alkaloid content had considerably increased. They reported that by over-expressing the genes (PMT, tropinone reductase and h6h) of scopolamine biosynthesis, up to four fold enhancement in scopolamine production was recorded (Palazon et al., 2008; Fig. 8.4).

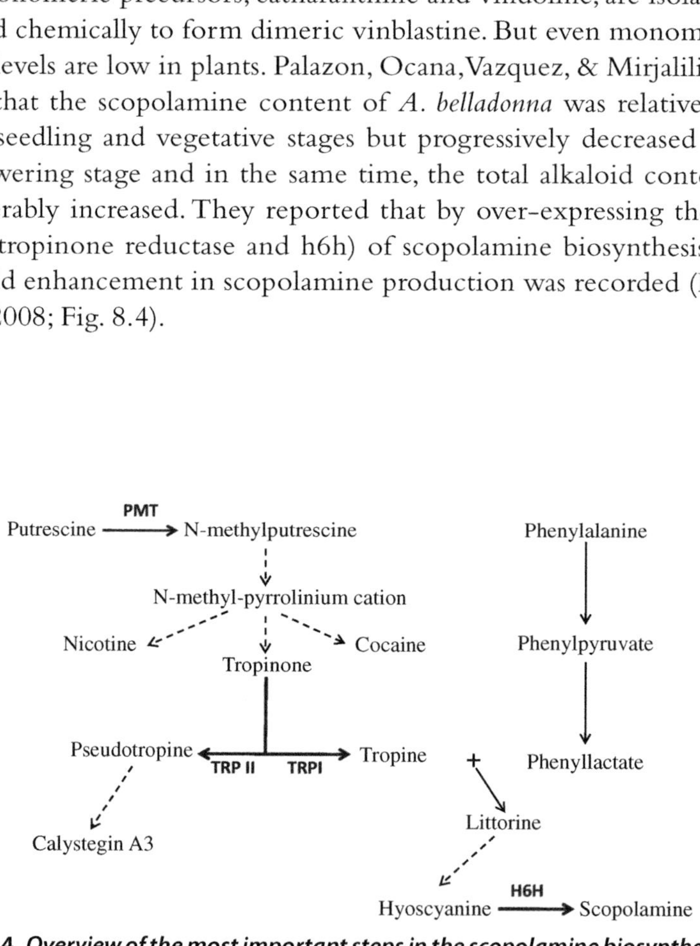

Figure 8.4 ***Overview of the most important steps in the scopolamine biosynthetic pathway.*** The enzymes over-expressed in scopolamine-producing hairy root cultures are in *blue. Broken arrows* indicate multiple steps. *PMT,* Putrescine N-methyltransferase; *TRP I,* Tropinone reductase I; *TRP II,* Tropinone reductase II, *H6H,* hyoscyamine -6ß-hydroxylase. *(Palazon et al., 2008)*

Metabolic engineering of phenolic compounds

Production of new antibiotics by manipulating genes encoding enzymes for a single pathway

The thioesterase activity of polycyclic polyketide synthases (PKSs) catalyzes the termination of the growing polyketide chain by cyclising the molecule through formation of a lactone ring. Repositioning the thioesterase gene of 6-deoxyerythronolide B synthase from the end of module 6 to the end of module 2 terminates the growing polyketide chain after two rounds of acid addition (Cortes et al., 1995). The resulting product is a triketide six-member lactone ring instead of the normal 14-member lactone ring of 6-deoxyerythronolide B. The result substantiates the earlier suggestion that the "multienzyme components of the modular PKSs are capable of functioning independently" (Cortes et al., 1995). Furthermore, the observation suggests a general approach to genetically control the length of polyketide chains and thereby the sizes of the macrocyclic polyketide ring.

Rational design of new antibiotics by mixing genes from different pathways

New polycyclic polyketides can be generated by recombining the genes for the biosynthesis of different polyketides. Results from recombining genes encoding several different metabolic pathways provided information to formulate a set of rules for the design of polycyclic polyketides (McDaniel, Ebert-Khosla, Hopwood, & Khosla, 1995). Chain lengths of the polyketide chains are determined by the minimal PKS. The genes for minimal PKSs and the corresponding chain lengths are *act* for a 16-carbon octaketide, *fren* for an 18-carbon nonaketide, *tcm* for a 20-carbon decaketide, and *whiE* for a 24-carbon dodecaketide. Only the genes for the ketosynthase and chain length factor need to come from the same PKS gene cluster because the acyl carrier protein does not affect chain length. The *act* ketoreductase catalyzes the reduction of the polyketide chain assembled by all of the minimal PKSs indicated earlier. Normally, the *act* ketoreductase catalyzes the reduction of the C9 carbonyl; however, C7 reductions can occur sometimes. Reduction of the ketone at C7 indicates that the first ring cyclization occurs between C7 and C9. Reduction of the ketone at C7 indicates that the first ring cyclization occurs between C5 and Cl0. First ring aromatization depends on the source of the aromatase because aromatases from different PKS gene clusters catalyze the reaction with different polyketide chain length substrates (Charles, 1996). Two groups of aromatases have been recognized

(Alvarez, Fu, Khosla, Hopwood, & Bailey, 1996). The first group includes aromatase from *act*, *gra*, *frem* and *gris*. The proteins exhibit sequence similarity between the first and second half of the amino acid sequence. The second group includes aromatases from *whiE*, *tcm*, *sch* and *cur* and does not contain a duplication of similar amino acid sequence. The first group aromatizes the first ring of the reduced polyketide chains and favours cyclization between C7 and C12. The second ring of polyketide chains that are not reduced and favor cyclization between C9 and C14 (Alvarez et al., 1996).

Role of transcription factors in metabolic engineering

Early work on pathway transcription factors in plants began with the discovery of the maize flavonoid pathway regulators COLORLESS1 (C1) and RED (R) (Latchman, 2003; Goff et al., 1990). Within a few years of their discovery, C1 and R were shown to induce flavonoid gene expression and anthocyanin accumulation in transgenic plants (Goff et al., 1990). These pioneering studies demonstrated the potential of transcription factors for the manipulation of complex metabolic pathways in plants. Fuelled by the advent of genomics, our understanding of mechanisms that govern the transcriptional regulation of metabolic pathways is expanding with increasing speed. Discovery of the transcription factors involved is accelerating (Table 8.2), and more experiments aim to exploit them to modulate the production of targeted metabolites.

The traditional approach to engineering plant metabolic pathways has been to target single slow or regulated steps that may limit pathway flux. Although this approach can be successful (Mahmoud & Croteau, 2001; Falco et al., 1995), identifying rate-limiting steps is often difficult (Dong, Braun, & Grotewold, 2001) and the level of end-product accumulation can be limited by more than one enzymatic activity. For example, greatly increased phytoene synthase activity in transgenic tomato plants that over-expressed the bacterial gene CRTB did not translate into a proportional elevation of carotenoid content, suggesting that one or more additional pathway steps may be limiting (Fraser et al., 2002). Furthermore, whereas the over-expression of one gene was sufficient to greatly induce beta-carotene accumulation in canola seeds (Shewmaker, Sheehy, Daley, Colburn, & Ke, 1999), three different genes had to be over-expressed to produce high beta-carotene "golden rice" because low levels of early pathway enzyme activities had to be supplemented (Ye et al., 2000). In such cases, an alternative may be to target transcriptional regulators of the pathway rather than pathway genes

Table 8.2 Transcription factors that control secondary metabolism in plants.

Plant transcription factor	Metabolite class	Mammalian homolog[a]	Function in mammals	DNA-binding domain
C1	Anthocyanins	c-MYB	Cell cycle	MYB[b]
P	Phlobaphenes			
TT2	Condensed Tannins			
PAP1	Anthocyanins			
AtMYB4	Sinapate Esters			
CrBPF1	Alkaloids			
R	Anthocyanins	c-MYC	Cell cycle	bHLH
TT8	Condensed Tannins			
CrMYC2	Alkaloids			
ORCA2	Alkaloids	None		AP2/ERF
ORCA3	Alkaloids			
CrGBF1	Alkaloids	CREB	Long-term memory, T-cell development, blood pressure	bZIP
CrGBF2	Alkaloids			

AP2/ERF, APETALA2/ethylene responsive factor; *At*, *Arabidopsis thaliana*; *bHLH*, basic helix-loop-helix; *bZIP*, basic leucine zipper; *Cr*, *Catharanthus roseus*; *CrBPF1*, priwinkle homolog of the MYB-like transcription-factor BPF1; *CREB*, cAMP response –element binding protein; *CrGBF*, G-box-binding factor; *CrMYC2*, MYC-type basic helix-loop-helix Transcription factor; *ORCA*, octadecanoid-responsive Catharanthus AP2-domain protein; *PAP1*, production of anthocyanin pigment 1; *TT*, transparent testa.

[a] Because there are many mammalian homolog for each class, only one example is given. This corresponds to the first mammalian homolog identified for each plant transcription factor family.

[b] The DNA-binding domains of mammalian MYB proteins contain three repeats. In plants, most MYB proteins contain two repeats although a minority, including CrBPF1, contain a single repeat.

themselves. The potential benefit of using transcription factors to modify flux through a metabolic pathway was highlighted by two recent studies that aimed to increase the concentration of health-beneficial flavonoids in tomato. In one experiment, Muir et al. (2001) constitutively over-expressed CHI, which encodes chalcone isomerase, an early flavonoid pathway enzyme that is expressed at low levels in tomato fruit. In a second experiment, Bovy et al. (2002) over-expressed the maize anthocyanin regulators Leafcolor (Lc) and C1 in a fruit-specific manner. As anticipated, each of these approaches resulted in increased flux through the flavonoid pathway in the fruit, although rather unexpectedly, fruit flavonols rather than anthocyanins increased. In a striking contrast, over-expression of CHI enhanced flavonol production only in the peel, whereas Lc and C1 caused an increase throughout the fruit. The reason for this discrepancy was that the genes that encode pathway enzymes upstream and downstream of CHI are active in the peel but not in the flesh. Together, these results underline that, in contrast to a pathway gene, a pathway activator can induce the accumulation of metabolites in a tissue in which most of the relevant enzymatic activities are normally insufficient. Before this recent tomato study, over-expression of R or of Lc and C1 and some of their orthologues had been shown to strongly induce anthocyanin accumulation in several species (Grotewold et al., 1998; Goldsbrough, Tong, Yoder, & Tong, 1996; Mooney et al., 1995 & Lloyd, Walbot, & Davis, 1992). These results demonstrated the potential of using transcription factors to increase flux through complex plant pathways. This potential was confirmed by experiments involving other pathway activators, in particular those controlling other branches of the flavonoid pathway. For example, P, which is responsible for the accumulation of red phlobaphene pigments in the pericarp of maize kernels, was over-expressed in maize suspension cells and shown to induce the accumulation of C-glycosyl flavones and flavan-4-ols (Grotewold et al., 1998). The recent discovery of the proanthocyanidin (condensed tannin) pathway gene regulators TRANSPARENT TESTA2 (TT2) and TT8 has opened the possibility of manipulating this branch of the flavonoid pathway (Nesi, Jond, Debeaujon, Caboche, & Lepiniec, 2001; Nesi et al., 2000). Over-expression of TT2 was recently shown to be sufficient to induce the accumulation of condensed tannin in tissues in which TT8 is expressed (Nesi et al., 2001). This result suggests that concurrent over-expression of TT2 and TT8 could be the key to controlling proanthocyanidin accumulation in other tissues. In a study targeted at yet another branch of the pathway, Yu et al. (2003) exploited the effect of R and C1 on early flavonoid pathway gene expression in soybean. They used CRC, a fusion of C1 and R, rather than separate gene constructs to boost substrate availability for isoflavone synthase (IFS), which catalyses the first committed step into isoflavonoid biosynthesis. To divert flux away from anthocyanins and flavonols, the pathway was further engineered by silencing the gene that encodes flavanone 3-hydroxylase. Remarkably, the resulting transgenic plants produced four times more isoflavone in seed than did control plants (Yu et al., 2003). In related studies, a combination of transcriptional activator and IFS over expression formed part of strategies that led to the production of isoflavones in non-legume species (Yu et al., 2000). Recently Ma et al. (2009) isolated and characterized AaWRKY1an *Artemisia annua* transcription factor that regulate the expression of amorpha-4, 11-dinene synthase a key regulatory gene of artemisinin biosynthetic pathway. They also demonstrated that transient expression of AaWRKY1 activate the expression of genes (hmgr, ads, cyp71av1, and dbr2) of artemisinin biosynthesis pathway.

RNAi (RNA interference)-mediated gene silencing

Using antisense gene for blocking activity of an enzyme is widely accepted, but in case of enzymes encoded by multigenes, it sometimes fails to block the activity. RNAi technology provided an alternative to block the activity of such enzymes that are not only encoded each by a multigene family but are also expressed across a number of tissues and developmental stages, as was the case with codeinone reductase involved in the biosynthesis of morphinan alkaloid (Larkin et al., 2007).

RNAi (RNA interference)-mediated by hpRNA (hairpin RNA) has been used in gene silencing in many species of plants. Liu, Blount, Steele, & Dixon (2002) reported that hpRNA-mediated down-regulation of ghSAD-1 and ghFAD2-1, two-key enzymes in the fatty-acid-biosynthesis pathway in cotton (*Gossypium hirsutum*), elevated the stearic acid content (44% compared with a normal level of 2%), and oleic acid content (77% compared with a normal level of 15%) in cotton seeds. It was also reported that suppression of one key enzyme, CaMXMT1, involved in the caffeine-biosynthetic pathway through hpRNA-mediated interference in coffee (*Coffea* spp.), decreased obromine and caffeine accumulation efficiently (30%–50% of that normally found in the species) (Ogita, Uefuji, Morimoto, & Sano, 2004). In a study by Allen et al. (2004), a hpRNA construct containing sequences from multiple cDNAs of genes in the codeine reductase gene family in the opium poppy (*Papaver somniferum*) was used to silence these enzymes in the pathway. In the transgenic plants, the non-narcotic alkaloid (*S*)-reticuline, which occurs upstream of codeine in the pathway, accumulated at the expense of morphine, codeine, opium and thebaine. There was another example in tomato [*Solanum lycopersicum* (formerly *Lycopersicon esculentum*] (Davuluri et al., 2005), the hp (hairpin) construct was used to suppress an endogenous photomorphogenesis regulatory gene, DET1, driven by a fruit-specific promoter. DET1 was degraded and the carotenoid and flavonoid content were increased while all other traits for fruit quality remained unchanged in transgenic plants compared with that in wild-type tomato. In a recent study, artemisinin content was enhanced upto 2–3 fold

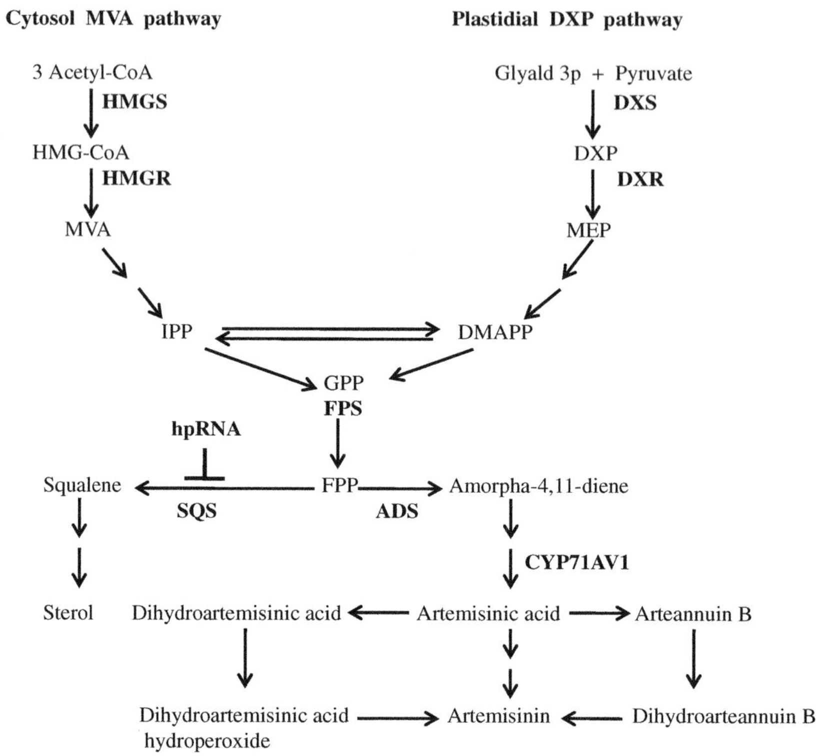

Figure 8.5 ***RNAi mediated suppression of SQS gene.*** *(Zhang et al., 2009).*

in transgenic *A. annua* plants by suppressing the expression of SQS (squalene synthase), a key enzyme of sterol pathway (a pathway competitive with that of artemisinin biosynthesis) by using hairpin-RNA-mediated RNAi (RNA interference) technique (Zhang et al., 2009) (Fig. 8.5). Thus, in recent years, RNAi technology has become an important tool for accelerating the breeding of aromatic and medicinal plants, where a conventional mutation breeding approach failed (Allen et al., 2004).

Combinatorial metabolic engineering for the production of plant therapeutic compounds

Combinatorial metabolism of terpenoids

Naturally occurring terpenoids are produced in small quantities, and thus, their purification results in low yields. Further, the complex structure of these molecules makes chemical synthesis challenging and often uneconomical due to poor yields. Metabolic engineering of these pathways in a common industrial biological host (*E. coli*) and (*Saccharomyces cerevisiae*) offers an attractive alternative to extractions from plants or chemical synthesis for producing large quantities of these complex molecules.

Metabolic engineering of artemisinin biosynthetic pathway

Application of combinatorial metabolic engineering for the production of artemisinin and its precursors, particularly in *Cichorium intybus* L. and microbes, has been adapted very recently (de Kraker et al., 2003; Bertea et al., 2005; Martin, Pitera, Wither, Newman, & Keasling, 2003; 2006; Ro et al., 2006). The production of precursors of artemisinin, such as, amorpha-4, 11-diene and artemisinic acid especially in *E. coli* and yeast has become a prime example of the capabilities of this parts-list-and-systems-design approach to microbial genetic engineering.

Keasling's group has developed a base technology for production of amorphadiene in *E. coli* (Martin et al., 2003). Bacteria already contain the MEP pathway for production of IPP/DMAPP, but they lack the MVA pathway. Instead, they added a truncated MVA pathway from *Saccharomyces cerevisiae* that was coupled to ADS in *E. coli* resulting in good bacterial growth and high-level production of amorphadiene estimated upto 100 μg in 12 h. Keasling's work is important because these engineered *E. coli* strains can serve as platform hosts for the production of essentially any terpenoid for which the biosynthetic genes are available. More recently, Teoh, Polichuk, Reed, Nowak, & Covello (2006) have isolated the next enzyme in the artemisinin pathway, a cytochrome P450 enzyme (CYP71AV1); this enzyme appears to catalyze the next three steps in artemisinin biosynthesis, an enzymatic function also confirmed by Keasling's group (J. Keasling, personal communication). Once cloned into a bacterial host and after optimization of the culture conditions, it should be possible to produce very large quantities of a close precursor to artemisinin in *E. coli*, thus making this important drug readily available in much larger quantities than previously thought possible. The concept of *E. coli* as a host cell producing sesquiterpenoids out of the endogenous pool of farnesyl diphosphate (FDP) has been investigated (Martin, Yoshikuni, & Keasling, 2001). This strategy has been combined with engineering of genes from the mevalonate dependent isoprenoid pathway, which resulted in an *E. Coli* strain producing 6.5 g/mL amorpha-4,11-diene (calculated as caryophyllene equivalent) from acetyl-CoA after supplementation of 15 g/L glucose (Tsuruta et al., 2009). Recently, attempts to use *S. cerevisiae* for the production of artemisinin precursors have been described. The expression of the amorphadiene synthase gene in yeast using plasmids and chromosomal integration led to the production of respectively 600 and 100 μg/ amorpha-4, 11-diene after 16 days batch cultivation (Lindahl et al., 2006). Using a *S. cerevisiae* strain containing an engineered MVA pathway coupled with the genes encoding amorphadiene synthase and CYP71AV1 the production of artemisinic acid up to 100 mg/l has been reported (Ro et al., 2006) (Fig. 8.6). This strain transported the artemisinin precursor outside the yeast cell, which makes purification of the product less complex. Artemisininc acid can be used for the semi-synthesis of artemisinin, but to lower the costs for production of the drug bioprocessing must be optimized (Liu et al., 1998).

Dafra Pharma International NV and Plant Research International (PRI) have initiated new research to produce artemisinin *via* genetically modified chicory plants. In studies carried out at Wageningen, the complete biosynthetic pathway of artemisinin was resolved (de Kraker et al., 2003 and Bertea et al., 2005; Fig. 8.7). In addition, the Wageningen group, headed by Prof. Harro Bouwmeester and Dr. Maurice Franssen, demonstrated that chicory enzyme(s) normally involved in the biosynthesis of the bitter sesquiterpene lactones in chicory, were capable of performing reactions required for the biosynthesis of artemisinin (de Kraker et al., 2003). The group of Prof. Bouwmeester has been tried to produce the chemical precursor for artemisinin (dihydroartemisinic acid) in the roots of chicory *via* a diversion of the biosynthesis of bitter compounds. On the other hand, the group of Prof. Bouwmeester has shown in a wide range of plant species that diversion of the biosynthesis of terpenes can be carried out very efficiently (Kappers et al., 2005). Moreover, they also demonstrated that up to 40 kg ha^{-1} dihydroartemisinic acid can be produced using genetically modified chicory.

Metabolic engineering of paclitaxel biosynthetic pathway

Paclitaxel was first isolated from *Taxus brevifolia* (pacific yew tree) Taxaceae in the sixties from last century (Wani, Taylor, Wall, Coggon, & McPhail, 1971), and its derivative Taxotere1 was clinically introduced 30 years later for the treatment of mainly ovarian and breast cancers. Isolation from the *T. brevifolia* bark is a problem, because of the low yield (500 mg kg^{-1}). Facing the high demand, various *Taxus* species are endangered in China and India. Croteau and coworkers isolated and identified several genes from differrent *Taxus* species that are responsible for steps in the biosynthesis and built a basis for today's combinatorial biosynthesis in a heterologous microorganism. Today, all genes have been cloned into *E. coli* and activity screening confirmed

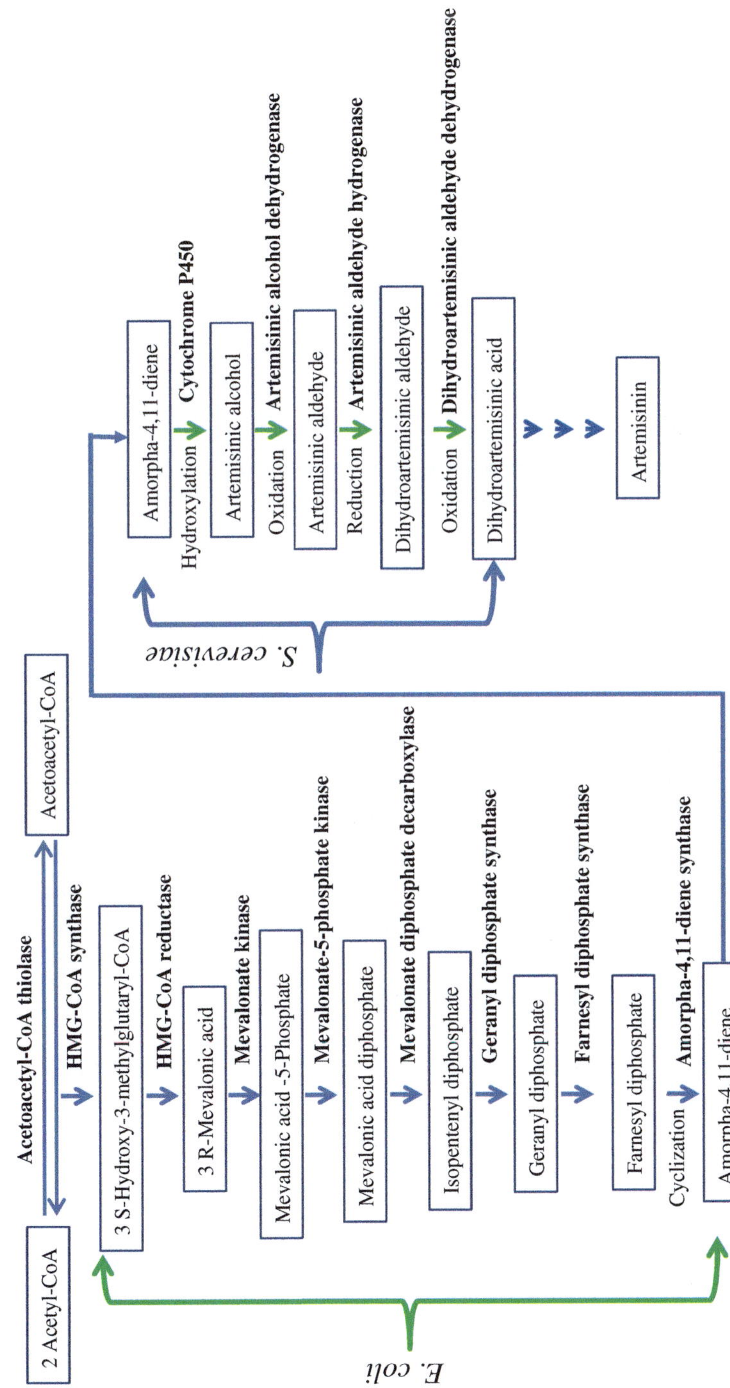

Figure 8.6 *Combinatorial biosynthesis of artemisinin starting from acetyl coenzyme A.*

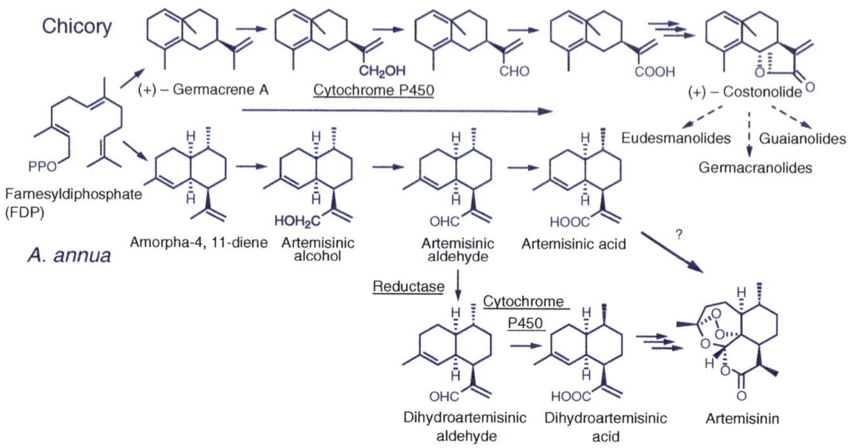

Figure 8.7 ***Biosynthetic routes of bitter sesquiterpene lactones in chicory and artemisinin in*** **Artemisia annua.** *(Bertea et al., 2005 and de Kraker et al., 2003).*

the function of isolated enzymes (Walker & Croteau, 2001, Long & Croteau, 2005, Jennewein, Rithner, Williams, & Croteau, 2001b, Walker, Schoendorf, & Croteau, 2000 & Chau & Croteau, 2004). The first intermediate, taxadiene can now be produced in *E. coli*. Co-expression of the taxadiene synthase from *T. brevifolia* (Wildung & Croteau, 1996) with a geranylgeranyl diphosphate synthase isolated from *Erwinia herbicola* (Math, Hearst, & Poulter, 1992), isopentenyl diphosphate synthase from *Schizosaccharomyces pombe* (Hahn & Poulter, 1995) and the endogenous deoxyxylulose 5-phosphate synthase from *E. coli* resulted in a production of 1.3 mg taxadiene per liter of cell culture (Huang, Roessner, Croteau, & Scott, 2001). This non-optimized system proved the principle of genetically engineering *E. coli* for the heterologous production of taxanes by combining enzymatic biosynthetic steps derived from several different organisms.

Metabolic engineering of alkaloid biosynthetic pathway

Combinatorial biosynthesis of alkaloids is known for a few examples like vincristine, vinblastine, ajmaline (Warzecha et al., 2000 Warzecha, Gerasimenko, Kutchan, & Stockigt, 2000), and morphine from plants and rebeccamycin and staurosporine from *Streptomyces albus* (Sanchez, Mendez, & Salas, 2006; Sanchez et al., 2006). The compounds mentioned have in common that a rather long biosynthetic pathway (30 enzymes for monoterpenoide indole alkaloids like vincristine and more than 17 enzymes for morphine) has to be elucidated and transferred into a heterologous host. In this chapter the combinatorial biosynthesis of morphine as benzylisoquinoline alkaloid and Vinca alkaloids as monoterpenoide indole alkaloids as examples for recent research strategies are discussed in detail.

Benzylisoquinoline alkaloids

Morphine is the most important member of the group of benzylisoquinoline alkaloids and is a natural product with high medicinal significance. Also other benzylisoquinoline alkaloids are medicinally important. Like morphine, codeine is used as an analgetic. Berberine and sanguinarine are used as antimicrobials and others as muscle relaxants such as parpaverine and (+)-tubocurarine. The morphine biosynthesis consists of 17 steps in *P. somniferum*, Papaveraceae, and has almost completely been elucidated. In the biosynthesis a key intermediate is (S)-norcoclaurine, that is biosynthesised by condensation of dopamine and 4-hydroxyphenylacetaldehyde (4-HPAA). The catalysing enzyme (S)-norcoclaurine synthase has recently been identified from *Thalictrum flavum* and cloned in *E. coli* (Samanani, Liscombe, & Facchini, 2004, Samanani & Facchini, 2002). Berberine as a second representative for benzylisoquinoline alkaloids is isolated from different plants, but is mostly associated with *Chelidonium majus*, Papaveraceae, and responsible for thecolor and the antimicrobial activity of the yellow latex and the plant extract (Da Cunha, Fechinei, Guedes, Barbosa-Filho, & Da Silva, 2005). With the exception of the oxidase leading to the quartenary nitrogen, all enzymes are known like the key step for bridging from (S)-reticulin to (S)-scoulerin (Dittrich and Kutchan, 1991), the introduction of a methyl group to give (S)-tetrahydrocolumbamin (Takeshita et al., 1995), and resulted in the building of a methylenedioxy ring (Ikezawa et al., 2003). Therefore we can expect that in the near future the successful combinatorial biosynthesis of berberine in a heterologous host will be tested.

Vinca alkaloids

The biosynthesis of vincristine and vinblastine is complex and is shown in for the early phase starting from geraniol to strictosidine (Vander Heijden, Jacobs, Snoeijer, Hallared, & Verpoorte, 2004). In the early phase tryptophan and secologanin as terpenoid precursors are condensed to form strictosidine as an important branching intermediate for also other alkaloids. In this short part of the entire route, already seven enzymes and corresponding genes are involved. From these seven genes four of these have been cloned in *E. coli* (Vander et al., 2004; Kutchan, Bock, & Dittrich, 1994). For the whole biosynthesis at least 30 biosynthetic and two known regulatory genes are involved, which encode around 35 intermediates. Furthermore, intracellular trafficking of intermediates between seven compartments must also be considered, what can be considered as major challenge in combinatorial biosynthesis (Vander et al., 2004). The tryptophan decarboxylase (TDC) and strictosidine synthase (STR) were the first two genes from *C. roseus* cloned from the monoterpenoid indole pathway into *S. cerevisiae* (Fig. 8.8). In the past single genes of the biosynthesis have been expressed in different heterologous organisms (Whitmer, 1999). The cDNA coding for STS from *R. serpentina* has previously been expressed in *E. coli* and in insect cells and was found to convert secologanin and tryptamine into strictosidine (Whitmer, 1999). After feeding the precursors tryptamine and secologanin, strictosidine and its aglycon were biosynthesised in *S. cerevisiae* as a new heterologous host. When strictosidine glucosidase was additionally over-expressed in the recombinant host *S. cerevisiae* carrying the tryptophan decarboxylase and strictosidine synthase gene a sufficient amount of strictosidine was formed (Geerlings et al., 2001). Besides in microbial hosts the mentioned genes of the early biosynthesis have also been cloned into *Nicotina tabacum*. The

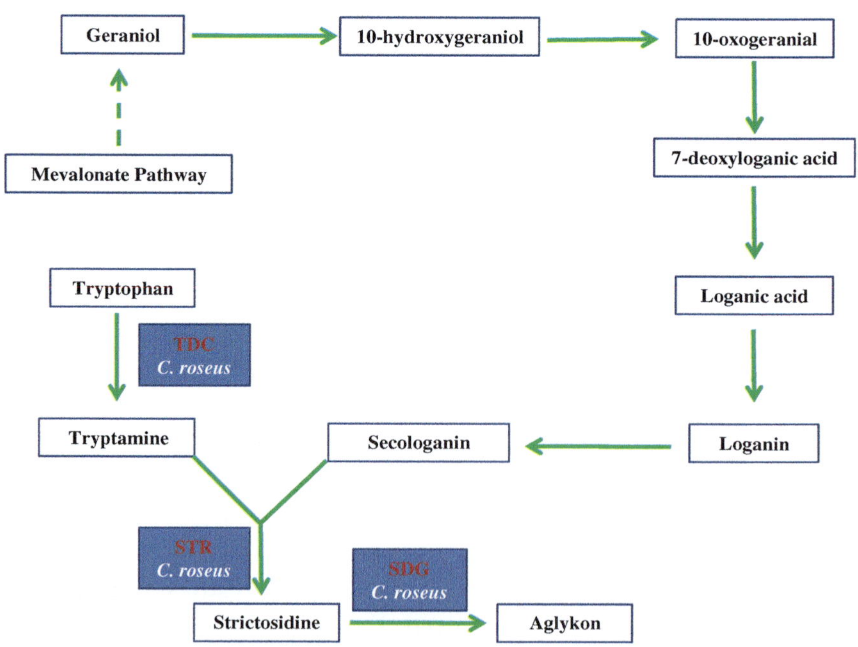

Figure 8.8 ***Early biosynthesis of* Vinca *alkaloids in* S. cerevisiae.** The *broken arrow* indicates multiple steps. *(Van der Heijden et al., 2004; Geerlings et al., 2001).*

major drawback however is the disability to hydrolyse strictosidine glucoside because *N. tabacum* does not possess specific glucosidases (Hallard et al., 1997). Later, strictosidine glucosidase (SDG) has also been successfully inserted and expressed in suspension culture of tobacco cells (Zarate, Dirks, van der, & Verpoorte, 2001). The strictosidine glucosidase protein in *N. tabacum* was present in the same high molecular weight complexes as known before in *C. roseus*.

Combinatorial metabolism of phenolic natural products

Flavonoids

Flavonoids represent a very important group of plant natural products. They are considered as health promoting substances in the human diet for their antioxidant, antiasthmatic, antiblood-clotting and anticancer activities. Flavonoids are exclusively produced in plants and found in almost all studied species in the plant kingdom. Flavonoids are produced via the so-called phenylpropanoid pathway, in which phenylalanine ammonia lyase (PAL) deaminates phenylalanine or tyrosine yielding cinnamic acid. The biosynthetic route on the enzymatic and genetic level has been elucidated in the past (Winkel-Shirley, 2001; Harborne & Williams, 2000). The biosynthesis starts with L-phenylalanine that is metabolized to cinnamic acid derivatives, which condensates with malonyl-CoA to a chalcone. In the biosynthesis cinnamic acid is hydroxylated by cinnamic-4-hydroxylases (C4H) to *para*-4-hydroxy-cinnamic acid, activated by 4-coumarate/cinnamate coenzyme A, coupled with 3 malonyl-CoA units and converted by chalcone synthase (CHS) to a chalcone derivate as first-committed precursor for the flavonoid biosynthesis. Chalcones are converted to flavonoids by a ring-closing step forming the heterocyclic C ring by chalcone isomerases. Naringenin is a chalcone and key intermediate leading to isoflavones, to condensed tannin precursors and, via different hydroxylation, glycosylation, prenylation and alkylation steps, to more than 600 known flavonoids (Harborne & Williams, 2000). Recent publications have documented the production of pinocembrin, naringenin and chrysin, apigenin, galangin, kaempferol and dihydrokaempferol in recombinant *E. Coli* BL21 (DE3). Because the main genes for flavonoids are missing in *E. coli*, recombinant plasmids (pUC, pET) containing the genes of interest have been constructed (Hwang, Kaneko, Ohnishi, & Horinouchi, 2003; Kaneko, Hwang, Ohnishi, & Horinouchi, 2003) (Table 8.1). These artificial gene clusters contain up to three genes from microorganisms or plant origin (*Glycyrrhiza echinata*, *Petroselinum crispum* and *Citrus sinensis*). Expression of all genes encoding the flavonoid biosynthesis up to the level of naringenin was successful, but only limited amounts of flavonoids were detected. To overcome this problem, the production of the essential precursor malonyl-CoA was increased by overexpression of the acetyl-Coa carboxylase from *Corynebacterium glutamicum*. In the recent study carried by (Miyahisa et al., 2005a; Miyahisa et al., 2005b, further biosynthetic genes have been introduced to modify the oxygenation pattern of flavonoids leading to kaempferol and apigenin. The published work is of high interest, because for the first time a nearly complete biosynthetic pathway from plants was established in a heterologous microorganism. In the future it will be of interest to investigate whether further enzymes modifying flavones and flavonols like glycosylation, prenylation or *O*-methylation, can be integrated.

Polyketide

Polyketides are secondary metabolites isolated from bacteria, fungi, plants and animals. Polyketides are usually biosynthesized through the decarboxylative condensation of malonyl-CoA derived extender units in a similar process to fatty acid synthesis (Robinson, 1991). Polyketides are structurally a very diverse family of natural products with diverse biological activities and pharmacological properties such as polyketide antibiotics, antifungals, cytostatics, anticholesterolemics, antiparasitics, coccidiostatics, animal growth promoters, and natural insecticides are in commercial use.

The 6-deoxyerythronolide B synthase gene cluster from *Saccharopolyspora erythraea* was engineered for expression in *Streptomyces coelicolor.* The recombinant bacteria synthesized a derivative of 6-deoxyerythronolide B that contained a methyl side chain instead of an ethyl side chain (Kao, Katz, & Khosla, 1994). The recombinant bacteria efficiently incorporated fed precursors into 6-deoxyerythronolide B (Cane, Luo, Khosla, Kao, & Katz, 1995). Both observations demonstrate expression of a heterologous gene to produce a functional enzyme. In an extension of these studies, recombinant 6-deoxyerythronolide B synthase was used for in vitro synthesis of 6-deoxyerythronolide B (Pieper, Luo, Cane, & Khosla, 1995). Synthesis of 6-deoxyerythronolide B, an important intermediate in erythromycin A biosynthesis, in a heterologous host reaffirms the prospect that genes for antibiotic synthesis may be cloned from the natural producing organisms and expressed in new hosts that may have been engineered or naturally offer special advantages for genetic manipulation to enhance commercial production and recombination of genes for related pathways to produce new polyketide natural products. Bedford and coworkers (Bedford, Schweizer, Hopwood, &

Khosla, 1995) cloned the gene for the biosynthesis of 6-methylsalicylic acid from the fungus *Penicillium patulum*. They inserted the gene into a regulatable *Streptomyces* expression cassette and introduced the recombinant plasmid into *Streptomyces coelicolor*, which had been engineered to possess the ability to synthesize polycyclic polyketide compounds. The design of this scheme enhanced the probability that the product of the synthase would be made, because the host has the known capacity to produce sufficient quantities of the precursors for the biosynthesis of polyketides. An advantage of this approach may be engineering one host to produce high levels of precursors that may feed the synthesis of several different products that may be assembled from similar precursors. Therefore, pathways for a variety of products from diverse species may be expressed at high levels in a single engineered host background. Particularly noteworthy is the fact that *Penicillium patulum* is a eukaryotic organism and *Streptomyces coelicolor* is a prokaryotic organism. Indeed the recombinant organism accumulated 6-methylsalicylic acid, thereby demonstrating the expression of functional 6-methylsalicylic acid synthase in *Streptomyces coelicolor*. Examples abound where proteins from eukaryotic organisms adopt insoluble and inactive conformations when they are expressed in prokaryotic organisms. It will be interesting to learn from additional examples of expression of eukaryotic genes specifying enzymes for antibiotic synthesis in prokaryotic host organisms whether the eukaryotic enzymes for antibiotic synthesis are generally active in antibiotic-producing bacterial hosts. The general success of this approach would provide the opportunity to incorporate the rich diversity of eukaryotic antibiotic synthesis into bacterial hosts. Expression of genes for kalafungin biosynthesis on a recombinant plasmid generated a new antibiotic even though the experiment was not an example of heterologous gene expression (Kakinuma et al., 1995). *Streptomyces tanashiensis* produces the antibiotic kalafungin. However, *Streptomyces tanashiensis* carrying a recombinant plasmid with the genes specifying kalafungin biosynthesis and 5 kilobases of plasmid DNA from the stability region of Scp2 synthesized a new antibiotic, tetrahydrokalafungin.

Stilbenes

Stilbenes are another class of phenylpropanoids that have given much attention, particularly resveratrol, which is used as an antioxidant (Jang et al., 1997) and for its ability to increase the lifespan of yeast, fruit fly, round worm (Howitz et al., 2003) and fish (Valenzano et al., 2006) by activating siruin deacetylases (Kaeberlein et al., 2005). *S. cerevisiae* reportedly produced resveratrol from endogenous phenylalanine using a combination of PAL,

4CL, and stilbene synthase (STS) (Zhang et al., 2006). Furthermore, *E. coli* expressing 4CL and STS converted supplemented cinnamic acids into their respective stilbenes at much higher levels (Katsuyama, Funa, Miyahisa, & Horinouchi, 2007 and Beekwilder et al., 2006).

Conclusions and summary

Several research groups have made significant progress in the metabolic engineering of terpenoids, alkaloids and phenolics during the past few years. Several studies of a range of plant species have shown that a high level of production of terpenoids, including modified products, can be obtained using metabolic engineering. The engineering of sesquiterpenes has been much less successful even though the precursor for the sesquiterpenes, FDP, is a ubiquitous molecule that supposedly is present in relatively large quantities, if only because it is the precursor for sterol biosynthesis. In 1995, Joseph Chappell discussed the possibility that metabolic channelling might occur in isoprenoid biosynthesis at a sub-cellular level. This might also occur in isoprenoid biosynthesis, which could explain why FDP (which is produced in relatively large amounts for sterol biosynthesis) is not or hardly available to introduced sesquiterpene synthases. Perhaps this obstacle to successful sesquiterpene metabolic engineering can be solved by artificial channelling, using fusion constructs of two or more enzymes from a pathway. Work with a fusion construct of FDP synthase and the sesquiterpene synthase epi-aristolochene synthase produced in *Escherichia coli* shows that such fusion constructs can be functional.

Transcription factors offer much promise for the manipulation of metabolic pathways because of their ability to control both multiple pathway steps and cellular processes that are necessary for metabolite accumulation. This is particularly true of complex pathways whose component enzymes are poorly characterized, such as those of secondary metabolism. Because of the intricacies of the transcriptional regulation of metabolism in plants, however, it will be important to enhance our knowledge of the transcriptional networks that respond to the activities of transcription factors.

Genetic and metabolic engineering has the potential to design a process to yield high quality, less expensive and/or completely novel products. In addition, mathematical and systematic treatment of the effect of these genetic perturbations provides more detailed analysis and insight. The combined knowledge is often capable of pointing to a direction for possible genetic changes in order to achieve a desired goal. Metabolic engineering principles have been used to improve biocatalysts either to increase product yield or to synthesize new products in areas such as recombinant protein, amino acid and antibiotic production processes.

Acknowledgment

One of the authors Mauji Ram would like to acknowledge Ipca Laboratories Ltd., Ratlam (M.P.), India for financial assistance in the form of Senior Research Executive.

References

Aharoni, A., Ashok, P., Francel, W. A., Cinzia, M. B., Robert, S., Zhongkui, S., Maarten, A. J., Wilfried, S., & Harro, J. B. (2004). Gain and loss of fruit flavor compounds produced by wild and cultivated Strawberry species. *The Plant Cell, 16*, 3110–3131.

Aharoni, A., Ashok, P., Stephan, D., Frans, G., Willem, J. K., Francel, W. A., harrie, a. V., Maarten, A. J., Wilfried, S., & Harro, J. B. (2004). Terpenoid metabolism in wild-type and transgenic Arabidopsis plants. *The Plant Cell, 15*, 2866–2884.

Allen, R. S., Millgate, A. G., Chitty, J. A., Thisleton, J., Miller, J. A., Fist, A. J., Gerlach, W. L., & Larkin, P. J. (2004). RNAi-mediated replacement of morphine with the non-narcotic alkaloid reticuline in *Opium poppy Nat. Biotechnology, 22*, 1559–1566.

Alvarez, M. A., Fu, H., Khosla, C., Hopwood, D. A., & Bailey, J. E. (1996). Engineered biosynthesis of novel polyketides: properties of the whiE aromatase/cyclise. *Nature Biotechnology, 14*, 335–338.

Aquil, S., Husaini, A. M., Abdin, M. Z., & Rather, G. N. (2009). Overexpression of HMG-CoA reductase gene leads to enhanced artemisinin biosynthesis in transgenic *A. annua* L. plants. *Planta Medica, 75*, 1–6.

Baiza, A. M., Quiroz-Moreno, A., Ruiz, J. A., & Victor, M. L. V. (1999). Genetic stability of hairy root cultures of *Datura stamonium. Plant Cell Tissue and Organ Culture, 59*, 9–17.

Bedford, D. J., Schweizer, E., Hopwood, D. A., & Khosla, C. (*1995*). Expression of a functional fungal polyketide synthase in the bacterium *Streptomyces coelicolor* A3 (2). *Journal of Bacteriology, 177*, 4544–4548.

Beekwilder, J., Wolswinkel, R., Jonker, H., Hall, R., de Vos, C. H., & Bovy, A. (2006). Production of resveratrol in recombinant microorganisms. *Applied and Environment Microbiology, 72*, 5670–5672.

Bertea, C. M., Freije, J. R., van der Woude, H., Verstappen, F. W., Perk, L., & Marquez, V. (*2005*). Identifi cation of intermediates and enzymes involved in the early steps of artemisinin biosynthesis in *Artemisia annua. Planta Medica, 71*, 40–47.

Besumbes, O., Sauret, G. S., Phillips, M. A., Imperial, S., Rodriguez, C. M., & Boronat, A. (2004). Metabolic engineering of isoprenoid biosynthesis in Arabidopsis for the production of taxadiene, the first committed precursor of Taxol. *Biotechnology and Bioengineering, 88*, 168–175.

Botella-Pavia, P., Besumbes, O., Phillips, M. A., Carretero-Paulet, L., Boronat, A., & Rodríguez-Concepción, M. (2004). Regulation of carotenoid biosynthesis in plants: evidence for a key role of hydroxymethylbutenyl diphosphate reductase in controlling the supply of plastidial isoprenoid precursors. *Plant Journal, 40*, 188–199.

Bovy, A., de Vos, R., Kemper, M., Schijlen, E., Almenar Pertejo, M., Muir, S., Collins, G., Robinson, S., Verhoeyen, M., & Hughes, S. (2002). High-flavonol tomatoes resulting from the heterologous expression of the maize transcription factor genes LC and C1. *Plant Cell, 14*, 2509–2526.

Cane, D. E., Luo, G., Khosla, C., Kao, C. M., & Katz, L. (1995). Erythromycin biosynthesis. Highly efficient incorporation of polyketide chain elongation intermediates into 6-deoxyerythronolide B in an engineered Streptomyces host. *Journal of Antibiotics (Tokyo), 48*, 647–651.

Carrari, F., Urbanczyk-Wochniak, E., Willmitzer, L., & Fernie, A. R. (*2003*). Engineering central metabolism in crop species: learning the system. *Metabolic Engineering.*, *5*, 191–200.

Chang, M. C. Y., Eachus, R. A., Trieu, W., Ro, D. K., & Keasling, D. J. (2007). Engineering E. coli for production of functionalized terpenoids using plant P450s. *Nature Chemical Biology*, *3*, 274–277.

Chappell, J., Wolf, J., Proulx, R. C., & Saunders, C. (1995). Is the reaction catalyzed by 3-hydroxy-3-methylglutaryl coenzyme A reductase a rate-limiting step for isoprenoid biosynthesis in plants? *Plant Physiology.*, *109*, 1337–1343.

Charles, L. H. (1996). Metabolic engineering of polyketide biosynthesis. *Current Opinion in Biotechnology*, 7, 560–562.

Charlwood, B. V., & Pletsch, M. (2002). Manipulation of natural product accumulation in plants through genetic engineering. *Journal of Herbs Spices and Medicina Plants*, *9*, 139–151.

Chau, M., & Croteau, R. (2004). Molecular cloning and characterization of a cytochrome P450 taxoid 2a-hydroxylase involved in Taxol biosynthesis. *Archives of Biochemistry and Biophysics*, *427*, 48–57.

Chen, D. H., Ye, H. C., & Li, G. F. (2000). Expression of a chimeric farnesyl diphosphate synthase gene in *Artemisia annua* L. transgenic plants via *Agrobacterium tumefaciens*-mediated transformation. *Plant Science*, *155*, 179–185.

Chen, F., Tholl, D., D'Auria, J. C., Farooq, A., Pichersky, E., & Gershenzon, J. (2003). Biosynthesis and emission of terpenoid volatiles from Arabidopsis flowers. *Plant Cell*, *15*, 481–494.

Chen, F., Ro, D. K., Petri, J., Gershenzon, J., Bohlmann, J., Pichersky, E., & Tholl, D. (2004). Characterization of a root-specific *Arabidopsis* terpene synthase responsible for the formation of the volatile monoterpene 1, 8-cineole. *Plant Physiology*, *135*, 1956–1966.

Chen, S., Li, H., & Liu, G. (2006). Progress of vitamin E metabolic engineering in plants. *Transgenic Research*, *15*, 655–665.

Cortes, I., Wiesmann, K. E., Roberts, G. A., Brown, M. J., Staunton, J., & Leadlay, P. F. (1995). Repositioning of a domain in a modular polyketide synthase to promote specific chain cleavage. *Science*, *268*, 1487–1489.

Da Cunha, E.V., Fechinei, I. M., Guedes, D. N., Barbosa-Filho, J. M., & Da Silva, M. S. (2005). Protoberberine alkaloids. *Alkaloids of Chemical Biology*, *62*, 1–75.

Davuluri, G. R., van Tuinen, A., Fraser, P. D., Manfredonia, A., Newman, R., Burgess, D., Brummell, D. A., King, S. R., Palys, J., & Uhlig, J. (2005). *Nature Biotechnology*, *23*, 890–895.

de Kraker, J., Schurink, M., Franssen, M. C. R., König, W. A., Groot, A., & Bouwmeester, H. J. (2003). Hydroxylation of sesquiterpenes by enzymes from chicory (*Cychorium intybus* L.) roots. *Tetrahedron*, *59*, 409–418.

Diemer, F., Caissard, J. C., Moja, S., Chalchat, J. C., & Jullien, F. (2001). Altered monoterpene composition in transgenic mint following the introduction of 4S-limonene synthase. *Plant Physiology and Biochemistry*, *39*, 603–614.

Dittrich, H., & Kutchan, T. M. (1991). Molecular cloning, expression, and induction of berberine bridge enzyme, an enzyme essential to the formation of benzophenanthridine alkaloids in the response of plants to pathogenic attack. *Proceedings of National Academy of Science USA*, *88*, 9969–9973.

Dixon, R. A. (*2005*). Engineering of plant natural product pathways. *Current Opinion in Plant Biology*, *8*, 329–336.

Dixon, R. A. (2001). Natural products and plant disease resistance. *Nature*, *411*, 843–847.

Dixon, R. A., & Steele, C. L. (1999). Flavonoids and isoflavonoids – a gold mine for metabolic engineering. *Trends in Plant Science*, *4*, 394–400.

Dong, X., Braun, E. L., & Grotewold, E. (2001). Functional conservation of plant secondary metabolic enzymes revealed by complementation of Arabidopsis flavonoid mutants with maize genes. *Plant Physiology*, *127*, 46–57.

Dudareva, N., Cseke, L., Blanc, V. M., & Pichersky, E. (1996). Evolution of floral scent in Clarkia: novel patterns of S-linalool synthase gene expression in the *C. breweri* flower. *Plant Cell*, *8*, 1137–1148.

Estevez, J. M., Cantero, A., Reindl, A., Reichler, S., & León, P. (2001). 1-Deoxy-D-xylulose-5-phosphate synthase, a limiting enzyme for plastidic isoprenoid biosynthesis in plants. *Journal of Biological Chemistry*, *276*, 22901–22909.

Facchini, P. J. (2001). Alkaloid biosynthesis in plants: Biochemistry, cell biology, molecular regulation, and metabolic engineering applications. *Annual Reviews of Plant Physiology and Plant Molecular Biology*, *52*, 29–66.

Facchini, P. J., Park, S. U., Bird, D. A., & Samanani, S. (2000). Toward the metabolic engineering of benzylisoquinoline alkaloid biosynthesis in opium poppy and related species. *Recent Research and Development in Phytochemistry*, *4*, 31–47.

Falco, S. C., Guida, T., Locke, M., Mauvais, J., Sanders, C., Ward, R. T., & Webber, P. (1995). Transgenic canola and soybean seeds with increased lysine. *Biotechnology*, *13*, 577–582.

Farmer, W. R., & Liao, J. C. (2001). Precursor balancing for metabolic engineering of lycopene production in *Escherichia coli*. *Biotechnology Program*, *17*, 57–61.

Ferreira, J. F. S., & Duke, S. O. (1997). Approaches for maximising biosynthesis of medicinal plant secondary metabolites. *AgBiotechnology News and Information*, *9*, 309N–315N.

Fong, R., Vroom, J. A., Hu, Z., Hutchinson, C. R., Huang, J., Cohen, S. N., & Kao, C. M. (2007). Characterization of a large, stable, high-copy-number *Streptomyces* plasmid that requires stability and transfer functions for heterologous polyketide overproduction. *Applied and Environmental Microbiology*, *73*(4), 1296–1307.

Forkmann, G., & Martens, S. (2001). Metabolic engineering and applications of flavonoids. *Current Opinion in Biotechnology*, *12*, 155–160.

Fraser, P. D., Romer, S., Shipton, C. A., Mills, P. B., Kiano, J. W., Misawa, N., Drake, R. G., Schuch, W., & Bramley, P. M. (2002). Evaluation of transgenic tomato plants expressing an additional phytoene synthase in a fruit-specific manner. *Proceedings of National Academy of Science USA*, *99*, 1092–1097.

Galili, G., Galili, S., Lewinsohn, E., & Tadmor, Y. (2002). Genetic, molecular and genomic approaches to improve the value of plant foods and feeds. *Critical Reviews in Plant Science*, *21*, 167–204.

Gantet, P., & Memelink, J. (2002). Transcription factors: tools to engineer the production of pharmacologically active plant metabolites. *Trends in Pharmacological Science*, *23*, 563–569.

Geerlings, A., Redondo, F. J., Contin, A., Memelink, J., van der, H. R., & Verpoorte, R. (2001). Biotransformation of tryptamine and secologanin into plant terpenoid indole alkaloids by transgenic yeast. *Applied Microbiology and Biotechnology*, *56*, 420–424.

Goff, S. A., Klein, T. M., Roth, B. A., Fromm, M. E., Cone, K. C., Radicella, J. P., & Chandler, V. L. (1990). Transactivation of anthocyanin biosynthetic genes following transfer of B regulatory genes into maize tissues. *EMBO Journal*, *9*, 2517–2522.

Goldsbrough, A. P., Tong, Y., Yoder, J. I., & Tong, Y. S. (1996). Lc as a non-destructive visual reporter and transposition excision marker gene for tomato. *Plant Journal*, *9*, 927–933.

Grotewold, E., Chamberlin, M., Snook, M., Siame, B., Butler, L., Swenson, J., Maddock, S., Clair, G. S., & Bowen, B. (1998). Engineering secondary metabolism in maize cells by ectopic expression of transcription factors. *Plant Cell*, *10*, 721–740.

Hahn, F. M., & Poulter, C. D. (1995). Isolation of Schizosaccharomyces pombe isopentenyl diphosphate isomerase cDNA clones by complementation and synthesis of the enzyme in *Escherichia coli*. *Journal of Biological Chemistry*, *270*, 11298–11303.

Hallard, D.,Van der Heijden, R.,Verpoorte, R., Cardoso, M. I. L., Pasquali, G., Memelink, J., & Hoge, J. H. C. (1997). Suspension cultured transgenic cells of *Nicotiana tabacum* expressing tryptophan decarboxylase and strictosidine synthase cDNAs from *Catharanthus roseus* produce strictosidine upon secologanin feeding. *Plant Cell Reports, 17*, 50–54.

Han, J. L., Liu, B.Y., Ye, H. C., Wang, H., Li, Z. Q., & Li, G. F. (2006). Effects of overexpression of the edogenous farnesyl diphosphate synthesis on the artemisinin content in *Artemisia annua* L. *Journal of Integrated Plant Biology, 48*(4), 482–487.

Harborne, J. B., & Williams, C. A. (2000). Advances in flavonoid research since. *Phytochemistry, 55*, 481–504.

Harker, M., Hellyer, A., John, C. C., Lanot, A., & Safford, R. (2003). Co-ordinate regulation of sterol biosynthesis enzyme activity during accumulation of sterols in developing rape and tobacco seed. *Planta, 216*, 707–715.

Hohn, T. M., & Ohlrogge, J. B. (1991). Expression of a fungal sesquiterpene cyclase gene in transgenic tobacco. *Plant Physiology, 97*, 460–462.

Howitz, K.T., Bitterman, K. J., Cohen, H.Y., Lamming, D.W., Lavu, S., Wood, J. G., Zipkin, R. E., Chung, P., Kisielewski, A., & Zhang, L. L. (2003). Small molecule activators of sirtuins extend *Saccharomyces cerevisiae* lifespan. *Nature, 425*, 191–196.

Hranueli, D., Cullum, J., Basrak, B., Goldstein, P., & Long, P. F. (2005). Plasticity of the Streptomyces genome-evolution and engineering of new antibiotics. *Current Medicinal Chemistry, 12*, 1697–1704.

Huang, Q., Roessner, C. A., Croteau, R., & Scott, A. I. (2001). Engineering *Escherichia coli* for the synthesis of taxadiene, a key intermediate in the biosynthesis of taxol. *Bioorganic and Medicinal Chemistry, 9*, 2237–2242.

Hwang, E. I., Kaneko, M., Ohnishi, Y., & Horinouchi, S. (2003). Production of plant-specific flavanones by *Escherichia coli* containing an artificial gene cluster *Appl. Applied Environmental Microbiology, 69*, 2699–2706.

Ikezawa, N., Tanaka, M., Nagayoshi, M., Shinkyo, R., Sakaki, T., Inouye, K., & Sato, F. (2003). Molecular cloning and characterization of CYP719, a methylenedioxy bridge-forming enzyme that belongs to a novel P450 family, from cultured *Coptis japonica* cells. *Journal of Biological Chemistry, 278*, 38557–38565.

Jang, M., Cai, L., Udeani, G. O., Slowing, K.V., Thomas, C. F., Beecher, C.W., Fong, H. H., Farnsworth, N. R., Kinghorn, A. D., & Mehta, R. G. (1997). Cancer chemopreventive activity of resveratrol, a natural product derived from grapes. *Science, 275*, 218–220.

Jennewein, S., & Croteau, R. (2001). Taxol: biosynthesis, molecular genetics, and biotechnological applications. *Applied Microbiology and Biotechnology, 57*, 13–19.

Jennewein, S., Rithner, C. D., Williams, R. M., & Croteau, R. B. (2001b). Taxol biosynthesis: taxane 13a-hydroxylase is a cytochrome P450-dependent monooxygenase. *Proceedings of National Academy of Science USA, 98*, 13595–13600.

Julsing, M. K., Koulman, A., Herman, J. W., Wim, J. Q., & Oliver, K. (2006). Combinatorial biosynthesis of medicinal plant secondary metabolites. *Biomolecular Engineering, 23*, 265–279.

Kaeberlein, M., McDonagh, T., Heltweg, B., Hixon, J., Westman, E. A., Caldwell, S. D., Napper, A., Curtis, R., DiStefano, P. S., & Fields, S. (2005). Substrate-specific activation of sirtuins by resveratrol. *Journal of Biological Chemistry, 280*, 17038–17045.

Kakinuma, S., Ikeda, H., Takada, Y., Tanaka, H., Hopwood, D. A., & Omura, S. (*1995*). Production of the new antibiotic tetrahydrokalafungin by transformants of the kalafungin producer *Streptomyces tanashiensis*. *Journal of Antibiotics (Tokyo), 48*, 484–487.

Kao, C. M., Katz, L., & Khosla, C. (1994). Engineered biosynthesis of a complete macrolactone in heterologous host. *Science, 265*, 509–512.

Kaneko, M., Hwang, E. I., Ohnishi, Y., & Horinouchi, S. (2003). Heterologous production of flavanones in *Escherichia coli*: potential for combinatorial biosynthesis of flavonoids in bacteria *J. Industrial Microbiology and Biotechnology, 30*, 456–461.

Kappers, I. F., Aharoni, A., Herpen, T. W. J. M., Luckerhoff, L. L. P., Dicke, M., & Bouwmeester, H. J. (2005). Genetic engineering of terpenoid metabolism attracts bodyguards to *Arabidopsis*. *Science, 309*, 2070–2072.

Krasnyanski, S., May, R. A., Loskutova, A., Ball, T. M., & sink, K. C. (1999). Transformation of the limonene synthase gene into peppermint (*Mentha piperita* L.) and preliminary studies on the essential oil profiles of single transgenic plants. *Theory of Applied Genetics, 99*, 676–682.

Kutchan, T. M., Bock, A., & Dittrich, H. (1994). Heterologous expression of the plant proteins strictosidine synthase and berberine bridge enzyme in insect cell culture. *Phytochemistry, 35*, 353–360.

Katsuyama, Y., Funa, N., Miyahisa, I., & Horinouchi, S. (2007). Synthesis of unnatural flavonoids and stilbenes by exploiting the plant biosynthetic pathway in *Escherichia coli*. *Chemical Biology, 14*, 613–621.

Larkin, P. J., Miller, J. A. C., Allen, R. S., Chitty, J. A., Gerlach, W. L., Kutchan, S. F. T. M., & Fist, A. J. (2007). Increasing morphinan alkaloid production by over-expressing codeinone reductase in transgenic *Papaver somniferum*. *Plant Biotechnology Journal, 5*, 26–37.

Latchman, D. S. (2003). *Eucaryotic Transcription Factors, edn4*. San Diego: Academic Press.

Leonard, E., Lim, K. H., Saw, P. N., & Koffas, M. A. (2007). Engineering central metabolic pathways for high-level flavonoid production in *Escherichia coli*. *Applied and Environmetal Microbiology, 73*, 3877–3886.

Lindahl, A. L., Olsson, M. E., Mercke, P., Tollbom, O., Schelin, J., Brodelius, M., & Brodelius, P. E. (2006). Production of the artemisinin precursor amorpha- 4,11-diene by engineered Saccharomyces cerevisiae. *Biotechnology Letters, 28*, 571–580.

Liu, C. J., Blount, J. W., Steele, C. L., & Dixon, R. A. (2002). Bottlenecks for metabolic engineering of isoflavone glycoconjugates in Arabidopsis. *Proceedings of National Academy of Science USA, 99*, 14578–14583.

Liu, C. Z., Wang, Y. C., Ouyang, F., Ye, H. C., & Li, G. F. (1998). Production of artemisinin by hairy root cultures of *Artemisia annua* L. in bioreactor. *Biotechnology Letters, 20*, 265–268.

Lloyd, A. M., Walbot, V., & Davis, R. W. (1992). Arabidopsis and *Nicotiana anthocyanin* production activated by maize regulators R and C1. *Science, 258*, 1773–1775.

Long, R. M., & Croteau, R. (2005). Preliminary assessment of the C13-side chain 20-hydroxylase involved in taxol biosynthesis. *Biochemical and Biophysical Research Communications, 338*, 410–417.

Lucker, J., Bouwmeester, H. J., Schwab, W., Blaas, J., Plas, L. H. W., & Verhoeven, H. A. (2001). Expression of Clarkia S-linalool synthase in transgenic petunia plants results in the accumulation of S-linalyl-b-Dglucopyranoside. *Plant Journal, 27*, 315–324.

Lucker, J., Bouwmeester, H. J., Schwab, W., Blaas, J., Plas, L. H. W., & Verhoeven, H. A. (2004). Increased and altered fragrance of tobacco plants after metabolic engineering using three monoterpene synthases from lemon. *Plant Physiology, 134*, 510–519.

Lucker, J., Bouwmeester, H. J., Blaas, J., Plas, L. H. W., & Verhoeven, H. A. (2004). Metabolic engineering of monoterpen biosynthesis: two-step production of (c)-trans-isopiperitenol by tobacco. *Plant Journal, 39*, 135–145.

Ma, D., Pu, G., Lei, C., Ma, L., Wang, H., Guo, Y., Chen, J., Du, Z., Wang, H., Li, G., Ye, H., & Liu, B. (2009). Isolation and characterization of AaWRKY1, an *Artemisia annua* transcription factor that regulate the Amorpha-4, 11-diene synthase gene, a key gene of artemisinin biosynthesis. *Plant and Cellular Physiology, 50*, 1246–2161.

Mahmoud, S. S., & Croteau, R. (2001). Metabolic engineering of essential oil yield and composition in mint by altering expression of deoxyxylulose phosphate reductoisomerase and menthofuran synthase. *Proceedings of National Academy of Science USA, 98*, 915–8920.

Mahmoud, S. S., William, M., & Croteau, R. (2004). Co-suppression of limonene-3-hydroxylase in peppermint promotes accumulation of limonene in the essential oil. *Phytochemistry, 65*, 547–554.

Martin, V. J., Pitera, D. J., Wither, S. T., Newman, J. D., & Keasling, J. D. (2003). Engineering a mevalonate pathway in *Escherichia coli* for production of terpenoids. *Nature Biotechnology, 21*, 796–802.

Martin, V. J., Yoshikuni, Y., & Keasling, J. D. (2001). The in vivo synthesis of plant sesquiterpenes by *Escherichia coli. Biotechnology and Bioengineering, 75*, 497–503.

Math, S. K., Hearst, J. E., & Poulter, C. D. (1992). The crtE gene in Erwinia herbicola encodes geranylgeranyl diphosphate synthase. *Proceedings of National Academy of Science USA, 89*, 6761–6764.

McDaniel, R., Ebert-Khosla, S., Hopwood, D. A., & Khosla, C. (1995). Rational design of aromatic polyketide natural products by recombinant assembly of enzymatic subunits. *Nature, 375*, 549–554.

Memerlink, J., Verpoorte, R., & Kijne, J. W. (*2001*). ORCAanisation of jasmonate-responsive gene expression in alkaloid metabolism. *Trends in Plant Science, 6*, 212–219.

Minas, W., Brünker, P., Pauli, T. K., & James, E. B. (1998). Improved erythromycin production in a genetically engineered industrial strain of *Saccharopolyspora erythraea. JT Biotechnology Program, 14*(4), 561–566.

Miyahisa, I., Funa, N., Ohnishi, Y., Martens, S., Moriguchi, T., & Horinouchi, S. (2005a). Combinatorial biosynthesis of flavones and flavonols in *Escherichia coli. Applied Microbiology and Biotechnology*, 1–6.

Miyahisa, I., Kaneko, M., Funa, N., Kawasaki, H., Kojima, H., Ohnishi, Y., & Horinouchi, S. (2005b). Efficient production of (2S)-flavanones by *Escherichia coli* containing an artificial biosynthetic gene cluster. *Applied Microbiology and Biotechnology, 68*, 498–550.

Mooney, M., Desnos, T., Harrison, K., Jones, J., Carpenter, R., & Coen, E. (1995). Altered regulation of tomato and tobacco pigmentation genes caused by the delila gene of Antirrhinum. *Plant Journal*, 7, 333–339.

Moore, B. S., Kalaitzis, J. A., & Xiang, L. (2005). Exploiting marine actinomycete biosynthetic pathways for drug discovery. *Antonie Van Leeuwenhoek, 87*, 49–57.

Muir, S. R., Collins, G. J., Robinson, S., Hughes, S., Bovy, A., Ric De Vos, C. H., van Tunen, A. J., & Verhoeyen, M. E. (2001). Overexpression of petunia chalcone isomerase in tomato results in fruit containing increased levels of flavonols. *Nature Biotechnology, 19*, 470–474.

Nesi, N., Debeaujon, I., Jond, C., Pelletier, G., Caboche, M., & Lepiniec, L. (2000). The TT8 gene encodes a basic helix-loop-helix domain protein required for expression of DFR and BAN genes in *Arabidopsis siliques. Plant Cell, 12*, 1863–1878.

Nesi, N., Jond, C., Debeaujon, I., Caboche, M., & Lepiniec, L. (2001). The Arabidopsis TT2 gene encodes an R2R3 MYB domain protein that acts as a key determinant for proanthocyanidin accumulation in developing seed. *Plant Cell, 13*, 2099–2114.

Ogita, S., Uefuji, H., Morimoto, M., & Sano, H. (2004). *Plant and Molecular Biology, 54*, 931–941.

Ohara, K., Tomomi, U., Tsuyoshi, e., Fumihiko, S., & Kazufumi, Y. (2003). Limonene production in tobacco with Perilla limonene synthase cDNA. *Journal of Experimental Botany, 54*, 2635–2642.

Paine, J. A., Shipton, C. A., Chaggar, S., Howells, R. M., Kennedy, M. J., Vernon, G., Wright, S. Y., Hinchliffe, E., Adams, J. L., Silverstone, A. L., & Drake, R. (2005). Improving the nutritional value of golden rice through increased pro-vitamin A content. *Nature Biotechnology, 23*, 482–487.

Palazon, J., Ocana, A. N., Vazquez, L. H., & Mirjalili, M. H. (2008). Application of metabolic engineering to the production of Scopolamine. *Biomolecule, 13*, 1722–1742.

Peebles, C.A.M., (2009). Metabolic engineering of the terpenoid indole alkaloid pathway of Catharanthus roseus hairy roots. PhD thesis, Rice University.

Pfeifer, B. A., & Khosla, C. (2001). Biosynthesis of polyketides in heterologous hosts. *Microbiology and Molecular Biology Reviews, 65*, 106–118.

Pfeifer, B., Hu, Z., Licari, P., & Khosla, C. (2002). Process and metabolic strategies for improved production of *Escherichia coli*-derived 6-deoxyerythronolide B. *Applied and Environmental Microbiology, 68*, 3287–3292.

Pichersky, E., & Gang, D. R. (2000). Genetics and biochemistry of secondary metabolites in plants: an evolutionary perspective. *Trends in Plant Science, 5*, 439–445.

Pieper, R., Luo, G., Cane, D. E., & Khosla, C. (*1995*). Cell-free synthesis of polyketides by recombinant erythromycin polyketide syntheses. *Nature, 378*, 263–266.

Ro, D. K., Paradise, E. M., Ouellet, M., Fisher, K. J., Newman, K. L., Ndungu, J. M., Ho, K. A., Eachus, R. A., Ham, T. S., Kirby, J., Chang, M. C., Withers, S. T., Shiba, Y., Sarpong, R., & Keasling, J. D. (2006). Production of the antimalarial drug precursor artemisinic acid in engineered yeast. *Nature, 440*, 940–943.

Robbins, M. P., Paolocci, F., Hughes, J. W., Turchetti, V., Allison, G., Arcioni, S., Morris, P., & Damiani, F. (2003). Sn, a maize bHLH gene, modulates anthocyanin and condensed tannin pathways in Lotus corniculatus 2003. *Journal of Experimental Botany, 54*, 239–248.

Roberts, S. C. (2007). Production and engineering of terpenoids in plant cell culture. *Nature of Chemical Biology, 3*, 387–395.

Robinson, J. A. (1991). Polyketide synthase complexes: their structure and function in antibiotic biosynthesis. *Philosophical Transactions of the Royal Society London B Biological Sciences, 332*(1263), 107–114.

Sa, G., Mi, M., He-chun, Y., Ben-ye, L., Guo-feng, L., & Kang, C. (2001). Effects of ipt gene expression on the physiological and chemical characteristics of *Artemisia annua* L. *Plant Science, 160*(4), 691–698.

Samanani, N., & Facchini, P. J. (2002). Purification and characterization of norcoclaurine synthase the first committed enzyme in benzylisoquinoline alkaloid biosynthesis in plants. *Journal Biological Chemistry, 277*, 33878–33883.

Samanani, N., Liscombe, D. K., & Facchini, P. J. (2004). Molecular cloning and characterization of norcoclaurine synthase, an enzyme catalyzing the first committed step in benzylisoquinoline alkaloid biosynthesis. *Plant Journal, 40*, 302–313.

Sanchez, C., Mendez, C., & Salas, J. A. (2006). Engineering biosynthetic pathways to generate antitumor indolocarbazole derivatives. *Journal of Industrial Microbiology and Biotechnology, 33*, 560–568.

Sanchez, C., Zhu, L., Brana, A. F., Salas, A. P., Rohr, J., Mendez, C., & Salas, J. A. (*2005*). Combinatorial biosynthesis of antitumor indolocarbazole compounds. *Proceedings of National Academy of Science USA, 102*, 461–466.

Sandmann, G., Römer, S., & Fraser, P. D. (*2006*). Understanding carotenoid metabolism as a necessity for genetic engineering of crop plants. *Metabolic Engineering, 8*, 291–302.

Schaller, H. B., Grausem, P., Benveniste, M.L., Chye, C.T.T., Song, Y.H., & Chua, N.H., (1995). Expression of the Hevea brasiliensis (H.B.K.) Mull. Arg. 3-hydroxy-3-methylglutaryl-coenzyme A reductase 1 in tobacco results in sterol overproduction. Plant Physiology 109: 761–770.

Shewmaker, C. K., Sheehy, J. A., Daley, M., Colburn, S., & Ke, D. Y. (1999). Seed-specific overexpression of phytoene synthase: increase in carotenoids and other metabolic effects. *Plant Journal, 20*, 401–412.

St-Pierre, B., Vazquez-Flota, F. A., & De Luca, V. (*1999*). Multicellular compartmentation of *Catharanthus roseus* alkaloid biosynthesis predicts intracellular translocation of a pathway intermediate. *Plant Cell, 11*, 887–900.

Sweetlove, L. J., Last, R. L., & Fernie, A. R. (*2003*). Predictive metabolic engineering: a goal for systems biology. *Plant Physiology, 132*, 420–425.

Takeshita, N., Fujiwara, H., Mimura, H., Fitchen, J. H., Yamada, Y., & Sato, F. (1995). Molecular cloning and characterization of S-adenosyl-L-methionine: scoulerine-9-O-methyltransferase from cultured cells of *Coptis japonica*. *Plant Cell Physiology, 36*, 29–36.

Teoh, K. H., Polichuk, D. R., Reed, D.W., Nowak, G., & Covello, P. S. (2006). *Artemisia annua* L. Asteraceae) trichome-specific cDNAs reveal CYP71AV1, a cytochrome P450 with a key role in the biosynthesis of the antimalarial sesquiterpene lactone artemisinin. *FEBS Letters, 580*, 1411–1416.

Tsuruta, H., Paddon, C. J., Eng, D., Lenihan, J. R., Horning, T., Anthony, L. C., Regentin, R., Keasling, J. D., Renninger, N. S., & Newman, J. D. (2009). High-level production of amorpha-4,11-Diene, a precursor of the antimalarial agent artemisinin, in *E. coli. PLoS ONE, 4*, e4489.

Valenzano, D. R., Terzibasi, E., Genade, T., Cattaneo, A., Domenici, L., & Cellerino, A. (2006). Resveratrol prolongs lifespan and retards the onset of age-related markers in a short-lived vertebrate. *Current Biology, 16*, 296–300.

Vander Heijden, R., Jacobs, D. I., Snoeijer, W., Hallared, D., & Verpoorte, R. (2004). The Catharanthus alkaloids: pharmacognosy and biotechnology. *Current Medical Chemistry, 11*, 607–628.

Vannelli, T., Wei Qi, W., Sweigard, J., Gatenby, A. A., & Sariaslani, F. S. (2007). Production of p-hydroxycinnamic acid from glucose in *Saccharomyces cerevisiae* and *Escherichia coli* by expression of heterologous genes from plants and fungi. *Metabolic Engineering, 9*, 142–151.

Walker, K., & Croteau, R. (2001). Taxol biosynthetic genes. *Phytochemistry, 58*, 1–7.

Walker, K., Schoendorf, A., & Croteau, R. (2000). Molecular cloning of a taxa-4(20), 11(12)-dien-5a-ol-O-acetyl transferase cDNA from Taxus and functional expression in Escherichia coli. *Archives of Biochemistry and Biophysics, 374*, 371–380.

Wallaart, T. E., Bouwmeester, H. J., Hille, J., Poppinga, L., & Maijers, N. C. (2001). Amorpha-4,11-diene synthase: cloning and functional expression of a key enzyme in the biosynthetic pathway of the novel antimalarial drug artemisinin. *Planta, 212*, 460–465.

Wani, M. C., Taylor, H. L., Wall, M. E., Coggon, P., & McPhail, A. T. (1971). Plant antitumor agents VI The isolation and structure of taxol, a novel antileukemic and antitumor agent from *Taxus brevifolia. Journal of the American Chemical Society, 93*, 2325–2327.

Wang, E., Rui, W., Joseph, D., John, H. L., Susheng, G., & George, J. W. (2001). Suppression of a P450 hydroxylase gene in plant trichome glands enhances natural-product-based aphid resistance. *Nature Biotechnology, 19*, 371–374.

Warzecha, H., Gerasimenko, I., Kutchan, T. M., & Stockigt, J. (2000). Molecular cloning and functional bacterial expression of a plant glucosidase specifically involved in alkaloid biosynthesis. *Phytochemistry, 54*, 657–666.

Weber, T., Welzel, K., Pelzer, S., Vente, A., & Wohlleben, W. (2003). Exploiting the genetic potential of polyketide producing streptomycetes. *Journal of Biotechnology, 106*, 221–232.

Whitmer, S. (1999). Aspects of terpenoid indole alkaloid formation by transgenic cell lines of *Catharanthus roseus* overexpressing tryptophan decarboxylase and strictosidine synthase Thesis LACDR, Leiden University, Leiden. *The Netherlands*.

Wildung, M. R., & Croteau, R. (1996). A cDNA clone for taxadiene synthase, the diterpene cyclase that catalyzes the committed step of taxol biosynthesis. *Journal of Biological Chemistry, 271*, 9201–9204.

Winkel-Shirley, B. (2001). Flavonoid biosynthesis a colorful model for genetics, biochemistry, cell biology and biotechnology. *Plant Physiology, 126*, 485–493.

Ye, X., Al-Babili, S., Kloti, A., Zhang, J., Lucca, P., Beyer, P., & Potrykus, I. (2000). Engineering the provitamin A (beta-carotene) biosynthetic pathway into (carotenoid-free) rice endosperm. *Science, 287*, 303–305.

Yu, O., Jung, W., Shi, J., Croes, R. A., Fader, G. M., McGonigle, B., & Odell, J. T. (2000). Production of the isoflavones genistein and daidzein in non-legume dicot and monocot tissues. *Plant Physiology, 124*, 781–794.

Yu, O., Shi, J., Hession, A. O., Maxwell, C. A., McGonigle, B., & Odell, J. T. (2003). Metabolic engineering to increase isoflavone biosynthesis in soybean seed. *Phytochemistry, 63*, 753–763.

Yun, D. J., Hashimoto, T., & Yamada, Y. (*1992*). Metabolic engineering of medicinal plants: transgenic *Atropa belladonna* with an improved alkaloid composition. *Proceedings of National Academy of Science USA, 89*, 11799–11803.
Zarate, R., Dirks, C., van der, H. R., & Verpoorte, R. (2001). Terpenoid indole alkaloid profile changes in *Catharanthus pusillus* during development. *Plant Science, 160*, 971–977.
Zhang, L., Ding, R., Chai, Y., Bonfill, M., Moyano, E., Caldentey, K. M. O., Xu, T., Pi, Y., Wang, Z., Zhang, H., Kai, K., Liao, Z., Sun, X., & Tang, K. (2004). Engineering tropane biosynthetic pathway in *Hyoscyamus niger* hairy root cultures. *Proceedings of National Academy of Science USA, 101*, 6786–6791.
Zhang, Y., Li, S. Z., Li, J., Pan, X., Cahoon, R. E., Jaworski, J. G., Wang, X., Jez, J. M., Chen, F., & Yu, O. (2006). Using unnatural protein fusions to engineer resveratrol biosynthesis in yeast and mammalian cells. *Journal of the American Chemical Society, 128*, 13030–13031.
Zhang, L., Jing, F., Li, F., Li, M., Wang, Y., Wang, G., Sun, X., & Tang, K. (2009). Development of transgenic *Artemisia annua* (Chinese wormwood) plants with an enhanced content of artemisinin, an effective anti-malarial drug, by hairpin-RNA mediated gene silencing. *Biotechnology Application of Biochemistry, 52*, 199–207.
Zimmermann, M. B., & Hurrell, R. F. (2002). Improving iron, zinc, and vitamin A nutrition through plant biotechnology. *Current Opinion of Biotechnology, 13*, 142–145.
Zook, M., Johnson, K., Hohn, T., & Hammerschmidt, R. (1996). Structural characterization of 15-hydroxytrichodiene, a sesquiterpenoid produced by transformed tobacco cell suspension cultures expressing a trichodiene synthase gene from *Fusarium sporotrichioides*. *Phytochemistry, 43*, 1235–1237.

CHAPTER 9

CRISPR/Cas9-mediated genome editing in medicinal and aromatic plants: developments and applications

Peerzada Arshid Shabir
Department of Botany, Government Degree College, Sopore, Jammu and Kashmir, India

Introduction

Plants are the "green chemical factories" with a wide range of metabolites that are either directly assimilated by humans or used as raw material in different industries. The safe application of these organic materials leads to more popularity of plant-derived medicines in the recent years (Soltani Howyzeh, Sadat Noori, Shariati, & Niazian, 2018) For a long time, classical biotechnology techniques have been exploited in order to obtain a multitude of beneficial traits like, from unraveling the entire metabolic pathway to increased metabolite and nutrient production, etc. However, conventional genetic engineering methods are not suitable for adding large-concerted changes into a plant, because the random insertion of DNA construct into one or more loci in one or more chromosomes can lead to undesirable effects. Therefore, these methods are not efficient to change the entire metabolic pathway in plants (Naqvi et al., 2010). Recent advances in genome engineering technologies are sparking a new revolution in biological research. Using next-generation sequencing (NGS) techniques, researchers can obtain the whole genomic and transcriptomic information of medicinal plants. These genomic and transcriptomics data can be combined with proteomics and metabolomics data to produce unidentified natural products of medicinal plants (Zhao et al., 2018).

The application of targeted genome-editing methods of transcription activator-like endonucleases (TALENs), zinc-finger nucleases (ZFNs), and CRISPR-Cas9 systems, subsequent to NGS analysis, can introduce synthetic biology to genetic and metabolic engineering in medicinal plants. It can create new varieties of medicinal plants that contain the

Medicinal and Aromatic Plants: Expanding their Horizons through Omics
http://dx.doi.org/10.1016/B978-0-12-819590-1.00009-4

desired-chemical compounds and/or new bio-products (Pouvreau, Vanhercke, & Singh, 2018). Protein-based editing tools, including TALENs and ZFNs have been powerful tools for manipulating genomes at the transcriptional level in the past few years. Both are artificial fusion proteins comprising an engineered DNA-binding domain fused to the nonspecific nuclease domain of the restriction enzyme FokI, and they have been used successfully in many organisms including plants (Palpant & Dudzinski, 2013). The latest ground-breaking technology for genome editing is based on RNA-guided engineered nucleases, which already hold great promise due to their simplicity, efficiency and versatility. The most widely used system is the type II clustered regularly interspaced short palindromic repeat (CRISPR)/Cas9 (CRISPR-associated) system from Streptococcus pyogenes (Jinek et al., 2012). CRISPR-Cas9 system has emerged as a revolutionary genome-engineering technology in clinical and agriculture research and there are increasing reports of application of CRISPR/Cas9 system in medicinal plants in recent years. (Weeks, Spalding, & Yang, 2015; Ma & Liu, 2016).

New modifications in CRISPR-Cas9 are being explored for novel applications such as silencing of multiple genes, simultaneously allowing characterization of gene families prevalent in plants. This will also provide a way to generate new variety of plants with multiple beneficial traits. Further, CRISPR-Cas9 system under the control of tissue-specific or inducible promoters can help in spatial or temporal genome altercations (Wang et al. 2015). Therefore, CRISPR-Cas9 has become an efficient, easy, and quick tool for raising improved plants with both addition of better traits and removal of undesirable features (Belhaj, Chaparro-Garcia, Kamoun, & Nekrasov, 2013). This system has also become a new tool of synthetic plant biology for designing minimal plant cell, such as engineering cells lacking nonessential components and division. These minimal cells can then be deployed as a basal system for production of novel biological systems (Bortesi & Fischer, 2015).

CRISPR-Cas9 mechanism

CRISPR is an acronym for clustered regularly interspaced short palindromic repeats and Cas9 is a nuclease associated with CRISPRs. This CRISPR system was first discovered in 1987 as a defense system in *Escherichia coli* against viruses. Later, they were found in 40% of sequenced bacterial genomes and 90% of archaea. A series of experiments involving bioinformatics tools unveiled various CRISPR/Cas components and their function in providing adaptive immunity to bacterial cells. In fact, scientists discovered that the CRISPR loci typically consist of a clustered set of

CRISPR-associated (Cas) genes and the signature CRISPR array—a series of 21–47 bp repeat sequences, (with an average of 32 bp—direct repeats) interspaced by 25–40bp variable sequences (spacers), which were consecutively repeated at regular intervals in the bacterial genome (Wang et al., 2015). In a given species, the nucleotides found in the repeats are highly conserved, and the spacers are also uniform in length, but with a highly variable sequence content. In 2005, sequencing and computational analyses indicated that the sequences of many CRISPR spacers are derived from extra-chromosomal DNA elements, with most being similar to bacteriophages and conjugative plasmids. A functional CRISPR/Cas locus is thus composed of two distinguishable components: an array of short-repeats interspersed with spacers derived from genomic sequences of an invading phage, and an operon of Cas genes (Jinek et al., 2012). Whereas Cas genes are translated into proteins, most CRISPR arrays are first transcribed as a single RNA before subsequent processing into shorter CRISPR RNAs (crRNAs), which direct the nucleolytic activity of certain Cas enzymes to degrade target nucleic acids. Overall, the adaptive immunity provided CRISPR/Cas systems can be divided into three stages. The first stage is CRISPR spacer acquisition. This occurs just following the entry of an invading virus or plasmid; specific fragments or protospacer sequences from the virus or plasmid are acquired and integrated into the CRISPR array between two adjacent repeats at the proximal end of a CRISPR locus. Duplication of the repeat sequence adjacent to the leader sequence is also required, creating a new repeat-spacer unit. Stage two involves CRISPR expression (Zhao et al., 2018). The CRISPR arrays, including the spacers, are transcribed during subsequent encounters with invasive DNA and are processed into small interfering CRISPR RNAs (crRNAs) by nucleases. Stage three is CRISPR interference. Mature crRNA-containing spacers and Cas proteins assemble to form stable CRISPR ribonucleoprotein complexes (RNPs) that scan the invading elements for a sequence complementary to the crRNA spacer. If the crRNA spacer matches the invasive nucleic acid perfectly, it initiates the cleavage of the invading DNA. Therefore, CRISPR/Cas systems provide elegant, effective, and dynamic immunity against invading genetic elements. At the same time, several clusters of CRISPR-associated (Cas) genes were identified to be well-conserved and typically adjacent to the repeat elements (Jansen et al., 2002), serving as a basis for the eventual classification of three different types of CRISPR systems (types I–III) (Haft, Selengut, Mongodin, & Nelson, 2005). Types I and III CRISPR loci contain multiple Cas proteins, now known to form complexes with crRNA to facilitate the recognition and destruction of target nucleic acids (Hale et al., 2009). In contrast, type II system originating from *Streptococcus pyogenes* can direct the cleavage of target DNA sites with a single Cas9 protein, thus making them the simplest of the three types. Cas9 is a large, multifunctional protein possessing two nuclease domains: one at the N-terminus called a RuvC-like domain, and the other in the middle section called the HNH domain, each cleaves one strand of a double- stranded DNA. Under natural conditions, Cas9 is inactive. It is activated when combined with the crRNA. The Cas9-crRNA complex scans a DNA double strand for protospacer-adjacent motifs (PAMs-the trinucleotide NGG) using Watson–Crick pairing between crRNA and targeted DNA. Once anchored at the proper PAMs, the HNH nuclease domain cleaves the RNA –DNA hybrid, while RuvC cleaves the other strand to form a double-strand break (DSB). In 2011, a trans-encoded small CRISPR RNA (tracrRNA) partially complementary to the repeat sequences of crRNA precursor transcripts was uncovered in the human pathogen *S. pyogenes* (type II CRISPR/Cas system) (Deltcheva et al., 2011). The tracrRNA directs pre-crRNA processing into mature crRNAs. This suggests that the CRISPR/Cas9 (type II system) is composed of at least three components—Cas9 endonuclease, CRISPR RNA (crRNA), and trans-activating crRNA (tracrRNA). In 2012, it was shown that in the type II DNA interference mechanism the mature crRNA and tracrRNA form a dual tracrRNA:crRNA structure that directs the Cas9 endonuclease to introduce DSB in the invasive target DNA (Jinek et al., 2012.) To simplify this system, Doudna and Charpentier (2014) engineered the dual tracrRNA:crRNA into a single guide RNA (sgRNA), which could also direct purified Cas9 to generate sequence- specific DSB in vitro. The sgRNA has two critical features: a 20-nucleotide sequence at the 5′ end that determines the specificity of the DNA target site through Watson-Crick base pairing, and a hairpin structure at the 3′-end that binds to Cas9, mimicking the dual tracrRNA:crRNA structure required to direct Cas9 to cleave target DNAs. Therefore, current CRISPR-Cas9 systems are modifications of bacterial CRISPR-Cas9 mechanism, comprising of modified Cas9 endonuclease and a sgRNA. Changing the 20-nucleotide sequence at the 5′-end of the sgRNA will enable the cleavage of any DNA of interest as long as there is a PAM (NGG) adjacent to the sequence. This seminal work highlights the potential to exploit the CRISPR/Cas9 system as an RNA-programmable genome-editing tool, and indeed it lays the foundation for CRISPR/Cas9-mediated genome editing. In practice, a synthetic sequence-specific nuclease is designed to recognize the chosen genomic site and is transfected into the cell, where it creates a double-strand DNA break (DSB) at the site. Once a DSB is generated, nonhomologous end-joining (NHEJ) or homology-directed repair (HDR) DNA repair mechanisms are initiated. HDR can accurately repair DSBs using the homologous sequence flanking a DSB or an exogenously supplied DNA "donor template" as the template. However, NHEJ repair is error-prone and frequently causes insertions or deletions around the sites of DNA breaks, leading to gene knockout. In eukaryotic cells, DSBs are preferentially repaired by NHEJ events, thus providing a promising strategy for research on plant functional genomics and crop improvement. In the past decade, ZFNs and TALENs have been successfully used in a variety of organisms and all three genome-editing technologies, ZFNs, TALENs, and CRISPR/Cas are able to induce DSBs at specific sites in the genome, which might be repaired by NHEJ or HDR that results in gene mutations at the target site. However, in contrast, ZFNs and TALENs are based on the protein-guided recognition mechanism and several publications have reported successful modifications in native plant genes, but these methods are relatively tedious, expensive and with only a small number of genes modified to date (Chen & Gao, 2014). Furthermore, delivery of genome engineering reagents into plant cells is a major barrier to the effective use of these technologies for creating novel traits (Baltes et al. 2014) and thus has not been widely adopted by the plant-research community until now. A more recently developed CRISPR/Cas system, which recognizes target DNA according to Watson-Crick base pairing between its guide RNA(s) and DNA, is the simplest one to implement and quickly became the most popular and powerful tool for genome engineering. The advanced CRISPR/Cas9 technology not only provides a molecular tool for investigating biological questions in depth, but also enables the development of innovative and practical applications of biology (Doudna & Charpentier, 2014). Immediately after its early use to edit the genomes of animals and bacteria (Cong et al., 2013; Hwang et al. 2013), its efficacy was demonstrated in the model plant systems of Arabidopsis, rice, sorghum, and tobacco (Feng et al., 2013; Li et al., 2013; Mao et al., 2013; Nekrasov, Staskawicz, Weigel, Jones, & Kamoun, 2013; Shan et al., 2013; Xie & Yang, 2013). Since these initial studies, CRISPR/Cas9 for targeted genome editing has become a mainstream method used by thousands of laboratories around the world. It has been applied easily, rapidly, and efficiently to edit endogenous genes in the variety of cell types and organisms and large number of genome-editing papers using CRISPR/Cas9 have been published.

Applications

With its high efficiency and simplicity, CRISPR/Cas9 and its modified versions have been widely explored in various organisms with many applications—gene mutation, gene expression repression or activation, and epigenome editing. In plants, the application of CRISPR/Cas9 is just emerging. Application of CRISPR/Cas9 in plant genome editing was first reported in 2013 targeting *Arabidopsis* (Jiang et al., 2013), and since then the CRISPR/Cas9 system has been successfully applied in many other plants, including *Sorghum bicolor* (Jiang et al., 2013), *Triticum aestivum* (Shan et al., 2013), *Oriza sativa* (Xie & Yang, 2013), *Nicotiana benthamiana* (Nekrasov et al., 2013), *Citrus sinensis* (Jia & Wang, 2014), *Nicotiana tabacum* (Baltes, Gil-Humanes, Cermak, Atkins, & Voytas, 2014), *Zea mays* (Liang, Zhang, Chen, & Gao, 2014), *Glycine max* (Jacobs, LaFayette, Schmitz, & Parrott, 2015), *Lactuva sativa* (Woo et al., 2015), *Medicago truncatula* (Michno et al., 2015), *Populus tremula* (Zhou, Jacobs, Xue, Harding, & Tsai, 2015), *Solanum tuberosum* (Wang et al., 2015), *Brassica oleracea* (Lawrenson et al., 2015), and *Cucumis sativus* (Chandrasekaran et al., 2016). Moreover, the modified genes generated by CRISPR/Cas9 show stable inheritance and expression for several generations and thus have been widely used to create various mutants and become a routine tool in plant laboratories around the world. To date, site-specific single gene knockout mutants, multiplex genome targeted knockout mutants, fragment deletions, gene replacements, and targeted gene insertions have been created using CRISPR/Cas9. The use of CRISPR-Cas9 system has also been extended to understand transcriptional regulation (Cheng et al., 2013), epigenetic modification (Hu et al. 2014), microscopic visualization of specific genome loci (Chen et al., 2013), epigenome editing (Morita et al., 2016), CRISPR-mediated activation (Lowder et al., 2018), interference (Tang et al., 2017), and knock in experiments (Wang, Sheng, Wen, & Du, 2017).

Plant secondary metabolites, with their great chemical diversity, varied biological functions, and pharmacological activities constitute interesting and important research. Understanding the origins and vast diversification of plant metabolism has been a long-standing goal in plant biology. Considering the complexity of the biosynthetic and regulatory process associated with the production of secondary metabolites, complete knowledge of the biosynthetic pathway including the various intermediates and enzymes that are involved is critical for the success of metabolic engineering (Zhao et al., 2018). The identification of genes, involved in the natural product synthesis by clustered or nonclustered biosynthetic pathways, is first priority of plant synthetic biologist to develop an artificial circuit for building a complex system that is competent to perform the desirable function in controlled manner. Till date, very little is known about biosynthetic genes of clinically used existing drugs which prevent access to engineered hosts for their production. However, recently discovered genome editing tool CRISPR/Cas9 technology has been precisely successful in identification of target genes for manipulation and tells about the possible consequences of such manipulation. The technique is becoming a popular strategy for optimizing metabolic pathways to improve the production of target metabolites. This molecular elucidation involves several approaches and several steps starting with the identification of genes or regulatory mechanisms such as transcription factors that control the secondary metabolite pathways up to the cloning and precise editing of the genes involved (Tang et al., 2017). For instance, CRISPR/Cas9system approach is being widely applied in down regulation or functional knockout of the genes to decrease the production of a certain unwanted group of compounds and increase the concentration of a desired secondary metabolite. The basic principle behind these approaches is knocking out an enzymatic step in the particular pathway by reducing the level of corresponding mRNA or protein. For example, CRISPR/Cas9 system was used effectively to control the production of bioactive alkaloids in the opium poppy where it was used to regulate benzylisoquinoline alkaloids (BIAs) metabolism and biosynthesis by knocking-out the 4′OMT2 gene (Alagoz, Gurkok, Zhang, & Unver, 2016). In *Dioscorea zingiberensis*, CRISPR/Cas9 system has been applied for targeted mutagenesis in Dzfps gene that led to reduction in the activity of farnesyl pyrophosphate synthase (FPS) enzyme and, subsequently, the content of squalene was 1.6 times less than wild- type plants. Recently, metabolic pathways and their dynamics producing dhurrin, a product of amino acid L-tyrosine that provides resistance against flea beetle, have been successfully characterized in Sorghum bicolor (Laursen et al., 2016) using CRISPR/Cas9 technology to make them emendable for biopharmaceuticals production. Metabolite engineering of this pathway has been carried out by nuclear genome transformation in Arabidopsis and tobacco (Tattersall et al., 2001) and organelle transformation in *N. tabacum* ensuring plasticity of this pathway (Gnanasekaran et al., 2016). The results of these recent studies show the significance of CRISPR/Cas9 system in medicinal plants, which can simply and purposefully change the chemical profile of useful medicinal plants and can also provide an alternative production system for PSM compounds for commercial exploitation.

Precise control of endogenous target gene expression is one of the core components of plant synthetic biology. Basically, this phenomenon has broad spectrum applications, from acceleration and suppression of single gene to switching ON/OFF multiple genes. To achieve desired transcription of endogenous genes, synthetic regulatory units are required and these regulatory units are synthetic promoters, activators, and suppressors. Earlier, to alter the transcription of target genes, a natural DNA-binding domain of TFs linked with short stretch of peptide that suppresses transcription has been used (Tang et al., 2017). Recently, identification and characterization of RNA-guided Cas9 nucleases open new avenues for desired manipulation in plant genome. Cas9 protein comprises of two major domains, RuvC and HNC with catalytic activity. Nuclease activity of the Cas9 protein can be inactivated by precise point mutation in catalytic domain. These Disabled nucleases can be used to regulate gene expression because they can still bind to their target DNA sequence. This variant of Cas9, known as dead Cas9 (dCas9), is fused with cytidine deaminase enzyme and can be guided at specific sites for base editing (Lowder et al., 2015). This protein is unable to cut DNA, but it can still be recruited to specific DNA sequences by gRNAs. Inactive Cas9 (dCas9) protein has potential to suppress the transcription activity by challenging the transcription machinery (Qi et al., 2013; Bikard et al., 2013). Along with this, dCas9 can be fused with the transactivation or trans-repression domain of a transcription factor and the precise and reversible transcriptional control of target genes becomes possible (Gilbert et al., 2013). Because the production of secondary metabolites is often dependent on the networking of several biosynthetic pathways, it is not surprising that a particular intermediate becomes rate limiting because the intermediate may be utilized by other competing pathways. Targeting this rate limiting through regulation of the enzymes associated with conversion of this intermediate toward the desired pathway for metabolite synthesis has been used as a strategy in many secondary metabolite synthesis pathways. For example, Piatek et al. (2015) modulated the transcription of both a reporter construct and the endogenous PDS gene in *N. benthamiana*, fusing the dCas9 C-terminus to the EDLL domain and the TAL activation domain to generate transcriptional activators and to the SRDX domain from the ERF transcription factor to generate a repressor. They observed that transcriptional activity was influenced by the position of the gRNA with respect to the transcriptional start site, as well as the nature of the target strand (sense or antisense). Zhou et al. (2018) applied the dCas9 for suppression of rosmarinic acid synthase gene (SmRAS) in *Salvia miltiorrhiza.*

The researchers designed the sgRNA that target RAS gene and the authors reported that in the hairy root extract of CRISPR/ Cas9-obtained mutants, the contents of RA and LAB compounds were decreased, whereas the volume of salvianic acid A sodium (SAAS) and sodium salt form of DHPL were increased. Even naked dCas9 without any effector domains has been shown to repress both synthetic and endogenous genes through the steric blocking of transcription initiation and elongation, although the degree of repression was modest in mammalian cells (Qi et al., 2013). The use of dCas9 for specific gene regulation provides an alternative approach in species that currently lack controllable expression systems. Multiple gRNAs targeting the same promoter also demonstrate synergistic effects, indicating that tuning the level of transcriptional control is possible using this approach (Piatek et al., 2015). In Addition to Its simple operation, Cas9 protein can also be guided by several gRNAs targeting different promoters and thus inactivating or editing multiple sites simultaneously (Qi et al., 2013). The adaptability of Cas9/gRNA to be directed to any DNA sequence makes them more amenable to fuse with any kind of effector domain to modulate genomic structure and function. An artificial DNA-binding domain can also be associated with effector domains such as transposases, DNA and histone methyl transferases, and acetyl transferases for modulation of epigenetic activities (Gaj, Gersbach, & Barbas, 2013). Though, epigenetic engineering is at nascent stage in plant systems. Still, researchers tried to confirm whether CRISPR/Cas9-mediated transcriptional activator would be able to reverse silencing effect of methylation. To confirm the hypothesis, dCas9 was fused with transcriptional activator domain VP64 (Beerli, Segal, Dreier, & Barbas, 1998). Under normal plant growth conditions, an imprinting gene, fertilization-independent seed 2 (AtFIS2), is silenced in vegetative tissue because of the methylation of regulatory region (Jullien, Kinoshita, Ohad, & Berger, 2006). The expression of AtFIS2 was considerably high in all transgenic plants harboring dCas9-VP64 construct. Thus, a dCas9 linked with transcriptional activator domain can recognize methylated DNA and drastically activate silenced genes in plants (Lowder et al., 2015).

Conclusion

CRISPR/Cas9 technology has several attractive features, including high-efficiency, ease of use, versatility, and the capability for multiplexed modifications; thus, it has become the most promising genome editing tool and boasts of a promising future in making the desired mutation in plants. At present, the CRISPR-Cas system has been tested in only a few medicinal plants, due to lack of availability of sufficient sequence information in many MAPs, but we firmly believe that intense further research will harness the full potential of applying the CRISPR-Cas system in other medicinal plants to identify the genes and enzymes that are involved in the biosynthetic pathway of different secondary metabolites. Recent progress demonstrates that the CRISPR/Cas9 technology is becoming the ultimate molecular tool for genome engineering and the forte of gene editing in plants including medicinal plants has been radically changed by CRISPR-Cas9 technology. In complex metabolic pathways where the modification of a single gene is not desirable, multiplex genome editing through the targeted knockout of a few genes, or the up- or down-regulation of several genes simultaneously, will create valuable agronomic traits in target plants. With a broader suite of CRISPR tools, we can expect that complex traits will be modified at will in the near future. The CRISPR- Cas9 thus holds the promising potential to reshape biosynthetic pathway in heterologous medicinal plants with artificially designed and precise controlled genetic circuit to maximize production of biopharmaceuticals. Nonetheless, to have a much greater impact on medicinal plant biology, further efforts are needed to optimize the CRISPR/Cas9 protocols for making it more user-friendly and freely accessible for research and practical applications.

References

Alagoz, Y., Gurkok, T., Zhang, B., & Unver, T. (2016). Manipulating the biosynthesis of bioactive compound alkaloids for next-generation metabolic engineering in opium poppy using CRISPR-Cas 9 genome editing technology. *Scientific Report*, *3*, 1–6.

Baltes, N. J., Gil-Humanes, J., Cermak, T., Atkins, P. A., & Voytas, D. F. (2014). DNA replicons for plant genome engineering. *The Plant Cell*, *26*, 151–163.

Beerli, R. R., Segal, D. J., Dreier, B., & Barbas, C. F. (1998). Toward controlling gene expression at will: specific regulation of the erbB-2/HER-2 promoter by using polydactyl zinc finger proteins constructed from modular building blocks. *Proceedings of the National Academy of Sciences of the United States of America*, *95*(25), 14628–14633.

Belhaj, K., Chaparro-Garcia, A., Kamoun, S., & Nekrasov, V. (2013). Plant genome editing made easy: targeted mutagenesis in model and crop plants using the CRISPR/Cas system. *Plant Methods*, *9*, 1–39.

Bikard, D., Jiang, W., Samai, P., Hochschild, A., Zhang, F., & Marraffini, L. A. (2013). Programmable repression and activation of bacterial gene expression using an engineered CRISPR-Cas system. *Nucleic Acids Research*, *41*(15), 7429–7437.

Bortesi, L., & Fischer, R. (2015). The CRISPR/Cas9 system for plant genome editing and beyond. *Biotechnology Advances.*, *33*, 41–52.

Chandrasekaran, J., Brumin, M., Wolf, D., Leibman, D., Klap, C., Pearlsman, M., et al. (2016). Development of broad virus resistance in non–transgenic cucumber using CRISPR/Cas9 technology. *Molecular Plant Pathology*, *17*, 1140–1153.

Chen, K., & Gao, C. (2014). Targeted genome modification technologies and their applications in crop improvements. *Plant Cell Reports*, *33*, 575–583.

Chen, S., Oikonomou, G., Chiu, C. N., Niles, B. J., Liu, J., Lee, D. A., et al. (2013). A large-scale in vivo analysis reveals that TALENs are significantly more mutagenic than ZFNs generated using context- dependent assembly. *Nucleic Acids Research*, *41*, 2769–2778.

Cheng, A. W., Wang, H., Yang, H., Shi, L., Katz, Y., Theunissen, T. W., et al. (2013). Multiplexed activation of endogenous genes by CRISPR-on, an RNA-guided transcriptional activator system. *Cell Research*, *23*, 1163–1171.

Cong, L., Ran, F. A., Cox, D., Lin, S., Barretto, R., & Habib, N. (2013). Multiplex genome engineering using CRISPR/Cas systems. *Science*, *339*, 819–823.

Deltcheva, E., Chylinski, K., Sharma, C. M., Gonzales, K., Chao, Y., Pirzada, Z. A., et al. (2011). CRISPR RNA maturation by trans-encoded small RNA and host factor RNase III. *Nature*, *31*, 602–607.

Doudna, J. A., & Charpentier, E. (2014). Genome editing. The new frontier of genome engineering with CRISPR-Cas9. *Science*, *346*(6213), 1258096.

Feng, Z., Zhang, B., Ding, W., Liu, X., Yang, D. L., Wei, P., Cao, F., Zhu, S., Zhang, F., Mao, Y., & Zhu, J. K. (2013). Efficient genome editing in plants using a CRISPR/Cas system. *Cell Research*, *23*(10), 1229–1232.

Gaj, T., Gersbach, C. A., & Barbas, C. F. (2013). III, ZFN, TALEN, and CRISPR/Cas-based methods for genome engineering. *Trends in Biotechnology*, *31*(7), 397–405.

Gilbert, L. A., Larson, M. H., Morsut, L., Liu, Z., Brar, G. A., & Torres, S. E. (*2013*). CRISPR-mediated modular RNA-guided regulation of transcription in eukaryotes. *Cell*, *154*, 442–451.

Gnanasekaran, T., Karcher, D., Nielsen, A. Z., Martens, H. J., Ruf, S., Kroop, X., Olsen, C. E., Motawie, M. S., Pribil, M., Moller, B. L., & Bock, R. (2016). Transfer of the cytochrome P450-dependent dhurrin pathway from *Sorghum bicolor* into *Nicotiana tabacum* chloroplasts for light-driven synthesis. *Journal of Experimental Botany*, *67*(8), 2495–2506.

Haft, D. H., Selengut, J., Mongodin, E. F., & Nelson, K. E. (2005). A guild of 45 CRISPR-associated (Cas) protein families and multiple CRISPR/Cas subtypes exist in prokaryotic genomes. *PLoS Computational Biology*, *1*(6), 1–60.

Hale, C. R., Zhao, P., Olson, S., Duff, M. O., Graveley, B. R., & Wells, L. (2009). RNA-guided RNA cleavage by a CRISPR RNA-Cas protein complex. *Cell*, *139*, 945–956.

Hu, Z., Yuan, T., Ye, X., Peipei, C., Wei, S., Yuhua, S., Licheng, G., Haibo, H., Chao, X., Shilin, C., & Xiuqiao, Z. (2014). Rapid identification and verification of indirubin-containing medicinal plants. *Evidence Based Complement. Alternative Medicine*, *2015*, 1–15.

Hwang, W. Y., Fu, Y., Reyon, D., Maeder, M. L., Tsai, S. Q., & Sander, J. D. (2013). Efficient genome editing in zebrafish using a CRISPR-Cas system. *Nature Biotechnology*, *31*, 227–229.

Jacobs, T. B., LaFayette, P. R., Schmitz, R. J., & Parrott, W. A. (2015). Targeted genome modifications in soybean with CRISPR/Cas9. *BMC Biotechnology*, *15*, 15–31.

Jansen, R., Embden, J. D., Gaastra, W., & Schouls, L. M. (2002). Identification of genes that are associated with DNA repeats in prokaryotes. *Mol. Microbiol.*, *43*, 1565–1575.

Jia, H., & Wang, N. (2014). Targeted genome editing of sweet orange using Cas9/sgRNA. *PLoS One*, *9*(4), 1–6.

Jiang, W. Z., Zhou, H. B., Bi, H. H., Fromm, M., Yang, B., & Weeks, D. P. (2013). Demonstration of CRISPR/Cas9/sgRNA-mediated targeted gene modification in Arabidopsis, tobacco, sorghum and rice. *Nucleic Acids Research*, *41*(20), e188.

Jinek, M., Chylinski, K., Fonfara, I., Hauer, M., Doudna, J. A., & Charpentier, E. (2012). A programmable dual-RNA-guided DNA endonuclease in adaptive bacterial immunity. *Science*, *337*, 816–821.

Jullien, P. E., Kinoshita, T., Ohad, N., & Berger, F. (2006). Maintenance of DNA methylation during the *Arabidopsis* life cycle is essential for parental imprinting. *The Plant Cell*, *18*(6), 1360–1372.

Laursen, T., Borch, J., Knudsen, C., Bavishi, K., Torta, F., Martens, H. J., Silvestro, D., Hatzakis, N. S., Wenk, M. R., Dafforn, T. R., & Olsen, C. E. (2016). Characterization of a dynamic metabolon producing the defense compound dhurrin in sorghum. *Science*, *354*, 890–893.

Lawrenson, T., Shorinola, O., Stacey, N., Li, C., Ostergaard, L., Patron, N., Uauy, C., & Harwood, W. (2015). Induction of targeted, heritable mutations in barley and *Brassica oleracea* using RNA-guided Cas9 nuclease. *Genome Biology*, *16*(1), 258–1258.

Li, J. F., Norville, J. E., Aach, J., McCormack, M., Zhang, D., & Bush, J. (2013). Multiplex and homologous recombination-mediated genome editing in *Arabidopsis* and *Nicotiana benthamiana* using guide RNA and Cas9. *Nature Biotechnology*, *31*, 688–691.

Liang, Z., Zhang, K., Chen, K., & Gao, C. (2014). Targeted mutagenesis in Zea mays using TALENs and the CRISPR/Cas system. *Journal of Genetics and Genomics*, *41*, 63–68.

Lowder, L. G., Paul, J. W., Baltes, N. J., Voytas, D. F., Zhang, Y., Zhang, D., Tang, X., Zheng, X., Hsieh, T. F., & Qi, Y. (2015). A CRISPR/Cas9 toolbox for multiplexed plant genome editing and transcriptional regulation. *Plant Physiology*, *169*(2), 971–985.

Lowder, L. G., Zhou, J., Zhang, Y., Malzahn, A., Zhong, Z., Hsieh, T. F., Voytas, D. F., Zhang, Y., & Qi, Y. (2018). Robust transcriptional activation in plants using multiplexed CRISPR-Act 2.0 and mTALE-act systems. *Molecular Plant*, *11*(2), 245–256.

Ma, X., & Liu, Y. G. (2016). CRISPR/Cas9-based genome editing systems and analysis of targeted genome mutations in plants. *Hereditas*, *38*, 118–125.

Mao, Y., Zhang, H., Xu, N., Zhang, B., Gou, F., & Zhu, J. K. (2013). Application of the CRISPR-Cas system for efficient genome engineering in plants. *Molecular Plant*, *6*, 2008–2011.

Michno, J. M., Wang, X., Liu, J., Curtin, S. J., Kono, T. J., & Stupar, R. M. (2015). CRISPR/Cas mutagenesis of soybean and *Medicago truncatula* using a new web-tool and a modified Cas9 enzyme. *GM crops & food*, *6*, 243–252.

Morita, S., Noguchi, H., Horii, T., Nakabayashi, K., Kimura, M., Okamura, K., Sakai, A., Nakashima, H., Hata, K., Nakashima, K., & Hatada, I. (2016). Targeted DNA demethylation *in vivo* using dCas9-peptide repeat and scFv-TET1 catalytic domain fusions. *Nature Biotechnology*, *34*, 1060–1065.

Naqvi, S., Farre, G., Sanahuja, G., Capell, T., Zhu, C., Christou, P., Farre, G., & Christou, P. (2010). When more is better: multigene engineering in plants. *Trends in Plant Science*, *15*, 48–56.

Nekrasov, V., Staskawicz, B., Weigel, D., Jones, J. D., & Kamoun, S. (2013). Targeted mutagenesis in the model plant *Nicotiana benthamiana* using Cas9 RNA-guided endonuclease. *Nature Biotechnology*, *31*, 691–693.

Palpant, N. J., & Dudzinski, D. (2013). Zinc finger nucleases: looking toward translation. *Gene Therapy*, *20*, 121–127.

Piatek, A., Ali, Z., Baazim, H., Li, L., Abulfaraj, A., & Al-Shareef, S. (2015). RNA-guided transcriptional regulation in planta via synthetic dCas9-based transcription factors. *Plant Biotechnology Journal*, *13*, 578–589.

Pouvreau, B., Vanhercke, T., & Singh, S. (2018). From plant metabolic engineering to plant synthetic biology: the evolution of the design/ build/test/learn cycle. *Plant Science*, *273*, 1–10.

Qi, L. S., Larson, M. H., Gilbert, L. A., Doudna, J. A., Weissman, J. S., & Arkin, A. P. (2013). Repurposing CRISPR as an RNA-guided platform for sequence-specific control of gene expression. *Cell*, *152*, 1173–1183.

Shan, Q., Wang, Y., Li, J., Zhang, Y., Chen, K., & Liang, Z. (2013). Targeted genome modification of crop plants using a CRISPR-Cas system. *Nature Biotechnology*, *31*, 686–688.

Soltani Howyzeh, M., Sadat Noori, S. A., Shariati, J. V., & Niazian, M. (2018). Essential oil chemotype of iranian ajowan (*Trachyspermum ammi* L.). *Journal of Essential Oil Bearing Plants*, *21*(1), 273–276.

Tang, X., Lowder, L. G., Zhang, T., Malzahn, A. A., Zheng, X., Voytas, D. F., Zhong, Z. H., Chen, Y.Y., Ren, Q. R., Li, Q., Kirkland, E. R., Zhang, Y., & Qi, Y. P. (2017). A CRISPR-Cpf1 system for efficient genome editing and transcriptional repression in plants. *Nature Plants*, *3*, 17018.

Tattersall, D. B., Bak, S., Jones, P. R., Olsen, C. E., Nielsen, J. K., Hansen, M. L., Hoj, P. B., & Moller, B. L. (*2001*). Resistance to an herbivore through engineered cyanogenic glucoside synthesis. *Science*, *293*, 1826–1828.

Wang, L. J., Sheng, M.Y., Wen, P. C., & Du, J.Y. (2017). Morphological, physiological, cytological and phytochemical studies in diploid and colchicine-induced tetraploid plants of *Fagopyrum tataricum* (L.) Gaertn. *Botanical Studies*, *58*, 1–2.

Wang, S., Zhang, S., Wang, W., Xiong, X., Meng, F., & Cui, X. (2015). Efficient targeted mutagenesis in potato by the CRISPR/Cas9 system. *Plant Cell Reports*, *34*, 1473–1476.

Wang, Z. P., Xing, H. L., Dong, L., Zhang, H.Y., Han, C.Y., & Wang, X. C. (2015). Egg cell-specific promoter-controlled CRISPR/Cas9 efficiently generates homozygous mutants for multiple target genes in Arabidopsis in a single generation. *Genome Biology*, *16*, 144.

Weeks, D. P., Spalding, M. H., & Yang, B. (2015). Use of designer nucleases for targeted gene and genome editing in plants. *Plant Biotechnology Journal*, *14*, 483–495.

Woo, J. W., Kim, J., Kwon, S. I., Corvalán, C., Cho, S. W., Kim, H., et al. (2015). DNA-free genome editing in plants with preassembled CRISPR-Cas9 ribonucleoproteins. *Nature Biotechnology*, *33*(11), 1162–1164.

Xie, K., & Yang, Y. (2013). RNA-guided genome editing in plants using a CRISPR–Cas system. *Molecular plant*, *6*, 1975–1983.

Zhao, Y., Li, L., & Zheng, G. (2018). CRISPR/dCas9-mediated multiplex gene repression in Streptomyces. *Biotechnology Journal*, *13*, 1800121.

Zhou, X., Jacobs, T. B., Xue, L. J., Harding, S. A., & Tsai, C. J. (2015). Exploiting SNPs for biallelic CRISPR mutations in the outcrossing woody perennial Populus reveals 4-coumarate: CoA ligase specificity and redundancy. *New Phytologist*, *208*, 298–301.

Zhou, Z., Tan, H., Li, Q., Chen, J., Gao, S., Wang, Y., et al. (2018). CRISPR/Cas9-mediated efficient targeted mutagenesis of RAS in *Salvia miltiorrhiza*. *Phytochemistry*, *148*, 63–70.

CHAPTER 10

Proteomics research in aromatic plants and its contribution to the nutraceuticals and pharmaceutical outcomes

Jameel R. Al-Obaidi
Kuala Lumpur, Malaysia

Introduction

Medicinal and aromatic plants (MAPs) have been deeply rooted and associated with human civilization and evolution from time immemorial (Kala, 2000, 2005; Niero, Cechinel Filho, & Yunes, 2018). The historical and cultural acceptability and reliability of MAPs may place this precious natural resource in high-demand (Daniel, 2016; Pundir & Jain, 2015). Apart from the use of existing knowledge and information of MAPs, there are a number of stages in MAPs segment from the collection of plants information to breeding and cultivation, harvesting and extraction of active compounds, value addition, and processing and drug development. Medicinal and fragrant plants occupy a substantial role in human cultures that have assisted to improve the wellbeing of people since early times (Fokunang et al., 2011). Preliminary records on use of medicinal and aromatic plants were found in various parts of the world, such as China, Greece, Middle East, and India, representing that these early civilizations used native fragrant and medicinal plants to improve their health in their own distinct ways before ideas were shared. Medicinal and aromatic plants, however, continue to influence human life, culture, and history (Inoue, Hayashi, & Craker, 2017). Plant-based extracts or secondary metabolite affects the physiological processes of different living organisms (Vaghasiya, Dave, & Chanda, 2011). Therefore, the major attention of the researchers is to improve their specific products (metabolite) which is useful due to its medicinal, pharmaceutical or aromatic properties. Since these are important aspects in the medicinal field of science, researchers should find the approaches for its improvement and more

Medicinal and Aromatic Plants: Expanding their Horizons through Omics
http://dx.doi.org/10.1016/B978-0-12-819590-1.00010-0

effective medicinal plants breeding methods (Nunes & Albuquerque, 2018). Based on previous reports, about three-quarter of the human population is consuming products with medicinal herbs origin with a quarter of prescribed medicines are derived from wild-medicinal plants (Alamgir, 2017; De Silva, 1997; Ekor, 2014). Phytotherapy is based on the use of those plant-based remedies and medicinal products for the purpose of avoidance and handling illnesses. The quality of each final product is only guaranteed by the use of raw supplies of high quality, the defined process of manufacturing, and validated equipment. Keeping a high control on the production process of herbal drugs and herbal isolates such as plant extracts and essential oils is done according to the requirements of pharmaceuticals preparation requirements (Benzie & Wachtel-Galor, 2004). Notably, medicinal and aromatic plants and their derivatives can serve as starting materials for pharmaceutical research and drug development. Current areas of utilization constitute powerful drivers for the usage of these natural resources (Atanasov et al., 2015). The elevated demands, together with the restricted accessibility, and possible exhaustion of these plant resources, make it required to increase efforts to obtain more knowledge regarding research and development and production (Agrawal, Tsay, Shyur, Wu, & Wang, 2017). There are thousands of plant species worldwide with medicinal and aromatic properties (Nunes & Albuquerque, 2018), the improved understanding of the growth and the expansion of herbal medicines, contributing to the search for plant-based active substances to develop drugs and increase beneficial options for healthcare professionals (Pan et al., 2013). Enhancement of the productivity and quality of these natural plant products through conventional breeding is still a challenge. However, recent improvements in plant "omics" research have generated knowledge leading to a better understanding of the complexity of biochemistry involved in the biosynthesis of these plant secondary metabolites. This omics research also concerned the identification and isolation of genes involved in different steps of a number of metabolic pathways (Van Emon, 2016). It is a known fact that isolated plant products and active materials played a major role in the development of modern drug development. Many of the extracted compounds are still used nowadays, or they have functioned as an example for the production of a large number of drugs (Pan et al., 2013). Over time, the use of herbal drugs and other natural products have developed on the basis of both progressive and unsuccessful experiences. The collected rich experiences have progressively established into traditional medicine in different parts of the world (Yuan, Ma, Ye, & Piao, 2016). The process of extract active compounds and its derivatives from medicinal and aromatic plant considered a complex process and require highly sophisticated methods with respect to aromatic and medicinal plant science. In respect to that, omics platforms such as proteomics techniques is a very important in the development of aromatic and medicinal plant research, in this chapter we review the studies involved in plant metabolic pathways that produce bioactive compounds, the growth of these plants, biological and non-biological stress and also the effectiveness of those plants and their compounds in vivo and in vitro toward the contribution to human health.

Proteomics studies on essential oil from aromatic plants

Cloves consider important aromatic plant with many therapeutic properties, essential oil from the plant proof to disturb proteins such as lipopolysaccharide and serine protease of the food poisoning bacteria *Campylobacter jejuni* which might give the oil antibacterial properties and can be used to control this foodborne illness (Kovács et al., 2016). Australian tea tree *Melaleuca alternifolia* essential oil is known to have antimicrobial activities (Carson & Riley, 1995), in a high-throughput iTRAQ proteomic study, showed that the treatment of the oil of the necrotrophic fungus Botrytis cinerea. The oil affecting metabolism pathways of the fungus by differentiation of more than 700 proteins (Xu, Shao, Wei, Xu, & Wang, 2017). Green cardamom (*Elettaria cardamomum*) is an aromatic spice planted mainly in South Asia for medicinal potential uses besides the main use in many Asian cuisines (Sengupta & Bhattacharjee, 2009). Extract from cardamom showed anti-proliferative activities toward human oral carcinoma cell, protein-protein interaction in that study showed the involvement of PP1-HDAC2-p53 and ERK1/2-p53 pathways in-bisabolene-induced mitochondrial apoptosis, this phosphoproteomic-based research useful for emerging anti-cancer medicines from γ-bisabolene (Jou et al., 2015).

Proteomic of growth and cultivation of medicinal and aromatic plants

The family Zingiberaceae considered one of the well-studied groups of the plant with great beneficial medicinal and nutraceuticals potential benefits to human health (Kumar, Asish, Sabu, & Balachandran, 2013). Turmeric is one of the plants that is widely used in many dishes especially in Asia, the effect of storage on the plant rhizomes has been studied using proteomic platform which showed sweet potato storage protein such as sporamin were highly expressed during the dormant stage (Chokchaichamnankit et al., 2009).

Curcumin is the main active compound from turmeric with anticancer and antiinflammatory properties proteomic study revealed that curcumin binds to different proteins including structural, enzymes and proteins involved in apoptosis pathway during interaction with mouse brain (Firouzi et al., 2014). Another proteomic study on the same direction showed the ability of the curcumin to contribute to anti-cancer activity on colon cancer by binding to hundreds of proteins (Wang et al., 2016) and curcumin-induced signaling during interaction with oral cancer cell line (Sathe et al., 2016). Not only the direct contribution of turmeric to human health has been investigated but also the studies on plant disease have been investigated such as proteomic study against phytopathogenic fungi *Fusarium graminearum* which shown the ability of the plant extract to disrupt cell membrane and inhibit of ergosterol synthesis and respiratory chain (Chen et al., 2018). *Curcuma comosa* is another plant in the same family with medicinal properties. A proteomic study has also been conducted on this ginger-like plant which revealed lectins and antioxidant proteins that is related to their activity found in other Zingiberaceae plant species (Boonmee, Srisomsap, Chokchaichamnankit, Karnchanatat, & Sangvanich, 2011). In the same family ginger *Zingiber officinale* also it is widely used as a spice in food and beverages. Similar, many of the family species have antioxidant and anticancer activities (Y. Li, Tran, Duke, & Roufogalis, 2012). Changes in the protein in ginger from both China and Ghana were correlated with the metabolite changes in both varieties which indicate that metabolism-linked enzymes are accountable for the difference within ginger species (Yin et al., 2018). Fennel flower seed (black seed) is a plant with cosmetic and pharmaceutical products due to its various biological activities, the seed oil has potential anticancer and antidiabetic activity (Jan, Ahmad, Rehman, Gani, & Khaqan, 2019). Proteomic research conducted on the plant leaves explained the plant tolerance against drought resistance, proteins such as phosphoribulokinase RuBisCO and glyceraldehyde-3-phosphate dehydrogenase founded to have a contribution in the plant tolerance (Khodadadi et al., 2017; Khodadadi et al., 2019). Proteomic study on the plant revealed several seeds storage proteins, proteins acting as catalytic activity enzymes and transport and binding proteins that are responsible for synthesis and storage of different components, such as oils (Alanazi et al., 2016). Fenugreek (*Trigonella foenum-graecum* L.) is an annual seed spice widely distributed around the world. The wild or cultivated fenugreek is commonly used as a traditional food or medicine due to anti-diabetic, hypocholesterolaemic, and anti-microbial effects (Snehlata & Payal, 2012). Proteomic signature of fenugreek indicated changes in the photosynthetic proteins related to defense upon treatment with methyl jasmonate and cholesterol (Ciura et al., 2017). Another study on this plant investigated the effect of planting time on the growth of the herb (Mostafaie et al., 2018).Vanilla species considered a very important aromatic plants group with some potential medicinal benefits such as anticancer, antimicrobial and antianxiety beside the main use for the plant for natural flavoring (Bythrow, 2005). In a gel-proteomics study on the species *Vanilla planifolia*, it has been found that both vanillin precursor, coumaric acid, and glucoside, produced at the early stage of the plant shooting differentiation (Palama et al., 2010). Callus differentiation of the same species have also been investigated at the molecular level whereas transport, metabolism, transcription and to a more extend stress proteins were expressed (Tan, Chin, Liddell, & Alderson, 2013). Gel-free proteomics revealed that another species from this Vanilla species could be a good alternative source for the production of natural vanilla by identifying proteins linked to vanillin production such ACC synthase, 2 chalcone-flavanone isomerase and vanillin synthase (Lopes et al., 2019). One of the most well-studied (especially at the molecular level) medicinal plant genus is Mentha (Maffei, Bertea, & Mucciarelli, 2007). Many proteomic types of research have been conducted on this taxon due to its importance in the food industry as long as its medicinal usage such as the study on *Mentha spicata* leaf. The study identified a few proteins responsible for terpenoid and phenylpropanoids synthesis (Champagne & Boutry, 2013). Saffron (*Crocus sativus*) one of the priciest aromatic plants used in traditional medicine to treat cough and asthma (Gohari, Saeidnia, & Mahmoodabadi, 2013). Proteomic analysis of the plant grown under Cadmium (Cd) contamination whereas many proteins related to energy, heat-shock, metabolism, signaling and defense and heat shock showed differentially expression profile compared to the control samples which indicated a disruption in the photosynthesis and chloroplast function of the plant leave (Rao, Lv, & Yang, 2017). Somatic embryogenesis process in saffron affected the molecular process of the plant, changes in protein groups similar to those changes affected by cadmium such as heat shock protein and energy proteins (Sharifi, Ebrahimzadeh, Ghareyazie, Gharechahi, & Vatankhah, 2012). However, despite these earlier studies with efficient protein extraction protocols for the plant leaves (Hurkman & Tanaka, 1986) and callus (Carpentier et al., 2005), a more recent study compare protein extraction methods from leaves and floral parts for proteomic analysis, the study concluded that NP-40 extraction method is the best to serve the purpose (Mehraj, Kamili, Nazir, Haq, & Balkhi, 2018). Key proteins that involve in the differentiation of flower bud were detected and identified by Zhang and his group (Zhang, Wei, Li, Jiang, & Zhang, 2016). The researchers found gene regulatory proteins involved in the biosynthetic components of saffron flower stigmas. Chaperonin and anti-oxidant related proteins could help in industrial selective breeding of the plant. To differentiate between saffron stigmas and other plant adulterants a proteomic study conducted to seek for protein-based bio-marker such as phosphoenolpyruvate carboxylase, α-1,4-glucan-protein synthase and glyceraldehydes-3-phosphate dehydrogenase considering the age and the geographical origin of the saffron (Paredi et al., 2016). American ginseng is a traditional medicinal plant with potential healing effects on a few types of cancer and Parkinson's disease (Lei et al., 2017). Hybrid ginseng leaves proteins related to carbohydrate catabolism and nitrogen fixation was linked the heterosis of the hybrid biomass and ginsenoside production in a proteomic study conducted by Lei and his group (Lei et al., 2017). The American ginseng (*Panax quinquefolius*) differ in application and its medicinal properties with the Asian ginseng (*Panax ginseng*) which considered when of the well-studied medicinal and aromatic plant at the molecular level, therefore a study at the molecular level to differentiate between the two species was necessary. A proteomic study conducted to tackle this issue was identified proteins to differentiate not only the two species but also to differentiate diverse parts such as main roots, skin, and the rhizome of the same species (Lum et al., 2002). A later study used the proteomic platform to differentiate roots, leaves and cultured hairy roots which revealed tissue-specific proteins existed in leaves and main roots but not the cultured hairy root (S. I. Kim et al., 2003). Another recent proteomic on the hairy root materials were investigating the changes in the proteins during the cultivation of Red sage hairy roots identified more than 800 proteins with 5 novel Cytochromes P450 (CYPs) and another 5 novel alkaloids enzyme that believed to be responsible tanshinones synthesis which itself responsible for anti-oxidant and anti-inflammatory properties (Contreras, Leroy, Mariage, & Wattiez, 2019). Another comparison made at the proteome level, but this time between the Asian ginseng and what know as Indian ginseng *Withania somnifera* with proteins identified from both roots with RNA processing and defense mechanism (Nagappan et al., 2012). Even though the genome sequencing only was done recently, 81 proteins from ginseng fruits linked to the electron enzymes and ATPase with potential contributors to the antioxidant activities were identified (S. W. Kim et al., 2016). Changes related in the metabolites happening during the development and the growth of ginseng captured in a 2DE-based proteomic study shown proteins related to energy metabolism, ginsenosides synthesis, and antioxidation activity will be accumulating in a higher concentration in the older ginseng fruit (R. Ma et al., 2016). At the post "ginseng genome" era, the proteomics research takes a new direction with high-throughput platforms used to identify more proteins like the study which identified more than 800 proteins during heat stress research on Asian ginseng leaves which showed common symptoms or respond to heat stress such as the overexpression of heat shock proteins, RNA transport, and decrease in the photosynthetic activity related proteins (S. W. Kim et al., 2019). Celery (*Apium graveolens*) considered an important aromatic and medicinal plant cultivated widely worldwide and consumed for its fresh leaves for direct consumption or as spices and cooking additives (Al-Asmari, Athar, & Kadasah, 2017). As the leave considered the most important part of this herb proteomic study conducted on this plant were focusing on the leaves such as the study on plant leave upon temperature stress where 71 proteins have been identified (Huang et al., 2017). The proteins were related to amino acid and protein synthesis and antioxidant increased in their expression during the cold condition, however, the chlorophyll content showed a decrease under heat and cold temperature.

Proteomics and aromatic/medicinal plant disease

Zingiber zerumbet is a wild, tropical pharmaceutical herb that shows a high degree of tolerance to diseases affecting cultivated ginger, proteomic study on the plant detected proteins from the Barley stripe mosaic virus together with the plant proteins which confirm the infection of the plant and reflect changes in defense, developmental processes, and secondary metabolite enzymes (Mahadevan, Jaleel, Deb, Thomas, & Sakuntala, 2015). Investigating the protein changes during rhizome rot caused by *Pythium aphanidermatum* also conducted on *Curcuma longa*. The study revealed the contribution of 12 proteins that are differentially expressed which found to be contributed to the plant defense, which seems to be boosted by the biological control agent *Pseudomonas fluorescens* (Prabhukarthikeyan et al., 2017). A proteomic approach was used to study the changes in the leaf proteome profile of the wild mint *Mentha arvensis* infected with a phytopathogenic fungus *Alternaria alternata* (Sinha & Chattopadhyay, 2011), the result of that study showed proteins related to protein-protein communication between the plant and the fungus, however, the 45 identified proteins showed that the early response was not enough to overcome the infection. The same group did compare the proteomic changes between the transgenic and wild mint during interaction with *A. alternata* fungus (Sinha et al., 2013). The study showed significant differences in the protein expression in the transgenic plants, besides that, the study revealed that transgenic plant show more abundance in proteins related to defense and stress, redox, metabolism proteins, cell signaling, and translation regulation. A more recent study on the same plant-disease interaction 21 differentially expressed proteins from leaf membrane including channel and receptors proteins such as AKT1-like potassium and membrane trafficking protein like ADP ribosylation factor, one-third of those proteins found to be linked to defense against fungal infection (Datta, Kumar, & Chattopadhyay, 2018). Cultivation of garlic hampered by the fertility issue, for that reason a transcriptomic and proteomic study was conducted comparing male sterile and the fertile garlic flowers reveled protein biomarkers such as ADP-ribosylation related to cell death pollen abortion (Shemesh-Mayer et al., 2015). Label-free proteomics were conducted on the black pepper plant during Phytophthora challenge that identified earlier more than 500 proteins with carbon fixation and sugar metabolism function (Mahadevan et al., 2016) and later peptidomic study that showed a group of antimicrobial peptides (AMPs) as essential immunity factors against phytopathogenic infection (Umadevi, Soumya, George, & Anandaraj, 2018). The same group recently investigated Trichoderma induced protein bio-marker on the black pepper, those proteins involved in the lignin synthesis, isoflavonoids and ethylene pathways, proteins like subtilisin-like proteas, NmrA protein, bisphosphate aldolase protein, and isoflavone reductase was significantly up-regulated during the tri-interaction (the plant-the pathogenic fungus and the biocontrol) worth further investigations in the future to discover their role in the black pepper tolerance (Umadevi & Anandaraj, 2019).

Proteomics medicinal and pharmaceutical properties

Researchers investigated the anti-cancer properties of fenugreek on CNS T cell lymphoma cells, the study revealed a group of proteins with interest in the therapeutic anticancer activity with the consideration of the possible differences of plant medicinal activity based on geographical origin (Alsemari et al., 2014). Despite the medical importance, proteomic also helped researchers to investigate the potential allergens from the plant like those found in peanut (Faeste, Christians, Egaas, & Jonscher, 2010). *Rosmarinus officinalis* L. (rosemary) is a medicinal plant native to the Middle East and Southern Europe. The plant is known for its antioxidant, anti-inflammatory, antimicrobial and potential antitumor (de Oliveira, Camargo, & de Oliveira, 2019). Rosemary extract was shown in a proteomic study anticancer activity against colon cancer, the extract changed hundreds of proteins that elicit adaptive responses to alleviate the stress (Valdés, Artemenko, Bergquist, García-Cañas, & Cifuentes, 2016). The extract also showed inhibition in the growth of xenograft tumor growth by altering RNA modification and amino acid synthesis (Valdés et al., 2017). Asian ginseng can change the protein expression in the striatum of 1-methyl-4-phenyl-1,2,3,6-tetrahydropyridine (MPTP)-injected mouse model with 30 proteins changed by ginseng administration after 3 days, those proteins were linked to energy and mitochondrial respiratory suggesting a neuroprotective effect of Asian ginseng (D. Kim et al., 2016). Another neuroprotective compound named oxyphylla A obtained from the fruit of *Alpinia oxyphylla* were discovered using an LC-MS/MS proteomic-based research (G. Li et al., 2016). Potential of medicinal properties of alliums such as garlic (*Allium sativum*) has been extensively used for a long time for its ability to boost immunity activity besides its antibacterial and antifungal characteristic (Iciek, Kwiecień, & Włodek, 2009). Fructose-arginine from garlic extract found to be involved in the anti-inflammatory respond in garlic extracts by altering the protein expression of proteins related to oxidative stress in B-12 murine cell line (H. Zhou et al., 2014). Besides that, Garlic anticancer activity well explained when a proteomic study conducted on a human T lymphocyte cell (Jurkat T-cell) and mouse fibroblast (L-929), the Allicin produced during garlic wounding or crashing found to change hundreds of proteins in these two cell lines causing thioallylation and with its ability to decrease enolase activity and increase zinc release (Iciek et al., 2009). Another compound derived from garlic, diallyl trisulfide, were tested against gastric cancer cell line (BGC823) and it found that many proteins in the cell line changes were related to apoptotic pathways which might be another evidence on the potential anticancer activity of garlic and its derivatives (Shao et al., 2005). Tea leaves considered a very important source of antibacterial and antioxidant-related compounds with many health benefit properties (Naveed et al., 2018). Molecular-based color changing in tea leaves were investigated using 2DE proteomic study comparing the changes in the proteins expressed in purple leave (young) to the green leave (mature), the study showed the importance of chalcone-related enzymes to the purple-shoot phenotype and those enzymes role in anthocyanin synthesis and tea flavor improvement (Zhou, Chen, Lee, Li, & Sun, 2017). Another proteomic study related to color development in tea leaves were conducted on white leave cultivars where more than 400 proteins identified during color development stages related to acetylation (Y.-X. Xu et al., 2017). Moreover, Li and his group also identified a group of proteins related to albinism in the plant leaves and linked to chloroplast damage (Q. Li et al., 2011). Different tea varieties (yellow and multicolor leaves) were also compared at the proteome level with proteins related to photosynthesis, cytochrome, and electron transport were identified, those proteins revealed a relationship between the low level of photosynthesis and the abnormality in chloroplast development (C. Ma et al., 2016). Seasonal changes related to ever-growing winter cultivar tea were studied comparing the proteins expressed in a high level in winter compared to the spring season, proteins such as RUBISCO and ATP synthase were highly accumulated in winter as a signal of normal growth ability under low-temperature stress (Liu et al., 2017). The recent proteomic and transcriptomic study revealed that flavonoids level inversely proportional to the leave maturity, and that particularly contributed by the high expression level of proteins related to shikimic acid and flavonoid (Wu et al., 2019). The drought stress considered one of the abiotic stress that affected the tea production where abscisic acid plays a very important role in improving protein transport, carbon metabolism and resistant protein (L. Zhou et al., 2014). Red sage (*Salvia miltiorrhiza*) is a herb with medicinal properties know to be used as Chinese alternative medicine (Lin & Hsieh, 2010) with claims on improving the blood circulation and antioxidative properties. Proteomic research used to support such theories by examining the changes in the rat smooth muscle cell line (A10) exposed to red sage extract, the research showed many protein changes (such as vimentin up-regulation) related to cell proliferation and reduction of reactive oxygen species (Hung, Wang, & Pan, 2010), however there is still a big need for clinical research to prove these claims. Together with NMR-based metabolomics platform, proteomic techniques helped in identifying and characterizing a novel peptide called roseltide from the medicinal plant *Hibiscus sabdariffa* (roselle plant) which considered knottin-type neutrophil elastase inhibitor with medicinal properties against neutrophil elastase-associated diseases (Loo et al., 2016).

Conclusion

Metabolites from medicinal and aromatic plants have contributed for a long time in the discovery and development of drugs and health supplements due to their relative safety to human consumption and their high-efficiency.

Herbal medicinal and natural-based pharmaceutical science and technology have vastly develop during the last few decades. So far, only a few medicinal plant species have complete genomic information, however, proteomic techniques can also be applied to species without genomic information. Proteomics techniques found to be useful to clarify the regulatory pathways underlying important medicinal and aromatic plant traits, and proteomic databases will contribute to the establishment of technology to regulate, enhance ,and improve plant productivity, reduce disease susceptibility and most importantly provide evidence for the medicinal and pharmaceutical effect to human health.

References

Agrawal, D. C., Tsay, H. S., Shyur, L. F., Wu, Y. C., & Wang, S.Y. (2017). *Medicinal Plants and Fungi: Recent Advances in Research and Development:*. Singapore: Springer.

Al-Asmari, A. K., Athar, M. T., & Kadasah, S. G. (2017). An updated phytopharmacological review on medicinal plant of Arab region: *Apium graveolens* Linn. *Pharmacognosy Reviews, 11*(21), 13–18 doi: 10.4103/phrev.phrev_35_16.

Alamgir, A.N. M. (2017). Cultivation of Herbal Drugs, Biotechnology, and In Vitro Production of Secondary Metabolites, High-Value Medicinal Plants, Herbal Wealth, Herbal, Trade, In A.N.M., Alamgir, (Ed.), *Therapeutic Use of Medicinal Plants and Their Extracts: Volume 1: Pharmacognosy* (pp. 379-452). Cham: Springer International Publishing.

Alanazi, I. O., Benabdelkamel, H., Alfadda, A. A., AlYahya, S. A., Alghamdi, W. M., Aljohi, H. A., & Masood, A. (2016). Proteomic Analysis of the Protein Expression Profile in the Mature Nigella sativa (Black Seed). *Applied Biochemistry and Biotechnology, 179*(7), 1184–1201. doi: 10.1007/s12010-016-2058-z.

Alsemari, A., Alkhodairy, F., Aldakan, A., Al-Mohanna, M., Bahoush, E., Shinwari, Z., & Alaiya, A. (2014). The selective cytotoxic anti-cancer properties and proteomic analysis of *Trigonella Foenum-Graecum. BMC Complementary and Alternative Medicine, 14*(1), 114. doi: 10.1186/1472-6882-14-114.

Atanasov, A. G., Waltenberger, B., Pferschy-Wenzig, E. -M., Linder, T., Wawrosch, C., Uhrin, P., & Stuppner, H. (2015). Discovery and resupply of pharmacologically active plant-derived natural products: a review. *Biotechnology Advances, 33*(8), 1582–1614. doi: 10.1016/j.biotechadv.2015.08.001.

Benzie, I., & Wachtel-Galor, S. (2004). Herbal Medicine: An Introduction to its History, Usage, Regulation, Current Trends, and Research Needs. In I. F. Benzie, & S. W.-G. (Eds.), *Herbal Medicine: Biomolecular and Clinical Aspects, Second Edition* (pp. 78–89). Florida CRC Press.

Boonmee, A., Srisomsap, C., Chokchaichamnankit, D., Karnchanatat, A., & Sangvanich, P. (2011). A proteomic analysis of *Curcuma comosa* Roxb. rhizomes. *Proteome Science, 9*, 43–143. doi: 10.1186/1477-5956-9-43.

Bythrow, J. D. (2005). Vanilla as a medicinal plant. *Seminars in Integrative Medicine, 3*(4), 129–131 doi: https://doi.org/10.1016/j.sigm.2006.03.001.

Carpentier, S. C., Witters, E., Laukens, K., Deckers, P., Swennen, R., & Panis, B. (2005). Preparation of protein extracts from recalcitrant plant tissues: an evaluation of different methods for two-dimensional gel electrophoresis analysis. *Proteomics, 5*(10), 2497–2507.

Carson, C., & Riley, T. (1995). Antimicrobial activity of the major components of the essential oil of Melaleuca alternifolia. *Journal of Applied Bacteriology, 78*(3), 264–269.

Champagne, A., & Boutry, M. (2013). Proteomic snapshot of spearmint (*Mentha spicata* L.) leaf trichomes: A genuine terpenoid factory. *Proteomics, 13*(22), 3327–3332. doi: 10.1002/pmic.201300280.

Chen, C., Long, L., Zhang, F., Chen, Q., Chen, C., Yu, X., & Long, Z. (2018). Antifungal activity, main active components and mechanism of Curcuma longa extract against i. *PLOS ONE, 13*(3), e0194284. doi: 10.1371/journal.pone.0194284.

Chokchaichamnankit, D., Subhasitanont, P., Paricharttanakul, N., Sangvanich, P., Svasti, J., & Srisomsap, C. (2009). Proteomic alteration during dormant period of *Curcuma Longa* Rhizomes. *Journal of Proteomics & Bioinformatics, 2*, 380–387. doi: 10.4172/jpb.1000098.

Ciura, J., Bocian, A., Kononiuk, A., Szeliga, M., Jaromin, M., & Tyrka, M. (2017). Proteomic signature of fenugreek treated by methyl jasmonate and cholesterol. *Acta Physiologiae Plantarum, 39*(5), 112. doi: 10.1007/s11738-017-2416-7.

Contreras, A., Leroy, B., Mariage, P. -A., & Wattiez, R. (2019). Proteomic analysis reveals novel insights into tanshinones biosynthesis in Salvia miltiorrhiza hairy roots. *Scientific Reports, 9*(1), 5768. doi: 10.1038/s41598-019-42164-3.

Daniel, M. (2016). *Medicinal plants: chemistry and properties:*. Enfield, New Hampshire, USA CRC press.

Datta, R., Kumar, D., & Chattopadhyay, S. (2018). Membrane proteome profiling of Mentha arvensis leaves in response to Alternaria alternata infection identifies crucial candidates for defense response. *Plant Signaling & Behavior, 13*(4), e1178423. doi: 10.1080/15592324.2016.1178423.

de Oliveira, J. R., Camargo, S. E. A., & de Oliveira, L. D. (2019). *Rosmarinus officinalis* L. (rosemary) as therapeutic and prophylactic agent. *Journal of Biomedical Science, 26*(1), 5. doi: 10.1186/s12929-019-0499-8.

De Silva, T. (1997). Industrial utilization of medicinal plants in developing countries. In G. Bodeker (Ed.), *Medicinal plants for forest conservation and health care* (pp. 34–44). Rome: Food and Agriculture Organization of the United Nations.

Ekor, M. (2014). The growing use of herbal medicines: issues relating to adverse reactions and challenges in monitoring safety. *Frontiers in Pharmacology, 4*, 177–1177. doi: 10.3389/fphar.2013.00177.

Faeste, C. K., Christians, U., Egaas, E., & Jonscher, K. R. (2010). Characterization of potential allergens in fenugreek (*Trigonella foenum-graecum*) using patient sera and MS-based proteomic analysis. *Journal of proteomics, 73*(7), 1321–1333. doi: 10.1016/j.jprot.2010.02.011.

Firouzi, Z., Lari, P., Rashedinia, M., Ramezani, M., Iranshahi, M., & Abnous, K. (2014). Proteomics screening of molecular targets of curcumin in mouse brain. *Life Sciences, 98*(1), 12–17 doi: https://doi.org/10.1016/j.lfs.2013.12.200.

Fokunang, C., Ndikum, V., Tabi, O., Jiofack, R., Ngameni, B., Guedje, N., & Kechia, F. (2011). Traditional medicine: past, present and future research and development prospects and integration in the National Health System of Cameroon. *African Journal of Traditional, Complementary and Alternative Medicines, 8*(3), 284–295.

Gohari, A. R., Saeidnia, S., & Mahmoodabadi, M. K. (2013). An overview on saffron, phytochemicals, and medicinal properties. *Pharmacognosy reviews*, 7(13), 61–66. doi: 10.4103/0973-7847.112850.

Huang, W., Ma, H. -Y., Huang, Y., Li, Y., Wang, G. -L., Jiang, Q., & Xiong, A. -S. (2017). Comparative proteomic analysis provides novel insights into chlorophyll biosynthesis in celery under temperature stress. *Physiologia Plantarum, 161*(4), 468–485. doi: 10.1111/ppl.12609.

Hung, Y. -C., Wang, P. -W., & Pan, T. -L. (2010). Functional proteomics reveal the effect of Salvia miltiorrhiza aqueous extract against vascular atherosclerotic lesions. *Biochimica et Biophysica Acta (BBA) - Proteins and Proteomics, 1804*(6), 1310–1321 doi: https://doi.org/10.1016/j.bbapap.2010.02.001.

Hurkman, W. J., & Tanaka, C. K. (1986). Solubilization of plant membrane proteins for analysis by two-dimensional gel electrophoresis. *Plant Physiology*, *81*(3), 802–806.

Iciek, M., Kwiecień, I., & Włodek, L. (2009). Biological properties of garlic and garlic-derived organosulfur compounds. *Environmental and Molecular Mutagenesis*, *50*(3), 247–265. doi: 10.1002/em.20474.

Inoue, M., Hayashi, S., & Craker, L.E. (2017). Culture, History and Applications of Medicinal and Aromatic Plants in Japan *Aromatic and Medicinal Plants-Back to Nature*: InTechOpen, DOI: 10.5772/66505.

Jan, K., Ahmad, M., Rehman, S., Gani, A., & Khaqan, K. (2019). Effect of roasting on physicochemical and antioxidant properties of kalonji (Nigella sativa) seed flour. *Journal of Food Measurement and Characterization*.doi: 10.1007/s11694-019-00052-4.

Jou, Y. -J., Chen, C. -J., Liu, Y. -C., Way, T. -D., Lai, C. -H., Hua, C. -H., & Lin, C. -W. (2015). Quantitative phosphoproteomic analysis reveals γ-bisabolene inducing p53-mediated apoptosis of human oral squamous cell carcinoma via HDAC2 inhibition and ERK1/2 activation. *Proteomics*, *15*(19), 3296–3309. doi: 10.1002/pmic.201400568.

Kala, C. P. (2000). Status and conservation of rare and endangered medicinal plants in the Indian trans-Himalaya. *Biological Conservation*, *93*(3), 371–379 https://doi.org/10.1016/S0006-3207(99)00128-7.

Kala, C. P. (2005). Current Status of Medicinal Plants Used by Traditional Vaidyas in Uttaranchal State of India. *Ethnobotany Research and Applications*, 3–12.

Khodadadi, E., Fakheri, B. A., Aharizad, S., Emamjomeh, A., Norouzi, M., & Komatsu, S. (2017). Leaf proteomics of drought-sensitive and-tolerant genotypes of fennel. *Biochimica et Biophysica Acta (BBA)-Proteins and Proteomics*, *1865*(11), 1433–1444.

Khodadadi, E., Hashiguchi, A., Fakheri, B. A., Aharizad, S., Emamjomeh, A., Norouzi, M., & Komatsu, S. (2019). Differences in fennel seed responses to drought stress at the seed formation stage in sensitive and tolerant genotypes. *Journal of Plant Biochemistry and Biotechnology*, *28*(1), 35–49. doi: 10.1007/s13562-018-0461-y.

Kim, D., Jeon, H., Ryu, S., Koo, S., Ha, K. -T., & Kim, S. (2016). Proteomic analysis of the effect of Korean Red Ginseng in the striatum of a Parkinson's disease mouse model. *PLOS ONE*, *11*(10), e0164906. doi: 10.1371/journal.pone.0164906.

Kim, S. I., Kim, J. Y., Kim, E. A., Kwon, K. -H., Kim, K. -W., Cho, K., & Park, Y. M. (2003). Proteome analysis of hairy root from Panax ginseng C. A. Meyer using peptide fingerprinting, internal sequencing and expressed sequence tag data. *Proteomics*, *3*(12), 2379–2392. doi: 10.1002/pmic.200300619.

Kim, S.W., Gupta, R., Lee, S. H., Min, C.W., Agrawal, G. K., Rakwal, R., & Kim, S.T. (2016). An integrated biochemical, proteomics, and metabolomics approach for supporting medicinal value of panax ginseng fruits. *Frontiers in Plant Science*, 7(994), doi: 10.3389/fpls.2016.00994.

Kim, S.W., Gupta, R., Min, C.W., Lee, S. H., Cheon, Y. E., Meng, Q. F., & Kim, S.T. (2019). Label-free quantitative proteomic analysis of Panax ginseng leaves upon exposure to heat stress. *Journal of Ginseng Research*, *43*(1), 143–153 doi: https://doi.org/10.1016/j.jgr.2018.09.005.

Kovács, J. K., Felső, P., Makszin, L., Pápai, Z., Horváth, G., Ábrahám, H., & Schneider, G. (2016). Antimicrobial and virulence-modulating effects of clove essential oil on the foodborne pathogen *Campylobacter jejuni*. *Applied and Environmental Microbiology*, *82*(20), 6158–6166. doi: 10.1128/AEM. 01221-16.

Kumar, K. M. P., Asish, G. R., Sabu, M., & Balachandran, I. (2013). Significance of gingers (Zingiberaceae) in Indian System of Medicine - Ayurveda: an overview. *Ancient science of life*, *32*(4), 253–261. doi: 10.4103/0257-7941.131989.

Lei, X., Wang, Z., Song, J., Liang, S., Yao, L., Hou, Z., & Wang, Y. (2017). Comparative Proteomic Analysis of Panax ginseng CA Meyer× Panax quinquefolius L. Leaves and Parental Lines Active Ingredients from Aromatic and Medicinal Plants. In H. A.

El-Shemy (Ed.), *Active Ingredients from Aromatic and Medicinal Plants*. IntechOpen DOI: 10.5772/66511.

Li, G., Zhang, Z., Quan, Q., Jiang, R., Szeto, S. S., Yuan, S., & Chu, I. K. (2016). Discovery, synthesis, and functional characterization of a novel neuroprotective natural product from the fruit of alpinia oxyphylla for use in Parkinson's disease through LC/MS-based multivariate data analysis-guided fractionation. *Journal of Proteome Research, 15*(8), 2595–2606.

Li, Q., Huang, J., Liu, S., Li, J., Yang, X., Liu, Y., & Liu, Z. (2011). Proteomic analysis of young leaves at three developmental stages in an albino tea cultivar. *Proteome Science, 9*(1), 44. doi: 10.1186/1477-5956-9-44.

Li, Y., Tran, V. H., Duke, C. C., & Roufogalis, B. D. (2012). Preventive and Protective Properties of Zingiber officinale (Ginger) in Diabetes Mellitus, Diabetic Complications, and Associated Lipid and Other Metabolic Disorders: A Brief Review. *Evidence-based complementary and alternative medicine: eCAM, 2012*, 516870–1516870. doi: 10.1155/2012/516870.

Lin, T. -H., & Hsieh, C. -L. (2010). Pharmacological effects of Salvia miltiorrhiza (Danshen) on cerebral infarction. *Chinese medicine, 5*, 22–122. doi: 10.1186/1749-8546-5-22.

Liu, S., Gao, J., Chen, Z., Qiao, X., Huang, H., Cui, B., & Liu, J. (2017). Comparative proteomics reveals the physiological differences between winter tender shoots and spring tender shoots of a novel tea (*Camellia sinensis* L.) cultivar evergrowing in winter. *BMC Plant Biology, 17*(1), 206–1206. doi: 10.1186/s12870-017-1144-x.

Loo, S., Kam, A., Xiao, T., Nguyen, G. K. T., Liu, C. F., & Tam, J. P. (2016). Identification and characterization of Roseltide, a knottin-type Neutrophil elastase inhibitor derived from *Hibiscus sabdariffa. Scientific Reports, 6*, 39401 doi: 10.1038/srep39401 https://www.nature.com/articles/srep39401#supplementary-information.

Lopes, E. M., Linhares, R. G., de Oliveira Pires, L., Castro, R. N., Souza, G. H. M. F., Koblitz, M. G. B., & Macedo, A. F. (2019). Vanilla bahiana, a contribution from the Atlantic Forest biodiversity for the production of vanilla: a proteomic approach through high-definition nanoLC/MS. *Food Research International, 120*, 148–156 doi: https://doi.org/10.1016/j.foodres.2019.02.038.

Lum, J. H. -K., Fung, K. -L., Cheung, P. -Y., Wong, M. -S., Lee, C. -H., Kwok, F. S. -L., & Lo, S. C. -L. (2002). Proteome of oriental ginseng panax ginseng C. A. Meyer and the potential to use it as an identification tool. *Proteomics, 2*(9), 1123–1130 doi: 10.1002/1615-9861(200209)2:9 < 1123::AID-PROT1123 > 3.0.CO;2-S.

Ma, C., Cao, J., Li, J., Zhou, B., Tang, J., & Miao, A. (2016). Phenotypic, histological and proteomic analyses reveal multiple differences associated with chloroplast development in yellow and variegated variants from *Camellia sinensis. Scientific Reports, 6*, 33369 doi: 10.1038/srep33369 https://www.nature.com/articles/srep33369#supplementary-information.

Ma, R., Sun, L., Chen, X., Mei, B., Chang, G., Wang, M., & Zhao, D. (2016). Proteomic analyses provide novel insights into plant growth and ginsenoside biosynthesis in forest cultivated panax ginseng (*F. Ginseng*). *Frontiers in Plant Science*, 7, 1–11. doi: 10.3389/fpls.2016.00001.

Maffei, M., Bertea, C., & Mucciarelli, M. (2007). Anatomy, physiology, biosynthesis, molecular biology, tissue culture and biotechnology of mint essential oil production. In B. M. Lawrence (Ed.), *Mint: the genus Mentha* (pp. 41–85). Boca Raton: CRC Press.

Mahadevan, C., Jaleel, A., Deb, L., Thomas, G., & Sakuntala, M. (2015). Development of an efficient virus induced gene silencing strategy in the non-model wild ginger-Zingiber zerumbet and investigation of associated proteome changes. *PLOS ONE, 10*(4), e0124518. doi: 10.1371/journal.pone.0124518.

Mahadevan, C., Krishnan, A., Saraswathy, G. G., Surendran, A., Jaleel, A., & Sakuntala, M. (2016). Transcriptome- assisted label-free quantitative proteomics analysis reveals novel insights into Piper nigrum—Phytophthora capsici Phytopathosystem. *Frontiers in Plant Science*, 7(785), doi: 10.3389/fpls.2016.00785.

Mehraj, S. S., Kamili, A. N., Nazir, R., Haq, E., & Balkhi, H. M. (2018). Comparative evaluation of extraction methods for total proteins from *Crocus sativus* L (Saffron). *Saudi Journal of Biological Sciences*, *25*(8), 1603–1608 doi: https://doi.org/10.1016/j.sjbs.2016.04.011.

Mostafaie, A., Kahrizi, D., Mohammadi, M., Yari, K., Rostami, H., Yaghotipoor, A., & Mostafaie, H. (2018). Effect of planting time and vermicompost on the proteomic pattern of fenugreek (*Trigonella foenum*-graecum). *Cellular and Molecular Biology (Noisy-le-Grand, France)*, *64*(9), 46–51.

Nagappan, A., Karunanithi, N., Sentrayaperumal, S., Park, K. -I., Park, H. -S., & Lee, D. H. S Natesan##S. (2012). Comparative root protein profiles of Korean ginseng (Panax ginseng) and Indian ginseng (Withania somnifera). *The American Journal of Chinese Medicine*, *40*(01), 203–218.

Naveed, M., BiBi, J., Kamboh, A. A., Suheryani, I., Kakar, I., Fazlani, S. A., & XiaoHui, Z. (2018). Pharmacological values and therapeutic properties of black tea *(Camellia sinensis*): a comprehensive overview. *Biomedicine & Pharmacotherapy*, *100*, 521–531 doi: https://doi.org/10.1016/j.biopha.2018.02.048.

Niero, R., Cechinel Filho, V., & Yunes, R. A. (2018). Medicinal Plants and Phytomedicines. In V. Cechinel Filho (Ed.), *Natural Products as Source of Molecules with Therapeutic Potential: Research & Development, Challenges and Perspectives* (pp. 1–33). Cham: Springer International Publishing.

Nunes, A. T., & Albuquerque, U. P. (2018). South American Biodiversity and Its Potential in Medicinal and Aromatic Plants. In U. P. Albuquerque, U. Patil, & Á. Máthé (Eds.), *Medicinal and Aromatic Plants of South America: Brazil* (pp. 3–15). Dordrecht: Springer Netherlands.

Palama, T. L., Menard, P., Fock, I., Choi, Y. H., Bourdon, E., Govinden-Soulange, J., & Kodja, . (2010). Shoot differentiation from protocorm callus cultures of Vanilla planifolia (Orchidaceae): proteomic and metabolic responses at early stage. *BMC Plant Biology*, *10*(1), 82. doi: 10.1186/1471-2229-10-82.

Pan, S. -Y., Zhou, S. -F., Gao, S. -H., Yu, Z. -L., Zhang, S. -F., Tang, M. -K., & Ko, K. -M. (2013). New perspectives on how to discover drugs from herbal medicines: cam's outstanding contribution to modern therapeutics. *Evidence-based Complementary and Alternative Medicine : eCAM*, *2013*, 627375–1627375. doi: 10.1155/2013/627375.

Paredi, G., Raboni, S., Marchesani, F., Ordoudi, A. S., Tsimidou, Z. M., & Mozzarelli, A. (2016). Insight of saffron proteome by gel-electrophoresis. *Molecules*, *21*(2), doi: 10.3390/molecules21020167.

Prabhukarthikeyan, S., Manikandan, R., Durgadevi, D., Keerthana, U., Harish, S., Karthikeyan, G., & Raguchander, T. (2017). Bio-suppression of turmeric rhizome rot disease and understanding the molecular basis of tripartite interaction among *Curcuma longa*, *Pythium aphanidermatum* and Pseudomonas fluorescens. *Biological Control*, *111*, 23–31.

Pundir, R. K., & Jain, P. (2015). Mechanism of Prevention and Control of Medicinal Plant-Associated Diseases. In D. Egamberdieva, S. Shrivastava, & A. Varma (Eds.), *Plant-Growth-Promoting Rhizobacteria (PGPR) and Medicinal Plants* (pp. 231–246). Cham: Springer International Publishing.

Rao, J., Lv, W., & Yang, J. (2017). Proteomic analysis of saffron (*Crocus sativus* L.) grown under conditions of cadmium toxicity. *Bioscience Journal*, *33*(3), .

Sathe, G., Pinto, S. M., Syed, N., Nanjappa, V., Solanki, H. S., Renuse, S., & Chatterjee, A. (2016). Phosphotyrosine profiling of curcumin-induced signaling. *Clinical Proteomics*, *13*(1), 13. doi: 10.1186/s12014-016-9114-0.

Sengupta, A., & Bhattacharjee, S. (2009). Cardamom (Elettaria cardamomum) and Its Active Constituent, I, 8-cineole. In *Molecular Targets and Therapeutic Uses of Spices: Modern Uses for Ancient Medicine* (pp. 65–85). Singapore: World Scientific.

Shao, J., Zhao, K., Li, N., Xu, N., Guo, R., Li, S., & Lu, Y. (2005). A proteomic investigation into a human gastric cancer cell line BGC823 treated with diallyl trisulfide. *Carcinogenesis, 27*(6), 1222–1231. doi: 10.1093/carcin/bgi306.

Sharifi, G., Ebrahimzadeh, H., Ghareyazie, B., Gharechahi, J., & Vatankhah, E. (2012). Identification of differentially accumulated proteins associated with embryogenic and non-embryogenic calli in saffron (*Crocus sativus* L.). *Proteome Science, 10*(1), 3–13. doi: 10.1186/1477-5956-10-3.

Shemesh-Mayer, E., Ben-Michael, T., Rotem, N., Rabinowitch, H. D., Doron-Faigenboim, A., Kosmala, A., & Kamenetsky, R. (2015). Garlic (*Allium sativum* L.) fertility: transcriptome and proteome analyses provide insight into flower and pollen development. *Frontiers in Plant Science, 6*, 271–1271. doi: 10.3389/fpls.2015.00271.

Sinha, R., Bhattacharyya, D., Majumdar, A. B., Datta, R., Hazra, S., & Chattopadhyay, S. (2013). Leaf proteome profiling of transgenic mint infected with Alternaria alternata. *Journal of Proteomics, 93*, 117–132 doi: https://doi.org/10.1016/j.jprot.2013.01.020.

Sinha, R., & Chattopadhyay, S. (2011). Changes in the leaf proteome profile of *Mentha arvensis* in response to Alternaria alternata infection. *Journal of proteomics, 74*(3), 327–336 doi: https://doi.org/10.1016/j.jprot.2010.11.009.

Snehlata, H. S., & Payal, D. R. (2012). Fenugreek (*Trigonella foenum-graecum* L.): an overview. *International Journal of Current Pharmaceutical Research, 2*(4), 169–187.

Tan, B. C., Chin, C. F., Liddell, S., & Alderson, P. (2013). Proteomic analysis of callus development in *Vanilla planifolia* Andrews. *Plant Molecular Biology Reporter, 31*(6), 1220–1229. doi: 10.1007/s11105-013-0590-3.

Umadevi, P., & Anandaraj, M. (2019). Proteomic analysis of the tripartite interaction between black pepper, *Trichoderma harzianum* and Phytophthora capsici provides insights into induced systemic resistance mediated by *Trichoderma* spp. *European Journal of Plant Pathology, 154*, 607–620. doi: 10.1007/s10658-019-01685-3.

Umadevi, P., Soumya, M., George, J. K., & Anandaraj, M. (2018). Proteomics assisted profiling of antimicrobial peptide signatures from black pepper (*Piper nigrum* L.). *Physiology and Molecular Biology of Plants, 24*(3), 379–387. doi: 10.1007/s12298-018-0524-5.

Vaghasiya, Y., Dave, R., & Chanda, S. (2011). Phytochemical analysis of some medicinal plants from western region of India. *Research Journal of Medicinal Plants, 5*(5), 567–576.

Valdés, A., Artemenko, K. A., Bergquist, J., García-Cañas, V., & Cifuentes, A. (2016). Comprehensive proteomic study of the antiproliferative activity of a polyphenol-enriched rosemary extract on colon cancer cells using nanoliquid chromatography–orbitrap MS/MS. *Journal of Proteome Research, 15*(6), 1971–1985. doi: 10.1021/acs.jproteome.6b00154.

Valdés, A., García-Cañas, V., Pérez-Sánchez, A., Barrajón-Catalán, E., Ruiz-Torres, V., Artemenko, K. A., & Cifuentes, A. (2017). Shotgun proteomic analysis to study the decrease of xenograft tumor growth after rosemary extract treatment. *Journal of Chromatography A, 1499*, 90–100 doi: https://doi.org/10.1016/j.chroma.2017.03.072.

Van Emon, J. M. (2016). The omics revolution in agricultural research. *Journal of agricultural and food chemistry, 64*(1), 36–44. doi: 10.1021/acs.jafc.5b04515.

Wang, J., Zhang, J., Zhang, C. -J., Wong, Y. K., Lim, T. K., Hua, Z. -C., & Lin, Q. (2016). In situ proteomic profiling of curcumin targets in HCT116 colon cancer cell line. *Scientific Reports, 6*, 22146–122146. doi: 10.1038/srep22146.

Wu, L. -Y., Fang, Z. -T., Lin, J. -K., Sun, Y., Du, Z. -Z., Guo, Y. -L., & Ye, J. -H. (2019). Complementary iTRAQ proteomic and transcriptomic analyses of leaves in tea plant (*Camellia sinensis* L.) with different maturity and regulatory network of flavonoid biosynthesis. *Journal of Proteome Research, 18*(1), 252–264. doi: 10.1021/acs.jproteome.8b00578.

Xu, J., Shao, X., Wei, Y., Xu, F., & Wang, H. (2017a). iTRAQ proteomic analysis reveals that metabolic pathways involving energy metabolism are affected by tea tree oil in *Botrytis cinerea*. *Frontiers in Microbiology, 8*, 1989–11989. doi: 10.3389/fmicb.2017.01989.

Xu, Y. -X., Chen, W., Ma, C. -L., Shen, S. -Y., Zhou, Y. -Y., Zhou, L. -Q., & Chen, L. (2017b). Proteome and acetyl-proteome profiling of *Camellia sinensis* cv. 'Anji Baicha' during periodic albinism reveals alterations in photosynthetic and secondary metabolite biosynthetic pathways. *Frontiers in Plant Science*, *8*(2104.), doi: 10.3389/fpls.2017.02104.

Yin, X., Wang, S. -L., Alolga, R. N., Mais, E., Li, P., Yang, P., & Qi, L. -W. (2018). Label-free proteomic analysis to characterize ginger from China and Ghana. *Food Chemistry*, *249*, 1–7 doi: https://doi.org/10.1016/j.foodchem.2017.12.062.

Yuan, H., Ma, Q., Ye, L., & Piao, G. (2016). The traditional medicine and modern medicine from natural products. *Molecules*, *21*(5), 559.

Zhang, H., Wei, Q., Li, C., Jiang, C., & Zhang, H. (2016). Comparative proteomic analysis provides insights into the regulation of flower bud differentiation in *Crocus Sativus* L. *Journal of Food Biochemistry*, *40*(4), 567–582. doi: 10.1111/jfbc.12254.

Zhou, H., Qu, Z., Mossine, V.V., Nknolise, D. L., Li, J., Chen, Z., & Gu, Z. (2014a). proteomic analysis of the effects of aged garlic extract and its fruarg component on lipopolysaccharide-induced neuroinflammatory response in microglial cells. *PLOS ONE*, *9*(11), e113531. doi: 10.1371/journal.pone.0113531.

Zhou, L., Xu, H., Mischke, S., Meinhardt, L. W., Zhang, D., Zhu, X., & Fang, W. (2014b). Exogenous abscisic acid significantly affects proteome in tea plant (*Camellia sinensis*) exposed to drought stress. *Horticulture Research*, *1*, 14029. doi: 10.1038/hortres.2014.29.

Zhou, Q., Chen, Z., Lee, J., Li, X., & Sun, W. (2017). Proteomic analysis of tea plants (*Camellia sinensis*) with purple young shoots during leaf development. *PLOS ONE*, *12*(5), e0177816. doi: 10.1371/journal.pone.0177816.

CHAPTER 11

Fluxes of nutrients in mycorrhiza: what has fluxomics taught us in the plant-fungus interaction?

Jesús Antonio Salazar-Magallón[a], Arturo Huerta de la Peña[a], Hebert Jair Barrales-Cureño[b]

[a]Unit in Development for Research and Technology Transfer in Biological Control, Postgraduate College, San Pedro Cholula, Puebla, México

[b]Natural Science Divsion, Puebla State Intercultural University, Huehuetla, Puebla, México

Introduction

The term mycorrhiza is used to describe an association formed by a group of fungi (mycelium) that, when having contact with the roots of the plants, can wrap them forming a mantle and penetrate them intercellularly through the cells of the cortex, as in the case of the ectomycorrhiza or, as in the case of the arbuscular mycorrhiza, penetrate the root, but no mantle is formed. At the same time, the hyphae branch out into the ground, forming an extensive network of hyphae capable of interconnecting, underground, the roots of plants of the same or different species (Parniske, 2008). This network of mycelium allows, under certain conditions, a free flow of nutrients, which can be measured through a science called fluxomics, toward the host plants and between the roots of the interconnected plants, which suggests that the mycorrhiza establishes a great union under the soil between plants that, at first sight, could seem distant and without any relation. Thus, mycorrhiza offers the host plant and the ecosystem different benefits in terms of survival and functioning (Gutjahr & Parniske, 2013). This association occurs in approximately 90% of the plants, so it is located in all the ecosystems of the world and, therefore, in different latitudinal gradients. In addition, it is important to note that there are fungi that can be found in several types of soil and climates, having a global distribution pattern, which indicates that they are, apparently, adapted to different habitats; the physical and chemical factors of the soil can restrict its distribution, so that mycorrhizal associations can be considered cosmopolitan and general. However, depending on

Medicinal and Aromatic Plants: Expanding their Horizons through Omics
http://dx.doi.org/10.1016/B978-0-12-819590-1.00011-2

the environment and the interacting species, the participants can be facultative or obligated (Finlay, 2008).

As shown in Table 11.1, the current classification was proposed by Harley and Smith in 1983, and was endorsed by Smith and Read in 1997, where seven different types of mycorrhizae are recognized, considering both their structural characteristics and the taxonomic group of the fungus or the plant involved and the morphological alterations experienced by the parties in the development of the new structure (Schüler, Schwarzott, & Walker, 2001). It is important to recognize that the mycorrhiza is not just the fungus, as is generally believed, it is rather an interesting process in which a structure is formed that allows the mutual exchange of water and nutrients through the interaction of certain specialized parts of at least two individuals (plant and fungus) and, like many other interactions between species, is result of a long evolutionary history (Sanders & Croll, 2010).

During this interaction process a whole range of levels of "benefit" can be presented for each of the parties, depending on the conditions in which the interaction is established. We can consider that under certain conditions (environmental, physiological, or genetic) both parties "benefit" (Camarena-Gutiérrez, 2012). Considering all the aforementioned, it can be proposed that measuring the flow of nutrients (fluxomics) can be understood in broad terms, as during this interaction process the hyphae of the fungus act as an extension of the root, increasing its exploration surface in the substrate, which gives the plant greater opportunity to absorb water and essential minerals and which has lasted for very long period of time.

Mycorrhizal fungi and their impact on medicinal plants

The consumption of medicinal plants throughout the world is growing, in addition, these plants are now cultivated for high-economic performance and have great importance in terms of food, cosmetics, and pharmaceuticals (Sharma, Prasad, & Shrivastav, 2015). Recent research has shown that mycorrhizal fungi can interact with some medicinal plants (Zeng et al., 2013). Mycorrhizal fungi promote the growth of medicinal plants especially in arid and semi-arid regions by increasing water and providing nutrients, especially nitrogen and phosphorus from the soil to the plant (Baum, El-Tohamy, & Gruda, 2015, Aroca & Ruiz-Lozano, 2009). On the other hand, mycorrhizal fungi provide protection to plants against the attack of phytopathogens (Jung, Martinez-Medina, López-Raez, & Pozo, 2012) and can generate important changes in the physiology of plants (Armada,

Table 11.1 Classification of mycorrhizal fungi depending on the plant that colonizes and their interaction with the host.

Mycorrhiza classification	Host	Description
Ectomycorrhiza	Some species of: Pinaceae, Fagaceae y Betulaceae, Salicaceae, Tiliaceae, Rosaceae, Leguminosae y Juglandaceae.	It is an interaction in which the hyphae of a fungus penetrate the secondary roots of the plant to develop, surrounding the cells of the radical cortex, and form an intercellular network called the Hartig network, in addition to a mycelial layer (set of hyphae that make up the body or talus of the fungus) on the outside of the root, called the mantle. The fungi involved in this association are mainly members of the Basidiomycotin group.
Arbuscular mycorrhiza	Different species of legumes, citrus, papaya, avocado, apple, mango, strawberry and peach.	It is an obligatory association for the fungi that form it, but not for plants. In this case Hartig's net or mantle is not formed, and is characterized because hyphae penetrate the root, enter the cells and can form two types of structures. Its main characteristic is the structure called arbuscular, which originates near the vascular cylinder of the plant through numerous successive dichotomous branches of a hypha and has the function of transferring nutrients from and to the plant. The second structure is called a vesicle, and may or may not be present, depending on the fungus. The arbuscular mycorrhizal fungi belong to the group of Zygomycetes.
Mycorrhizae of orchids or orchidoid endomycorrhizae	Orchids	In this case the plant (orchid) is very dependent on the fungus, since it stimulates the germination of its seeds and the initial growth of the seedling. In its seedling stage, orchids are chlorophyll (they do not present chlorophyll) and therefore saprobic (they do not produce their own food), so they depend directly on the contributions of carbon compounds and nutrients provided by the fungus. The fungi that form it are from the Basidiomycotin group.

(*Continued*)

Table 11.1 Classification of mycorrhizal fungi depending on the plant that colonizes and their interaction with the host. (*Cont.*)

Mycorrhiza classification	Host	Description
Ericoid mycorrhiza	All species of the families *Ericaceae* (except for the genera *Arbutus* and *Arctostaphylos*), *Epacridaceae* and *Empetraceae*	In this case, the fungus also penetrates the radical cells, but it is distinguished because the plant involved is generally of the order Ericales (commonly called heather), although this type of mycorrhiza has also been observed in some bryophytes (group of mosses), and the fungus is from the Ascomycotina group.
Ectendomycorrhiza	Coniferous plants of the genus *Pinus*	This type of mycorrhiza is special, since it has characteristics of the ectomycorrhizae (Hartig network and mantle), but simultaneously presents a certain degree of intracellular penetration, as in the endomycorrhizas. In some cases, the mantle is not formed, but always the Hartig network. This interaction occurs mainly between fungi of the Basidiomycotina and Ascomycotina groups.
Arbotoid mycorrhiza	Occurs in plants of the genera *Arctostaphylos*, *Arbutus* and *Pyrola*,	It is a type of ectendomycorrhiza, since it is observed that simultaneously the fungus penetrates the radical cells of the plant and forms the network of Hartig. The associated fungi are always Basidiomycotina; species of the genera *Hebeloma, Laccaria, Poria, Rhizopogon, Pisolithus, Thelephora, Piloderma, Cenococcum* and *Lactarius* have been reported. Generally, the fungi that form arbotoid mycorrhiza can form ectomycorrhiza if they interact with plants of the genus *Pinus.*
Monotropoid mycorrhiza	Associated plants are of the genus *Sarcodes, Pterospora* and *Monotropa*	It has been observed that the fungus that forms monotropoid mycorrhiza is able to colonize the roots of nearby trees (one to two meters) mainly of the Pinus and Picea genera, and transport nutrients from the tree to the chlorophyll plants.

Azcón, López-Castillo, Calvo-Polanco, & Ruiz-Lozano, 2015, Talaat & Shawky, 2012). In recent years, the positive impacts of mycorrhizal fungi have been reported in medicinal plants, especially in their growth, the physiology, content and quality of essential oils and the mitigation of environmental stress conditions (Aslani, Hassani, Rasouli-sadaghiani, Esmailpour, & Rohi, 2014, Zolfaghari, Nazeri, Sefidkon, & Rejali, 2013, Scagel & Lee, 2012). A large part of these studies evaluated the inoculation of plants with a single species of mycorrhizal fungi. However, some reports have shown that autochthonous mycorrhizae are more effective in the cultivation of medicinal plants than non-native mycorrhizae (Yassen et al., 2016). The use and application of autochthonous mycorrhizae are beneficial because they are species-adapted to the local climatic conditions, the soil and the study area (Symanczik, Courty, Boller, Wiemkem, & Al-Yahyaei, 2015). Therefore, the selection of more efficient mycorrhizal fungi is a fundamental requirement in the programs of biological fertilization, so it is important to know the differences in the levels of compatibility between the host and the autochthonous mycorrhiza (Engelmoer & Kiers, 2015).

What is fluxomics?

The total flow of a metabolic pathway (the rate of change of production of the final metabolite) is considered by the speed at which the enzymes and transporters of the pathway catalyze reactions or improve transport processes. Then, the flow of a biosynthetic route is the variable that serves as an indicator of the speed of all its individual components. Therefore, fluxomics is the discipline of omics that deals with the analysis of metabolic fluxes (Sanford, Soucaille, Whited, & Chotan, 2002). Unlike other omics, such as transcriptomics, proteomics, or metabolomics, which provide a static view of biological systems, fluxomics provides information on the dynamics of the system, that is, its evolution over time. Thus, for example, if transcriptomics, proteomics, and metabolomics are quantified in a cell line, information is obtained on the transcripts, proteins and metabolites present in the sample at the time of the experiment, but if fluxomics is quantified it is possible to infer how the system over time, for example, how long it will take to duplicate the cell line. Moreover, it is considered that fluxoma, the set of metabolic fluxes in a metabolic system is a direct manifestation of the metabolic phenotype (Winter & Krömer, 2012). Also, unlike transcriptomics, proteomics or metabolomics that can be quantified directly, fluxoma can only be estimated indirectly by integrating measurements of production and consumption of metabolites, transcriptomics, proteomics or metabolomics data into mathematical models and / or distribution maps of ^{13}C in different metabolites obtained from experiments in which cells are incubated with substrates marked with ^{13}C (Niittylae, Chaudhuri, Sauer, & Frommer, 2009).

How is flow analyses carried out in the metabolic pathways?

There are several methods to determine the flow of nutrients in the various metabolic pathways of an organism, among which kinetic models and stoichiometric models are prominent where substrates marked with ^{13}C are used. Based on this principle, different algorithms have been performed to maximize the consistency between the measures of gene expression and the activity or inactivity of the reactions simulated by the model. In addition, algorithms have also been developed to integrate metabolomics information in these models. These algorithms are based on the principle that if a metabolite is detected under the study conditions, it implies that some of the reactions in which it participates must be active (Klein & Heinzle, 2012).

Kinetic models

In kinetic models, a metabolic system is described by a system of first-order differential equations. Thus, for each metabolite a differential equation is formulated that defines its variation over time. These are constructed by adding the flows produced by each metabolite and subtracting the flows that consume it based on the stoichiometry of the reactions defined in the metabolic network (Jamshidi & Palsson, 2008). Here, the flow of a reaction is a function of the concentration of those metabolites that participate or regulate the reaction. The mathematical function, which expresses how the flow of a particular reaction depends on the concentrations of metabolites, is called the kinetic equation. In reactions catalyzed by enzymes, kinetic equations usually reflect the total amount of enzyme, its affinity for substrates and its degree of inhibition or activation (Costa, Hartmann, & Vinga, 2016). The complexity and number of parameters associated with a kinetic equation depends on the detail with which you want to model the reaction. To obtain the maximum correspondence between the predictions of the model and the experimental data, an adjustment of the parameters of the kinetic equations is normally made. The numerical resolution of the differential equations system allows us to predict how metabolite concentrations evolve and, by extension, the flows (Saa & Nielsen, 2017).

Stoichiometric models

In stoichiometric models, the metabolic system is described as a linear system of equations and inequations with metabolic fluxes as variables. They are based on the assumptions that the study system is in steady state, that is, in a state in which metabolite concentrations are constant over time (Feng et al., 2002). This implies that the sum of the flows of the reactions that produce a certain metabolite must be equal to the sum of the flows of reactions that consume this same metabolite, that is, that the flows of production and consumption for all metabolites are balanced. Additionally, you can also define an upper and lower limit for each flow. These limits can be used to incorporate experimental data, allowing restricting the flows of consumption and production of metabolites to experimentally determined values (Fleming et al., 2010). The system of equations and inequations together with the upper and lower limits for the flows serves to define a space of solutions, that is, possible flow values. To select the best solutions within this space is usually defined a biologically desirable objective in the system, for example, the maximization of one or more flows (Stalidzans, Seiman, Peebo, Komasilovs, & Pentjuss, 2018). The biological objective set will depend on the system studied. Thus, in a highly proliferative system, such as a bacterium or tumor cell, it is usually assumed that "the goal" is to maximize growth. This objective is implemented through maximizing the production of the biomass components of the organism, giving the different components a weight proportional to their abundance. The result is optimal distribution of flux (fluxomic), that is to say, that maximizes the production of biomass and therefore the bacterial or tumoral proliferation. This approach is known as flux balance analysis (Curran et al., 2012).

Fluxomics assisted by ^{13}C

Regardless of whether kinetic models or stoichiometric models are used, a challenge in quantifying flows are the degrees of freedom associated with the large number of branches and cycles contained in a metabolic network (Quek, Wittmann, Nielsen, & Krömer, 2009). The use of substrates marked with ^{13}C, a stable isotope of carbon, provides the means to reduce this uncertainty. This is possible because the conversion of substrates into products through different metabolic pathways results in characteristic brand patterns in intermediates and metabolic products (Niedenführ, Wiechert, & Nöh, 2015). For example, pyruvate can be incorporated into the Krebs cycle through both pyruvate dehydrogenase (PDH) and pyruvate carboxylase (PC). If pyruvate is marked, for example due to the incubation of the cells under study with ^{13}C-labeled glucose, these two reactions result in a different brand pattern in glutamate. In this way, if after incubation with glucose marked with ^{13}C, the mark pattern in glutamate is quantified, the relative activity of the PDH and PC can be inferred (Crown & Antoniewizc, 2013).

To integrate the experimental data of the brand, the flow distribution that best reproduces the experimentally determined brand patterns in different metabolites of the studied system is sought. For this, kinetic or stoichiometric models are used, coupled to a model that calculates the isotopomer distribution and simulates the propagation of the brand through the system (Heux et al., 2015). A key factor to consider is whether the brand is in a stationary isotopic state or not. In an isotopic steady state, the abundances of isotopomers and mass isotopomers can be calculated as a function of the distribution of flows and the mark on the substrates, making it possible to simplify the problem. However, the distribution of isotopomers may have a high transition time, particularly in those metabolites that are found in high amounts inside cells, such as glycogen, and this time may be longer than the time of the experiment. In these cases, more complex mathematical approaches must be used (Krömer, Quek, & Nielsen, 2009).

Benefits of the interaction: fluxomic of nutrients between the plant and the fungus

The economic importance of the mycorrhizal association lies in the harmonic relationship of nutritional support that is established between both organisms, with the bidirectional flow of nutrients (Ferrol, Barea, & Azcon-Aguilar, 2002). For three decades it has been shown that the fungus absorbs, mainly, P from the soil (Chiu, Liu, & Harrison, 2001) and transports it to the plant (Solaiman & Saito, 2001), and from this one moves a series of carbon compounds toward the fungus (Bago et al., 2003). For this reason, the plant, although it is capable of growing independently, generally has greater development when it is colonized by the mycorrhizal fungus; above all, in conditions of low-levels of nutrients in the soil, characteristic of tropical soils and affecting agricultural productivity. The structure of the arbuscular mycorrhiza (AM), represents one of the absorbent organs of the subsoil of most plants in nature (Harrison, 1997) and has been one of the oldest and most prosperous strategies that have developed the root systems of plants for establishment of reciprocal benefit with microorganisms (Remy et al., 1994).

The AM fungi obtain carbon from the host plant (Pfeffer, Douds, Bécard, & Schachar-Hill, 1999), in the form of hexose molecules, which is converted to lipids (triacylglycerol and carbohydrates-glycogen- in the intraradical mycelium, which are translocated to the extraradical mycelium and from which the structural carbohydrate chitin and the carbohydrates from the trehalose and glycogen stores are synthesized (Bago et al., 2003) .The movement of the phosphate ion -from the soil to the root- through the network of external hyphae, initiates with the absorption of the soil solution by phosphate transporters H + -ATPase (located in the extraradical hyphal), followed by the conversion to short-chain polyphosphates that are translocated to the intraradical structures through mobile vacuoles of the fungus and the subsequent hydrolysis of the same by the phosphatases located in the intraridial hyphae (Ferrol et al., 2002) Then, the P is transferred from the fungus to the interfacial apoplasts. by flow, which is taken (by plant) by membrane transporters, in such a way that they generate growth increase, health, and resistance to stress, particularly for mycorrhized plants under limiting nutrient conditions (Pearson & Jakobsen, 1993).

Fluxomics of carbon

The movement and speed of transport of carbon in the plant-fungus relationship has been studied by several authors (Bago et al., 2003; Simard et al. 2003; Zhu & Miller, 2003).The fluxes of carbon in most cases is carried out from the plant toward the fungus, however, the movement of the fungus back to the plant has been described only in very special cases where the plant has certain restrictions of addition of carbon, for example, in achlorophyllous plants (Bidartondo et al., 2002) Currently, the mechanisms of how these carbon movements occur in this symbiotic system have not been very well understood, however, in this part of the work,We will briefly describe the studies carried out by several authors, detailing carbon fluxes in mycorrhizal fungi (Fig. 11.1).

As a first step, trehalose plays a fundamental role in the transport of carbon from the plant to the fungus, as has been demonstrated in previous studies, this molecule participates in the germination of the spore, it has been observed that trehalose concentrations decrease and they are synthesized again during the symbiotic state, and trehalose functions as a buffer for intracellular glucose (Pfeffer et al., 1999, Shachar-Hill et al., 1995). According to the studies carried out by Bago, Pfeffer, Zipfel, Lammers, & Shachar-Hill (2002) and Simard et al. (2003) when the germinated spores are exposed to hexose, this hexose enters the cell and is degraded; this is

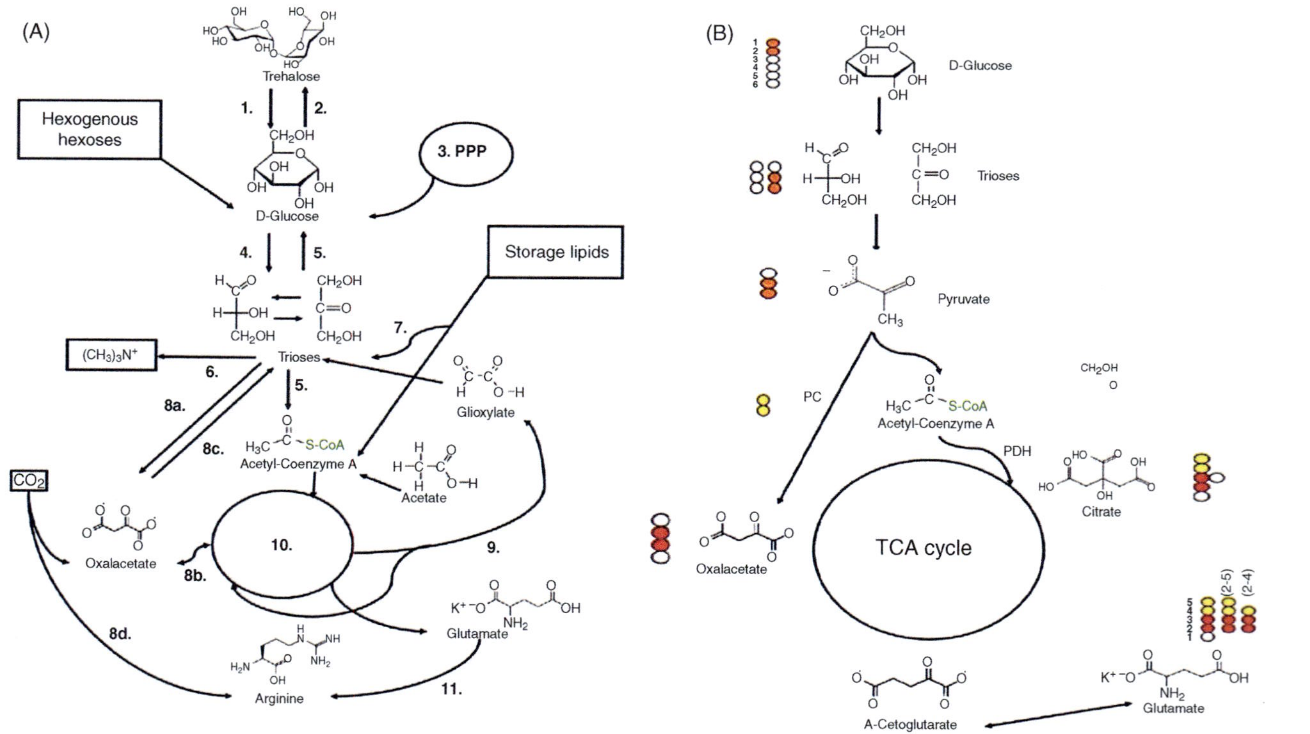

Figure 11.1 ***Simplified diagrams of carbon fluxes in the metabolic pathways of mycorrhizal fungi.*** **(A)** the enzymatic reactions as well as the products obtained during the metabolism in mycorrhizal fungi are shown: *(1)* synthesis of trehalose from glucose phosphate and UDP-glucose; *(2)* lysis of trehalose by trehalase; (3) Alternative metabolic pathway known as hexose monophosphate; *(4)* Gluconeogenesis; *(5)*- Glycolysis; *(6)* Non-photosynthetic carbon metabolism; *(7)* β-oxidation; *(8)* dark fixation of CO_2 through pyruvate carboxylase to oxaloacetate *(8a)* or carbomyl phosphate synthetase *(8d)*; *(9)* glyoxylate cycle; *(10)*- Krebs cycle; *(11)* Arginine synthesis. **(B)** representative diagram of the carbons marked in the hexoses that enter the different metabolic pathways, the circles represent carbon atoms, specifically the gray ones represent ^{12}C, the red circles represent ^{13}C that has entered the cycle of Krebs through pyruvate carboxylase (PC), the yellow circles represent the ^{13}C that has entered the Krebs cycle through pyruvate dehydrogenase (PDH) and the brown circles represent the ^{13}C that have not yet entered the Krebs cycle.

carried out in an asymbiotic state (intraradical) via glycolysis. Also, in studies using nuclear magnetic resonance with glucose isotopes it was demonstrated that the extra-radical hyphae a symbiotic state can take and use the hexoses inside the root (Bago et al., 2003).

Bécard and Piché, (1989) proved with their research that CO_2 favors asymbiotic growth, also suggest that CO_2 can be a source of carbon in anabolic processes. On the other hand, based on the observations made by Johnson et al. (2002), demonstrated a significant addition of CO_2 through pyruvate decarboxylase in the Krebs cycle. Finally, Lake, Ostle, Rangel-Castro, & Jonhson (2006) reported that dark fixation of CO_2 as part of gluconeogenesis does not favor a net balance of C, because there is a carboxylation/decarboxylation in the Krebs cycle. In a study conducted, the importance of carbon fluxes through the glyoxylate cycle in mycorrhizal fungi was demonstrated. Through the results shown by a nuclear magnetic resonance, they interpreted that this cycle participates for the generation of hexose, whether the triacylglycerols enter directly into this cycle or through the decomposition of fatty acids by β-oxidation. These observations were made mainly on the germination of the spores, because the largest amount of energy reserved in the spores are not carbohydrates but accumulated fatty acids.

There are routes that intervene in the fluxes of carbon in the plant-fungus interaction. One of these routes is that of hexose monophosphate or pentose phosphate, which uses a hexose to generate sugars of 5 carbons, necessary to make NADPH, and synthesize nucleotides and nucleic acids. Therefore, its objectives are to produce NADPH, in addition, it can be divided into two phases, the first is oxidative and the reactions are irreversible, and the second is non-oxidative and the reactions are reversible. It is known that PPP is active in fungi and its action is consistent with recent observations in the symbiotic state (Bago et al. 1999; Pfeffer et al. 1999). Another way in which carbon flows are involved is in the synthesis of amino acids, where the metabolic precursors of amino acids are also metabolic intermediates of other important processes of metabolism. For example, ribose 5 phosphate is an intermediary of pentose metabolism; 3-phosphoglycerate, which in turn is the precursor of erythrose 4-phosphate, is an important intermediate of glycolysis; Phosphoenolpyruvate, one of the last intermediates of glycolysis, is also an amino acid precursor when it is combined with erythrose 4-phosphate; Pyruvate is the compound resulting from glycolysis. Likewise, two intermediates of the Krebs cycle, α-ketoglutarate and oxaloacetate, are also the two most important metabolic precursors of amino acids (Bago et al., 1999).

Fluxomics of nitrogen

Nitrogen is an element whose availability limits the optimal development of plants in all ecosystems. In this sense, the mycorrhizal fungi can absorb NH_4^+ and NO_3^-, in addition, they have the capacity to improve the access to the sources of organic nitrogen of the soil (Hodge, Campbell, & Fitter, 2001). This absorption, called translocation, made by the fungus becomes an important pathway for the absorption of nitrogen by the plant (Chen, Gu, Wang, Chen, & Xu, 2017). Currently, mechanisms of transfer of nitrogen from the fungus to the plant have been described (Chen et al., 2017; Fellbaum et al., 2012; Jin, Jiang, & Zang, 2011; Tian, Kasiborsky, Lammers, Bücking, & Shachar-Hill, 2010; Kasola, Workmaster, & Spada, 2007), where they determine the operation of some metabolic pathways in which nitrogen is translocated from the extratrarradic mycelium (ERM) to the intraradic mycelium (IRM) as arginine. At this point, we will describe two of the main mechanisms of nitrogen absorption, the first is carried out in an asymbiotic state, that is, when the spore germinates and a second model where the transport of NH_4^+ and NO_3^- ions is described from the soil until the IRM to later toward the plant.

Nitrogen fluxes in spore germination

In the absorption and assimilation of nitrogen, a continuous administration of energy and carbon skeletons is needed during the germination of spores (Fig. 11.2). Apparently, the sources of storage of nitrogen, such as proteins and carbohydrates as stored lipids that will later be lysed to produce amino acids during this phase. During germination, glycine, serine, asparagine, glutamate, and alanine are generally formed from nitrogen storage molecules (Avio & Giovannetti, 1998; Bago et al., 1999). Jin et al. (2011) reported that a metabolic pathway in the synthesis of glycine and serine is through transamination with glyoxylate, which causes the synthesis of glycine and the conversion of glycine to serine by the enzyme serine hydroxymethyltransferase (SHMT). At the same time, they showed that the primary source of glyoxylate is obtained through the conversion of isocitrate to succinate in the glyoxylate cycle, which is related to the metabolism of acetyl-CoA, which is caused by the breakdown of storage lipids in the spores through β-oxidation (Fig. 11.2). In the transamination reactions with glyoxylate, alanine, glutamate, serine and aspartate are used as the amino group donors. Therefore, all these metabolic reactions detail the ability of mycorrhizal fungal spores to degrade lipids as an energy source for carbohydrate and amino

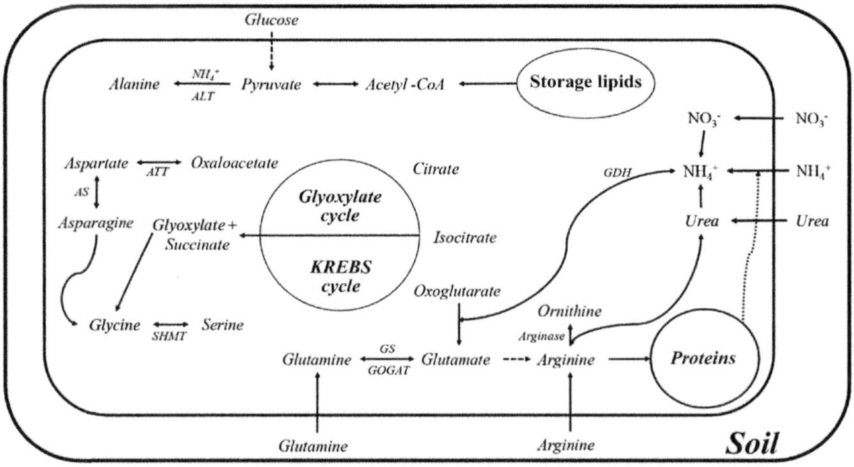

Figure 11.2 ***Representative model of metabolic pathways and nitrogen fluxes in spore germination through amino acid metabolism.*** *ALT*, alanine transaminase; *AAT*, aspartate aminotransferase; *AS*, asparagine synthase; *SHMT*, serine hydroxymethyl transferase; *GS*, glutamine synthetase; *GOGAT*, glutamate synthase; *GDH*, glutamate dehydrogenase; *NR*, nitrate reductase.

acid biosynthesis (Chen et al., 2017; Fellbaum et al., 2012; Jin et al., 2011; Tian et al., 2010; Kasola et al., 2007).

Also, during germination of the spores, they can incorporate exogenous sources of nitrogen to carry out an amino acid synthesis de novo (Gachomo et al., 2009). When the spores are exposed to NH_4^+, NO_3^- or urea gives rise to high concentrations of free amino acids, for example, arginine, and glutamate. As presented in Fig. 11.2, Glutamate is a key point in the metabolism of amino acids and Arginine seems to play a role in the storage and transport of Nitrogen that is assimilated during presymbiotic growth (Bago, Pfeffer, & Shachar-Hill, 2001). Some authors propose that the concentration of these amino acids was increased when it was supplemented with an inorganic source of Nitrogen, concluding that Novo amino acids can be synthesized (Jin et al., 2011; Gachomo et al., 2009, Bago et al., 2001). Therefore, based on the above evidence, amino acid synthesis does not limit presymbiotic growth, in addition, while fungi do not depend on exogenous nitrogen to germinate or develop, they can take advantage of available inorganic nitrogen sources to synthesize amino acids.

Nitrogen fluxes in the plant-fungus symbiosis

Mycorrhizal fungi can consume both organic and inorganic nitrogen from the soil and transport this nutrient to the plant, as shown by the experiments

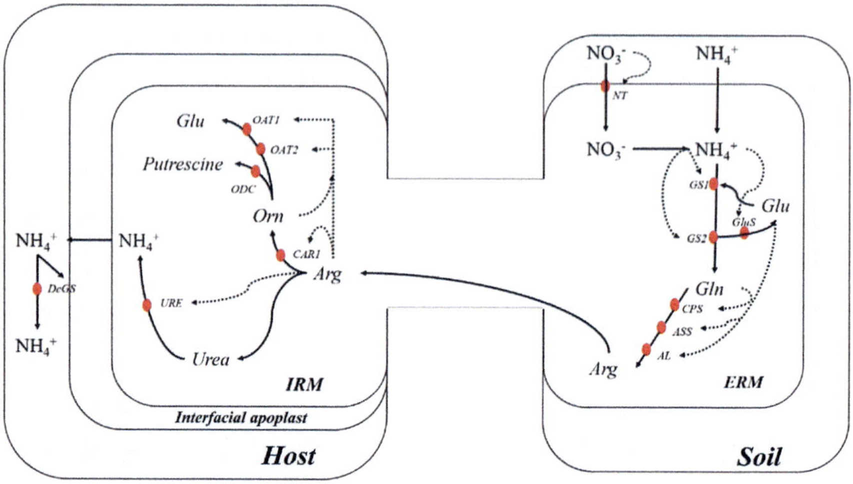

Figure 11.3 ***Representative diagram of the flow of nitrogen in the metabolic reactions of mycorrhizal fungi.*** Nitrogen is introduced from the ground to the host through metabolic conversions to Arginine, then it is transported to the intraradical mycelium, where it is again transported through a series of metabolic reactions, to finally be transferred and assimilated by the plant. *AL*; Arginosuccinate lyase; *ASS*, Argininosuccinate synthase; *CAR*, Arginase; *CPS*, Carbomyl-phosphate synthase; *GS*, Glutamine synthetase 1 and 2; *GluS*, Glutamate synthase; *NT;* Nitrate transporter; *OAT*, Ornithine aminotransferase; *URE*, Urease.

carried out by Govindarajulu et al. (2005), where the ERM exposed to NH_4^+, NO_3^- or urea marked with ^{15}N, metabolize this nitrogen and translocate it to the roots, as shown in Fig. 11.3 (Fellbaum et al., 2012). This nitrogen is translocated in the hyphae as arginine, however, it is quickly transformed into urea to finally be converted to the plant as NH_4^+, while the carbon skeletons generated during the splitting of the arginine are reincorporated again to the fungal groups (Bago et al. 2001;Govindarajulu et al. 2005). Now, we can understand how the external hyphae of the mycorrhizal fungi capture the inorganic nitrogen as NO_3^- and NH_4^+ and organic nitrogen as amino acids and transfer some, sometimes a large part of them to the plants (Bago, Vierheilig, Piché, & Azcón-Aguilar, 1996; Hawkins, Johansen, & George, 2000; Govindarajulu et al., 2005; Jin et al. 2005).

Some authors describe that NH_4^+ is the preferred nutrient in most circumstances (Read & Pérez-Moreno, 2003). The transfer of nitrogen to the host tends to be higher when NH_4^+ is administered instead of NO_3^- to the hyphae of mycorrhizal fungi, according to the data reported by Tanaka and Yano (2005), although it is likely that nitrogen is transferred from the

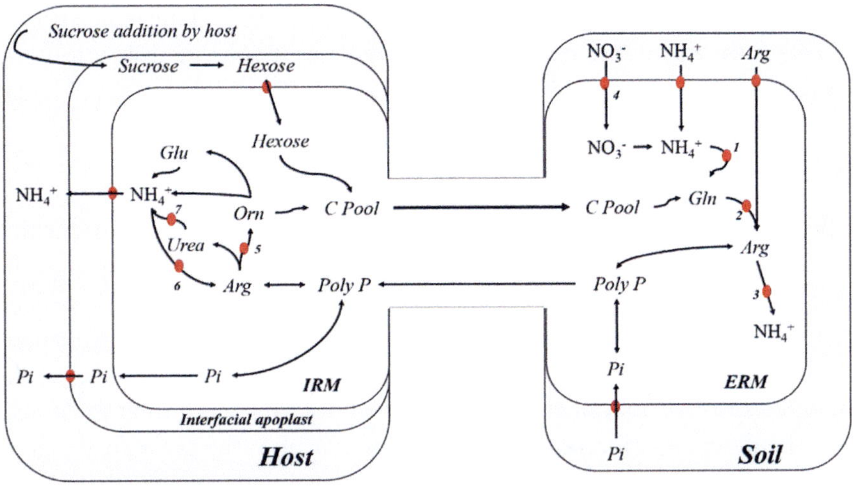

Figure 11.4 ***Representative model of flow of carbon, nitrogen and phosphorus in mycorrhizal fungi.*** *1-* GS and GluS ↑; *2-* CPS, ASS and AL ↑; *3-* CAR1 and URE ↓; *4-* NT↓; CAR ↑; CPS, ASS, AL and GS ↓; *7-* URE ↑.

extraradical hypha to arginine before transferring to the plant as NH_4^+ (Govindarajulu et al. 2005; Cruz et al. 2007).

Fluxomics of phosphorus

Several authors have described, as shown in Fig. 11.4, that phosphate is introduced into the cell by high-affinity phosphate transporters in the extraradical mycelium (Harrison & van Buuren, 1995; Solaiman & Saito, 2001). In later studies, it was shown that the phosphate may move within the fungus as polyphosphate (polyP) and once in the intraric hyphae, the long-chains are hydrolyzed, which facilitates the transfer to the host (Ohtomo and Saito 2005; Bago et al. 2002; Harrison 1999). Phosphate transfer of the fungus to the plant seems to occur mainly at the interface, although the expression of phosphate transporters around hyphae in Paris type coils has also been demonstrated (Karandashov, Nagy, Wegmuller, Amrhein, & Bucher, 2004). The activity of the ATPase plant is strongly expressed in the arbuscular membrane (Smith, Grace, & Smith, 2009) and phosphate accumulation as polyP strongly correlated with AM colonization (Ohtomo & Saito, 2005). Most importantly, a subfamily (subfamily 1 under the Pht1 family) of plant phosphate transporters (StPT4) is now known to be expressed only in colonized plants; the first was in *Solanum tuberosum* (Rausch et al., 2001), and subsequently they were identified in several other taxa (Javot, Penmetsa, Terzaghi, Cook, & Harrison, 2007). The acquisition of P through the symbiotic pathway down regulates the direct absorption of P by the plant (Smith et al., 2009).

Perspectives for the future and conclusions

In this chapter, we have described the fluxes of the main nutrients which are consumed and transported in the fungus-plant interaction through the study of fluxomics. By means of kinetic models and stoichiometric models made by several authors, each with its advantages and limitations, have been able to elucidate the metabolic pathways of each nutrient. Likewise, the principles of fluxomics of ^{13}C have also been described and their great utility in the determination of metabolic fluxes. For now, the great challenge of fluxomics for the next few years is to develop better approximations that are able to integrate kinetic models with genomic scale models, thus combining the advantages of these study techniques to obtain increasingly real and accurate data. Finally, it is worth noting that the current and future trend is to facilitate the free exchange of fluxomic data generated among the scientific community. The achievement of this objective requires the generation of international standards, databases of fluxomic information and electronic infrastructures that facilitate the dissemination of the obtained data.

References

Aroca, R., & Ruiz-Lozano, J. M. (2009). Induction of plant tolerance to semi-arid environments by beneficial soil microorganisms - a review. In E. Lichtfouse (Ed.), *Climate change, intercropping, pest control and beneficial microorganisms. Sustainable Agriculture Reviews* (pp. 121–135). (2). The Netherlands: Springer.

Armada, E., Azcón, R., López-Castillo, O. M., Calvo-Polanco, M., & Ruiz-Lozano, J. M. (2015). Autochthonous arbuscular mycorrhizal fungi and Bacillus thuringiensis from a degraded Mediterranean area can be used to improve physiological traits and performance of a plant of agronomic interest under drought conditions. *Plant Physiology and Biochemistry, 90*, 64–74 https://doi.org/10.1016/j.plaphy.2015.03.004.

Aslani, Z., Hassani, A., Rasouli-sadaghiani, M., Esmailpour, B., & Rohi, Z. (2014). Effects of arbuscular mycorrhizal (AM) fungi on essential oil content and nutrients uptake in basil under drought stress. *Journal of Medicinal Plants and By-products, 2*, 147–153.

Avio, L., & Giovannetti, M. (1998). The protein pattern of spores of arbuscular mycorrhizal fungi: comparison of species, isolates and physiological stages. *Mycological Research, 102*, 985–990.

Bago, B., Pfeffer, P. E., Abubaker, J., Jun, J., Allen, J. W., Brouillette, J., Douds, D. D., Lammers, P. J., & Shachar-Hill, Y. (2003). Carbon export from arbuscular mycorrhizal roots involves the translocation of carbohydrate as well as lipid. *Plant Physiology, 131*, 1496–1507 https://doi.org/10.1104/pp.102.007765.

Bago, B., Pfeffer, P. E., Zipfel, W., Lammers, P., & Shachar-Hill, Y. (2002). Tracking metabolism and imaging transport in arbuscular mycorrhizal fungi. *Plant and Soil, 244*, 189–197 https://doi.org/10.1023/A:1020212328955.

Bago, B., Pfeffer, P. E., & Shachar-Hill, Y. (2001). Could the urea cycle be translocating nitrogen in the arbuscular mycorrhizal symbiosis? *New Phytologist, 149*, 4–8 https://doi.org/10.1046/j.1469-8137.2001.00016.x.

Bago, B., Pfeffer, P. E., Doubs, D. D., Brouillette, J., Bécard, G., & Shachar-Hill, Y. (1999). Carbon metabolism in spores of the arbuscular mycorrhizal fungus glomus intraradices as revealed by nuclear spectroscopy magnetic resonance. *Plant Physiology, 121*, 263–271 https://doi.org/10.1104/pp.121.1.263.

Bago, B., Vierheilig, H., Piché, Y., & Azcón-Aguilar, C. (1996). Nitrate depletion and pH changes induced by the extraradical mycelium of the arbuscular mycorrhizal fungus *Glomus intraradices* grown in monoxenic culture. *New Phytologist, 133*, 273–280.

Baum, C., El-Tohamy, W., & Gruda, N. (2015). Increasing the productivity and product quality of vegetable crops using arbuscular mycorrhizal fungi: a review. *Scientia Horticulturae, 187*, 131–141 https://doi.org/10.1016/j.scienta.2015.03.002.

Bécard, G., & Piché, Y. (1989). Fungal growth stimulation by CO_2 and root exudates in the vesicular-arbuscular mycorrhizal symbiosis. *Applied Environmental Microbiology, 55*, 2320–2325.

Bidartondo, M. I., Redecker, D., Hijri, I., Wiemken, A., Bruns, T. D., Dominguez, L., Sersic, A., Leake, J. R., & Read, D. J. (2002). Epiparasitic plants specialized on arbuscular mycorrhizal fungi. *Nature, 419*, 389–392 https://doi.org/10.1038/nature01054.

Camarena-Gutiérrez, G. (2012). Plant-arbuscular mycorrhizal fungi interactions. *Revista Chapingo Serie Ciencias Forestales y del Ambiente., 18*(3), 409–421 http://doi:10.5154/r.rchscfa.2011.11.093.

Costa, R. S., Hartmann, A., & Vinga, S. (2016). Kinetic modeling of cell metabolism for microbial production. *Journal of Biotechnology, 219*(10), 126–141 https://doi.org/10.1016/j.jbiotec.2015.12.023.

Chen, A., Gu, M., Wang, S., Chen, J., & Xu, G. (2017). Transport properties and regulatory roles of nitrogen in arbuscular mycorrhizal symbiosis. *Seminars in Cell and Developmental Biology., 74*, 80–88 http://dx.doi.org/10.1016/j.semcdb.2017.06.015.

Chiu, T. J., Liu, H., & Harrison, M. J. (2001). The spatial expression patterns of a phosphate transporter (MtPT1) from *Medicago truncatula* indicate a role in phosphate transport at the root/soil interface. *Plant Journal, 25*, 281–293 https://doi.org/10.1046/j.1365-313x.2001.00963.x.

Curran, K. A., Crook, N. C., & Alper, H. S. (2012). Using flux balance analysis to guide microbial metabolic engineering. *Methods in Molecular Biology, 834*, 179–216 http://doi:10.1007/978-1-61779-483-4_13.

Crown, S. B., & Antoniewicz, M. R. (2013). Publishing ^{13}C metabolic flux analysis studies: a review and future perspectives. *Metabolic Engineering, 20*, 42–48 https://doi.org/10.1016/j.ymben.2013.08.005.

Cruz, C., Egsgaard, H., Trujillo, C., Ambus, P., Requena, N., Martins-Loucao, M. A., & Jakobsen, I. (2007). Enzymatic evidence for the key role of arginine in nitrogen translocation by arbuscular mycorrhizal fungi. *Plant Physiology, 144*, 782–792 https://doi.org/10.1104/pp.106.090522.

Engelmoer, D. J. P., & Kiers, E. T. (2015). Host diversity affects the abundance of the extraradical arbuscular mycorrhizal network. *New Phytologist, 205*, 1485–1491 doi: 10.1111/nph.13086.

Fellbaum, C. R., Gachomo, E. W., Beesetty, Y., Choudhari, S., Strahan, G. D., Pfeffer, P. E., Kiers, T. E., & Bücking, H. (2012). Carbon availability triggers fungal nitrogen uptake and transport in arbuscular mycorrhizal symbiosis. *Proceedings of the National Academy of Sciences (PNAS), 109*(7), 2666–2671 https://doi.org/10.1073/pnas.1118650109.

Feng, X., Page, L., Rubens, J., Chircus, L, Colletti, P., Pakrasi, H., & Tang, Y. (2002). Bridging the gap between fluxomics and industrial biotechnology. *Journal of Biomedicine and Biotechnology, 2010*, 1–13 http://doi:10.1155/2010/460717.

Ferrol, N., Barea, J. M., & Azcon-Aguilar, C. (2002). Mechanisms of nutrient transport across interfaces in arbuscular mycorrhizas. *Plant Soil, 244*, 231–237 https://doi.org/10.1007/978-94-017-1284-2_22.

Fleming, R. M. T., Thiele, I., Provan, G., & Nasheuer, H. P. (2010). Integrated stoichiometric, thermodynamic and kinetic modelling of steady state metabolism. *Journal of Theorical Biology, 264*(3), 683–692 http://doi:10.1016/j.jtbi.2010.02.044.

Finlay, R. D. (2008). Ecological aspects of mycorrhizal symbiosis: with special emphasis on the functional diversity of interactions involving the extraradical mycelium. *Journal of Experimental Botany, 59*, 1115–1126 http://doi:10.1093/jxb/ern059.

Gachomo, E., Alle, J. W., Pfeffer, P. E., Govindarajulu, M., Douds, D. D., Jin, H. R., Nagahashi, G., Lammers, P. J., Shachar-Hill, Y., & Bücking, H. (2009). Germinating spores of *Glomus intraradices* can use internal and exogenous nitrogen sources for de novo biosynthesis of amino acids. *New Phytologis., 184*, 399–411 http://doi:10.1111/j.1469-8137.2009.02968.x.

Govindarajulu, M., Pfeffer, P. E., Jin, H., Abubaker, J., Douds, D. D., Allen, J. W., Bücking, H., Lammers, P. J., & Shachar-Hill, Y. (2005). Nitrogen transfer in the arbuscular mycorrhizal symbiosis. *Nature, 435*, 819–823 https://doi.org/10.1038/nature03610.

Gutjahr, C., & Parniske, M. (2013). Cell and developmental biology of Arbuscular mycorrhiza symbiosis. *Annual Review of Cell and Developmental Biology, 29*, 593–617 https://doi.org/10.1146/annurev-cellbio-101512-122413.

Harrison, M. J. (1999). Molecular and cellular aspects of the arbuscular mycorrhizal symbiosis. *Annual Review of Plant Physiology and Plant Molecular Biology, 50*, 361–438.

Harrison, M. J. (1997). The arbuscular mycorrhizal symbiosis: an underground association. *Trends Plant Science, 2*, 54–60 https://doi.org/10.1016/S1360-1385(97)82563-0.

Harrison, M. J., & Van Buuren, M. L. (1995). A phosphate transporter from the mycorrhizal fungus *Glomus versiforme*. *Nature, 378*, 626–629 https://doi.org/10.1038/378626a0.

Hawkins, H. J., Johansen, A., & George, E. (2000). Uptake and transport of organic and inorganic nitrogen by arbuscular mycorrhizal fungi. *Plant and Soil, 226*, 275–285 https://doi.org/10.1023/A:1026500810385.

Hodge, A., Campbell, C. D., & Fitter, A. H. (2001). An arbuscular mycorrhizal fungus accelerates decomposition and acquires nitrogen directly from organic material. *Nature, 413*, 297–299 https://doi.org/10.1038/35095041.

Heux, S., Bergès, C., Millard, P., Portais, J. C., & Létisse, F. (2015). Recent advances in high-throughput ^{13}C-fluxomics. *Current Opinion in Biotechnology, 43*, 104–109 https://doi.org/10.1016/j.copbio.2016.10.010.

Karandashov, V., Nagy, R., Wegmuller, S., Amrhein, N., & Bucher, M. (2004). Evolutionary conservation of a phosphate transporter in the arbuscular mycorrhizal symbiosis. *Proceedings of the National Academy of Sciences (PNAS), 101*, 6285–6290.

Jamshidi, N., & Palsson, B. (2008). Formulating genome-scale kinetic models in the post-genome era. *Molecular Systems Biology, 4*(171), 1–10 http://doi:10.1038/msb.2008.8.

Jin, H. R., Jiang, d. H., & Zang, P. H. (2011). Effect of carbon and nitrogen availability on metabolism of aminoacids in germinating spores of arbuscular mycorrhizal fungi. *Pedosphere, 21*(4), 432–442 https://doi.org/10.1016/S1002-0160(11)60145-8.

Jin, H., Pfeffer, P. E., Douds, D. D., Piotrowski, E., Lammers, P. J., & Shachar-Hill, Y. (2005). The uptake, metabolism, transport and transfer of nitrogen in an arbuscular mycorrhizal symbiosis. *New Phytologist, 168*, 687–696 https://doi.org/10.1111/j.1469-8137.2005.01536.x.

Javot, H., Penmetsa, R. V., Terzaghi, N., Cook, D. R., & Harrison, M. J. (2007). A Medicago truncatula phosphate transporter indispensable for the arbuscular mycorrhizal symbiosis. *Proceedings of the National Academy of Sciences (PNAS), 104*, 1720–1725 https://doi.org/10.1073/pnas.0608136104.

Johnson, D., Leake, J. R., Ostle, N., Ineson, P., & Read, D. J. (2002). *In situ* $^{13}CO_2$ pulse-labelling of upland grassland demonstrates a rapid pathway of carbon flux from arbuscular mycorrhizal mycelia to the soil. *New Phytologist, 153*(2), 327–334 https://doi.org/10.1046/j.0028-646X.2001.00316.x.

Jung, S. C., Martinez-Medina, A., López-Raez, J. A., & Pozo, M. J. (2012). Mycorrhiza-induced resistance and priming of plant defenses. *Journal of Chemical Ecology, 38*, 651–664 https://doi.org/10.1007/s10886-012-0134-6.

Kasola, K. R., Workmaster, B., & Spada, P. (2007). Inoculation of cranberry (*Vaccinium macrocarpon*) with the ericoid mycorrhizal fungus *Rhizoscyphus ericae* increases nitrate influx. *New Phytologist*, *176*, 184–196 http://doi:10.1111/j.1469-8137.2007.02149.x.

Klein, S., & Heinzle, E. (2012). Isotope labeling experiments in metabolomics and fluxomics. *Systems Biology and Medicine*, *4*(3), 261–272 https://doi.org/10.1002/wsbm.1167.

Krömer, J., Quek, L. E., & Nielsen, L. (2009). ^{13}C-Fluxomics: a tool for measuring metabolic phenotypes. *Australian Biochemist*, *40*(3), 17–20.

Lake, J. R., Ostle, N. J., Rangel-Castro, J. I., & Jonhson, D. (2006). Carbon fluxes from plants through soil organisms determined by field $^{13}CO_2$ pulse-labelling in an upland grassland. *Applied Soil Ecology*, *33*(2), 152–175 https://doi.org/10.1016/j.apsoil.2006.03.001.

Niittylae, T., Chaudhuri, B., Sauer, U., & Frommer, W. B. (2009). Comparison of quantitative metabolite imaging tools and carbon-13 techniques for fluxomics. *Methods in Molecular Biology*, *553*, 355–372 http://doi:10.1007/978-1-60327-563-7_19.

Niedenführ, S., Wiechert, W., & Nöh, K. (2015). How to measure metabolic fluxes: a taxonomic guide for 13C fluxomics. *Current Opinion in Biotechnology*, *34*, 82–90 https://doi.org/10.1016/j.copbio.2014.12.003.

Ohtomo, R., & Saito, M. (2005). Polyphosphate dynamics in mycorrhizal roots during colonization of an arbuscular mycorrhizal fungus. *New Phytologist*, *167*, 571–578 http://doi:10.1111/j.1469-8137.2005.01425.x.

Parniske, M. (2008). Arbuscular mycorrhiza: the mother of plant root endosymbiosis. *Nature Reviews in Microbiology*, *6*, 763–775 https://doi.org/10.1038/nrmicro1987.

Pearson, J. N., & Jakobsen, Y. (1993). The relative contribution of hyphae and roots to phosphorus uptake by arbuscular mycorrhizal plants, measured by dual labelling with ^{32}P and ^{33}P. *New Phytologist*, *124*, 489–494.

Pfeffer, P. E., Douds, D. D., Bécard, G., & Schachar-Hill, Y. (1999). Carbon uptake and metabolism and transport of lipids in an arbuscular mycorrhiza. *Plant Physiology*, *120*, 587–598 https://doi.org/10.1104/pp.120.2.587.

Quek, L. E., Wittmann, C., Nielsen, L. K., & Krömer, J. O. (2009). OpenFLUX: efficient modelling software for 13C-based metabolic flux analysis. *Microbial Cell Factories*, *8*(25), 1–15 http://doi:10.1186/1475-2859-8-25.

Rausch, C., Daram, P., Brunner, S., Jansa, J., Laloi, M., Leggewie, G., Amrhein, N., & Bucher, M. (2001). A phosphate transporter expressed in arbuscule-containing cells in potato. *Nature*, *414*, 462–466 https://doi.org/10.1038/35106601.

Remy W., Taylor T. N., Hass H., & Kerp H. (1994). 400 millon year old vesicular-arbuscular mycorrhizae. (VAM). *Proceedings of the National Academy of Science (PNAS)*. USA 91:11841-11843. https://doi.org/10.1073/pnas.91.25.11841.

Read, D. J., & Pérez-Moreno, J. (2003). Mycorrhizas and nutrient cycling in ecosystems – a journey towards relevance? *New Phytologist*, *157*, 475–492 https://doi.org/10.1046/j.1469-8137.2003.00704.x.

Saa, P. A., & Nielsen, L. K. (2017). Formulation, construction and analysis of kinetic models of metabolism: A review of modelling frameworks. *Biotechnology Advances*, *35*(8), 981–1003 https://doi.org/10.1016/j.biotechadv.2017.09.005.

Sanders, I. R., & Croll, D. (2010). Arbuscular mycorrhiza: The challenge to understand the genetics of the fungal partner. *Annual Reviews in Genetics*, *44*, 271–292 http://doi:10.1146/annurev-genet-102108-134239.

Sanford, K., Soucaille, P., Whited, G., & Chotan, G. (2002). Genomics to fluxomics and physiomics — pathway engineering. *Current Opinion in Microbiology*, *5*, 318–322 https://doi.org/10.1016/S1369-5274(02)00318-1.

Shachar-Hill, Y., Pfeffer, P. E., Douds, D., Osman, S. F., Doner, L. W., & Ratcliffe, R. G. (1995). Partitioning of intermediate carbon metabolism in VAM colonized leek. *Plant Physiology*, *108*, 7–15 https://doi.org/10.1104/pp.108.1.7.

Sharma, P., Prasad, G., & Shrivastav, A. (2015). Evaluation of antimicrobial activity of some medicinal plants against most prevalent diabetic wound pathogens. *Research & Reviews: Journal of Medical Science and Technology*, *4*(1), 13–20.

Schüler, A., Schwarzott, D., & Walker, C. (2001). A new fungal phylum, the *Glomeromycota* phylogeny and evolution. *Mycological Research, 105*, 1413–1421 http://doi:10.1017/S0953756201005196.

Scagel, C. F., & Lee, J. (2012). Phenolic composition of basil plants is differentially altered by plant nutrient status and inoculation with mycorrhizal fungi. *HortScience, 47*, 660–671 https://doi.org/10.21273/HORTSCI.47.5.660.

Smith, F. A., Grace, E. J., & Smith, S. E. (2009). More than a carbon economy: nutrient trade and ecological sustainability in facultative arbuscular mycorrhizal symbioses. *New Phytologist., 182*, 347–358 https://doi.org/10.1111/j.1469-8137.2008.02753.x.

Simard S.W., Jones M. D., Durall D. M. (2003). Carbon and nutrient fluxes within and between mycorrhizal plants. In: van der Heijden M.G.A., Sanders I.R. (eds) Mycorrhizal Ecology. Ecological Studies (Analysis and Synthesis), vol 157. Springer, Berlin, Heidelberg. https://doi.org/10.1007/978-3-540-38364-2_2.

Solaiman, M. D. Z., & Saito, M. (2001). Phosphate efflux from intraradical hyphae of *Gigaspora margarita in vitro* and its implication for phosphorus traslocation. *New Phytologist, 151*, 525–533.

Stalidzans, E., Seiman, A., Peebo, K., Komasilovs, V., & Pentjuss, A. (2018). Model-based metabolism design: constraints for kinetic and stoichiometric models. *Biochemical Society Transactions, 46*(2), 261–267 http://doi:10.1042/BST20170263.

Symanczik, S., Courty, P. E., Boller, T., Wiemkem, A., & Al-Yahyaei, M. N. (2015). Impact of water regimes on an experimental community of four desert arbuscular mycorrhizal fungal (AMF) species, as affected by the introduction of a non-native AMF species. *Mycorrhiza, 25*, 639–647 https://doi.org/10.1007/s00572-015-0638.

Talaata, N. B., & Shawky, B. T. (2012). Influence of arbuscular mycorrhizae on root colonization, growth and productivity of two wheat cultivars under salt stress. *Archives of Agronomy and Soil Science, 58*, 85–100 https://doi.org/10.1080/03650340.2010.506481.

Tanaka, Y., & Yano, K. (2005). Nitrogen delivery to maize via mycorrhizal hyphae depends on the form of N supplied. *Plant, Cell and Environment, 28*, 1247–1254 https://doi.org/10.1111/j.1365-3040.2005.01360.x.

Tian, C., Kasiborsky, B., Lammers, P., Bücking, H., & Shachar-Hill, Y. (2010). Regulation of the nitrogen transfer pathway in the arbuscular mycorrhizal symbiosis: gene characterization and the coordination of expression with nitrogen flux. *Plant Physiology, 153*, 1175–1187 https://doi.org/10.1104/pp.110.156430.

Yassen, T., Yassen, K., Rahim, F., Wali, S., Ahmad, I., Begum, H., & Ghani, S. (2016). Arbuscular mycorrhizal fungi spores diversity and AMF infection in some medicinal plants of District Charsadda KPK. *Pure and Applied Biology, 5*(4), 1176–1182 http://dx.doi.org/10.19045/bspab.2016.50141.

Winter, G., & Krömer, J. O. (2012). Fluxomics – connecting 'omics analysis and phenotypes. *Environmental Microbiology* http://doi:10.1111/1462-2920.12064.

Zolfaghari, M., Nazeri, V., Sefidkon, F., & Rejali, F. (2013). Effect of arbuscular mycorrhizal fungi on plant growth and essential oil content and composition of *Ocimum basilicum* L. *Iranian Journal of Plant Physiology, 3*, 643–650 DOI: 10.22034/IJPP. 2013.540674.

Zeng, Y., Guo, L. P., Chen, B. D., Hao, Z. P., Wang, J. Y., Huang, L. Q., et al. (2013). Arbuscular mycorrhizal symbiosis and active ingredients of medicinal plants: current research status and prospectives. *Mycorrhiza, 23*, 253–265 DOI 10.1007/s00572-013-0484-0.

Zhu, Y., & Miller, R. M. (2003). Carbon cycling by arbuscular mycorrhizal fungi in soil–plant systems. *Trends in Plant Science, 8*(9), 407–409 https://doi.org/10.1016/S1360-1385(03)00184-5.

CHAPTER 12

Metabolomics of medicinal and aromatic plants: Goldmines of secondary metabolites for herbal medicine research

Amrina Shafi[a], **Insha Zahoor**[b,c]
[a]Department of Biotechnology, School of Biological Sciences, University of Kashmir, Srinagar, Jammu and Kashmir, India
[b]Drug Therapeutics and Neurobiology Lab, Department of Biotechnology and Bioinformatics Centre, University of Kashmir, Srinagar, Jammu and Kashmir, India
[c]Department of Neurology, Henry Ford Hospital, Detroit, MI, United States

Introduction

Plants are considered to be the main resource providers of diverse materials that are useful for a variety of purposes including food, energy, and medicine since ancient times. Plants have provided humans with medicines since the beginning of civilization, and the majority of the population in many Asian and African countries still depends on traditional herbal medicines for primary healthcare (Hosni, 2012). Presently, medicinal and aromatic plants (MAPs), preferred in many medicinal systems as they are renewable sources, are generally considered safer, and are readily available (Saxena, Saxena, Nema, Singh, & Gupta, 2013). MAPs are the sources of thousands of chemicals that possess their own functional benefits making the plants one of the most preferred sources of drugs development and synthesis (Strohl, 2000; Karunamoorthi, Jegajeevanram, Vijayalakshmi, & Mengistie, 2013). Further, on the basis of ethnomedicinal data, herbal medicine is not only easily accessible at low prices for primary healthcare, but also can serve as a valuable reservoir for pharmacological drug development (Padalia, 2012).

Presently, the world is facing issues of complex diseases and the need of the hour is to discover and develop novel drug leads. MAPs have been clinically explored and used as therapeutic sources, as they contain diverse secondary metabolites with therapeutic potential (Patwardhan & Vaidya, 2010; Mukherjee et al. 2012). However, the phytotherapeutic mechanism of action of valuable therapeutic compounds and novel bioactive substances needs to be understood. Humans have, though, relied on medicinal plants

Medicinal and Aromatic Plants: Expanding their Horizons through Omics
http://dx.doi.org/10.1016/B978-0-12-819590-1.00012-4

for centuries, yet a lot of important information about the molecular and cellular systems of the plants only started coming to light with the advent of modern biotechnological and bioinformatics techniques (omics). Conventional methods for plant-based drug discovery are much expensive and time-consuming (DiMasi, Hansen, & Grabowski, 2003). However, the application of computational approaches helps to catalyze the process of plant-based drug research (Harishchander, 2017). With the advent of these omic sciences, the role of various genes and proteins in the metabolic processes has been recently decoded (Van Emon, 2015). A recent development in bioinformatics resources brought a major change in current studies of MAPs including plant-based drug discovery. Combination of omic approaches is required, to understand the biological and other metabolic processes of medicinal plants, to analyze, and to interpret the data associated with it (Harishchander, 2017). Availability of curated databases of MAPs played a vital role in the area of plant-based drug discovery. Use of all omic data combined with chemometric analysis tools is termed as systems biology. Systems biology has helped is a complete understanding of a system or a physiological state by integrating different "omics" data sources for the analysis of networks and regulation (Ray et al., 2002). Out of the available omics strategies, metabolomics has resulted in the production of inclusive metabolite profiles providing a clear understanding of diagnostic changes in the levels of metabolites, leading to therapeutic monitoring of drug targets through the elucidation of metabolomics pathways.

Since the scientific basis of MAPs-based medicine is still poor in herbal medicine, researchers have raised public health-related issues on the usage of MAPs based on herbal medicines and on herb-drug interactions (Patwardhan & Vaidya, 2010; Mukherjee et al. 2012). Thus, modernization and development of herbal-based drugs require a comprehensive evaluation of metabolites and underlying metabolic pathways and elucidation of their mechanism of actions. Further, the potential of lesser known/under-explored MAPs needs additional research, so that these natural chemical goldmines can be explored and better utilized in future. Metabolomics is playing a key role in meeting the above-said challenging tasks and there has been ever-increasing employment of bioinformatics tools in the study of MAPs metabolomics. In this chapter, we have highlighted some of the recent advances in the field of bioinformatics particularly metabolomics and how they have helped in opening up new doors for understanding the basic biological aspects of the MAPs. The medicinal effect of MAPs is oriented towards the secondary plant metabolites, which have played an important role in alleviating several ailments. In modern medicine, they have provided lead compounds for the production of medications for treating various diseases from migraine up to cancer. In this chapter, we are presenting various databases of secondary plant metabolites and different metabolomics engineering strategies for enhancing secondary metabolite production in MAPs.

MAPs significance and secondary metabolites

Humans have exploited various natural resources for the cure and treatment of diseases and among these the MAPs have been the most reliable. MAPs and medicinal and aromatic crops (MAC) are high-value crops; the natural products obtained from these crops are low-volume high-value commodities that have numerous applications in various industries such as the flavor and fragrance, perfumery and cosmetics, pharmaceutical, aromatherapy, and as ingredients in various consumer products (Agyare, Obiri, Boakye, & Osafo, 2013). Various parts of the MAPs are used for medicinal purposes (Chanda, 2014), as they contain diverse primary and secondary metabolites that help in various biochemical perturbations after their administration (Agyare et al., 2013). Moreover, the modern pharmaceutical industry and medicinal system also rely on these phytochemicals as the natural ingredient or as the lead molecule for the generation of semi-synthetic and synthetic derivatives. Population explosion and the scarcity of modern medicines have resulted in an increased emphasis on this traditional healing system relying on plants (Leonti, 2013). Thus, MAPs serves as an important source of raw materials for traditional as well as modern medicines. About 80% of the world population relies on herbal medicines for primary health care needs as per the World Health Organization (WHO) (Akerele, 1993).

As secondary metabolites are an important source of therapeutic drugs or drug leads, MAPs also have the ability to synthesize a unique array of these phytochemicals as metabolites (Fiehn, 2002). Currently, the structures for various natural products are known, which represents 15% of the estimated 350,000 plant species (Stitt & Fernie, 2003). The botanical drugs that we use today are all illustrations of complex mixtures enriched in plant secondary metabolites and are classified as: alkaloids, glycosides, tannins, phenolic compounds, volatile oils, terpenoids, saponins, and steroids, etc. (Kamal & Khan, 2014). MAPs and MAP-derived compounds have become most attractive category for researchers because of the presence of these structurally diverse secondary metabolites which find a diverse array of applications from drugs to industrial and agricultural or environmental

(Thomas, Joy, Mathew, & Skaria, 2000). Thus due to broad structural diversity in secondary metabolites and their wide ranges of pharmacological activities; MAPs are considered as valuable and never-ending sources of novel chemical structures having therapeutic potential and therefore represents a "Chemical Goldmines" of novel products and applications (Zwenger & Basu, 2008). Thus isolation of active constituents form these "chemical goldmines" offer huge opportunities for their industrial utilization as valuable sources for new molecules for flavor and drug development.

Plants and their products have also been used as traditional medicines for the treatment of common ailments (Crozier, Jaganath, & Clifford, 2006); up to 70,000 species of plants are used in folk medicine (Farnsworth & Soejartto, 1991). MAPs and the usage of their products have been reported for traditional societies such as India, China, Africa, and in the developed countries like Europe, Australia, and North America. About 100,000 secondary metabolites have been isolated from higher plants (Afendi, Okada, & Yamazaki, 2012), nearly 7,500 plant species in India (Shankar & Majumdar, 1997) and about 1,000 medicinal plants in China, are commonly used in ethnomedicines (He & Sheng, 1997). This increased utilization of MAPs is due to ease of availability, minimal side effects, and lack of a need to get regulatory approvals for the plant-based natural products (Pandey & Tripathi, 2014).

Advances in chemistry and pharmacology have validated the claims of traditional medicines and have discovered active principles. About 30% of all US-FDA approved drugs introduced in the market are either natural products, botanical drugs, or semisynthetic analogues of a natural product (Newman & Cragg, 2012). The enormous biosynthetic potential of MAPs is yet to be exploited completely and biotechnology could be used to generate novel chemical compounds, with enhanced or newer bioactivities, through the intervention of the latest technologies. Processes like metabolic engineering and gene pharming have been successfully generating the required amounts of the metabolites from the plants leading ultimately to a faster drug discovery process (Webster, Teh, & Ma, 2016). To improve the safety and consistency of these chemical goldmines, additional focus is required to define the pharmacology, stability, and bioavailability of these products (Bent & Ko, 2004). Improved production of secondary metabolites, the study of biochemical pathways and associated enzymes, biosynthesis of the desired constituents, mode of action at target sites, the potential toxicity of constituents used in drug development are the exciting frontiers for future research.

Metabolomics: a component of the "OMICS" system

MAPs evolved the biosynthesis of a cornucopia of novel chemicals to survive and communicate in a complex ecological environment and can be considered as huge compound libraries as the chemical diversity of plants is enormous. The advances in a number of systems-biology disciplines (genomics, transcriptomics, metabolomics, and computation biology), fuelled by the decreasing costs for generating large-scale molecular data, has revolutionized our research approaches in the field of MAPs (Champagne & Boutry, 2013). Among omics science, metabolomics is an emerging and a part of the systems biology that is primarily concerned with the high-throughput snapshot of a metabolome at a given time point and under specific physiological conditions (Champagne & Boutry, 2013). Various modern technologies have been explored for assessment of plant secondary metabolites; metabolomics is a newer addition to the spectrum of omics approaches and is of interest for herbal medicine (Shyur & Yang, 2008). Both the primary and secondary metabolites are studied under the canopy of this discipline (Wolfender, Rudaz, Choi, & Kim, 2013). Metabolomics strategies may actually offer the most valuable and functional information that is crucial in system biology studies (Fig. 12.1). Actually, it is challenging because of the complexity of the diverse metabolic characteristics and abundances of molecules (Gomez-Casati, 2016). Metabolomics allows for a global assessment of a cellular state within the context of the immediate environment, taking into account gene expression, genetic regulation, altered kinetic activity and regulation of enzymes, and changes in metabolic reactions (Menedes, Kell, & Westerhoff, 1996). It differs from the classical or traditional targeted phytochemical analysis in various fundamental aspects, such as being a data-driven approach with predictive power that aims to assess all measurable metabolites.

Bioinformatics has helped in deciphering important information about various genetic players of MAPs and how they tend to relate to one another. Bioinformatics methods such as hierarchical cluster analysis, principal component analysis and others are employed to statistically process the massive amount of data, resulting from metabolomics profiling (Okada, Afendi, Altaf-Ul-Amin, Takahashi, & Nakamura, 2010). An appealing strategy is to interconnect the data obtained from omics to cellular interaction networks known as system biology, which may explain the activity of complex herbal mixtures in a comprising fashion. Therefore, systems biology is appreciated as an innovative discipline to study holistic phytotherapeutical approaches. There is a continuous quest for new molecules from MAPs, as the synthetic

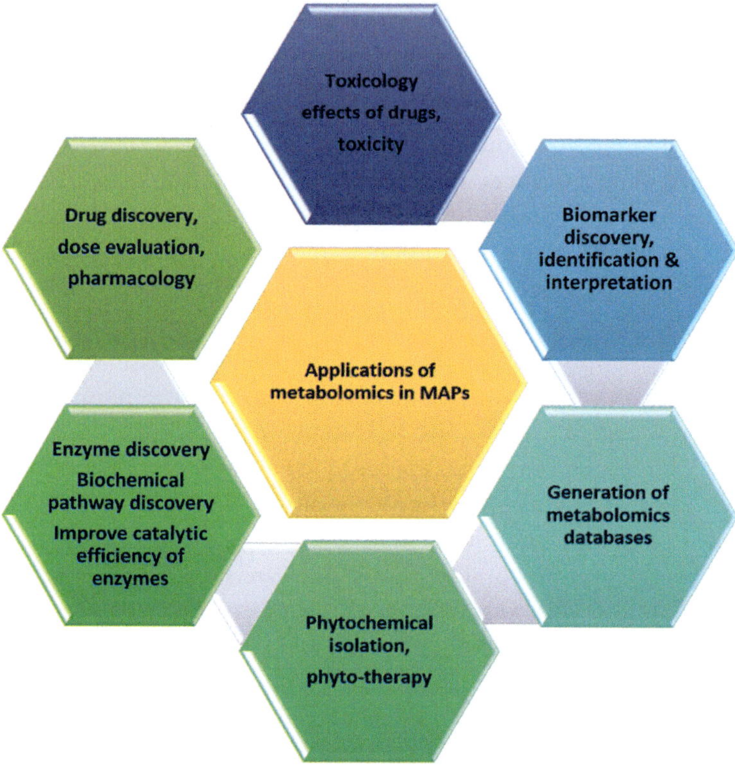

Figure 12.1 ***Applications of metabolomics in MAPs.***

libraries in the pharmaceutical industry have not yielded the expected number of lead molecules. Some of the applications of metabolomics are:

1. Metabolomics is a novel and a powerful analytical tool in herbal medicine research for comprehensive profiling of secondary metabolites (Ellis, Dunn, Griffin, Allwood, & Goodacre, 2007; Wolfender et al., 2013). The fast development of metabolomics technology provides us with a valuable opportunity to advance the studies of medicinal plants as more than 50,000 metabolites have been characterized from MAPs.
2. Metabolomics as a postgenomics tool is often regarded as offering distinct advantages when compared to other "-omics" technologies and one advantage of conducting metabolomics is that genomics information is not needed.
3. Metabolomics can be used to compare metabolite quantitative changes in medicinal materials between different ages, origins, organs, developmental stages, environmental cultivation, and culture conditions, and

processing methods. Recent advances in metabolomics have enabled rapid identification and quantification of yet unidentified metabolites (Wang, Wang, Li, Paradesi, & Brown, 2007).

4. It can help us in understanding the metabolic pathways for the production of these bioactive compounds, generate metabolic fingerprinting of MAPs for the authentication and quality control, classify medicinal plants, and establish a quantitative version of chemotaxonomic analysis to advance our knowledge of the evolutional relationship of medicinal plants (Fig. 12.1).
5. Metabolomics has become a powerful tool in drug discovery and development by identification and profiling of secondary metabolites in MAPs (Gahlaut, Vikas, Gothwal, Kulharia, & Chhillar, 2013). It may provide a systems biology approach for target compound analysis (TCA) in medicinal plants.
6. Multidisciplinary facets of metabolomics have been exploited including drug discovery and development, high-throughput screening for evaluation of plant drugs (Ulrich-Merzenich et al., 2007).
7. Metabolomics is the most imperative approach that provides metabolite profiles for studying biochemical networks and gives a clear understanding of the effects of the drug–herbal interventions (Patwardhan & Vaidya, 2010; Tagore, Chowdhury, & De, 2014). This metabolic data can be used for generating metabolic correlation networks.

Strategies for metabolomic analysis

Metabolomics is a rapidly developing technology and major approaches currently used in plant metabolomics research include metabolic fingerprinting, metabolite profiling, and targeted analysis (Shulaev, 2006). Some of the strategies used for the metabolomic study are described further:

Metabolomics: A holistic quantification and identification of all metabolites within a biological system, under a given set of conditions (Dunn & Ellis, 2005). This state is currently unrealizable, with any single or combination of metabolomic approaches.

Metabolic/Metabolite profiling: It is involved in identification and quantification of metabolites related through their metabolic pathway(s) or similarities in their chemistry and is aimed at a simultaneous measurement of all or a set of metabolites in a sample. There are two types of metabolite profiling that are used today: targeted and non-targeted (broad) (Fig. 12.2). The targeted analysis is fully calibrated and delivers an absolute quantification of

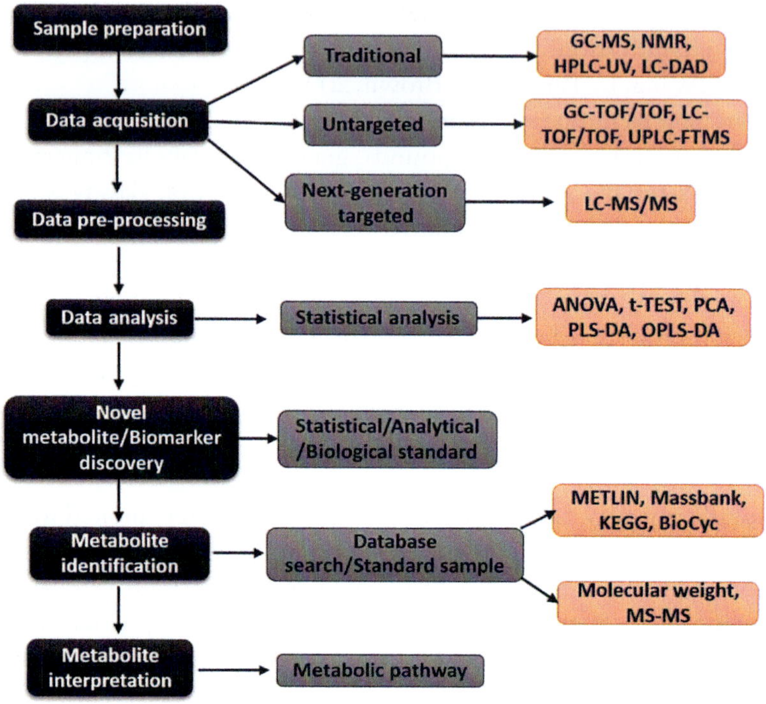

Figure 12.2 ***Metabolomics strategies in MAPs.***

the quantity of particular metabolite in a given sample. However, non-targeted approaches aim to provide information on as comprehensive as possible a selection of metabolites. Multiple analytical techniques can be used for metabolite profiling (Shulaev, 2006; Sumner, Mendes, & Dixon, 2003) like NMR, GC-MS, liquid chromatography-mass spectrometry (LC-MS), capillary electrophoresis–mass spectrometry (CE-MS), and FT-IR spectroscopy.

Metabolite fingerprinting: This is a rapid and high-throughput method, where global metabolite profiles are obtained from crude samples or simple cellular extracts. In general, metabolic fingerprinting is largely used to identify metabolic signatures or patterns associated with a particular stress response without identification or precise quantification of all the different metabolites in the sample (Giri et al., 2010). Pattern-recognition analysis is then performed on the data to identify features specific to a fingerprint. Fingerprinting can be performed with a variety of analytical techniques, including NMR (Krishnan et al. 2005), MS (Goodacre, York, Heald, & Scott, 2003), Fourier transform ion cyclotron resonance mass spectrometry or Fourier transforms infrared (FT-IR) spectroscopy (Johnson et al. 2003).

Metabolite footprinting: The measurement of metabolites secreted from the intracellular complement of a biological system into its extracellular medium or matrix. With the development of metabolomics, the terms metabolic fingerprint and metabolic footprint have now become commonly used. Particularly in cells type studies, metabolic fingerprinting refers directly to the whole set of intracellular metabolites (Lin, Wu, Tjeerdema, & Viant, 2007), while metabolic footprinting applies to a set of extracellular metabolites (Pope, MacKenzie, Defernez, Aroso, & Fuller, 2007).

Chemoprofiling: As the chemical composition of plants may vary to some extent and needs to be standardized to guarantee comparable therapeutic effects. A number of chromatographic fingerprinting analyses are known to disclose the detectable ingredients composition and concentration distribution (Yang, Zhao, Wang, Liang, & Zeng, 2010). Standard analytical techniques include thin-layer chromatography (TLC), high-performance liquid chromatography (HPLC) and capillary electrophoresis. Recently, novel technological developments became available for chemo-profiling, such as infrared spectroscopy, metabolic fingerprinting and quantitative determinations based on NMRspectra.

Metabolic fluxomics: It studies the overall flux of a metabolite between different regions (Niedenführ, Wiechert, & Nöh, 2015). It also estimates the complications in modeling intracellular transport of various primary and secondary metabolites. For fluxomic analysis, a radioactive isotope of carbon (^{13}C) is used in a cell system (Niedenführ et al., 2015). Algorithms based on differential equations have been developed to model the metabolic networks and study their interrelationships. Many programs employ a cellular modeling environment that mimics the in vivo conditions for carrying out the metabolic profiling and flux analysis studies within a plant cell (Boudon, Chopard, Ali, Gilles, & Hamant, 2015).

Novel, high-throughput and cost-effective analytical platforms for metabolite profiling

Natural product research requires new strategies to renovate traditional methodologies, which are too expensive and/or time-consuming (Kinghorn et al., 2008). Latest metabolomics tools are effective and specific for rapid identification and generating new knowledge of the pharmacological and toxic effects of the plant metabolite (Michel, Halabalaki, & Skaltsounis, 2013; Harvey, 2008). Varied analytical techniques are applied in metabolite profiling due to the huge-chemical diversity and variation of metabolites concentration (Theodoridis et al., 2008). Consecutively, new lead molecules are being developed through hyphenated and new analytical methods to improve the accuracy, consistency of plant-based preparations, and for stricter standardization of herbal drugs (Ulrich-Merzenich et al., 2007). The laboratory techniques generally employed for the purpose are the variants of chromatography (liquid chromatography, gas chromatography, or high-performance liquid chromatography) for separation purpose and a spectroscopic technique (UV-visible spectroscopy, NMR, or MS) for detection. Bioinformatics and cheminformatics tools are then employed for interpreting the outcomes in the forms of peaks. This fully characterized metabolite is then processed for further analysis like pharmacological screening. The methodology of metabolomics typically employs two principal analysis techniques (Fig. 12.2). The first technique is analytical chemistry, aimed at comprehensive, simultaneous, high-throughput and accurate metabolite analysis. Liquid chromatography (LC) (e.g., HPLC and UPLC) and gas chromatography (GC) analyses, as well as capillary electrophoresis are frequently employed for metabolite separation. HPLC coupled with a Photodiode Array Detector (HPLC-PDA) is usually used for secondary metabolite profiling (Hosni, Msaada, Ben Tâarit, & Marzouk, 2011). MS (mass-spectrometry) is also a popular analytical tool in metabolomics where metabolite identification is based on structural information. Although metabolite profiling and its interpretation by this method are complex due to limited spectral databases (Dunn, Broadhurst, Atherton, Goodacre, & Griffin, 2011), still MS is considered as one of the key tools for effective metabolite identification, usually based on a mass match of metabolites with the databases (Horai, Arita, Kanaya, Nihei, & Ikeda, 2010; Wishart, Jewison, Guo, Wilson, & Knox, 2013). An LC-MS-based metabolomic study has become a powerful analytical tool for the assessment of various secondary metabolites in herbal medicine (Mukherjee, Harwansh, & Bahadur, 2016). These secondary metabolites have been found to possess a broad range of therapeutic properties, including antioxidant, cardioprotective, and antihypertensive potential (Alamgeer, Malik, & Bashir, 2015). NMR spectroscopy is also frequently used in metabolomic studies (Cloarec, Campbell, Tseng, Braumann, & Spraul, 2007). NMR spectroscopy is a highly popular technique among phytochemists, for the structural elucidation and functional characteristics/information of metabolites of interest based on the interpretation of NMR spectral features (Dunn et al., 2011). NMR-based metabolomics has been shown to be very useful due to its untargeted and unbiased features. Scognamiglio, D'Abrosca, Esposito, & Fiorentino (2015) applied NMR spectroscopic technique for the evaluation of metabolite changes in seven aromatic Mediterranean plant species during different seasons. They targeted the detection and quantification of both primary and secondary metabolites and identified flavonoids (quercetin, apigenin, and kaempferol) and phenylpropanoid derivatives (rosmarinic and chlorogenic acid) as the principal polyphenols. Furthermore, NMR is highly reproducible it allows the contemporary identification and quantification of a large number of compounds, needs short times of analysis and allows the identification of unknown compounds, as it gives important structural information (Forseth & Schroeder, 2011). Thus, NMR-based metabolomics approaches extremely powerful for the profiling of crude extracts and the rapid identification of natural products (Nakabayashi & Saito, 2013).

A range of other analytical tools are also in use for metabolomics analysis, including GC-MS (gas chromatography-equipped with MS), GC × GC-MS (two-dimensional gas chromatography coupled with MS), FTIR (Fourier transform infrared spectroscopy), LC-MS (liquid chromatography coupled with MS) and CE-MS (capillary electrophoresis coupled with MS) (Theodoridis, Gika, & Wilson, 2008). Initially only GC was hyphenated to MS, and its usage for separation was limited to only volatile metabolites (Roux, Lison, Junot, & Heilier, 2011). Now, LC hyphenated to mass spectrometry (LC-MS) is a comparatively more powerful and emphatic tool and plays a significant contribution toward profiling of drug metabolism and bioactivation (Li et al. 2012). Recently another imperative tool that is, UPLC (ultra-performance liquid chromatography), has been introduced for the metabolomics research. UPLC-MS methods revealed improved potential toward the evaluation of differential metabolic pathway activities due to improved sensitivity and resolution (Lu et al., 2008). Zhao and Lin, (2014) highlighted the applications of UPLC-based metabolomics for biomarkers discovery in clinical chemistry. Cox, Oha, Keasling, Colson, & Hamann (2014) elaborated the significance of metabolomics for biomarkers characterization in natural product research and categorized metabolomics tools into five basic categories, that is, NMR, MS, LC × MS, GC × MS, and integrated analytical strategies. In a recent review, Chen et al. (2016) comprehensively described the applications of metabolomics for toxicological biomarker discovery related to natural product research. Recently, the technologies that are commonly used for global metabolome studies have been increasingly combined with multi-hyphenated techniques such as GC × GC-time-of-flight (TOF), GC-TOF-MS, and UPLC-quadrupole (Q)-TOF-MS to enable compound analysis using a wider range of metabolome perspectives (Okada, Nakamura, Kanaya, Takano, & Malla, 2009; Lee, Jung, Shin, Kim, & Moon, 2014). Shi, Cao, Wang, Aa, &

Duan (2016) have also described the applications of metabolomics techniques for efficacy and toxicological studies of traditional Chinese herbal medicines (TCHM). They highlighted that NMR and MS-based analytical methods are more commonly being used for TCHM research; however, the sensitivity and resolution of MS-based techniques (normally coupled with LC and GC) are much higher than that of NMR.

Metabolomics as a tool for quality evaluation of herbal products from MAPs

Natural products from MAPs have been the focus area for research in multidisciplinary fields for the development of templates of new chemical entities (NCEs). In order to rationalize the use of herbal products in different forms, more particularly the extracts/marketed product in therapy as is being used nowadays, a need-based and novel concept of chemo-profiling or metabolomics is gaining impetus. MAPs contain the diverse number of chemical components; the composition of metabolites is dynamically regulated by gene expression networks that change with time and cultivation conditions (Trethewey, 2004). To this end, a very powerful approach is metabolomics, the comprehensive analysis of the set of low molecular weight compounds of a biological system under a given condition (Kim, Choi, & Verpoorte, 2011). Plant metabolomics involved the measurement of genotypic and phenotypic changes in the cellular systems by analyzing the complete set of compounds present in plant cell (Gahlaut et al., 2013). For metabolomic analysis, standardization through the markers, and DNA fingerprinting, several analytical techniques are available for assuring consistent quality and efficacy of herbal medicine (Ellis et al. 2007; Efferth & Greten, 2012). Furthermore, for aromatic plants, a great effort has been devoted to the study of essential oils (Efferth & Greten, 2012) and recent research has shown MAPs as a source of bioactive compounds (Vallverdú-Queralt, Regueiro, Martínez-Húelamo, Alvarenga, & Leal, 2014).

The advantages of herbal medicines over synthetic and modern drugs include holistic and personalized approaches, the synergism of pharmacologically active constituents in herbs and minimal side effects in medication for complex diseases derived from multifactorial causes (Verpoorte et al., 2009). In other words, metabolites in plants interact with multiple targeted proteins in humans, which regulate gene expression and lead to a dynamical state change in both the human metabolome and human physiological activity. Progress in pharmacogenomics has provided considerable support for a more holistic view regarding diagnosis and treatment (Wang, Lamers, Korthout, van Nesselrooij, & Witkamp, 2005). Metabolomics has played a significant role in the elucidation of the therapeutic potential of these MAPs (Fig. 12.1) by identifying and quantifying a diverse group of secondary metabolites. MAPs-derived medicines have been recognized as representative multi-compound drugs and can be investigated by metabolomics technology. This study has given rise to special emphasis on phytomedicine research. It can be very useful in shifting the paradigm in drug discovery and development from natural resources. Metabolomics can be used as an effective platform to understand the phytochemical basis of such therapeutically active phytoconstituents (Williamson, 2001). In many cases of drug analysis, few of the plant secondary metabolites are barely detected due to the low-presence in plants and hence low therapeutic activity. But in case of MAPs and their formulations, the biological activities can be produced synergistically due to the presence of several constituents therein.

The advent of genomic and metabolomic technologies has now made it possible to bring the field of MAPs natural products into the 21st century and replace serendipitous and haphazard finding by rational design and discovery. Genomics-based "phytochemical arrays (genome, transcriptome, proteome, and metabolome)" have been established for measurement and analysis of several aspects including metabolite profiling in plants (Bino, Hall, Fiehn, & Kopka, 2004). Metabolomics is also widely applied in plant multi-omics studies and is considered to be an effective approach for comprehensively elucidating the biosynthetic flow of "gene-to-metabolite" on the molecular level and the correlations among gene expression, proteins and metabolites in systems biology and functional genomics. (Fukushima, Kusano, Redestig, Arita, & Saito, 2009; Saito, Hirai, & Yonekura-Sakakibara, 2007). MAPs are being considered as alternative sources of finding new chemical entities (NCEs) for drug discovery and development for example, Aspirin (a semi-derivative compound from the *Salix alba*) is a wonder drug being used to treat pain and other complications for many years (Newman & Cragg, 2012), secondary metabolites like paclitaxel (Taxol), camptothecin (irinotecan, topotecan) and podophyllotoxins (etoposide, teniposide) etc. have been reported to possess potential anticancer activity (Harvey, 2007). Metabolomic fingerprinting can be very helpful in the field of herbal medicine for drug discovery, systems biology, gene-function analysis and various diagnostic techniques through different modern hyphenated technologies. In addition, recent improvements in metabolomics technologies have incorporated the valuable tools of gene-function analysis, system biology, and diagnostic platforms. New therapeutic challenges and trends continually create an increasing need for more rapid scientific responses, and metabolomics is currently positioned to be an important tool for providing biomarkers for better diagnosis and prognosis.

In the current MAPs research, however, metabolomics has been independently applied to evaluation and discrimination based on metabolite profiles and fingerprints, and on marker metabolites. Herbal medicines including crude drugs and traditional medicines generally contain pharmacologically active constituents even if they are minor constituents in total metabolites. Metabolomics has also provided a fresh understanding of drug action and has the potential to identify biochemical pathways toward which new drugs might be directed. In the early 21st century, scientists increasingly focused on pharmaco-metabolomics, which enabled a broad look at metabolic responses to drugs and in turn stimulated new lines of investigation. Initially compounds were subjected to metabolism studies after their discovery and the metabolites were characterized by conventional spectral techniques after their isolation from biological matrices (Prasad, Garg, Takwani, & Singh, 2011). This success in drug discovery is linked to the high-chemical diversity of natural sources; nevertheless, the chemical variability and the vast number of metabolites (Wolfender, Marti, & Ferreira Queiroz, 2010). Comparative metabolomics strategy coupled with cell and gene-based assays were used for species classification and anti-inflammatory bioactivity validation of medicinal Echinacea species (Hou et al., 2010). More metabolic profiling has been conducted in the medicine (*Medicago truncatula*) which is also a model organism for legume biology (Barsch, Tellstrom, Patschkowski, Kuster, & Niehaus, 2006a; Barsch, Carvalho, Cullimore, & Niehaus, 2006b).

During recent years, metabolomic studies have been carried out worldwide by the scientific community to explore the metabolite profiles and metabolic pathways of different biofluids/herbal extracts (Meyer & Maurer, 2015; Zhang et al., 2015). TERPMED is a proposal devoted to plant terpenes, the largest and most diverse group of plant that will focus on sesquiterpene lactones and phenolic diterpenes because of the presence of distinct functional groups and their high potential as novel human drugs for treating cancer and neurological disorders. Classic genetic, molecular and metabolic engineering strategies (Bioballistics, *Agrobacterium tumefaciens* transformation, recombinant enzymes) for the production of natural compounds or even optimization of MAPs have expanded and improved productivity of plant-derived fine chemicals. Combining science and engineering in this research field was claimed as "Combinatorial Biosynthesis" and later as "Synthetic Biology." Thus, there is a possibility of application of the latest genetic engineering technologies for the exploitation of MAPs by increasing the level of desired natural products, gain insight into metabolic pathways even for new biosimilar chemicals.

Bioinformatics database resources for MAPs

Several specialized biological databases and web resources have been developed which provides updated information about MAPs with medicinal properties. Metabolomics, similar to transcriptomic and proteomics, generates huge volume of data that require specialized bioinformatics and data mining tools to gain knowledge (Tota et al., 2013). Metabolomics data handling, analysis and mining and its integration with other omics platforms have been dramatically improved in recent years because of the development of an array of publicly available bioinformatics tools (Fig. 12.3). In order for metabolomics data to be seamlessly integrated with other global molecular datasets that define the biological status of tissue(s), it needs to be organized and normalized in a standard format that enables cross-referencing with multiple datasets (Caspi, Foerster, Fulcher, Kaipa, & Krummenacker, 2008).

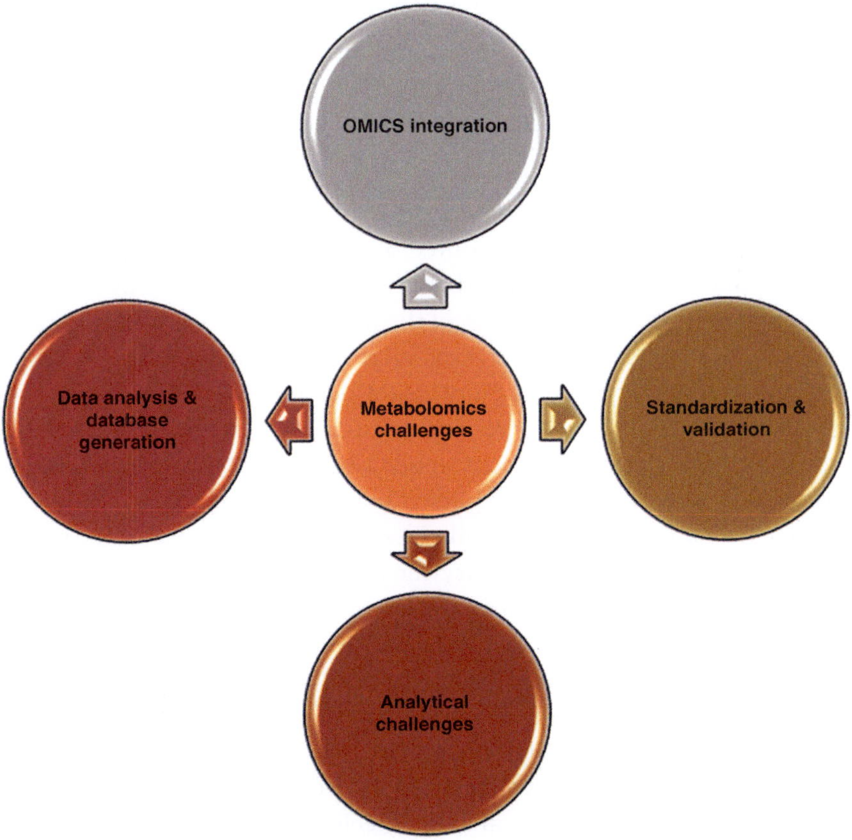

Figure 12.3 *Challenges faced by metabolomics in MAPs.*

The data can inform modeling of the metabolic networks of specialized plant products both by the researchers who deposited the data and by the broader research community. From a human health perspective, these data and techniques can enable bioengineering of MAPs both to produce larger quantities of medicinally-useful compounds as well as to produce new specialized compounds with targeted therapeutic potentials (Kale, Haug, Conesa, Jayseelan, & Moreno, 2016).

Here we describe various resources that will be useful for MAPs research (Table 12.1). Metabolite annotation and identification are highly dependent on the availability and quality of databases. For annotation, several programs can be used, including the Golm Metabolite Database, MassBank, METLIN, FiehnLib and Lipid Search. For NMR data, the Madison Metabolomics Consortium Database is also useful for annotating signals from primary metabolites in plants. In addition to this spectral database, there are also several compound databases including KNApSAcK, Kyoto Encyclopaedia of Genes and Genomes (KEGG) PubChem and LIPIDMAPS.

Conclusion and future perspectives

MAPs based products have always been touted as remedies to treat various symptoms and even today, a large part of the world population relies on herbal medicines as a major source of health care. In some rural areas, traditional medicines based on herbal drugs still remain the only source of health care. Even today almost 30% of the modern drugs we use are actually derived from natural products (an ever-increasing number of these are coming from MAPs) are now in the process of being approved for market. We are now at the beginning of a new phase in which integrative approaches of genomics and metabolomics are applied to the study the metabolism of MAPs. These approaches have begun to transform our understanding of at least two main aspects of herbal medicines: (1) the biosynthesis, and pathway regulation, of many plant secondary metabolites of medicinal importance and their exploitation for biomarker-driven drug discovery and development (2) the mechanism of action of many of these plant herbal components on human metabolism and health. Plant metabolomics provides a comprehensive understanding of the spectrum of phytochemical constituents of MAPs and this approach is being exploited in a wide range of applications including medical science, synthetic biology, etc. Advances in analytical chemistry, computation and biotechnology have led to the recent development of the latest methodologies for broad-metabolite profiling.

Table 12.1 Bioinformatics resources for secondary metabolites of medicinal plant.

S.No.	Database for metabolomics	Information	References
1.	InDiaMed (Indian Medicinal Plants for Diabetes)	Biochemical, chemical, geographical, and pharmacological information of the medicinal plants	Tota et al. (2013)
2.	IMPPAT (The Indian Medicinal Plants, Phytochemistry, and Therapeutics)	Information about phytochemical constituents of medicinal plants. The database includes 1742 medicinal plants, 9596 phytochemicals, and 1124 therapeutic uses, and nonredundant information about 9596 phytochemicals with standard chemical identifiers and structure, as well as 960 potential druggable phytochemicals	Mohanraj et al. (2017)
3.	MPD3	Information about phytochemicals, activities, structural, and test target of medicinal plants. It contains more than 5000 phytochemicals from around 1000 medicinal plants with 80 different properties, and 200-plus targets. Moreover, the database provides 632 genus and 1022 plant-based information including phytochemicals (7062), targets (271), and activities (91) records.	Mumtaz et al. (2017)
4.	Natural Products Alert (NAPRALERT)	The database provides information about natural products/secondary metabolites, including ethano-medical, pharmacological, and biochemical extracts	Loub et al. (1985)
5.	Indonesian medicinal plants database	Provides information about medicinal plants and three-dimensional (3D) structures of chemical compounds. The database contains 3825 species records with 16244 local names and 6776 records documented in 12,980 species-contents interaction along with 3D structures of 1412 chemical compounds from medicinal plants	Yanuar et al. (2011)

(Continued)

Table 12.1 Bioinformatics resources for secondary metabolites of medicinal plant (*Cont.*)

S.No.	Database for metabolomics	Information	References
6.	Super Natural II	Provides information on 326,000 natural compounds with corresponding two-dimensional structures and their physicochemical properties. It also facilitates pathways information related to synthesis and degradation of the natural products, in addition to their mechanism in relation to drugs with similar structure and their target	Banerjee et al. (2014)
7.	HerbMed	The database provides hyperlinked access to scientific data underlying the use of herbs for health	Wootton (2002)
8.	Medicinal Plant Metabolomics Resource (MPMR)	Framework for generating experimentally testable hypotheses about the metabolic networks that lead to the generation of specialized compounds, identifying genes that control their biosynthesis and establishing a basis for modeling metabolism in less studied species	Medicinal Plant Metabolomics Resource (2012)
9.	Medicinal Plant Consortium (MPC)	It provides transcriptomic and metabolomics resources for 14 key medicinal plants for the advancement of drug production and development	MPC (http://medicinalplantgenomics.msu.edu/contacts.shtml)
10.	HMDB	It enlists 42,000 metabolites and the number of lipid variants.	Wishart et al. (2009)
11.	METLIN	Database of metabolite MS spectra	Smith et al. (2005)
12.	MetaCyc	Database for the study of metabolic pathways	Caspi et al. (2008)
13.	Golm metabolite Database	Information about mass spectra from active metabolites quantified by GS-MS	Kopka et al. (2005)

S.No.	Database for metabolomics	Information	References
14.	MassBank	A first public repository of mass spectra of small chemical compounds for life sciences (<3000 Da), it contains 605 electron-ionization mass spectrometry	Horai et al. (2010)
15.	FiehnLib	GC/MS libraries based on time-of-flight mass spectrometry (GC-TOF) and on quadrupole mass spectrometry (GC-Quad) of over 1000 primary and secondary metabolites.	Tobias et al. (2009)
16.	Madison Metabolomics Consortium	A resource for metabolomics research based on NMR and MS	Cui et al. (2008)
17.	KNApSAcK	The database contains chemical diversity of metabolites which occurred through the evolution of species, accumulation and search of metabolite-species relationships	Shinbo et al. (2006)
18.	Medicinal Plant Genomics Resource	The database also contains metabolic profiling data (LC-MS), collected from several tissues of medicinal plants	Medicinal Plant Genomics Resource (2018) (http://medicinalplantgenomics.msu.edu/)
19.	LIPIDMAPS	Consortium developed for a number of online tools for performing tasks such as drawing lipid structures and predicting possible structures from (MS) data	Fahy et al. (2007)
20.	MetaboAnalyst	Comprehensive metabolomics data analysis, interpretation, and integration with other omics data	Chong et al. (2018)
21.	ASCA	A new method that can deal with complex multivariate datasets containing an underlying experimental design, such as metabolomics datasets	Smilde et al. (2005)

(Continued)

Table 12.1 Bioinformatics resources for secondary metabolites of medicinal plant (*Cont.*)

S.No.	Database for metabolomics	Information	References
22.	KaPPA-View	Transcriptomic and metabolomics database	Tokimatsu et al. (2005)
23.	MetNet	Software is designed to enable biologist to visualize, statistically analyze and model a metabolic and regulatory network map of Arabidopsis	Wurtele et al. (2003)
24.	CathaCyc	Metabolic pathway database built from metabolic and RNA-seq data of the plant *C. roseus*, a repository for genes, enzymes, reactions, and pathways of primary and secondary metabolism	Van Moerkercke et al. (2013)
25.	AraCyc	A database containing biochemical pathways of Arabidopsis	Zhang et al. (2005)
26.	MedicCyc	A Pathway database for *M. truncatula,* which currently features more than 250 pathways with related genes, enzymes and metabolites	Urbanczyk-Wochniak & Sumner (2007)
27.	Metabolights	This database is cross-species and contains information about metabolite structures, spectra and biological roles	Kale et al. (2016)
28.	PRIMe	Genomic and metabolomics data/NMR spectroscopy, GC-MS, LC-MS and CE-MS database	Akiyama et al. (2008)
29.	MapMan	A tool that displays large -data sets of metabolic pathways or other processes. METABOLITESCAVENGER groups hundreds of metabolites into pathways	Thimm et al. (2004)
30.	Metabolomics at Rothamsted	National centre for plant and microbial metabolites	MeT-RO (https://www.rothamsted.ac.uk/metabolomics)

Integration of different broad-range metabolite profiling into genomics approaches (system biology) driven by metabolome data will bring a change in our understanding of the regulation of metabolic networks and, therefore, facilitate improvements in our understanding of MAPs. As futuristic diagnostic tools, both metabolomics and proteomics can be the promising strategies for the efficient translation and understanding of metabolite–disease and protein-disease associations leading to a revolution in therapeutic monitoring. We see in this avalanche of knowledge both challenges and avenues for further research. We think there is an urgent need to develop faster, more informative and comprehensive analytical approaches for profiling and characterizing a larger number of metabolites; these challenges can be overcome also with the development of computational metabolomics strategies for metabolite annotation, de novo pathway reconstruction, and analysis of natural variation. We clearly recognize the long history and the potential of MAPs as a source of drug leads, but there is need of more intense scrutiny that should be conducted on herbal drugs–including rigorous studies on their chemical composition and clinical trials–before claims could be made in relation to their therapeutic efficacy.

References

Afendi, F. M., Okada, T., Yamazaki, M., et al. (2012). KNApSAcK family databases: integrated metabolite-plant species databases for multifaceted plant research. *Plant Cell and Physiology, 53*, e1.

Agyare, C., Obiri, D. D., Boakye, Y. D., & Osafo, N. (2013). Anti-inflammatory and analgesic activities of African medicinal plants. In V. Kuete (Ed.), *Medicinal Plant Research in Africa* (pp. 725–752). London: Elsevier.

Akerele, O. (1993). *Nature's medicinal bounty: don't throw it away* (pp. 390-395). In: World health, forum., World Health Organization, Switzerland,

Akiyama, K., Chikayama, E., Yuasa, H., Shimada, Y., Tohge, T., et al. (2008). PRIMe: a Web site that assembles tools for metabolomics and transcriptomics. *In Silico Biology,* 8, 0027.

Alamgeer, M. N., Malik, H., Bashir, S., et al. (2015). Cardiotonic and vasoconstriction effects of aqueous methanolic extract of *Paspalidium flavidum* L. *Pakistan Journal of Pharmaceutical Sciences, 28*, 437–441.

Banerjee, P., Erehman, J., Gohlke, B. O., Wilhelm, T., Preissner, R., & Dunkel, M. (2014). Super natural II-a database of natural products. *Nucleic Acids Research, 43*, D935–D939.

Barsch, A., Carvalho, H. G., Cullimore, J.V., & Niehaus, K. (2006b). GC-MS based metabolite profiling implies three interdependent ways of ammonium assimilation in Medicago truncatula root nodules. *Journal of Biotechnology, 127*, 79–83.

Barsch, A., Tellstrom, V., Patschkowski, T., Kuster, H., & Niehaus, K. (2006a). Metabolite profiles of nodulated alfalfa plants indicate that distinct stages of nodule organogenesis are accompanied by global physiological adaptations. *Molecular Plant Microbe Interaction, 19*, 998–1013.

Bent, S., & Ko, R. (2004). Commonly used herbal medicines in the United States: a review. *American Journal of Medicine, 116*, 478–485.

Bino, R. J., Hall, R. D., Fiehn, O., Kopka, J., et al. (2004). Potential of metabolomics as a functional genomics tool. *Trends in Plant Science*, *9*, 418–425.

Boudon, F., Chopard, J., Ali, O., Gilles, B., Hamant, O., et al. (2015). A computational framework for 3D mechanical modeling of plant morphogenesis with cellular resolution. *PLoS Computational Biology*, *11*(1), e1003950.

Caspi, R., Foerster, H., Fulcher, C. A., Kaipa, P., Krummenacker, M., et al. (2008). The MetaCyc database of metabolic pathways and enzymes and the BioCyc collection of pathway/genome databases. *Nucleic Acids Research*, *36*, D623–D631.

Champagne, A., & Boutry, M. (2013). Proteomics of non-model plant species. *Proteomics*, *13*, 663–673.

Chanda, S. (2014). Importance of pharmacognostic study of medicinal plants: an overview. *Journal of Pharmacognosy and Phytochemistry*, *2*(5), 69–73.

Chen, D. -Q., Chen, H., Chen, L., Tang, D. -D., Miao, H., & Zhao, Y. -Y. (2016). Metabolomic application in toxicity evaluation and toxicological biomarker identification of natural product. *Chemico Biology Interaction*, *252*, 114–130.

Chong, J., Soufan, O., Li, C., Caraus, I., Li, S., Bourque, G., Wishart, D. S., & Xia, J. (2018). MetaboAnalyst 4.0: towards more transparent and integrative metabolomics analysis. *Nucleic Acids Research*, *46*(W1), W486–W494.

Cloarec, O., Campbell, A., Tseng, L. H., Braumann, U., & Spraul, M. (2007). Virtual chromatographic resolution enhancement in cryoflow LC-NMR experiments *via* statistical total correlation spectroscopy. *Analytical Chemistry*, *79*, 3304–3311.

Cox, D. G., Oha, J., Keasling, A., Colson, K. L., & Hamann, M. T. (2014). The utility of metabolomics in natural product and biomarker characterization. *Biochimica et Biophysica Acta*, *1840*, 3460–3474.

Crozier, A., Jaganath, I. B., & Clifford, M. N. (2006). Phenols, Polyphenols and Tannins: An overview. In A. Crozier, M. N. Clifford, & H. Ashihara (Eds.), *Plant Second Metabolism Occurs in Structural Role of Human Diet* (pp. 1–24). Oxford: Blackwell Publishing Ltd.

Cui, Q., Lewis, I. A., Hageman, A. D., Anderson, M. E., Li, J., et al. (2008). Metabolite identification via the Madison Metabolomics Consortium Database. *Nature Biotechnology*, *26*, 162–164.

DiMasi, J. A., Hansen, R. W., & Grabowski, H. G. (2003). The price of innovation: New estimates of drug development costs. *Journal of Health Economics*, *22*, 151–185.

Dunn, W. B., & Ellis, D. I. (2005). Metabolomics: current analytical platforms and methodologies. *Trends in Analytical Chemistry*, *24*, 285–294.

Dunn, W. B., Broadhurst, D. I., Atherton, H. J., Goodacre, R., & Griffin, J. L. (2011). Systems level studies of mammalian metabolomes: the roles of mass spectrometry and nuclear magnetic resonance spectroscopy. *Chemical Society Reviews*, *40*, 387–426.

Efferth, T., & Greten, H. J. (2012). Medicinal and aromatic plant research in the 21st century. *Medicinal & Aromatic Plants*, *1*, e110.

Ellis, D. I., Dunn, W. B., Griffin, J. L., Allwood, J. W., & Goodacre, R. (2007). Metabolic fingerprinting as a diagnostic tool. *Pharmacogenomics*, *8*, 1243–1266.

Fahy, E., Sud, M., Cotter, D., & Subramaniam, S. (2007). LIPID MAPS online tools for lipid research. *Nucleic Acids Research*, *35*, W606–W612.

Farnsworth, N., & Soejartto, D. (1991). Global importance of medicinal plants. In O. Fiehn, V. Heywood, & H. Synge (Eds.), *Conservation of Medicinal Plants* (pp. 25–51). Cambridge: Cambridge University Press.

Fiehn, O. (2002). Metabolomics – the link between genotypes and phenotypes. *Plant Molecular Biology*, *48*, 155–171.

Forseth, R. R., & Schroeder, F. C. (2011). NMR-spectroscopic analysis of mixtures: from structure to function. *Current Opinion in Chemical Biology*, *15*, 38–47.

Fukushima, A., Kusano, M., Redestig, H., Arita, M., & Saito, K. (2009). Integrated omics approaches in plant systems biology. *Current Opinion in Chemistry and Biology*, *13*, 532–538.

Gahlaut, A., Vikas, D. M., Gothwal, A., Kulharia, M., Chhillar, A. K., et al. (2013). Proteomics and metabolomics: mapping biochemical regulations. *Drug Invention Today*, *5*, 321–326.

Giri, L., et al. (2010). Chromatographic and spectral fingerprinting standardization of traditional medicines: an overview as modern tools. *Research Journal of Phytochemistry*, *4*, 234–241.

Gomez-Casati, D. (2016). Metabolomics applications in plant biotechnology. *Metabolomics*, *6*, e146.

Goodacre, R., York, E.V., Heald, J. K., & Scott, I. M. (2003). Chemometric discrimination of unfractionated plant extracts analyzed by electrospray mass spectrometry. *Phytochemistry*, *62*, 859–863.

Harishchander, A. (2017). A review on application of bioinformatics in medicinal plant research. *Proteomics and Bioinformatics*, *1*, 000104.

Harvey, A. L. (2007). Natural products as screening resource. *Current Opinion in Chemistry and Biology*, *11*, 480–484.

Harvey, A. L. (2008). Natural products in drug discovery. *Drug Discovery Today*, *13*, 894–901.

He, S., & Sheng, N. (1997). Utilization and conservation of medicinal plants in China with special reference to Atractylodes lancea. In G. Bodeker, K. K. S. Bhat, J. Burley, & P. Vantomme (Eds.), *Medicinal plants for forest conservation and health care. Non-wood forest products* (pp. 109–111). (11). Rome: FAO.

Horai, H., Arita, M., Kanaya, S., Nihei, Y., Ikeda, T., et al. (2010). Massbank: a public repository for sharing mass spectral data for life sciences. *Journal of Mass Spectrometry*, *45*, 703–714.

Hosni, K. (2012). Medicinal and Aromatic Plants in Developing Countries. *Medicinal and Aromatic Plants*, *1*, e118.

Hosni, K., Msaada, K., Ben Tâarit, M., & Marzouk, B. (2011). Phenological variations of secondary metabolites from *Hypericum triquetrifolium* Turra. *Biochemical Systematics and Ecology*, *39*, 43–50.

Hou, C. C., Chen, C. H., Yang, N. S., Chen, Y. P., Lo, C. P., Wang, S.Y., et al. (2010). Comparative metabolomics approach coupled with cell-and gene-based assays for species classification and anti-inflammatory bioactivity validation of Echinacea plants. *Journal of Nutritional Biochemistry*, *21*, 1045–1059.

Johnson, H. E., Broadhurst, D., Goodacre, R., & Smith, A. R. (2003). Metabolic fingerprinting of salt-stressed tomatoes. *Phytochemistry*, *62*, 919–928.

Kale, N. S., Haug, K., Conesa, P., Jayseelan, K., Moreno, P., et al. (2016). MetaboLights: An Open-Access Database Repository for Metabolomics Data. *Current Protocols in Bioinformatics*, *53*, 1–18.

Kamal, A., & Khan, M. M. R. (2014). Phytochemical evaluation of some medicinal plants. *Indian Journal of Plant Science*, *3*(4), 5–8.

Karunamoorthi, K., Jegajeevanram, K., Vijayalakshmi, J., & Mengistie, E. (2013). Traditional medicinal plants a source of phytotherapeutic modality in resource-constrained health care settings. *Evidence-Based Complementary and Alternative Medicine*, *18*(1), 67–74.

Kim, H. K., Choi, Y. H., & Verpoorte, R. (2011). NMR-based plant metabolomics: where do we stand, where do we go? *Trends in Biotechnology*, *29*, 267–275.

Kinghorn, A., et al. (2008). Novel Strategies for the Discovery of Plant-Derived Anticancer Agents. *Pharmaceutical Biology*, *41*, 53–67.

Kopka, J., Schauer, N., Krueger, S., Birkemeyer, C., Usadel, B., et al. (2005). GMD@CSB.DB: the Golm metabolome database. *Bioinformatics*, *21*(8), 1635–1638.

Krishnan, P., Kruger, N. J., & Ratcliffe, R. G. (2005). Metabolite fingerprinting and profiling in plants using NMR. *Journal of Experimental Botany*, *56*, 255–265.

Lee, J., Jung, Y., Shin, J. H., Kim, H. K., Moon, B. C., et al. (2014). Secondary metabolite profiling of *Curcuma* species grown at different locations using GC/TOF and UPLC/Q-TOF MS. *Molecules*, *19*, 9535–9551.

Leonti, M. (2013). Traditional medicines and globalization: current and future perspectives in ethnopharmacology. *Frontiers in Pharmacology*, *4*, 92.

Li, F., Gonzalez, F. J., & Ma, X. (2012). LC–MS-based metabolomics in profiling of drug metabolism and bioactivation. *Acta Pharmaceutica Sinica B*, *2*, 118–125.

Lin, C. Y., Wu, H., Tjeerdema, R. S., & Viant, M. R. (2007). Evaluation of metabolite extraction strategies from tissue samples using NMR metabolomics. *Metabolomics*, *3*, 55–67.

Loub, W. D., Farnsworth, N. R., Soejarto, D. D., & Quinn, M. L. (1985). NAPRALERT: Computer handling of natural product research data. *Journal of Chemical Information and Computer Sciences*, *25*, 99–103.

Lu, X., Zhao, X., Bai, C., Zhao, C., Lu, G., & Xu, G. (2008). LC–MS-based metabonomics analysis. *Journal of Chromatography B*, *866*, 64–76.

Medicinal Plant Consortium (MPC). Available at http://medicinalplantgenomics.msu.edu/contacts.shtml) (Accessed, 2018).

Medicinal Plant Genomics Resource. Available at http://medicinal plantgenomics.msu.edu/ (Accessed May 23, 2018).

Medicinal Plant Metabolomics Resource. Available online: http://metnetdb.org/mpmr_public/ (Accessed on November 12, 2012).

Menedes, P., Kell, D. B., & Westerhoff, H.V. (1996). Why and when channelling can decrease pool size at constant net flux in a simple dynamic channel. *Biochimica et Biophysica Acta*, *1289*, 175–186.

MeT-RO "The National Centre for Plant and Microbial Metabolomics, Rothamsted Research", West Common, Harpenden, AL5 2JQ, UK, Available at https://www.rothamsted.ac.uk/metabolomics.

Meyer, G. M. J., & Maurer, H. H. (2015). In vivo, in vitrometabolism sudies of glaucine, a new herbal high by GC-MS, LCMS, LC-HR-MS, NMR techniques. *Toxichem Krimtech*, *82*, 323.

Michel, T., Halabalaki, M., & Skaltsounis, A. L. (2013). New concepts, experimental approaches, and dereplication strategies for the discovery of novel phytoestrogens from natural sources. *Planta Medicine*, *79*(7), 514–532.

Mohanraj, K., Karthikeyan, B. S., Vivek-Ananth, R. P., Chand, R. B., Aparna, S. R., et al. (2017). IMPPAT: A curated database of Indian medicinal plants, phytochemistry and therapeutics. *Scientific Reports*, 206–995.

Mukherjee, P. K., Harwansh, R. K., Bahadur, S., et al. (2016). Metabolomics of medicinal plants – a versatile tool for standardization of herbal products and quality evaluation of ayurvedic formulations. *Current Science*, *111*, 1624–1630.

Mukherjee, P. K., Nema, N. K., Venkatesh, P., & Debnath, P. K. (2012). Changing scenario for promotion and development of Ayurveda – way forward. *Journal of Ethnopharmacology*, *143*, 424–434.

Mumtaz, A., Ashfaq, U. A., ul Qamar, M. T., Anwar, F., Gulzar, F., et al. (2017). MPD3: a useful medicinal plants database for drug designing. *Natural Product Research*, *31*, 1228–1236.

Nakabayashi, R., & Saito, K. (2013). Metabolomics for unknown plant metabolites. *Anals of Bioanalytical Chemistry*, *405*(15), 5005–5011.

Newman, D. J., & Cragg, G. M. (2012). Natural products as sources of new drugs over the 30 years from 1981 to 2010. *Journal of Natural Products*, *75*, 311–335.

Niedenführ, S., Wiechert, W., & Nöh, K. (2015). How to measure metabolic fluxes: a taxonomic guide for 13 C fluxomics. *Current Opinion in Biotechnology*, *34*, 82–90.

Okada, T., Afendi, F. M., Altaf-Ul-Amin, M., Takahashi, H., Nakamura, K., et al. (2010). Metabolomics of medicinal plants: the importance of multivariate analysis of analytical chemistry data. *Current Computer Aided Drug Design*, *6*, 179–196.

Okada, T., Nakamura, Y., Kanaya, S., Takano, A., Malla, K. J., et al. (2009). Metabolome analysis of Ephedra plants with different contents of ephedrine alkaloids by using UPLC-Q-TOF-MS. *Planta Medicine*, *75*, 1356–1362.

Padalia, R. C. (2012). Medicinal and Aromatic plants: Chemical Goldmines. *Medicinal Aromatic plants, 1*, e113.

Pandey, A., & Tripathi, S. (2014). Concept of standardization, extraction and pre phytochemical screening strategies for herbal drug. *Journal of Pharmacogonics Phytochemistry, 2*(5), 115–119.

Patwardhan, B., & Vaidya, A. D. (2010). Natural products drug discovery: accelerating the clinical candidate development using reverse pharmacology approaches. *Indian Journal of Experimental Biology, 48*, 220–227.

Pope, G. A., MacKenzie, D. A., Defernez, M., Aroso, M. A., Fuller, L. J., et al. (2007). Metabolic footprinting as a tool for discriminating between brewing yeasts. *Yeast, 24*, 667–679.

Prasad, B., Garg, A., Takwani, H., & Singh, S. (2011). Metabolite identification by liquid chromatography-mass spectrometry. *Trend in Analytical Chemistry, 30*, 360–387.

Ray, L.B., Chong, L.D., & Gough, N.R. (2002). Computational biology: Sci STKE, (p. 10).

Roux, A., Lison, D., Junot, C., & Heilier, J. F. (2011). Applications of liquid chromatography coupled tomass spectrometry-based metabolomics in clinical chemistry and toxicology: a review. *Clinical Biochemistry, 44*, 119–135.

Saito, K., Hirai, M.Y., & Yonekura-Sakakibara, K. (2007). Decoding genes with coexpression networks and metabolomics - 'majority report by precogs'. *Trends in Plant Science, 13*, 36–43.

Saxena, M., Saxena, J., Nema, R., Singh, D., & Gupta, A. (2013). Phytochemistry of medicinal plants. *Journal of Pharmacogn Phytochemistry, 1*(6), 168–182.

Scognamiglio, M., D'Abrosca, B., Esposito, A., & Fiorentino, A. (2015). Metabolomics: an unexplored tool for allelopathy studies. *Journal of Allelochemical Interactions, 1*, 9–23.

Shankar, D., & Majumdar, B. (1997). Beyond the biodiversity convention: the challenges facing the biocultural heritage of India's medicinal plants. In , Bodeker, G., Bhat, K.K.S., Burley, J., Vantomme, P., (eds.), Medicinal plants for forest conservation and health care (pp 87-99). Non-wood forest products 11. FAO, Rome.

Shi, J., Cao, B., Wang, X. -W., Aa, J. -Y., Duan, J. -A., et al. (2016). Metabolomics and its application to the evaluation of the efficacy and toxicity of traditional Chinese herb medicines. *Journal of Chromatography B, 1026*, 204–216.

Shinbo, Y., Nakamura, Y., Altaf-Ul-Amin, M., Asahi, H., Kurokawa, K., et al. (2006). KNApSAcK: A Comprehensive Species-Metabolite Relationship Database. In K. Saito, R. A. Dixon, & L. Willmitzer (Eds.), *Plant Metabolomics, Biotechnology in Agriculture and Forestry* (vol 57). Berlin, Heidelberg: Springer.

Shulaev, V. (2006). Metabolomics technology and bioinformatics. *Brief Bioinformatics*, 7, 128–139.

Shyur, L. F., & Yang, N. S. (2008). Metabolomics for phytomedicine research and drug development. *Current Opinion in Chemical Biology, 12*, 66–71.

Smilde, A. K., Jansen, J. J., Hoefsloot, H. C., Lamers, R. J., van der Greef, J., & Timmerman, M. E. (2005). ANOVA-simultaneous component analysis (ASCA): a new tool for analyzing designed metabolomics data. *Bioinformatics, 21*(13), 3043–3048.

Smith, C. A., O'Maille, G., Want, E. J., Qin, C., Trauger, S. A., et al. (2005). METLIN: a metabolite mass spectral database. *Therapeutic Drug Monitoring, 27*, 747–751.

Stitt, M., & Fernie, A. R. (2003). From measurements of metabolites to metabolomics: an "on the fly" perspective illustrated by recent studies of carbon-nitrogen interactions. *Current Opinion in Biotechnology, 14*, 136–144.

Strohl, W. R. (2000). The role of natural products in a modern drug discovery program. *Drug Discovery Today, 5*, 39–41.

Sumner, L. W., Mendes, P., & Dixon, R. A. (2003). Plant metabolomics: large-scale phytochemistry in the functional genomics era. *Phytochemistry, 62*, 817–836.

Tagore, S., Chowdhury, N., & De, R. K. (2014). Analyzing methods for path mining with applications in metabolomics. *Gene, 534*, 125–138.

Theodoridis, G., Gika, H. G., & Wilson, I. D. (2008). LC-MS-based methodology for global metabolite profiling in metabonomics/ metabolomics. *Trends in Analytical Chemistry*, *27*, 251–260.

Thimm, O., Blaesing, O. E., Gibon, Y., Nagel, A., Meyer, S., et al. (2004). Mapman: a user-driven tool to display genomics datasets onto diagrams of metabolic pathways and other biological processes. *The Plant Journal*, *37*, 914–939.

Thomas, J., Joy, P. P., Mathew, S., & Skaria, B. P. (2000). Plant sources of aroma chemicals and medicines in India. *Chemical Industry Digest*, 104–108.

Tobias, K., Gert, W., Do, Y. L., Yun, L., Mine, P., Sevini, S., & Oliver, F. (2009). FiehnLib – mass spectral and retention index libraries for metabolomics based on quadrupole and time-of-flight gas chromatography/mass spectrometry. *Analytical Chemistry*, *81*(24), 10038–10048.

Tokimatsu, T., et al. (2005). KaPPA-View. A web-based analysis tool for integration of transcript and metabolite data on plant metabolic pathway maps. *Plant Physiology*, *138*, 1289.

Tota, K., Rayabarapu, N., Moosa, S., Talla, V., Bhyravbhatla, B., & Rao, S. (2013). InDiaMed: A comprehensive database of Indian medicinal plants for diabetes. *Bioinformation*, *9*, 378.

Trethewey, R. (2004). Metabolite profiling as an aid to metabolic engineering in plants. *Current Opinion in Plant Biology*, 7, 196–201.

Ulrich-Merzenich, G., Zeitler, H., Jobst, D., Panek, D., Vetter, H., & Wagner, H. (2007). Application of the 'omic' technologies in phytomedicine. *Phytomedicine*, *14*, 70–82.

Urbanczyk-Wochniak, E., & Sumner, L. W. (2007). MedicCyc: a biochemical pathway database for Medicago truncatula. *Bioinformatics*, *23*(11), 1418–1423.

Vallverdú-Queralt, A., Regueiro, J., Martínez-Húelamo, M., Alvarenga, J. F., Leal, R., et al. (2014). A comprehensive study on the phenolic profile of widely used culinary herbs and spices: rosemary, thyme, oregano, cinnamon, cumin and bay. *Food Chemistry*, *154*, 299–307.

Van Emon, J. M. (2015). The omics revolution in agricultural research. *Journal of Agricultural and Food Chemistry*, *64*(1), 36–44.

Van Moerkercke, A., Fabris, M., Pollier, J., Baart, G. J. E., Rombauts, S., et al. (2013). CathaCyc, a metabolic pathway database built from Catharanthus roseus RNA-seq data. *Plant and Cell Physiology*, *54*, 673–685.

Verpoorte, R., Crommelin, D., Gilissen, L. J. W. J., Schuitmaker, H., van der Greef, J., & Witkamp, R. F. (2009). A system view on the future of medicine: Inspiration form Chinese medicine? *Journal of Ethnopharmacology*, *121*, 479–481.

Wang, L. J., Wang, S. Z., Li, Y. H., Paradesi, M. S., & Brown, S. J. (2007). The model organism database for Triboliumcastaneum. *Beetle Base. Nucleic Acids Research*, *35*, D476–D479.

Wang, M., Lamers, R. J., Korthout, H. A., van Nesselrooij, J. H., Witkamp, R. F., et al. (2005). Metabolomics in the context of systems biology: bridging traditional Chinese medicine and smolecular pharmacology. *Phytotherapy Research*, *19*, 173–182.

Webster, G., Teh, A. Y. H., & Ma, J. K. C. (2016). Synthetic gene design–the rationale for codon optimization and implications for molecular pharming in plants. *Biotechnology and Bioengineering*, *114*(3), 492–502.

Williamson, E. M. (2001). Synergy and other interactions in phytomedicines. *Phytomedicine*, *8*, 401–409.

Wishart, D. S., Jewison, T., Guo, A. C., Wilson, M., Knox, C., et al. (2013). HMDB 3.0 – the human metabolome database in 2013. *Nucleic Acids Research*, *41*, D801–D807.

Wishart, D. S., Knox, C., Guo, A. C., Eisner, R., Young, N., et al. (2009). HMDB: a knowledgebase for the human metabolome. *Nucleic Acids Research*, *37*, D603–D610.

Wolfender, J. L., Rudaz, S., Choi, Y. H., & Kim, H. K. (2013). Plant metabolomics: from holistic data to relevant biomarkers. *Current Medicine Chemistry*, *20*, 1056–1090.

Wolfender, J. -L., Marti, G., & Ferreira Queiroz, E. (2010). Advances in Techniques for Profiling Crude Extracts and for the Rapid Identificationof Natural Products: Dereplication, Quality Control and Metabolomics. *Current Organic Chemistry*, *14*(16), 1808–1832.

Wootton, J. C. (2002). Development of HerbMed®: An interactive, evidence-based herbal database. *Advances in Phytomedicine*, *1*, 55–60.

Wurtele, E. S., Li, J., Diao, L., Zhang, H., Foster, C. M., et al. (2003). MetNet: Software to Build and Model the Biogenetic Lattice of *Arabidopsis*. *Computational and Functional Genomics*, *4*(2), 239–245.

Yang, H., Zhao, C., Wang, X., Liang, Y., Zeng, Y., et al. (2010). Chromatographic fingerprint investigation for quality evaluation and control of Shengui hair-growth tincture. *Planta Medicine*, *76*, 372–377.

Yanuar, A., Mun'im, A., Lagho, A. B., Syahdi, R. R., Rahmat, M., & Suhartanto, H. (2011). Medicinal plants database and three dimensional structure of the chemical compounds from medicinal plants in Indonesia. *International Journal of Computer Science Issue*, *8*, 180–183.

Zhang, P., Foerster, H., Tissier, C. P., Mueller, L. A., Paley, S., Karp, P. D., & Rhee, S.Y. (2005). MetaCyc and AraCyc: metabolic pathway databases for plant research. *Plant Physiology*, *138*, 27–37.

Zhang, Z. H., Wei, F., Vaziri, N. D., Cheng, X. L., Bai, X., Lin, R. C., & Zhao, Y.Y. (2015). Metabolomics insights into chronic kidney disease and modulatory effect of rhubarb against tubulointerstitial fibrosis. *Scientific Reports*, *5*, 1–17.

Zhao, Y. -Y., & Lin, R. -C. (2014). UPLC–MSE application in disease biomarker discovery: The discoveries in proteomics to metabolomics-mini review. *Chemico-Biological Interactions*, *215*, 7–16.

Zwenger, S., & Basu, C. (2008). Plant terpenoids: applications and future potentials. *Biotechnology and Molecular Biology Review*, *3*, 1–7.

CHAPTER 13

Exploration of biotechnological studies in low-calorie sweetener *Stevia rebaudiana*: present and future prospects

Shamshad A. Khan[a], Priyanka Verma[b], Laiq Ur Rahman[c], Varsha A. Parasharami[b]
[a]Department of Applied Biotechnology, College of Applied Sciences, Ministry of Education, Sur, Oman
[b]CSIR-National Chemical Laboratory, Division of Biochemical Sciences, Pune, Maharashtra, India
[c]Plant Biotechnology Division, Central Institute of Medicinal and Aromatic Plants (CSIR-CIMAP), Lucknow, Uttar Pradesh, India

Introduction

Increased uptake of junk food, excessive intake of sugar in form of beverages and various other forms, and tendency to be more lethargic has led to an unhealthy and nondisciplined life style. This abnormal routine and non-awareness toward our health are slowly leading people toward various deadly diseases like heart attack, obesity, high-blood pressure, hyper tension, and diabetes. Obesity and diabetes type II which are considered to be interlinked to each other are major causes of diseases in human kind and have shifted the focus toward sweetening compounds with low calories. This situation provided an ideal platform to artificial sweeteners like saccharin, sucralose and aspartame to take over from sugar related compound (Pol, Hohnova, & Hyotylainen, 2007). The dramatic increase in diabetic patients for last two decades has posed as one of the biggest and a major medical condition which the human kind is facing right now. The faster pace at which diabetes is spreading can be estimated by WHO's annual report which state that by the end of 2030 approximately 360 million people will be suffering from this deadly disorder and especially in India which is known as the "The Diabetic Capital of The World" nearly 70 million people will be diabetic till 2025. According to the research done by the Madras Diabetes Research Foundation in collaboration with International Diabetes Federation (IDF) it was found that compared to 50.8 million people suffering from type 2

Medicinal and Aromatic Plants: Expanding their Horizons through Omics
http://dx.doi.org/10.1016/B978-0-12-819590-1.00013-6

diabetes in 2010, the number of patients increased to 62.4 million people in 2011 (National Diabetes Audit 2012–13). As far as other countries are concerned the number of diabetic patients increased from 5.6 million to 20.9 million in the United States from 1980s to 2010 which shows a steep rise in diabetic case from 2.5% to 6.9% (Polonsky, 2012) and China faces a tough challenge as WHO data states that the economy of China will be hit by $ 558 billion in foreign national income due to diabetes and heart-related diseases. In the United Kingdom the conditions are even deteriorating as three million people are suffering from the disorder and a hike of 38% was noted since 2001–13 (Polonsky, 2012). According to WHOs annual 2016 report, in comparison to 108 million people suffering from diabetes Type1 in 1980s, the number shoots up to approximately 422 million people in 2014 which means that global prevalence of this disorder almost doubled itself since 1980 (WHO, 2016). Among all the products that are playing a major role in the rise of blood glucose leading to increase in diabetic numbers, one of the central cause is beverages containing sugar which directly affect the diabetes-enhancing parameters. Although efforts have been made previously to include nonsugar low-calorie molecules like saccharin, aspartame, sucralose, and other related molecules in to the daily used products but still consumption of these artificial sweetening compounds on daily basis has not been considered safe according to several studies performed. There are reports which state that the use of these synthetic sugar alternatives may lead to increase in brain tumors, bladder cancer and various other related diseases on their long-term uptake (Tandel, 2011). Taking a note from this recent scenario there is an urgent requirement to develop a low-calorie sweetening alternative which can be natural, safe and can be an alternative to sugar and other artificial sweetening agent with an efficacious and sound economic approach. The sweet glycosides from *S. rebaudiana* can be an alternate sugar option as these glycosides are sweeter than sucrose with less calorific value. Apart from it, these sweet glycosides have fewer side effects as compared to the regular use of low-calorie sweetening agents used nowadays.

Apart from it the vegetative propagation in *S. rebaudiana* is slow which affects the overall steviol glycoside profile of the plant. So there was an urgent requirement to develop a rapid clonal propagation method for fast and true to mother plant clones which can overcome the problem of slow vegetative propagation and seed-germination associated variations. With these associated problems this plant also lack the desirable characters such as high content of sweetening steviol glycosides (Rebaudioside A, stevioside) and through in vitro callus induction and regeneration of somaclones from them and their further clonal selection may lead to varieties with high-steviol glycoside profile with effective sweetening properties.

Stevia rebaudiana (Bertoni) (Family: Asteraceae) is a perennial shrub which is found domesticated in India since past 20 centuries (Sivaram & Mukundan, 2003). This plant system is grabbing lot of attention because of its natural high-sweetening, low-calorie properties. Mostly the leaf segment contain the steviol glycoside such as stevioside, rebaudioside A and their derivatives which are responsible for sweetening attributes of this plant (Khan, Rahman, Verma, & Shanker, 2016). Among the glyocosides of *S. rebaudiana*, rebaudioside A was found to be the most desirable of all the steviol glycosides as it leaves the least aftertaste bitterness as compared to other steviol glycosides (Megeji, Kumar, Singh, Kaul, & Ahuja, 2005) but recently in has been revealed that commercial scale up of steviol glycoside and the companies associated with them are more interested toward other steviol glycosides such as reb D, O, and M. Moreover it has been concluded that the use of mixture of steviol glycosides with a defined set of combinations are more desirable now as it has better taste and economically also affordable (Fry, 2016). The glycosylation of the steviol backbone increases the sweetening value of these compounds (Chaturvedi & Prakash, 2012). Most importantly the pattern of glycosylation heavily influences the taste perception of these intensely sweet glycosides. Interestingly studies have also shown that glycoside content depends upon the age of the leaves also as the older leaves were found sweeter that the young leaves. Earlier it was assumed that except the aerial parts (mostly leaves) no other parts of *S. rebaudiana* have shown to produce the sweet glycosides and the reason given for the absence of these glycosides in other parts such as roots is deprivation of chlorophyll (Bawane, Gopalakrishna, & Akki, 2012). Later on while studying the photoperiodic effects on the metabolism of steviol glycosides it was claimed that other plant parts may also contain some amount of steviol glycoside (Ceunen & Geuns, 2013). The Stevia based sweeteners have been increased up from 3500 tonnes (2013) to 10,000 tonnes (2017) according to market trends and further the estimated cost of market value of steviol glycosides have been reached up to US $275 million which is almost twice of the initial value (Leatherhead Food Research and Mintel, 2014).

As far as the market overview of the Stevia-based products are concerned the market of Stevia is divided in to two parts: Extract type based and on application based. The extract category also has three segments in the market namely powder, liquid and leaf segments. The forecasting of the Stevia market states that the expected reach of stevia based sweeteners will reach up to US$ 565.2 million till 2020 in comparison to its value of US$ 347.0 million in 2014 (http://www.futuremarketinsights.com). The improved taste, market support and versatility of the *Stevia* based products have made companies such as Coca- Cola, PesiCo Inc, Nestle S.A., Stevia Corp. and Cargill Inc. to invest at a higher scale (https://www.technavio.com: Global *Stevia* Market 2016–2020).

As far as the origin and history of Stevia is concerned it is assumed that sweet species, that is, *S. rebaudiana* was first introduced to Europe when Spain conquered the south America in 16thcentury but the original studies of this plant started in 20th century in the United Kingdom. Although there are more than eighty wild species of *Stevia* which have been reported (Table 13.1) only from North America and over two hundred species from South America but sweetening property has only been found in *S. rebaudiana* and some of its related species (Bawane et al., 2012). It was in year 1887 when this herb was given the name *Eupatorium rebaudianum* by Dr. Moises Santiago Bertoni after the name of Rebaudi a chemist who was considered the first person to study the chemical aspect of this plant. Initially the Guarani Indians residing in Paraguay, Brazil and Argentina were known to use *Stevia* as a sweetening agent in 800s but in 1987, Dr. Moises Santiago Bertoni (College of Agriculture, Asuncion) found this herb botanically and

Table 13.1 Stevia species and their characteristic features.

Stevia species	Characteristic features
Stevia ivifolia	Height around 2 ft., stem erect, leaves rhomboidal, flower heads white, arranged in fastigiated corymbs
Stevia ovata	Height around 2 ft, stem erect, leaves ovate, flower-heads white, arranged in compact fastigiated corymbs
Stevia purpurea	Height around 18 inch, stem erect,Leaves lanceolate, flower heads purple arranged in fastigiated corymbs
Stevia rebaudiana	Height 1–1½ fet stem puberulous leaves opposite, oblanceolate, flower heads very small, whitish in a corymb Leaves with sweet flavor
Stevia salicifolia	Height 18 inch leaves opposite, narrow, lanceolate, nearly or quite entire, very shortly stalked, almost connate. Flower-heads white in spreading corymbs.
Stevia serrata	Height 18 inch stem erect, branched, leaves alternate, almost glabrous, Flower heads white or pink, in fastigiated

introduced *S. rebaudiana* as a new variety and later in 1903 performing some experiment it was realized that sweetening property of *S. rebaudiana* was higher than sugar compounds (Midmore & Rank, 2002). Although efforts were first made to commercialize this crop initially in Paraguay way back in 1964 but a serious major effort was made in Japan when this crop was introduced by Sumida (1968) followed by its spread to the other countries. In the recent times China is leading the way in *Stevia* cultivation and other countries like Tanzania, Indonesia, Brazil, Canada, Korea and Mexico are also following the trends. Soon after the discovery of sweetening properties of this herb the larger scale plantation started way back in 1908 as prior to this time the *Stevia* plants were restricted to its natural habitats only. Soon after this discovery US department of Agriculture realized its medicinal properties and later in 1931, stevioside was isolated in form of white crystalline extract from the leaves of this plant by two French chemists who reported it to be responsible for the sweetening properties of *Stevia rebaudiana*. In between the year 1960–1970 Japanese enterprises realized the importance of steviosides and its derivatized products were included in consortium of food products (Saxena & Ming, 1988). From that time plantation has been taken up by many countries such as Indonesia, Brazil, US, Canada, Japan and Tanzania (Brandle & Telmer, 2007).

Steviol glycosides: importance and biosafety

The concentration of steviol glycosides are not the same throughout the plant and differ because of factors such as the conditions at which the cultivar is grown, cultivar to cultivar difference and the various other parameters. The stevioside ranges from 4%–20% of the dry weight of the leaves. In Stevia more than 30 steviol glycosides are present among which the major glycoside present are *S. rebaudiana* followed by rebaudioside and trace amounts of other glycosides such as steviolbioside 2, rebaudioside A4, B5, C6, D7, E8, F9, and dulcoside A10 (Fig. 13.1, Fig. 13.2) (Kennelly, 2002; Starrat, Kirby, Pocs, & Brandle, 2002). The *S. rebaudiana* plants and its content although have been reported to be used since centuries but the shortage of scientific validation has not allowed this important plant to be fully explored at commercial level. Among the glyocosides of *S. rebaudiana*, rebaudioside A is the most desirable of all the stevioside as it leaves the least aftertaste bitterness as compared to stevioside so the focus of the research on this plant is nowadays concerned on developing the varieties which are high on rebaudioside A content and less on stevioside content. The different approaches that have

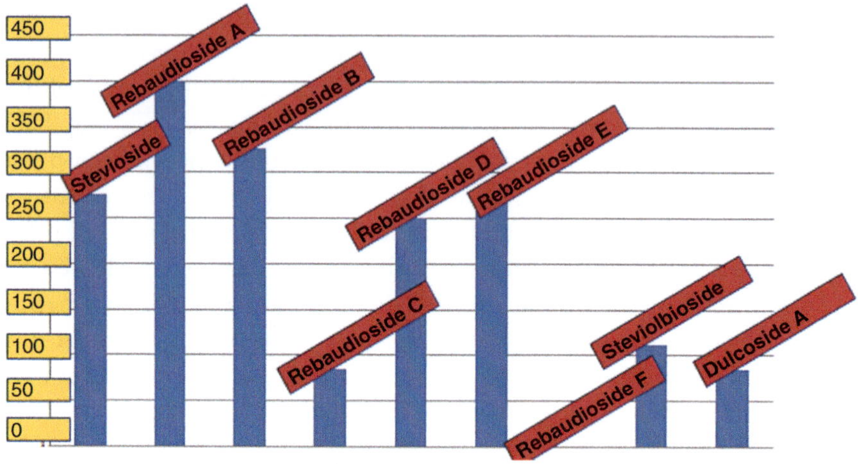

Figure 13.1 Relative sweetening strength (in fold) of different steviol glycosides and their sweetening capacity (in folds) compared to 0.4% sucrose solution.

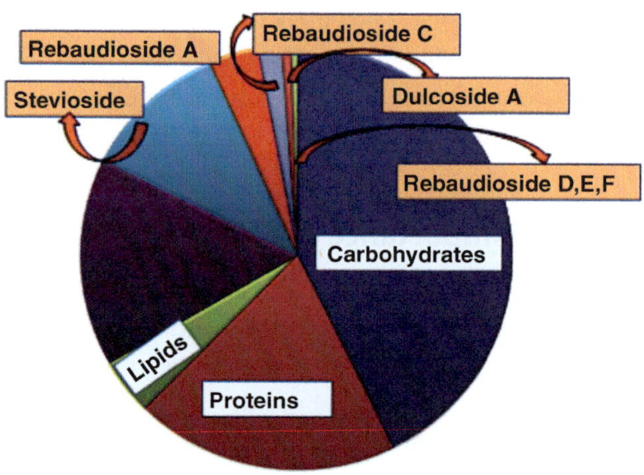

Figure 13.2 Total biochemical content (% leaf dry weight) in the *S.rebaudiana* plant.

been applied to improve the genetic traits of *S. rebaudiana* include transgenic approaches or the crossing among the varieties (Yadav et al., 2011). The studies on steviol glycosides have also shown that these glycosides are low-calorie sweetener because the steviol glycosides are not broken down by the digestive enzymes which results in low-glucose release in blood and a proper maintenance of blood glucose level in the body (Strauss, 1995). The daily intake of stevioside has been considered safe as different mutagenic studies have been performed for stevioside's safety assessments were found to be in safety range. Since time immemorial *Stevia* plant as a whole, its leaves and extract have been used by people as sweetening agent and there is not a single report which out rightly rejects the stevioside for their safe use. Although research have been previously done on stevioside component and leaf extracts as a whole for their antifertility studies as it was found that 5% solution of *Stevia* leaf extract may have a severe effect on sterility in both male and female rats but later on research investigation showed that oral consumption in a set range of this plant is safer for use (Yodyingyuad & Bunyawong, 1991). The daily intake of stevioside has been considered safe as different mutagenic studies have been performed for stevioside's safety assessments. The model organisms which were tested for the mutagenic activity of steviol glycosides includes sensitive strains of *B. subtilis*, *E. coli* WP2 uvrA/pKM101 and hamster but the studies does not show any significant proof of stevioside been mutagenic for human consumption (Geuns, 2003). According to mutagenic studies this fact can be established that *Stevia* and its steviol glycosides can be used as an alternate source of sweetening agent not only for diabetic patients but also for the normal person and the frequent use of these glycosides may lead to lowering the blood glucose level (Boeckh-Haebisch, 1992). The normal intake of steviol glycoside has been standardized at the range of 0–2 mg/kg body weight by World Health Organization for its safe and effective use (Beneford, DiNovi, & Schlatter, 2006) but later on it has been revised and now ADI is now fixed at 4 mg/kg BW as steviol equivalents not steviol glycoside by European Food Safety Authority (EFSA, 2010). The Food and Drug Association (FDA) of US considered the use of *Stevia* glycosides as safe as a natural sweetener in eatable products as well as beverages in 2008. The use of highly purified steviol glycosides have been permitted by European Union (EU) with the number (EU 1131, 2012) in December 2011. *Stevia* and its glycosides are known and familiar mostly for its low glycemic index and low calorific value but interestingly further investigations have shown that these glycosides have important activities other than being a low calorie sweetener. Previously the studies have been performed on the use of high concentration of steviosides and its antibacterial activities but this antibacterial activity was restricted to certain bacteria (Das et al., 1992) and this antibacterial property of steviosides play an important role in dental caries and its restrictions (Matsukubo & Takazoe, 2006). Interestingly steviosides have been known to play a role against diarrhea as these glycosides are recognized to lower down the muscle contractions of the intestine (Chatsudthipong & Muanprasat, 2009). Apart from these, various other aspect of the bioactivity of steviol glycosides have been studied such as its effect on human health and its biosafety, its antiviral activities, ability to maintain the blood glucose level and its effect on the renal function (Geuns, 2003). Recent report by Momtazi-Borojeni et al. (2017) has claimed that the sweetening agent such as steviol glycoside do not cause subacute or acute toxicity neither they have carcinogenic or mutagenic activities as assumed previously.Preclinical and clinical studies suggest therapeutic and pharmacological applications for stevia and their extracts because they are not toxic and exhibit several biological activities. This review presents the biological activity of *Stevia rebaudiana* Bertoni and their relationship to antidiabetic, anticariogenic, antioxidant, hypotensive, antihypertensive, antimicrobial, antiinflammatory, and antitumor activities.

Therapeutic and pharmacological studies on *S. rebaudiana* also suggested that compounds present in the plant are nontoxic and show a range of biological activities (Ruiz-Ruiz, Moguel-Ordonez, & Segura-Campos, 2017).

One of the major areas of research which is going to decide the fate of this low-calorie natural sweetener is its bio safety and human consumption studies as the assessment of biosafety is almost complete for steviol glycosides and countries like US, EU, Australia, and New Zealand are heading toward obtaining authorizations. More importantly the focal point of the research is more in the direction of monitoring of the use and absorption of steviol glycosides in the population after introduction of steviol glycosides in different foods. Although studies have shown that the steviol glycosides are not having adverse effect when tested at laboratory level but still there is a need to carry out a thorough validation process for oral consumption and its bio safety aspects. With the safe and effective use of these glycosides these low-calorie sweeteners can easily replace the sugar-derived products as *Stevia* plants are having the advantage of having a widespread abundance (Puri, Sharma, & Tiwari, 2011).

All over the world *Stevia* covers 32,000 ha for its cultivation and China covers the major part of it by contributing to 75% of this area. As far as India is concerned the total annual production is 600 ton with gradual increase in growing of the plants and the export demand steeply shooting up (Das, Gantait, & Mandal, 2011). The economic development of *Stevia* is vastly improving with its products slowly replacing the sugar as well as artificial substitutes from market. According to survey reports the crude extracts as well as the derived products from *Stevia* such as "reb A" have increased by 27% from the year 2010 to 2011 with an assumption that *Stevia* products shooting up to \$700m within coming 5 years. From business point of view *Stevia* cultivation can be considered as a good option provided the leaves are of optimum good quality. According to reports the tissue culture-raised plants were found to be more disease resistant compared to vegetative grown plants which are often prone to infection with fungal pathogens like *A. alternata*. With people's interest growing in low calorie products definitely *Stevia* products are going to make a contribution in economy of low-calorie natural sweeteners. One of the purified products of *Stevia* "reb A" already made a place in international sugar market with an extent of 25% as a safe and healthy sugar alternate. From economic point of view agricultural aspect of *Stevia* will play a major role in the future as the total stevioside content depends upon the leaf biomass that in a way depend on what strategies are to be adapted for efficient agricultural practices for *Stevia* cultivation. The previous works have suggested that the leaf biomass increases with the increasing planting density and steviol glycosides are found to be enhanced with the long day period (Metvier and Viana, 1979). A strict breeding program will have to be followed which includes harvesting methods, crop production system, handling method as well as seed and seedling development (Brandle & Telmer, 2007). Agronomically while at the higher altitude the cultivation of *Stevia* is annual but in its native places it is a polyannual herbaceous plant (Goettemoeller & Ching, 1999). The stem is short with elliptic-lanceolate leaves and roots have a tap-root system which further becomes diffused system in later stages of development. The height of the plants grows up to 0.60–0.80 m high (Andolfi, Ceccarini & Macchia, 2002). The longer exposure of light period (16 h) was found to enhance the glycoside content despite the fact that *Stevia* is a short day plant and lower temperature (below 20°C) have a deleterious effect on the content of steviol glycosides (Lavini et al., 2008). No preferred soil type is reported for *Stevia* cultivation as this plant grow on a long range of soil types starting from course textured sands, infertile, acid soils, neutral soil (pH6.5–7.5) to well-drained loams but has difficulty to grow on low drained soil or clay. *Stevia* is considered to be best suited to semi humid to subtropical conditions but the lower temperature tend to affect the roots of the plants thus killing of the plant, so the ideal temperature for growth of this plant is considered to be in the range of 15–30°C.

Biotechnological interventions: improvements in *S. rebaudiana*

The biochemical pathway elucidation in *S. rebaudiana* has long been a task to understand but the newly introduced techniques for analysis of genomic resources such as high throughput expressed sequence tags (ESTs) have provided a helping hand in understanding the complexity of the pathway. The sweet glycosides of *S. rebaudiana* are diterpenoids in nature and its biosynthetic pathway is quite similar to gibberellic acid pathway as both of these molecule have four steps common in their biosynthesis. The MEP pathway initiates the biosynthesis of steviol glycosides as out of sixteen steps in steviol glycoside synthesis, seven steps are common to MEP pathway. The hydroxylation of kaurenoic acid by kaurenoic acid 13-hydroxylase (SrKA13H) is the first step of diversion from gibberelic acid biosynthesisand toward steviol glycoside biosynthesis and considered as the first committed step in the direction of steviol glycoside biosynthesis (Brandle & Telmer, 2007). The leaves of *S. rebaudiana* are recognized at the site from where the pathway initiates and later after the synthesis of different steviol glycosides, they are been transported to different plant parts, however almost negligible concentration of these steviol glycosides are reported from stems and roots (Yadav & Guleria, 2012). This sixteen step biosynthesis of steviol glycosides starts with MEP pathway as major seven portion of the pathway is similar to MEP pathway. The plastids are the site from where the biosynthesis of steviol starts with 2-C-methyl-D-erythritol-4 phosphate (MEP) pathway. One molecule of DMAPP and three molecules of IPP combine to form geranylgeranyl diphosphate (GGDP). After sharing the initial seven steps with MEP pathway the steviol glycosides biosynthetic pathway shares its next four steps with gibberellic acid (GA) biosynthetic pathway (Bennett, Lieber, & Heftmann, 1967). In case of *Stevia*, the GGDP biosynthesis steps moves to steviol biosynthesis step by the involvement of four substrates namely Geranylgeranyl diphosphate which converts to Copalyl diphosphate in presence of enzyme copalyl diphosphate synthase (CPPS) which on subsequent action from enzyme kaurene synthase (KS) further gets converted to Kaurene. This kaurene gives rise to Kaurenoic acid in presence of the enzyme Kaurene Oxidase (KO) (Brandle & Telmer, 2007). The divergence in steviol glycoside pathway from gibberelic acid pathway comes from kaurene synthesis point when kaurene gets converted to kaurenoic acid in presence of enzyme kaurenoic acid hydroxylase (KAH) which gives rise to steviol which is considered the "pool" from which the other steviol glycosides are synthesized by further glucosylation or rhaminosylation (Fig. 13.3). The steviol glycoside biosynthesis is divided in to two parts—early phase and late phase and accordingly the genes are also divided in to early expressed genes and late expressed genes. The UDP-glycosyltransferases (UGTs) have the known function of transferring the sugar moieties from nucleotide sugar donor to the acceptor molecules

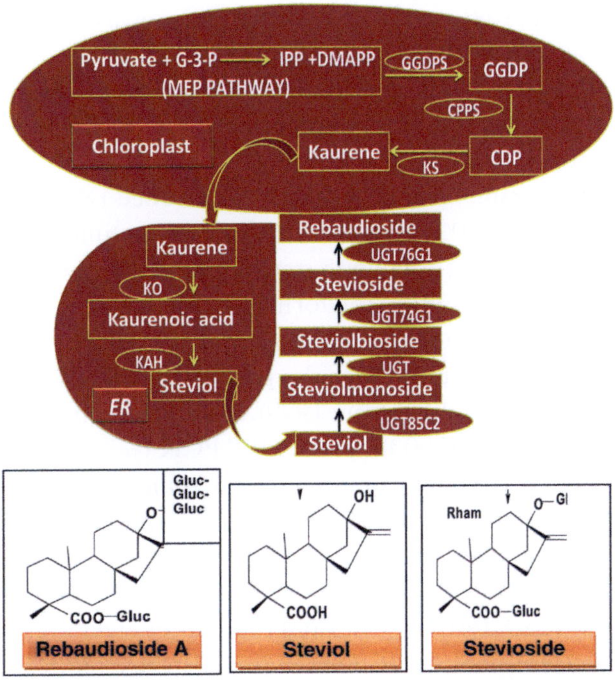

Figure 13.3 Biosynthesis of steviol glycosides and structure of three important steviol glycosides. *GGDPS*, geranylgeranyl diphosphate synthase; *CPPS*, copalyl diphosphate synthase; *KS*, kaurene synthase; *KO*, kaurene oxidase; *KAH*, kaurenoic acid hydroxylase; UGT85C2; 15 UGT; 16 UGT74G1; and 17 UGT76G1 (UDP-glucuronosyl transferases).

(Wang & Hou, 2009). Four of the important UGTs (*viz* UGT85C2, UGT, UGT74G1, and UGT76G1) mediate the steviol glycoside biosynthesis initiating from the backbone steviol molecules to the final step of rebaudioside A synthesis. The UDP-glycosyltransferases (UGTs) involved in the final steps are known to add glucose moiety to the steviol backbone to give rise to different steviol glycosides. These UGTs are encoding enzymes for glycosylation and belong to a single gene family which constitutes 69 various gene families. In case of UDP-glycosyltransferase-encoded by gene SrUGT74G1 plays an important role in glycosylation of steviolbiside at C-19 position which gives rise to stevioside. Further the glycosylation at C-13 position of stevioside converts it to rebaudioside A in presence of enzyme encoded by gene SrUGT76G1 .When the rhamnosylation occurs in places of glucosylation at the C13 glucose of the C2 position phenomenon results in formation of dulcoside A and the further glycosylation of it at C-3 position yields rebaudioside C formation. The perception of taste differs and depends upon position of glycosylation as these rhaminosylated glycosides are less sweet in taste compared to the glucosylated steviol glycosides (Tanaka, 1997). The DXS gene from *Stevia* has been cloned using RT-PCR as the sequences of these genes are highly conserved among various plant species and so the primer designing from this gene was an easy task (Totte et al., 2003). The complementation assay of this cloned DXS gene was done with *E.coli* strain MC4100 dxs::CAT to confirm the function of this gene (Totte et al., 2003). In the following steps the rearrangements of the DXP chain was done which results in the formation of 2C-methyl-D-erythritol-4-phosphate in the presence of the enzyme DXP reductoisomerase. This is considered as the committed step as the gene of this step is cloned in Mint and *Arabidopsis* (Lange and Croteau, 1999). The DXR gene was first cloned from *Stevia* by Totte et al. (2003) by using RT PCR technique and its activities was checked by complementation assay byusing *E. coli* strain MC4100 dxr::TET. The coming five steps of MEP pathway were characterized by cloning and characterizing the genes in *Arabidopsis* and followed by various other plant species which included *Stevia* also (Hsieh & Goodman, 2006).

One of the characterized genes responsible for steviol glycoside conversion from steviol pool, SrUGT85C2 has been assessed for its effect on gibberalic acid (GA_3) formation as well as chlorophyll synthesis as they all share the same MEP pathway. The study showed a decrease of 78%–83% in GA_3 and 64%–76% decrease in chlorophyll a and b when expressing the SrUGT85C2 gene in *Stevia* plants. This study clearly shows the role of SrUGT85C2 in down regulating the genes related to MEP pathway which leads to GA_3 and chlorophyll formation in transgenic *A. thaliana* plants (Guleria, Masand, & Yadav, 2015). Further the over expression of SrKA13H (ent-kaurenoic acid-13 hydroxylase) cDNA revealed the increment in steviol pools (1–3 μg g–1 DW). The transgenic *A. thaliana*plants also showed significant decrease in GA_3 concentrations (Guleria et al., 2015). The down regulation of ent-kaurene synthase, ent-kaurene oxidase, ent-kaurenoic acid hydroxylase, and three UDP dependent glycosyltransferases, UGT85C2, UGT74G1 and UGT76G1was performed by treating the plants with poly ethylene glycol which resulted in the considerable decrease in total steviol glycosides. This study clearly mentions the correlation between the regulation of these genes and their effect on the content of stevioside, rebaudioside, steviolbioside, rubusoside, and dulcoside A (Hajihashemi & Geuns, 2016). The genetic level manipulation in *Stevia* is a tough task as this plant is associated with genetic heterozygosity, self-incompatibility and lack of genomic information which leads to poor molecular prospection of this important plant therefore the old time techniques like plant breeding are adopted as the method of genetic level assessment which is associated with complexity and time taking efforts. As discussed before that UDP-glycotransferase family plays a major role in the final step of synthesis of SGs so most of the transgenic expression and manipulation at genetic level are related to over expression of the genes responsible for these UGTs. The transgenic over expression of SrUGT74G1 cDNA from *Stevia* was assessed in *A. thaliana* with the aim of producing steviol glycosides in the different plant system (Yadav & Guleria, 2012). The SrUGT74G1 gene expression in associated with the conversion of steviolbioside to stevioside but no accumulation of stevioside was reported in *A. thaliana* except the changes in GA_3 content and enhancement in catechin accumulation in transgenic plants. There in not much of study done for the genetic level regulation of SG biosynthesis but recently *Agrobacterium* mediated transient gene silencing through RNA interference has been checked for its effect on the silencing of the genes namely SrKA13H and three SrUGTs (SrUGT85C2, SrUGT74G1, and SrUGT76G1) (Yadav & Guleria, 2012) which are decisive in commitment toward steviol glycosides biosynthesis from GA pathway. This study was undertaken to understand the exact role and effect of blocking these genes on total steviol glycosides content of the plants which showed that SrKA13H and SrUGT85C2 play a key role in affecting the carbon flux for the synthesis of steviol glycoside biosynthesis or gibberalin biosynthesis but more importantly this study has hinted toward the existence of an alternate steviol glycoside biosynthesis route. The stevioside biosynthesis pathway include its steviol synthesis step which is further glycosylated to be converted in to these sweet glycosides such as stevioside and rebaudioside but the high accumulation of steviol at the longer periods are reported to be toxic to the plants. So there is a requirement to create variation in this important crop possibly at plant tissue culture level to generate a low steviol containing plants or the plants where the steviol content is further glycosylated as quickly as possible to the other compounds. As far as the role of in vitro studies of *Stevia* plants are concerned it has been a well established fact that this important shrub suffers from germination problems and genetic variability related issues with vegetative propagation also not been able to propagate this plant effectively (Misra et al., 2011). Apart from this multiplication associated problems some of the selective steviosides such as dulcoside (very low concentration in *Stevia*) are reported to show carcinogenic activities at laboratory level experiments. So there is an utmost requirement to deal with *Stevia* plants under in vitro level and develop the varieties which are rapidly propagative, high in rebaudioside content and zero in dulcoside composition (Madan et al., 2010).

In vitro studies in *S. rebaudiana*

Plant biotechnology is a field of high significance but not much of work has been done regarding production of compounds of commercial values. Constant variation in climate conditions and limitations regarding resource availability in field conditions makes plant an ideal system to be used as bio factories for constant production of variety of chemically and therapeutically important compounds. For the selection of desired cell types with useful variations plant cell culture have been considered as a most formidable technique (Wang & Wang, 2012). The *S. rebaudiana* plants with the above discussed importance needed to be cultivated at a larger scale but this plant is often found to be associated with low number of individual plants when grown vegetatively and on the other hand the seed propagated plants tend to show low germination % (Sivaram & Mukundan, 2003). The tissue culturing of this plant has to be performed with the foresight to improve the varieties with enhanced productivity of desired compounds. The initial in vitro studies in this plant have been performed related to media composition and effect of sucrose concentrations on overall growth of the plant (Table 13.2).

The ever increasing natural product demands are bending in the boundaries of tissue culture techniques so that the production of seconadary metabolites should meet the requirement.Somatic embryogenesis in *S. rebaudiana* is an important aspect of tissue culture studies as it provides the virtues of bipolar structure which contains the properties of a embryo of a zygote. More importantly the somatic embryo develops shoots and roots simultaneously which results in rapid growth of the plant and requires less time for acclimatization process. As the *S. rebaudiana*plants face the problem of low-germination percentage it necessitates the somatic embryogenesis studies in this plant system. The research performed of somatic embryogenesis in this plant system reveals that the plant growth regulators 2-4 D and 2i P work efficiently in combination to yield somatic embryos in *S. rebaudiana* (Naranjo, Betin, Trujillo, Posada, & Garces, 2015). Different types of amino acids such as glutamine and tryptophan also used to play an important role in somatic calli generation in combination with additives like casein hydrolysate, potato extract and yeast extract (Das and Mandal, 2010). Earlier Filho, Hashimoto, &Vieira (1993) found that high sucrose concentration also helps

Table 13.2 *In vitro* culture establishment studies carried out in *S. rebaudiana*

Explant	Media	Response	Days to induction	No. of shoot/ roots/explants	Reference
Leaves	MS + 2mg/L BAP	Shoot proliferation	42	43.9	Hassanen & Khalil, (2013)
In vitro plantlets	MS + 0.5 and 1.0 mg/L IBA	Root induction		4.5	
	MS + 1.0 mg/L IBA and 0.5 mg/L Alar.	Root induction		7.3	
Nodal Parts	MS + 1mg/LBAP + 0.5mg/L IBA	Shoot bud initiation		1.89	Laribi et al., (2012)
Callus culture	MS+ 0.5 BA + 1.0 2,4-D	Suspension cultures			Ferreira & Handro, (1988)
Suspension cultures	LS+ 0.5 BA +0.5 2,4D to form callus followed by medium with K (2.0 mg/L) + NAA (0.02 mg/L)	Callus culture followed by bud regeneration			
Leaves	MS + 1 BAP +2NAA	Shoot Bud Induction From Callus		5.0 ± 1.0	Khan et al. (2014)
Shoot Tips	B5+ 4mg/LBA + 0.8mg/L NAA	Shoot proliferation		28	Giridhar, Sowmya, Ramakrishna, & Ravishankar, (2010)
Stem Node Segment	MS +0.5 mg/L BA + 0.5 mg/LKn	Shoot multiplication		18.1	Alhady (2011)
In vitro plant	MS + 1.0 or 2.0 mg/L IBA	Rooting	42	7.1	
Shoot Tips	MS + 2mg/L Kn	Multiple Shoot Induction	15.58 ± 0.5	11	Das et al. (2011)
In vitro plants	MS	In vitro rooting			

(Continued)

Table 13.2 *In vitro* culture establishment studies carried out in *S. rebaudiana* (*Cont.*)

Explant	Media	Response	Days to induction	No. of shoot/ roots/explants	Reference
Leaf And Nodal Explants	MS+ 2 mg/L 2, 4-D	Callus Induction And Somatic Embryo Generation			Pande & Khetmalas, (2012)
Nodal Segments Leaf Explants	MS + 1mg/L BAP +2mg/L NAA MS + 3mg/L BAP +1mg/LIAA	Multiple Shoots Induction Callus Induction			Upadhyay et al. (2013)
Nodal Segments In vitro plants	MS + 1.5 mg/L BA + 0.5mg/L Kn MS +0.1 mg/IAA	Multiple Shoots Induction Rooting	5-12	8.75 ± 0.34 12.10	Ahmed et al. (2007)
Nodal Segments	MS + 2.5mg/L 2-4D 1.5 mg/L IAA	Shoot initiation Callus Induction			Patil (2013)
Shoot Tips In vitro plants	MS +1mg/L BA+ 0.5 mg/L IAA ½ Nitsch (N6) + 1.0 mg/L IAA	Shoot Inductions *In vitro* rooting	11.80 ± 0.80	16.20 ± 0.37	Anbazhagan, Kalpana, Rajendran, Natarajan, & Dhanavel (2010)
Root Tips	MS + 10.7 μM NAA	Adventitious Roots			Reis et al., (2011)
In vitro Derived Buds In vitro plants	MS + 1 mg/L BAP ½ MS + 0.4 mg/L NAA.	Multiple Shoot Regeneration Rooting	07	15.69 14.4 ± 3.41	Thiyagarajan & Venkatachalam (2012)
Auxiliary Buds	MS + 1.5 mg/L BAP + 0.1 mg/L NAA	Organogenesis			Mirniam, Roshandel, Otroshi, & Ebrahimi (2013)

Axillary Buds	MS + 1mg/L TDZ	Multiple Shoots		6.0	Lata, Chandra, Techen, El-Sohly, & Khan (2012)
Leaves	MS + 2mg/L BA+ 2mg/L Kn	Shoot Buds Induction		3.73 ± 0.88	Sreedhar et al. (2008)
Nodal segments	MS +0.5 mg/L BAP and 0.25 mg/L Kn MS + 1.0 mg/L IBA	Shoot formation Roots initiation		7.82 ± 0.7 30.12 ± 2.1	Razak, Ong, Yu, & Lau (2014)
Shoot tips containing axillary buds	MS + 2.0 mg/L BAP MS + 1.5mg/L TDZ MS+ 1.0 mg/L IBA	Shoot multiplication Multiple shoot formation Roots initiation		3.75 14.5 7.0	Ahmed et al. (2016)
Nodal and shoot-tip explants	MS + 2.0 mg/L BAP MS + 0.3 mg/L BAP + 0.3 mg/L Kin +0.1 mg/L NAA ½ MS+ 0.5 mg/L NAA	Shoot induction (98.0%) Shoot multiplication Rooting	4.7 30 7.4	34.9 21.2	Yadav et al. (2016)
Nodal explant	ER + 3.0mg/L BAP	Multiple shoot formation		4.11 ± 0.21	Mehta & Dadhich, (2015)
Nodal explant	WPM + 1.0 g/lYE; 100 μM CHI; 0.5 g/L CH; and 1.0 g/l ALG WPM + 0.5, 1.0, or 2.0 g/L PEC, 1.0 g/lYE,	Shoot multiplication Root initiation		3.90 ± 0.35 6.13 ± 0.33	Bayraktar et al. (2016)
Nodal explant	MS + 2.0 mg/L kinetin + 0.05% NaCl	Multiple shoot formation Roots initiation		3.01± 0.05	Gupta, Sharma, & Saxena (2016)

(*Continued*)

Table 13.2 *In vitro* culture establishment studies carried out in *S. rebaudiana* (*Cont.*)

Explant	Media	Response	Days to induction	No. of shoot/ roots/explants	Reference
Shoot apices	MS+ 18.09 μM 2,4-D + 7.38 2iP μM	Somatic embryo			Naranjo et al. (2015)
Leaf explant	MS + Yeast extract + Casein hydrolysate + Potato extract + 2.0 mg/L 2-4D + 0.5mg/L BAP	Embryogenic callus			Das and Mandal (2010)
Isolated protoplast	KM8P + 0.2 mg/L 2,4D) 1mg/L NAA+ 0.5 mg/L zeatin followed by shift to MS + 1mg/L BAP	Callus development Somatic embryo followed by plantlet regeneration			Lopez-Arellano et al. (2014)
Axillary buds	MS+ 2.0 mg/L 2, 4- D + 0.2mg/L BAP + 0.2mg/L TDZ followed by MS+ 1.0 mg/L 2, 4- D	Friable callus turned to somatic embryos			Banerjee & Sarkar, (2010)
Leaves	MS+ 0.25mg/L 2,4-D MS+ 1mg/L BA in a high sucrose medium (120 g/L).	Somatic Embryogenesis			Filho et al., (1993)
Floret explant	MS+ 9.05 μM 2,4-D	Embryogenic callus			Filho & Hattori (1997)
Nodal explant	MS+ 2.0 mg/L of NAA	Shoot multiplication			Gupta, Sharma, & Saxena (2015)
Leaf explant	MS+ 1.0 mg/L BA) + GA3(1.0mg/L) + Spd (1.0mg/L) 2, 4-D (2.0) + BA (2.0)+ Spd (2.0)	Shoot multiplication Callus induction		27.7 ± 1.3	Khalil et al., (2016)

Nodal explant	MS+ BAP (0.5 mg/L) + Kn (0.5 mg/L	Multiple shoot formation		17.5	Verma, Yadav, & Singh (2011)
Nodal segments	MS+ 0.5 mg dm^{-3} BAP MS+ 0.5 mg dm^{-3} IBA	Shoot proliferation Root induction	42	6.90 ± 2.65 7.90 ± 0.50	Tomaszewska-Sowa et al. (2016)
Shoot apex Node Leaf	MS+ BA (8.87 µm)+ IAA (5.71 µm) Half-strength MS medium supplemented with 4.90 µm IBA	Shoot induction Root induction		11.2 ± 0.8 10.0 ± 0.8 7.9 ±0.6	Sivaram and Mukundan (2003)
Microshoots	MS+ IBA 3.0+ CCC3.0	Root induction		25.33+ 1.5	Dey, Kundu, Bandyopadhyay, & Bhattacharjee, (2013)
Leaf explants	MS + 4.0 mg/L IBA+ 0.2 mg/L BAP MS + 1.0Kn + 0.2 NAA+ 40AS	Callus induction Shoot regeneration		12.51	Rathore, Yadav, Singh, & Singh, (2014)
In vitro plants	MS+ 0.5mg/L IBA.	Root induction	7-10		Preethi et al. (2011)
Leaf explants	MS + 1.0mg/L NAA+ 1.0 mg/L 2,4 2,4-D	Callus induction			Sharma et al., (2015)
Flower explants	MS+ 2.0 mg/L BA + 2.0 mg/L 2,4-D MS + 2.0 mg/L BA MS+ 2.0 mg/L BA	Callus induction Shoot organogenesis from callus Shoot induction and multiplication	30 30 21	21.6	Ahmad et al. (2011)

(Continued)

Table 13.2 *In vitro* culture establishment studies carried out in *S. rebaudiana* (*Cont.*)

Explant	Media	Response	Days to induction	No. of shoot/ roots/explants	Reference
Stem-tips					
	MS + 1.0 mg/L kin				
	Shoot multiplication	50–100 shoots			Labiri (2012)
Nodal segment	MS+ 0.1 mg/L Kin+	Direct shoot regeneration			Dwivedi et al. (2016)
Leaves	0.05 mg/L BAP				
Callus	MS+ 3.0mg/L BAP+ 4.0 mg/L NAA	Callus induction			
	Liquid MS+ 3.0 mg/L BAP+ 4.0 mg/L NAA + 1.0 mg/L ascorbic acid	Suspension culture			
Leaf explants	MS + 1BAP mg/L + 1 NAA mg/L+ 2.5 mg/L 2,4-D	Friable green callus			Javad et al. (2016)

in generation of somatic embryos in *S. rebaudiana* without callus interphase but these somatic embryos were not able to regenerate in to shoots. For efficient regeneration protocol establishment the protoplasts were also isolated from *S. rebaudiana* which further developed in to somatic embryo via callus interphase. These calli mediated-somatic embryos regenerated plantlets but there were no observable diffrenece in these protoplast regenerated plants in respect to growth pattern and morphology as compared to the control plants (Lopez-Arellano et al., 2014). The reduced concentration of 2, 4-D also effects the growth of somatic embryos positively as callus maintained on high 2, 4-D (2.0 mg/L) in combination with other growth hormones when shifted to low-concentration of 2, 4-D (1.0 mg/L) yielded somatic embryos (Banerjee & Sarkar, 2010). The effect of sucrose was checked for callus induction and somatic embryogenesis in *S. rebaudiana* and different concentration of sugar levels, that is, 1%–6% were checked on hypocotyls, intermodal, nodal and leaf segments. Among the sucrose concentrations 3% sucrose supplemented with 2, 4-D in MS medium was found suitable for somatic embryo generation. The indirect regeneration through callus culture has been seen as a method to create genetic variation among *S. rebaudiana* plant species. The dedifferentiated and unorganized calluses have also been a good source of adventitious shoot formation in *S. rebaudiana* plants (Khan, Rahman, Shankar, & Singh, 2014, Preethi et al., 2011). The indirect regeneration in many plants has been reported to improve its agronomic traits (Wang & Wang, 2012). Indirect regeneration through callus phase is considered an important step toward embryoids formation and somaclone generation (Dodds & Roberts, 1982). In *S. rebaudiana* the induction of callus and direct regeneration from leaf explants have been reported with a range of different combinations of auxins (NAA, IAA and 2-4D) and cytokinin (BA and Kinetin) and it was found that callus phase can be inducted on MS + 1 mg/L BAP + 2 mg/L NAA (Khan et al., 2014). Similar results were found by Preethi et al. (2011) while using MS+ 2 mg/L BA. Several reports on shoot bud initiation and for fast multiplication are available on *S. rebaudiana*. For faster and efficient multiplication of *S. rebaudiana* from nodal explants MS medium with 1 mg/L BAP was found suitable and for shoot bud initiation the combination of BAP and IBA (1 mg/L BAP and 0.5 mg/L IBA) was found ideal. For the faster multiple shoot proliferation the IBA alone or IAA in combination with BAP was found effective (Laribi, Rouatbi, Kouki, & Bettaieb, 2012). Hassanen and Khalil (2013) reported maximum number of shoot per explants with MS media supplemented with 2 mg/L BAP but the shoots were found to be very thin which lead to poor survival of plants at glasshouse level studies. Different growth retardant have also been tested on *S. rebaudiana* plants as the pre-acclimatization treatments for increasing the rooting efficiency. The growth retardants used were alar and cycocel and it resulted in efficient rooting process (Hassanen & Khalil, 2013). The different studies show that various media compositions have been assessed for rooting under in vitro raised plant and efficiency was found with MS media in full or half strength with combinations of IAA or IBA. The *Agrobacterium* based genetic transformation studies in *S. rebaudiana* plant system is restricted. The *A. rhizogenes*-mediated hairy root cultures aspects of *S. rebaudiana* were assessed and the hairy roots were established in half strength MS and White media by using *A. rhizogenes* strain 15834. The hairy roots were kept in dark for checking the presence of steviol glycosides. Among all the tested clones no presence of steviol glycosides were reported and further shifting of the hairy root cultures to kinetin and benzyladenine hormonal combinations induced callusing. These results were in accordance with earlier reports that steviosides are only present in the aerial parts especially leaves and absent in the lower segments of the plants (Yamazaki & Flores, 1991). The lack of presence of steviol glycosides can be attributed to the absence of chlorophyll in hairy root culture. Apart from hairy root cultures the adventitious roots of *Stevia* were also induced from root tips in MS media supplemented with 10.7 μM NAA in a roller bottle system and maintained for six months. The phytochemical analysis of these adventitious roots did not show the presence of any *Steviol* glycosides (Reis et al., 2011). Table 13.3 summarizes the steviol glycoside content reported in different tissues under in vitro and *ex vitro* conditions. For gene transfer studies and pathway analysis a direct regeneration protocol establishment from leaf explants was the requirement and Khan et al. (2014) provided it by establishing the direct regeneration protocol from leaf explants using EHA105 strain of *A. tumefaciens*. Apart from transgenic plants containing GUS gene, four different somaclone regenerant were also established from *S. rebaudiana* plants which were genetically different on the basis of ISSR profiling and the phytochemical analysis also showed the different level of glycosides in these variants (Khan et al., 2014). The effect of NaCl concentration (60, 90, and 120 mM) on *S. rebaudiana* plants were also checked and it was found that higher concentration of NaCl (120 mM) tends to negatively affect the total dry weight, chlorophyll content as well as the concentration of rebaudioside A and stevioside (Zeng, Chen, Li, Yi, & Wu, 2013). This study of effect of salinity on *S. rebaudina* states that plant can moderately tolerate the salt stress. Recently Khan et al. (2016) studied the affect of chemical and

Table 13.3 The different steviol glycosides content reported under in vitro and ex vitro conditions in *S. rebaudiana*.

	Steviol Glycoside				
Stevioside	**Rebaudioside A (%dw)**	**Dulcoside A (%dw)**	**Rebau-dioside C (%dw)**	**Rebau-dioside F (%dw)**	**Reference**
4%–13%dw	2–4	0.4–0.7			Wong (2011)
3.9 mg g-1 (DW *in vivo* leaves) 4.2 mg g-1 (DW *in vitro* shoots) 4.3 mg g-1 (DW callus) 4.6 mg g-1 (DW suspension culture) 4.1mg g-1(synthetic seed grown shoots)					Dwivedi et al. (2016)
1.34 ± 0.1 (Callus) 1.33 ± 0.07 (Suspension)	0.69 ± 0.03 3.85 ± 0.0				Gupta et al., (2014)
0.0039 g/g dw (Shoots) 0.0054 (g/g dw (Rooted shoots)					Dheeranupattana, Wangprapa, & Jatisatienr, (2008)
0.3%dw	0.6	0.9	14.4	2.7	Starrat et al., (2002)
3.703%dw					Shirwalkar et al. (2011)
6.83%–7.91%dw					Jadhao, Katekhaye, & Thorat (2011)
2.87%dw					Kaur (2011)
5.0%–10.0%dw	2.0–4.0	0.4–0.7	1.0–2.0	<1.0	Puri et al. (2011)
5.8 ± 1.3%dw	1.8 ± 1.2		1.3 ± 1.4		Gardana, Scaglianti, & Simonetti (2010)
7.01 ± 0.05%dw 9.23 ± 0.04%dw (In vitro)					Singh, Dwivedi, & Atri (2014)

(Continued)

Table 13.3 The different steviol glycosides content reported under in vitro and ex vitro conditions in *S. rebaudiana*.(*Cont.*)

	Steviol Glycoside				
Stevioside	**Rebaudioside A (%dw)**	**Dulcoside A (%dw)**	**Rebau-dioside C (%dw)**	**Rebau-dioside F (%dw)**	**Reference**
6.4 ± 0.89 %dw (Glasshouse plants)	3.2 ± 0.56				Khan et al. (2014)
	0.09				Rajasekaran, Ramakrishna, Udaya Sankar, Giridhar, & Ravishankar (2008)
5.40 ± 0.25%dw (In vitro)	3.42 ± 0.12 (In vitro)				Lata, Chandra, Techen, Wang, & Khan (2013)
8.2 ± 0.56% dw (EMS treated)	6.7 ± 0.34				Khan et al. (2016)
0.031 ± 0.009 %dw	0.040 ± 0.008 0.112 ± 0.014 (In vitro)	- 0.024 ± 0.001 (In vitro)	0.024 ± 0.005 0.020 ± 0.003 (In vitro)		Giridhar et al. (2010)
5-10%dw	2-4		1-2		Yadav et al. (2011)
14.69 %dw	0.55				Bayraktar et al. (2016)
3.4± 0.015 %dw (Suspension Culture) 0.06% dw (Morphogenic culture) 0.387% dw (Shoots regenerated from callus) 24.9, 21.6 and 30.2 mg/g (Field grown plants)					Bondarev, Reshetnyak, & Nosov (2001)

3.30± 0.28 (Etiolated in vitro regenerants) 0.06%dw (Morphogenic green callus) 25.40 mg/g (Field grown plants)					Ladygin et al. (2008)
82.48 µg/g					Aman, Hadi, Khalil, Zamir, & Ahmad (2013)
2.9 mg/g fresh weight (In vitro micro shoots)					Dey et al. (2013)
10.20%dw					Khalil et al. (2016)
1.72 %dw (Hairy root cultures) 1.79 mg/g (Field grown plant)					Pandey, Pandey, Pandey, Singh, & Banerjee (2016)
381.03 µg/g DW					Mathur & Shekhawat (2013)
1.34% (Elicited callus culture) 1.33 % (Elicited suspension culture)					Gupta et al. (2014)
9.6% (In vitro raised plants)					Karim et al. (2016)
1.37 %					Gupta et al. (2016)
1.47 % (Elicited callus culture)	0.44 ± 0.01%				Gupta et al. (2015)
1.47 ± 0.05					
8.75 %dw (Field grown plant)					Woelwer-Rieck, Lankes, Wawrzun, & Wust (2010)
5.8% (Field grown plant)					Gardana et al. (2010)

(*Continued*)

Table 13.3 The different steviol glycosides content reported under in vitro and ex vitro conditions in *S. rebaudiana*.(*Cont.*)

	Steviol Glycoside				
Stevioside	**Rebaudioside A (%dw)**	**Dulcoside A (%dw)**	**Rebau-dioside C (%dw)**	**Rebau-dioside F (%dw)**	**Reference**
4%–13% w/w	2%–4% w/w	0.4%–0.7% w/w	1%–2% w/w		Melis, Rocha, & Augusto, (2009)
33.87 mg/g					Javad et al., (2016)
40.0 mg/g Steviol glycoside (Russian leaf)					
6.4%–6.9% (w/w)	4.7%–5.0%	0.7 ± 0.1	1.4 ± 0.0	0.2 ± 0.0	Yucesan et al. (2016)
14.50 ± 0.83 mg/g Dry weight Leaves	80.86 ± 0.94				Zeng et al. (2013)
0.325 mg/g (Suspension culture)					Sharma et al., (2015)
40 mg/g FW (Leaves)					
Total steviol glycosides 5.8% (Callus culture) 4.9% (In vivo) 3.6% (In vitro)					Sivaram and Mukundan (2003)
381.03 μg/g DW (Suspension culture)					Mathur and Shekhawat (2013)

physical mutagen such as EMS and gamma radiation to create the variation in *S. rebaudiana* plants and found that the photosynthetic parameters such as chlorophyll content, stomatal conductance, and transpiration rate decreased in mutagen treated plants. On the other hand the rebaudioside A content in both the EMS and gamma-treated plant enhanced up to 1.5 and 2.0 folds.

It is very recently claimed that despite gibberalic acid having an important role in Steviol glycoside biosynthesis still there was no study performed on the effect of GA 4 and 7 on *S. rebaudiana*. It was found that under in vitro conditions effect GA4/7 has led to better leaf area formation, improved fresh to dry weight ration and interestingly concentration of rebaudioside A was found to be comparable to control plants when GA4 was used but decreased when GA7 was implemented (Pazuki, Aflaki, Yucesan, & Gurel, 2019).

The effect of different concentration of mannitol by adding them to media has also been recently studied and interestingly it was found that different mannitol concentrations effect the stevioside concentration differently. As per the study maximum accumulation of stevioside was found on 20 g/L treatment of mannitol but rebaudioside found its maximum presence on 30 g/L concentration of mannitol.

The protocols have been established for generating the suspension cultures from callus cultures and further regeneration of plants from the suspension-derived callus cultures. The friable type callus were suitable for generating suspension cultures in MS liquid medium with hormonal supplement of BA (0.5 mg/L) + 2,4-D (1.0 mg/L). Interestingly the suspension cultures which were derived from senescent calluses were showing variation in chromosome number but suspension derived from actively growing callus were diploid in nature. The regenerated plants from suspension cultures were having better characters such as plant vigour, chromosome numbers, rooting efficiency as well as morphological characters (Ferreira & Handro, 1988). The regeneration of suspension cultures was also reported by Sharma et al. (2015) and MS basal medium with combinations of BAP and ascorbic acid was used for regeneration of suspension cultures. Further the presence of total steviol glycoside content and stevioside content were also checked after elicitation with methyl jasmonate which resulted in maximum stevioside content.

The suspension cultures of *S. rebaudiana* have also been studied for the presence of steviol glycosides and protocols have been established for characterization and standardization of suspension cultures. Recently (Dwivedi et al. (2016)) studied different tissue culture aspects such as callus cultures, in vitro plants and synthetic seed derived plants and found that maximum stevioside content was in suspension cultures which show that single cell suspension could prove to be a alternate source for secondary metabolite production in *S. rebaudiana.* Plant growth regulators BAP and NAA in combination with ascorbic acid was effective in generating single cell suspension cultures (Dwivedi et al., 2016).

The callus cultures were used for inducing suspension cultures cell lines producing steviol glycosides and the dynamics of steviol glycoside production at different time interval have been assessed. It was revealed that the maximum stevioside production was at the starting of exponential phase of suspension culture with the maximum yield of 381.03 μg/g DW (Mathur & Shekhawat, 2013).The presence of salts such as NaCl and Na_2CO_3have also effected the concentration of total steviol glycosides as in case of callus and suspension culture the total steviol glycoside content have been increased in presence of NaCl and Na_2CO_3 up to 1.57% and 5.14%, respectively (Gupta, Sharma, & Saxena, 2014).

Transcriptomic-level studies in *S. rebaudiana*

Providing more insight to the steviol biosynthetic pathway Kim, Zheng, Liao, & Jang (2018) have claimed trichomes also has evident role to play in the metabolic pathway of this medicinal herb. The comparative transcriptomic analysis according to Kim et al. (2018) reveals biochemical specialization on leaves of *S. rebaudiana* for the production of diterpenoids. As per tissue specific chemical analysis it's reported that trichomes do not have a role to play in steviol glycoside accumulation as these SGs are stored in leaves but interestingly these trichomes works for accumulation of another type of diterpenoid which in labdane in nature such as agatholic acid and oxomanolyl oxide. The transcriptomic analysis with gene expression and phylogenetic analysis clearly mentioned separate gene members which are producing enzymes for other kind of labdane type diterpenoids which differ from enzymes which are responsible for synthesis of steviol glycoside. Additionally this particular report also states that in trichome two another homologs of copalyl diphosphate synthase (SrCPS) and kaurene synthase1 (SrKS1) are also found which are known by the name SrCPS2 and KS like (SrKSL) enzymes.

One of the major hindrance of transcripome level analysis of *S. rebaudiana* was addressed by Singh et al. (2017) which states that ontogeny dependent deviation has been an important factor which has created a major hindrance in SG biosynthesis level transcriptomic studies. To provide details of transcriptome level analysis at global level Singh et al. (2017) used global transcriptome sequencing approach with the help of assorted *de novo* transcripts and have successfully identified genes responsible for GAs biosynthesis as well as SGs biosynthesis. These analysis indicated toward a development based utilization of metabolic flux between GA and SGs biosynthesis as they both share the common pathway steps.

Novel microRNA and their target prediction in *S. rebaudiana* were identified by Mandhan and Singh (2015) which used high throughput small RNA sequencing. It resulted in the construction of s-RNA library with distinct sequences. It was reported that prediction for 12 novel mi RNA were made and UGTs and nucleotide sequences of *S. rebaudiana* were considered to be the sources from which these mi RNA are made. Recent transcriptomic analysis have been updating on the roles of trichomes in *S. rebaudiana* as comparative transcriptomic analysis revealed presence of new ditepenoid in this plant system. Global transcriptome sequencing analysis approaches have mentioned development-based utilization of metabolic flux between GA and SG biosynthesis. Apart from it high throughput small RNA sequencing data are also helping in construction of s RNA library.

Future prospects

The growing interest of people toward low-calorie sweetener and alarming rate at which diabetes is increasing is fluxing the interest toward an alternative sweetening option which is safe to use and *S. rebaudiana* fits in to this gap with its high-sweetening and low-calorie attributes. The current challenges posed to the production of these low-calorie sweetening source is the development of new varieties of *S. rebaudiana* plants which could have high concentration of stevioside and rebaudioside A. Moreover as discussed earlier, rebaudioside A is more desirable amongst all the steviol glycosides present in the plant as it leaves less bitter after taste as compared to stevioside and has better sweetening quotient. So recent works are taking these facts in to account and pathway studies in *S. rebaudiana* has been moving toward conversion of stevio pool to rebaudioside A which will require a better understanding and characterization of UGTs associated with these steps (Khan et al., 2014). The in vitro studies in recent times are concerned toward conversion of steviol pool to successive steviol glycosides as steviol pool accumulation in vacuoles leads to toxicity in plants itself.

The concentration of unwanted steviol glycosides such as dulcoside (Madan et al., 2010) is also of the aspect that needs to noted and blocking/diverting this secondary metabolite producing genes toward important glycosides would be a segment of the future studies concerned with *S. rebaudiana* plants. The somaclonal variation approach and generation of variants in *S. rebaudiana* plants could be a viable approach to generate plants with desired attributes. Apart from it in vitro tissue culturing in *S. rebaudiana* is of utmost requirement because this plant system suffers from low-germination problems. Adding to this, the vegetative propagation of this plant system is also not very effective (Misra et al., 2011). A better understanding of direct/indirect regeneration protocol, the use of *A. tumafaciens/A. rhizogenes* based genetic transformation studies for gene transfer and pathway exploration through gene over expression would be suitable approaches for improvement in this high-sweetening and low-calorie medicinal plant.

Acknowledgments

We gratefully acknowledge the financial support received from the Department of Science and Technology (DST) as DST-SERB Young Scientist Fellowship (Ministry/Department Letter No.YSS/2015/001417) Government of India, New Delhi to carry out this work and Director CSIR- NCL for providing all the necessary facilities.

References

Ahmed, M. B., Salahin, M., Karim, R., Razvy, M. A., Hannan, M. M., Sultana, R., Hossain, M., & Islam R (2007). An efficient method for *in vitro* clonal propagation of a newly introduced Sweetener Plant (*Stevia rebaudiana* Bertoni.) in Bangladesh. *American-Eurasian Journal of Scientific Research, 2*, 121–125.

Ahmed, B., Hossain, M., Islam, R., Saha, A. K., & Mandal, A. (2011). A review on natural sweetener plant stevia having medicinal and commercial importance. *Agronomics Glasnik, 73*, 75–91.

Alhady, MRAA (2011). Micropropagation of *Stevia rebaudiana* Bertoni. A new sweetening crop in Egypt. *Global Journal of Biotechnology and Biochemistry, 6*, 178–182.

Aman, N., Hadi, F., Khalil, S. A., Zamir, R., & Ahmad, N. (2013). Efficient regeneration for enhanced steviol glycosides production in *Stevia rebaudiana* (Bertoni). *Comptes Rendus Biologies, 336*, 486–492.

Anbazhagan, M., Kalpana, M., Rajendran, R., Natarajan, V., & Dhanavel, D. (2010). In vitro production of *Stevia rebaudiana* Bertoni. *Emirates Journal of Food and Agriculture, 22*, 216–222.

Andolfi, Ceccarini, L., & Macchia, M. (2002). Caratteristiche bioagronomiche di *Stevia rebaudiana. L'Informatore Agrario, 23*, 48–51.

Banerjee, M., & Sarkar, P. (2010). Somatic embryogenesis in*Stevia rebaudiana*Bertoni using differentconcentration of growth hormones. *International Journal of Plant Science, 5*, 284–289.

Bawane, A. A., Gopalakrishna, B., & Akki, K. S. (2012). An overview on *Stevia*: a natural calorie free sweetener. *International Journal of Advanced Pharmacy, Biology and Chemistry, 1*, 362–368.

Bayraktar, M., Naziri, E., Akgun, I. H., Karabey, F., Ilhan, E., Akyol, B., Bedir, E., & Gurel A (2016). Elicitor induced stevioside production, in vitro shoot growth, and biomass accumulation in micropropagated *Stevia rebaudiana*. *Plant Cell Tissue and Organ Culture, 127*, 289–300.

Beneford, D. J., DiNovi, M., & Schlatter, J. (2006). *Steviol Glycosides: Safety Evaluation of Certain Food Additives. WHO Food Additive Series*. Geneva: WHO pp 117-143.

Bennett, R. D., Lieber, E. R., & Heftmann, E. (1967). Biosynthesis of steviol from kaurene. *Phytochemistry, 6*, 1107–1110.

Boeckh-Haebisch EMA (1992). Pharmacological Trial of concentrated crude extract of *Stevia rebaudiana* (Bert) Bertoni in healthy volunteers. *Arquivos de Biologia e Tecnologia, 35*, 299–314.

Bondarev, N., Reshetnyak, O., & Nosov, A. (2001). Pecularities of diterpenoid steviol glycoside production in in vitro cultures of *Stevia rebaudiana* (Bert.). *Plant Science, 161*, 155–163.

Brandle, J. E., & Telmer, P. G. (2007). Steviol glycoside biosynthesis. *Phytochemistry, 68*, 1855–1863.

Ceunen, S., & Geuns, J. (2013). Glucose, sucrose, and steviol glycoside accumulation in *Stevia rebaudiana* grown under different photoperiods. *Biologia Plantarum, 57*(2), 390–394.

Chatsudthipong, V., & Muanprasat, C. (2009). Stevioside and related compounds: therapeutic benefits beyond sweetness. *Pharmacology Therapy, 121*, 41–54.

Chaturvedi, V. S. P., & Prakash, I. (2012). Spectral analysis and chemical studies of the sweet constituent, rebaudioside-A. *European Journal of Medicinal Plants, 2*, 57–65.

Das, A., & Mandal, N. (2015). Enhanced development of embryogenic callus in *Stevia rebaudiana* Bert by additive and amino acids. *Biotechnology, 9*, 368–372.

Das, A., Gantait, S., & Mandal, N. (2011). Micropropagation of an Elite-Medicinal plant: *Stevia rebaudiana* Bertoni. *International Journal of Agricultural Research, 1*, 40–48.

Das, S., Das, A. K., Murphy, R. A., Punwani, I. C., Nasution, M. P., & Kinghorn, A. D. (1992). Evaluation of the cariogenic potential of the intense natural sweeteners Stevioside and Rebaudioside A. *Caries Research, 26*, 363–366.

Dey, A., Kundu, S., Bandyopadhyay, A., & Bhattacharjee, A. (2013). Efficient micropropagation and chlorocholine chloride induced stevioside production of *Stevia rebaudiana* Bertoni. *Comptes Rendus Biologies, 336*, 17–28.

Dheeranupattana, S., Wangprapa, M., & Jatisatienr, A. (2008). Effect of sodium acetate on stevioside production of Stevia rebaudiana. *Acta Horticulture (ISHS), 786*, 269–272.

Dodds, J. H., & Roberts, L. W. (1982). *Experiments in plant tissue culture*. Cambridge, London, New York: Cambridge University Press.

Dwivedi, S., Alam, A., & Shekhawat, G. S. (2016). Antioxidant response of *Stevia rebaudiana* Bertoni (Angiosperms; Asteraceae) during developing phase of suspension cell culture. *Plant Science Today, 3*(2), 115–123.

European Food Safety Authority (2010). Annual Report 2010, Available from: https://www.efsa.europa.eu/sites/default/files/corporate_publications/files/ar10en.pdf.

Ferreira, C. M., & Handro, W. (1988). Production, maintenance and plant regeneration from cell suspension cultures of *Stevia rebaudiana* (Bert.). *Bertoni Plant Cell Reports*, 7, 123–126.

Filho, J. C. B., & Hattori, K. (1997). Embryogenic callus formation andhistological studies from*Stevia rebaudiana*(Bert.) bertoni floret explants. *Revista Brasileira de Fisiologia Vegetal, 9*, 185–188.

Filho, J. C. B., Hashimoto, J. M., & Vieira, L. G. E. (1993). Induction of somatic embryogenesis from leaf explants of *Stevia rebaudiana*. *Revista Brasileira de Fisiologia Vegetal, 5*, 51–53.

Fry J (2016) Stevia Innovation: Improved Leaf Extracts from Advanced Understanding of Taste, In Proceedings of the 9th Stevia symposium "From Field to Fork", Gothenburg, Sweden, ISBN: 978-90-74253-321, Euprint, Belgium. 33-40.

Gardana, C., Scaglianti, M., & Simonetti, P. (2010). Evaluation of steviol and its glycosides in Stevia rebaudiana leaves and commercial sweetener by ultra high performance liquid chromatography–mass spectrometry. *Journal of Chromatography A*, *1217*, 1463–1470.

Geuns, M. C. (2003). Stevioside. *Phytochemistry*, *64*, 913–921.

Giridhar, P., Sowmya, S. K., Ramakrishna, A., & Ravishankar, G. A. (2010). Rapid clonal propagation and stevioside profiles of *Stevia rebaudiana* Bertoni. *International Journal of Developmental Biology*, *4*, 47–52.

WHO Library Cataloguing-in-Publication Data (2016) Global report on diabetes.1. Diabetes Mellitus – epidemiology. 2. Diabetes Mellitus – prevention and control. 3. Diabetes, Gestational. 4. Chronic Disease. 5. Public Health. I. World Health Organization. ISBN 978 92 4 156525 7 (NLM classification:WK 810)© World Health Organization. https://apps.who.int/iris/bitstream/handle/10665/204871/9789241565257_eng.pdf?sequence=1

Global Stevia Market 2016-2020: https://www.technavio.com/report/global-food-stevia-market. Stevia Market: Global Industry Analysis and Opportunity Assessment 2014-2020, http://www.futuremarketinsights.com/reports/global-stevia-market.

Goettemoeller, J., & Ching, A. (1999). Seed Germination in Stevia rebaudiana. In J. Janick (Ed.), *Perspectives on new crops and new uses* (pp. 510–511). Alexandria, VA: ASHS Press.

Guleria, P., Masand, S., & Yadav, S. K. (2015). Diversion of carbon flux from 24ibberellins to steviol biosynthesis by over-expressing SrKA13H induced dwarfism and abnormality in pollen germination and seed set 24ibberel of transgenic *Arabidopsis*. *Journal of Experimental Botany*, *66*, 3907–3916.

Gupta, P., Sharma, S., & Saxena, S. (2014). Effect of salts (NaCl and $Na_2CO_{3)}$ on callus and suspension culture of Stevia rebaudiana for steviol glycoside (Natural sweetener) production. *Applied Biochemistry and Biotechnology Part A: Enzyme Engineering in Biotechnology*, *172*, 2894–2906.

Gupta, P., Sharma, S., & Saxena, S. (2015). Biomass yield and steviol glycoside production in callus and suspension culture of *Stevia rebaudiana* treated with proline and polyethylene glycol. *Applied Biochemistry and Biotechnology*, *176*, 863–874.

Gupta, P., Sharma, S., & Saxena, S. (2016). Effect of abiotic stress on growth parameters and steviol glycoside content in *Stevia rebaudiana* (Bertoni) raised in vitro. *Journal of Applied Research in Medicinal and Aromatic Plants* http://dx.doi.org/10.1016/j.jarmap.2016.03.004.

Hajihashemi, S., & Geuns, J. M. C. (2016). Gene transcription and steviol glycoside accumulation in *Stevia rebaudiana* under polyethylene glycol-induced drought stress in greenhouse cultivation. *FEBS Open Biology*, *6*, 937–944.

Hassanen, S. A., & Khalil, R. M. A. (2013). Biotechnological Studies for Improving of Stevia (*Stevia rebaudiana* Bertoni) *In vitro* Plantlets. *Middle-East Journal of Scientific Research*, *14*, 93–106.

Hsieh, M. H., & Goodman, H. M. (2006). Functional evidence for the involvement of *Arabidopsis* IspF homolog in the nonmevalonate pathway of plastid isoprenoid biosynthesis. *Planta*, *223*, 779–784.

Jadhao, D. B., Katekhaye, S. D., & Thorat, B. N. (2011). Improved RP-HPLC method for quantitative estimation of stevioside in *Stevia rebaudiana* Bertoni. *International Journal of Phytopharmacology*, *1*, 27–34.

Javad, S., Naz, S., Ilyas, S., & Aftab, A. (2016). Production of stevioside from callus and cell suspension cultures of Stevia rebaudiana(Bert.). *The Journal of Animal and Plant Science*, *26*(5), 1374–1382.

Karim, M. Z., Uesugi, D., Nakayama, N., Hossain, M. M., Ishihara, K., & Hamada, H. (2016). Identification of stevioside using tissue culture-derived Stevia (*Stevia rebaudiana*) leaves. *Biochemistry Insights*, *8*, 33–37.

Kaur, H. (2011). Chromatographic determination of stevioside in leaf parts of *in vitro* and *in vivo* regenerated plants of *Steviarebaudiana*. *International Journal of Natural Products and Research*, *4*, 44–48.

Kennelly, E. J. (2002). Sweet and non-sweet constituents of Stevia rebaudiana (Bertoni) Bertoni. In A. D. Kinghorn (Ed.), *Stevia, the genus Stevia. Medicinal and aromatic plants-industrial profiles* (pp. 68–85). London: Taylor and Francis.

Khalil, S. A., Kamal, N., Sajid, M., Ahmad, N., Zamir, R., Ahmad, N., & Ali, S. (2016). Synergism of polyamines and plant growth regulators enhanced morphogenesis, stevioside content, and production of commercially important natural antioxidants in *Stevia rebaudiana* Bert. *In Vitro Cellular and Developmental Biology–Plant*, *52*, 174–184.

Khan, S. A., Rahman, L., Shankar, K., & Singh, M. (2014). *Agrobacterium tumefaciens*-mediated transgenic plant and somaclone production through direct and indirect regeneration from leaves in *Stevia rebaudiana* with their glycoside profile. *Protoplasma*, *251*, 661–670.

Khan, S. A., Rahman, L., Verma, R., & Shanker, K. (2016). Physical and chemical mutagenesis in *Stevia rebaudiana*: variant generation with higher UGT expression and glycosidic profile but with low photosynthetic capabilities. *Acta Physiology Plant*, *38*, 4.

Kim, M. J., Zheng, J., Liao, M. H., & Jang, I. (2018). Overexpression of SrUGT76G1 in Stevia alters major steviol glycosides composition towards improved quality. *Plant Biotechnology Journal*, *17*(6), Available from: https://doi.org/10.1111/pbi.13035.

Ladygin, V. G., Bondarev, N. I., Semenova, G. A., Smolov, A. A., Resetnayk, O.V., & Nosov, A. M. (2008). Chloroplast Ultrastructure, Photosynthetic Apparatus Activities and Production of Steviol Glycosides in Stevia rebaudiana in vivo and in vitro. *Biology Plant*, *52*, 9–16.

Lange, B. M., & Croteau, R. (1999). Isopentenyl diphosphate biosynthesis via a mevalonate-independent pathway: isopentenyl monophosphate kinase catalyzes the terminal enzymatic step. *Proceedings of National Academy of Science USA*, *96*(24), 13714–9.

Laribi, B., Rouatbi, N., Kouki, K., & Bettaieb, T. (2012). *In vitro* propagation of *Stevia rebaudiana* Bertoni. A non-caloric sweetener and antidiabetic medicinal plant. *International Journal of Medicinal and Aromatic Plant*, *2*, 333–339.

Lata, H., Chandra, S., Techen, N., El-Sohly, M., & Khan, I. (2012). Determination of genetic stability of micropropagated plants of *Stevia rebaudiana* using inter-simple sequence repeat (ISSR) markers. *Planta Medica*, *78*, 1–6.

Lata, H., Chandra, S., Techen, N., Wang, Y. H., & Khan, I. A. (2013). Molecular analysis of genetic fidelity in micropropagated plants of *Stevia rebaudiana* Bert. Using ISSR marker. *American Journal of Plant Science*, *4*, 964–971.

Lavini, A., Riccardi, M., Pulvento, C., Luca, S. D., Scamosci, M., & Andria, R. (2008). Yield, quality and water consumption of Stevia rebaudiana Bertoni grown under different irrigationregimes in Southern Italy. *Italian Journal of Agronomics/Riverian Agronomics*, *2*, 135–143.

Leatherhead Food Research and Mintel. (2014). Stevia set to steal intense sweetener market share by 2017. *Mintel and Leatherhead Food Research* http://www.mintel.com/press-centre.

Lopez-Arellano, M., Dhir, S., Albino, Nilmarie Colon, Santiago, A., Morris, T., & Dhir, S. K. (2014). Somatic embryogenesis and plantlet regeneration from protoplast culture of *Stevia rebaudiana*. *British Biotechnology Journal*, *5*, 1–12.

Madan, S., Ahmad, S., Singh, G. N., Kohli, K., Kumar, Y., Singh, R., & Garg, M. (2010). *Stevia rebaudiana* (Bert.) Bertoni-A Review. *Indian Journal of Natural Products and Resources*, *1*, 267–286.

Mandhan, V., & Singh, K. (2015). Identification of novel microRNAs and their target prediction in *Stevia rebaudiana*. *Transcriptomics*, *3*, 1 DOI: 10.4172/2329-8936.1000107.

Mathur, S., & Shekhawat, G. S. (2013). Establishment and characterization of *Stevia rebaudiana* (Bertani) cell suspenson culture: An in vitro approach for production of stevioside. *Acta Physiology of Plant*, *35*, 931–939.

Mathur, S., & Shekhawat, G. S. (2013). Establishment and characterization of *Stevia rebaudiana* (Bertoni) cell suspension culture: an in vitro approach for production of stevioside. *Acta Physiology of Plant, 35*, 931–939.

Matsukubo, T., & Takazoe, I. (2006). Sucrose substitutes and their role in caries prevention. *International Dental Journal, 56*, 119–130.

Megeji, N. W., Kumar, J. K., Singh, V., Kaul, V. K., & Ahuja, P. S. (2005). Introducing *Stevia rebaudiana*. A natural Zero-Calorie sweetener. *Current Cellular Science, 88*, 801–804.

Mehta, J., & Dadhich, L. K. (2015). Micropropagation of *Stevia rebaudiana* Bertoni. In different kind of basal medium. *International Journal of Pure and Applied Bioscience, 3*, 264–270.

Melis, M. S., Rocha, S. T., & Augusto, A. (2009). Steviol effect, a glycoside of *Stevia rebaudiana*, on glucose clearance in rats. *Brazilian Journal of Biology, 69*, 371–374.

Metivier, J., & Viana, A. M. (1979). Determination of microgram quantities of stevioside from leaves of *Stevia rebaudiana* Bertoni by two dimensional thin layer chromatography. *Journal of Experimental Botany, 30*, 805–810.

Midmore DJ, Rank AH (2002) A new rural industry- Stevia to replace imported chemical sweeteners, RIRDC web publication, Project No. UCQ-16A.

Mirniam, A., Roshandel, P., Otroshi, M., & Ebrahimi, M. (2013). A Novel Protocol for *Stevia rebaudiana* Bert Regeneration. *Journal of Advanced Laboratory and Research Biology, 4*, 15–22.

Misra, H., Soni, M., Silawat, N., Mehta, D., Mehta, B. K., & Jain, D. C. (2011). Antidiabetic activity of medium-polar extract from the leaves of *Stevia rebaudiana* Bert. (Bertoni) on alloxan-induced diabetic rats. *Journal of Pharma Bioallied Science, 3*, 2.

Momtazi-Borojeni, A. A., Esmaeili, S. A., Abdollahi, E., & Sahebkar, A. (2017). A review on the pharmacology and toxicology of steviol glycosides extracted from S. rebaudiana. *Current Pharmaceutical Design, 23*, 1616–1622.

Naranjo, E. J., Betin, O. F., Trujillo, A. J. U., Posada, R. C., & Garces, L. A. (2015). Effectof genotype on the in vitro regeneration of *Stevia rebaudiana via* somatic embryogenesis. *Acta Biology Colombia, 21*, 87–98.

Pande, S., & Khetmalas, M. (2012). Effect of concentration of sucrose on callus induction & somatic embryogenesis of anti-diabetic plant: *Stevia Rebaudiana. Asian Journal of Biochemistry and Pharmacology Research, 4*, 2231–2560.

Pandey, H., Pandey, P., Pandey, S. S., Singh, S., & Banerjee, S. (2016). Meeting the challenge of stevioside production in the hairy roots of Stevia rebaudiana by probing the underlying process. *Plant Cell Tissue and Organ Culture* Doi:10.1007/s11240-016-1020-7.

Patil, N. M. (2013). *In vitro* culture of *Stevia rebaudiana* var Bertoni. *Journal of Chemical Biology and Physical Science, 3*, 2706–2711.

Pazuki, A., Aflaki, F., Yucesan, B., & Gurel, S. (2019). Effects of cytokinins, gibberellic acid 3, and gibberellic acid 4/7 on in vitro growth, morphological traits, and content of steviol glycosides in *Stevia rebaudiana. Plant Physiology and Biochemistry., 137*, 154–161.

Pol, J., Hohnova, B., & Hyotylainen, T. (2007). Characterisation of *Stevia rebaudiana* by comprehensive two-dimensional liquid chromatography time-of-flight mass spectrometry. *Journal of Chromatography, 11*, 85–92.

Polonsky, K. S. (2012). The past 200 years in diabetes. *New England Journal of Medicine, 367*, 1332–1340.

Preethi D, Sridhar TM, Naidu CV (2011) Efficient Protocol for Indirect Shoot Regeneration from Leaf Explantsof *Stevia rebaudiana* (Bert.) An Important Calorie Free Bio sweetener Journal of Phytology 3: 56-60.

Puri, M., Sharma, D., & Tiwari, A. K. (2011). Downstream processing of stevioside and its potential applications. *Biotechnology Advances, 29*, 781–791.

Rajasekaran, T., Ramakrishna, A., Udaya Sankar, K., Giridhar, P., & Ravishankar, G. (2008). Analysis of predominant steviosides in *Stevia rebaudiana* bertoni by liquid chromatography/electro spray ionization-mass spectrometry. *Food Biotechnology, 22*, 179–188.

Rathore, S., Yadav, K., Singh, N., & Singh, S. K. (2014). Comparative study on callus induction, proliferation and plantlets regeneration in two cultivars of *Stevia rebaudiana* bertoni – The only non caloric natural sweetener. *Pertanika Journal of Tropical Agriculture Science, 37*, 499–508.

Razak, U. N. A. A., Ong, C. B., Yu, T. S., & Lau, L. K. (2014). *In vitro* micropropagation of *Stevia rebaudiana* Bertoni in Malaysia. *Brazilian Archives of Biological Techology., 57*, 23–28.

Reis, R. V., Borges, A. P. P. L., Chierrito, T. P. C., Souto, E. R., Souza, L. M., Iacomini, M., Oliveira, A. J. B., & Goncalves, R. A. C. (2011). Establishment of adventitious root culture of *Stevia rebaudiana* Bertoni in a roller bottle system. *Plant Cell Tissue and Organ Culture, 106*, 329–335.

Ruiz-Ruiz, J. C., Moguel-Ordonez, Y. B., & Segura-Campos, M. R. (2017). Biological activity of Stevia rebaudiana Bertoni and their relationship to health. *Critical Reviews in Food Science and Nutrition, 57*, 2680–2690.

Saxena, N. C., & Ming, I. S. (1988). Preliminary harvesting characteristics of Stevia. *Physical Properties of Agricultural Material and Products, 3*, 299–303.

Sharma, N., Gauchan, D. P., Dhaka, A., Luitel, A., Shakya, S., & Shakya, R. (2015). Establishment of regenerative callus, cell suspension system and molecular characterization of stevia rebaudiana bertoni for the production of stevioside in in vitro. *International Journal for Research in Applied Science & Engineering Technology, 3*, ISSN: 2321–9653.

Shirwalkar, A., Parmar, V., Bhagat, J., & Khan, S. (2011). Identification and estimation of stevioside in the commercial samples of stevia leaf and powder by HPTLC and HPLC. *International Journal of Pharmaceuticals and Life Science, 2*, 1050–1058.

Singh, G., Singh, G., Singh, P., Parmar, R., Paul, N., Vashist, R., Swarnkar, M. K., Kumar, A., Singh, S., Singh, A. K., Kumar, S., & Sharma, S. K. (2017). Molecular dissection of transcriptional reprogramming of steviol glycosides synthesis in leaf tissue during developmental phase transitions in Stevia rebaudiana Bert. *Scientific Report*, 7, 11835 doi: 10.1038/s41598-017-12025-y.

Singh, P., Dwivedi, P., & Atri, N. (2014). In vitro shoot multiplication of Stevia and assessment of stevioside content and genetic fidelity of the regenerates. *Sugar Technology, 16*, 430–439.

Sivaram, L., & Mukundan, U. (2003). In vitro culture on *Stevia rebaudiana. In Vitro Cellular and Developmental Biology– Plant, 39*, 520–523.

Sreedhar, R. V., Venkatachalam, R., Thimmaraju, N., Bhagyalakshmi, M. S., Narayan, L., & Ravishankar, G. A. (2008). Direct organogenesis from leaf explants of *Stevia rebaudiana* and cultivation in bioreactor. *Biology Plantarum, 52*, 355–360.

Starrat, A. N., Kirby, C. W., Pocs, R., & Brandle, J. E. (2002). Rebaudioside F, a diterpene glycoside from *Stevia rebaudiana. Phytochemistry, 59*, 367–370.

Strauss, S. (1995). The perfect sweetener? *Technology Review, 98*, 18–20 Whitaker J Sweet justice: FDA relents on stevia. Human Events 51:11.

Sumida, T. (1968). Reports on Stevia introduced from Brazil as a new sweetness resource in Japan (English summary). *Journal of Cent Agriculture and Experimental Station, 31*, 1–71.

Tanaka, O. (1997). Improvement of taste of natural sweeteners. *Pure Applied Chemistry, 69*, 675–683.

Tandel, K. R. (2011). Sugar substitutes: Health controversy over perceived benefits. *Journal of Pharmacology and Pharmacotherapeutics, 2*, 236–243.

Thiyagarajan, M., & Venkatachalam, P. (2012). Large scale In vitro propagation of *Stevia rebaudiana* (bert) for commercial application: Pharmaceutically important and antidiabetic medicinal herb. *Industrial Crops and Products, 37*, 111–117.

Tomaszewska-Sowa, M., Figas, A., & Sawilska, A. K. (2016). A non caloric sweetener *Stevia rebaudiana* bertoni – tissue cultured plantlets fororganic farming and home gardening. *Infrastructure and Ecology of Rural Areas.* doi: http://dx.medra.org/10.14597/infraeco.2015.4.3.0911261-1273.

Totte, N., Van den Ende, W., Damme, V., Compernolle, F., Baboeuf, I., Geuns, J.M.C. (2003) Cloning and heterologous expression of early genes in gibberellins and steviol biosynthesis *via* the methylerythritol phosphate pathway in *Stevia rebaudiana* Canadian Journal of Botany 81: 517-522

Upadhyay S, Sharma S, Kumar R (2013) *In vitro* Morphological, Biochemical and Microbial Studies on Elite Clones of *Stevia rebaudiana* for Enhanced Production of Stevioside International Journal of Traditional and Herbal Medicine 1: 6-12.

Verma, S., Yadav, K., & Singh, N. (2011). Optimization of the protocols for surface sterilization, regeneration and acclimatization of *Stevia rebaudiana* Bertoni. *American – Eurasioan Journal of Agriculture and Environment Science, 11*, 221–227.

Wang, J., & Hou, B. (2009). Glycosyltransferases: key players involved in the modification of plant secondary metabolites. *Frontiers of Biology in China, 4*, 39–46.

Wang, Q. M., & Wang, L. (2012). An evolutionary view of plant tissue culture: somaclonal variation and selection. *Plant Cell Reports, 31*, 1535–1547.

Woelwer-Rieck, U., Lankes, C., Wawrzun, A., & Wust, M. (2010). 1st Improved HPLC method for the evaluation of the major steviol glycosides in leaves of Stevia rebaudiana. *European Food and Research Technology, 231*, 581–588.

Wong JSJ (2011). Determination of steviol glycosides from *Stevia* plant(*Stevia rebaudiana*) grown in different areas of Sabah. Thesis (Bachelor's degree of food science), University of Malasiya, Sabah-ID code 6177.

Yadav, A. K, Singh, S., Dhyani, D., & Ahuja, P. (2011). A review on the improvement of *Stevia [Stevia rebaudiana* (Bertoni)] Canadian Journal of Plant Science 91: 1–27.

Yadav, S. K., & Guleria, P. (2012). Steviol glycosides from Stevia: Biosynthesis pathway review and their application in foods and medicine. *Critical Reviews of Food Science and Nutrition, 52*, 988–998.

Yadav A, Subhah K, Poonia AK, Yadav IS, & Yadav RC. (2016) An efficient micropropagation protocol for *Stevia rebaudiana* Medicinal Plants. International Journal of Phytomedicnes and Related Industries 8:65-73.

Yamazaki, T., & Flores, H. E. (1991). Examination of steviol glycoside production by hair root and shoot cultures of *Stevia rebaudiana*. *Journal of Natural Products, 54*, 986–992.

Yodyingyuad, V., & Bunyawong, S. (1991). Effect of stevioside on growth and reproduction. *Human Reproduction, 6*, 158–165.

Yucesan, B., Buyukgoçmen, R., Mohammed, A., Sameeullah, M., Altug, C., Gurel, S., & Gurel, E. (2016). An efficient regeneration system and steviol glycoside analysis of *Stevia rebaudiana* Bertoni, a source of natural high-intensity sweetener. *In Vitro Cellular and Developmental Biology–Plant* DOI 10.1007/s11627-016-9765-6.

Zeng, J., Chen, A., Li, D., Yi, B., & Wu, W. (2013). Effects of salt stress on the growth, physiological responses, and glycoside contents of *Stevia rebaudiana* Bertoni. *Journal of Agriculture and Food Chemistry, 61*, 5720–5726.

CHAPTER 14

Saffron (*Crocus sativus* L.): phytochemistry, therapeutic significance and omics-based biology

Deepu Pandita
Government Department of School Education, Jammu, Jammu and Kashmir, India

Introduction

Crocus sativus L. (Saffron) belongs to family Iridaceae and is generally cultivated in Europe, Asia, and America (Shahi, Assadpour & Jafari, 2016). It is an aromatic and medicinal plant species of prime economic importance (Rios, Recio, Giner, & Manez, 1996; Ferrence & Bendersky, 2004). Saffron has a typical aroma which is produced by the phytochemical apocarotenoid safranal. With a price tag of up to € 30,000/kg, saffron is the priciest spice on globe and at times even exceeds gold price (Mystery solved—biologists explain the genetic origins of the saffron crocus, 2019) and is used since antiquity for its aroma and the ability to give dishes and textiles a golden-yellow hue (Negbi, 1999). Saffron is harvested from the flowers of the saffron crocus (*Crocus sativus*), mainly the stigma. Some 150,000–200,000 flowers need to be harvested by hand, to get 1 kg of saffron. Each flower has three stigmas which are also manually separated and then dried. These threads of stigma constitute the saffron spice (Mystery solved—biologists explain the genetic origins of the saffron crocus, 2019). Yearly around 200 tonnes of saffron stigma threads are harvested manually worldwide. Saffron's cultivation, manual harvesting, flower picking, and stigma separation makes this golden spice costly (Aytekin & Acikgoz, 2008; Molina, Valero, Navarro, Guardiola, & García-Luis, 2005), so even little boost in the yield and/or quality can enhance profitability. In 14th century, 14-week-long "Saffron War of Balsthal, was ignited when 363 kg (today's market prices > US$ 500,000) shipment of saffron on the way to town of Basel was captured and robbed from the Swiss merchants by nobles (Willard, 2002).

Medicinal and Aromatic Plants: Expanding their Horizons through Omics
http://dx.doi.org/10.1016/B978-0-12-819590-1.00014-8

Saffron is a triploid male-sterile plant harboring eight chromosome triplets (2n = 3x = 24) and propagates only vegetatively (Brighton, 1977; Mathew, 1977) through corms which are bulb-like structures on the stem of the saffron plants. Saffron is only vegetatively propagated by corms and all plants are clones with low, if any, genetic diversity (Fluch, Hohl, Stierschneider, Kopecky, & Kaar, 2010). It has a genome size of 1C = 3.45 Gbp (Brandizzi & Grilli Caiola, 1998). A consequence of this form of reproduction is that there is no room for improving saffron quality by crossing different cultivars. Thus, all modern saffron plants are genetically nearly identical. Saffron is under vegetative cultivation from daughter bulbs worldwide for more than 3500 years. There has been controversy on the probable parent species of the saffron for almost 100 years. Because of which the changes could not be made in the crocus genome by new breeding methods. Knowing the origin, in particular the originating plant species would enable saffron breeders to use new genotypes to broaden the diversity of the saffron crocus. However, breeding and genetic improvement of the cultivated saffron crocus as a crop is not possible, because of its male sterility caused by triploidy.

Despite its economic relevance, the genome and chromosomes genomic studies of the saffron crocus are relatively rare (Fluch et al., 2010; Alsayied, Fernandez, Schwarzacher, & Heslop-Harrison, 2015; Larsen, Orabi, Pedersen, & Ørgaard, 2015). But very recently this puzzle has been solved. The saffron crocus has descended from one wild species *Crocus cartwrightianus* found in Greece. Through saffron crocus genome sequencing and comparative chromosome analysis (Fluorescence in situ hybridization, FISH) of different crocus species, Dresden biologists were able to demonstrate that genomes of two *Crocus cartwrightianus* individuals with slight chromosomal differences had got fused, concluding the centuries-long search for the parental origin of this legendary plant (Schmidt, Heitkam, Liedtke, Schubert, & Menzel, 2019). The wild crocus species *C. cartwrightianus* from Greece was known as the sole primogenitor of the modern saffron plant, and the area in the vicinity of the Greek capital Athens as the region where it evolved (Origin of the saffron crocus traced back to Greece, 2019). Nemati, Blattner, Kerndorff, Erol, & Harpke (2018) performed single nucleotide polymorphism analyses of an intron of the nuclear topoisomerase gene, and genotyping-by-sequencing, and found that saffron and *C. cartwrightianus* show very high allele similarity concluding that *C. cartwrightianus* genotypes formed the autotriploid saffron. The researchers report that saffron crocus has descended from one species, the wild species *Crocus cartwrightianus*, found in Greece. Through genome sequencing and comparative chromosome analysis via fluorescence in situ hybridization (FISH) of different crocus species, the biologists demonstrated that genomes of two *Crocus cartwrightianus* individuals with slight chromosomal differences are fused. These findings conclude the centuries-long search for the origins of this plant (Mystery solved—biologists explain the genetic origins of the saffron crocus, 2019). Schmidt et al. (2019) demonstrated by comparative FISH to saffron and closely related *Crocus* species that *C. sativus* is an autotriploid species solely derived from heterogeneous cytotypes of *C. cartwrightianus*. Phylogenetic and population assignment analysis accompanied by chloroplast polymorphisms indicated that the *C. cartwrightianus* population in the vicinity of Athens is the most analogous to *C. sativus* suggestive of the crop being an autotriploid that evolved in Attica by combining two different genotypes of *C. cartwrightianus*. Triploid sterility and vegetative propagation prevented afterwards segregation of the favorable traits of saffron, resulting in worldwide cultivation of a unique clonal lineage (Nemati, Harpke, Gemicioglu, Kerndorff, & Blattner, 2019). The contributions of *C. cartwrightianus* to the formation of saffron were proposed previously in different studies (Grilli-Caiola, Caputo, & Zaier, 2004; Zubor, Suranyi, Gyori, Borbely, & Prokisch, 2004; Frizzi, Miranda, Pantani, & Tammaro, 2007; Petersen, Seberg, Thorsøe, Jørgensen, & Mathew, 2008; Ahrazem, Trapero, Gomez, Rubio-Moraga, & Gomez-Gomez, 2010; Harpke, Meng, Kerndorff, Rutten, & Blattner, 2013; Larsen et al., 2015).

Saffron life cycle

Saffron is an autumn flowering plant. The growth pattern of *Crocus sativus* is different from other crops and it can be divided into three stages, namely, flowering (Fig. 14.1), vegetative stage, and formation of corms. Under Indian conditions, flowering occurs during autumn (October through November), followed by vegetative stage throughout winter, and the formation of replacement corms at the base of the shoots. The corm of *Crocus sativus* is an underground stem which is a compact mass of starch resembling scaly leaves covered with closely reticulated sheath known as tunics (Fig. 14.1F–G). The shape of the corm varies from flattened to ovoid or sub-globose (Srivastava, 1963). Each newly formed corm has 1 or 2 apical buds (from which new leaves, floral axis, and 1 or 2 daughter corms are produced) and 4–7 secondary buds, placed irregularly in a spiral form in the lower portion. Each mother corm generates 1–3 purple flowers having perianth of six violet tepals (Fig. 14.1A–C). Depending on the size of mother corm, each mother corm produces 2–3 principal daughter corms from apical buds and

Figure 14.1 Crocus sativus L: (A) Saffron flowers showing flower parts. (B) Saffron flowers showing pollinator with pollen grains adhered to body. (C) Two saffron flowers produced from single corm. (D) and (E) Dry saffron stigma. (F) and (G) Corms.

several corms from lateral buds (Mathew, 1982). Corms have the tunic cover of parallel fibers and are well-adapted to tolerate cold winters and hot summers (Agayev & Zarifi, 2010). Flowers have an underground tubular ovary on central pistil and a thin style of 9–10 cm length divided into three dark-red branches stigmas, each one up to 30–40 mm long. The flower usually appears in autumn (September–November) and is marked by its large perianth consisting of 6 petaloid tepals, which is tube-like and slender, funnel-shaped. The six segments are almost equal in form and size, though the inner ones are invariably somewhat shorter than the outer ones and are concave, narrow, and oblong. The flowers are hysteranthous to sub-hysteranthous (Dhar, 2000). Three stamens are epiperianthous (Fig. 14.1A–C). The filaments are short and free and anthers are yellow in color, long, and basifixed (Khan, 1996). Gynoecium comprise of ovary, style, and stigma (Khan, 1996). The style is thread-like and branches into three style arms, that is, stigmas, which are extended and protrude from the perianth; they are tubular, characteristic reddish or orange red in color, and subclavate (Fernández, 2004). Commercial saffron is the dried-red stigma and attached yellowish style. *Crocus sativus* L. has 2.0–3.2 cm long three stigmata separated from each other by hand after harvesting (Modaghegh, Shahabian, Esmaeili, Rajbai, & Hosseinzadeh, 2008; Tarantilis, Beljebbar, Manfait, & Polissiou, 1998). The dried stigma constitutes the saffron of commerce (Fig. 14.1D–E).

Origin, history, and folk uses of saffron

There are various schools of thought which believe that the word Saffron has either originated from the Arabic word "*Zafaran*" meaning yellow, is a fascinating spice steeped in rich history (Grilli Caiola & Canini, 2010) or from the French term "*Safran,*" which is derived from the Latin word "*Safranum.*" *Safranum* is also allied to Italian *Zafferano* and Spanish *Azafran.* Saffron is known as "*Kum Kum*" and "*Kesar*" in Sanskrit and "*Koung*" in Kashmiri (Husaini et al., 2010). Saffron was scientifically named as "*Crocus sativus* var. *officinalis*" by Linnaeus (1762) (Grilli Caiola & Canini, 2010). The genus name *Crocus* has been derived from the Greek word "*Corycus,*" which is the name of a region in Cilcia of eastern Mediterranean. Today saffron is the name both for the *C. sativus* plant and spice acquired from its dried stigmas as well. Saffron is endemic to Mediterranean region with its origin from Greece, Asia Minor, and/or Persia (Vavilov, 1951; Bowels, 1952; Skrubis, 1989). The records of saffron cultivation date back to 1500–2500 BC in India, China, Iran, Greece, Mediterranean basin, and Eastern Europe (Negbi, 1989; Beiki, Keifi, & Mozafari, 2010). Saffron was widely grown in East and Mediterranean basin since Late Bronze Age (Zohary, 1994; Negbi, 1999). Saffron has been cultivated as a spice for at least 3500 years in Egypt and Middle East (Fernández, 2004). Saffron originated in southern Attica and probably was domesticated there. Saffron was afterward propagated vegetatively from corms (Brighton, 1977; Mathew, 1982; Fernández, 2004). Since at least 2300 years, farmers selected and replanted the corms of the saffron plants with extra-long, reddish, and more aromatic stigmas (Douglas & Perry, 2003). Infact, the accumulation of crocin, picrocrocin, and safranal in saffron's stigmas gives unique color, flavor, and aroma to the plant, respectively.

From Minoan frescos it is apparent that man already used wild saffron in the southern Aegean > 3600 years ago. Theophrastus for the first time described the cultivation and propagation of triploid saffron by corms (350 BCE–287 BCE) in Historia Plantarum (Negbi & Negbi, 2002). The Romans introduced saffron into Great Britain, while the Arabs brought saffron to Spain (The Royal Horticultural Society, 2003). There is no record as to when the cultivation of saffron was started in Jammu and Kashmir (India). Historical accounts of saffron cultivation in Kashmir dates back to 550 AD. It is believed that saffron was first spread to India by Persian rulers by transplanting the saffron cultivars in the Persian Empire (Dalby, 2002) or after Persia occupied Kashmir, Persian saffron was grown there around 500 BC (McGee, 2004). According to a Buddhist legend, saffron plantation in Kashmir back dates to 5th century BC (Dalby, 2002). So, saffron growing in Kashmir has origin from Persia (Singh & Dhar, 1976). Legends support the view that saffron was grown at Padampore (today's Pampore, Kashmir), India. *Rajatarangini* written by Kalhana in the 12th century which is the oldest historical treatise of Kashmir, records that saffron was there in Kashmir even prior to the reign of King Lalitaditya in 750 AD (Husaini et al., 2010). In Kashmir, saffron grows under temperate climatic conditions on "Karewas" which are uplands with lacustrine deposits at an altitude of 1585–1677 m above mean sea level (Kanth, Khanday, & Tabassum, 2008). Frost enhances the robustness and fragrance in saffron plants. In Kashmir (India), Afghanistan, Iran, and Pakistan, saffron production is the main source of income for many farmers, since the saffron booms in soils that cannot be utilized for agriculture.

Historical and archeological evidences suggest the use of *C. sativus* to 2500–1500 BC (Negbi, 1989; Ferrence & Bendersky, 2004). In ancient Egypt and Rome saffron was used as a dye, in drugs and perfumes and culinary purposes (Abdullaev, 1993; Abdullaev, 2002). *C. sativus* and

C. cartwrightianus have been employed as spice and dye for over 3500 years (Fernández, 2004; Grilli Caiola & Canini, 2010). According to Greek mythology, Zeus slept on a bed of saffron. Further, Saffron was used as a royal dye, perfume in saloons, courts, theatres and bathrooms in Greece (about 2000–146 BC) (Behmanesh, 1959; Munby, 1992; Blois & Spek, 2005) and was later extended to common people (Abrishami, 1987; Dadkhah, Ehtesham, & Fekrat, 2003). The use of *C. sativus* stigmas is shown in frescos from Crete and Santorini in 3600 years ago. Saffron was used in Iran to produce various energizing drinks and with cinnamon and orange jam (Abrishami, 1997). Even today Saffron is used as a healthy additive to add flavor, color, smell, and aroma to Iranian food such as chelow kabab (Asalgoo, Jahromi, Meftahi, & Sahraei, 2015; Behnia, 1991). It is used as dye for clothes (Bamford, 2006) and imparts color and flavor to food additives (Sampathu, Shivshankar, & Lewis, 1984). It is also used as the main foodstuff in preparing various saffron rice puddings, Zoolbia and Halva (Behnia, 1991; Moshiri et al., 2006). Assyrians and Babylonians (2nd BC) used saffron for cure of dyspnea, head problems, female issues of menstruation, delivery, and painful urination (Blois & Spek, 2005). Cleopatra is believed to (69–30 BC) (Wilson, 2006) bath in milk and saffron (Behnia, 1991). *Crocus sativus* was used as a gift with gold by Iranians who considered it as a spiritual item to scent the body of dead people, to document the prayers and rulers' orders and to paint the books (Abrishami, 1997). Since times immemorial, *Crocus sativus* has found use as a spice, dye, against rheumatism, and alcohol addictions, scarlet fever, smallpox, colds, asthma, eye, and heart diseases, tumor and cancer (Hartwell, 1982; Zareena, Variar, Cholar, & Bongirwar, 2001), respiratory decongestant, anodyne, emmenagogue, aphrodisiac, antidepressant, diaphoretic, antispasmodic, expectorant and sedative, painkiller during delivery and for "lady's malaise, as an aphrodisiac, antispasmodic, expectorant (Yu-Zhu et al., 2008), antispasmodic, antidepressant in Persian traditional medicine (Akhondzadeh et al., 2005), antiapoptotic and anticarcinogenic (Fernández-Sánchez et al., 2012) and in treatment of kidney and urinary disorders in the tribal communities of Ladakh, India (Ballabh, Chaurasia, Ahmed, & Singh, 2008). In accordance with the "Doctrine of Signatures, European practitioners acquire the yellowish hue of saffron as a sign of its medicinal use against jaundice (Darling Biomedical Library, 2002; *Saffron*, UCLA). In recent years, besides the *Crocus sativus* use as spice, medicinal, and therapeutic properties have also been documented in it with low doses (Winterhalter & Straubinger, 2000; Abdullaev and Frankel, 1999).

Distribution and production of saffron

The centre of diversity of genus *Crocus* is Balkans and Turkey and its species habitats in meadows, scrub, and woodland show wide distribution from Western Europe and North-western Africa (Rubio-Moraga et al., 2009; Harpke et al., 2015) through Near East, Central Asia to western China with the centre of diversity on the Balkan Peninsula and in Asia Minor (Mathew 1982). Almost all saffron grows in a belt bounded by the Mediterranean in the west and mountainous Kashmir in the east. Saffron is a high value low volume spice that grows throughout Mediterranean Europe and Western Asia between 10° west and 80° east longitudes and 30 to 50° north latitudes (Yasmin & Nehvi, 2013). All other continents except Antarctica produce smaller amounts. In 1991, Some 300 t (300,000 kg) of whole threads and powder are gleaned yearly, (Katzer, 2010) of which 50 t (50,000 kg) is top-grade "coupe" saffron (Negbi, 1999).

The first cultivation fields of *Crocus sativus* were near Zagros and Alvand mountains for the period of the Kingdom of Media (708–550 BC) (Shabani, 2001; Blois & Spek, 2005; Pirnia, 2006), but how they used *Crocus sativus* is unknown (Abrishami, 1997; Dadkhah Ehtesham et al., 2003; Abrishami, 2004). It is believed that ancient Persians were the first who cultivated saffron rather than using only the wild flower (Dadkhah Ehtesham et al., 2003). Today, Saffron is cultivated in Iran, Spain, India, Greece, Morocco, Italy, Turkey, France, Switzerland, Israel, Pakistan, Azerbaijan, China, Egypt, United Arab Emirates, Japan, and few attempts have been made to introduce this spice in Australia, New Zealand, the United States, Argentina, and Chile (Fernández, 2004). Saffron is cultivated in Iran ancient times and exports saffron of high quality and distinctive properties to other countries. The world's total annual saffron production was 300 t/yr in year 2004, out of which Iran contributed 80% from an area of 50,000 ha (Fernández, 2004). Khorasan province (Iran) accounts for 46,000 ha of area under saffron cultivation. Iran is the largest producer accounting for almost 80% of the total world production (Ahmad et al., 2011). Spain (10%–12%) is the second largest saffron producer followed by India (3.3%), Greece (2.0%), and Morocco (0.3%) (Keify & Beiki, 2012). Iran is by far the world's most important producer:—in 2005 it grossed some 230 tonnes (230,000 kg) of dry threads, or 93.7% of the year's global total mass; much of the Iranian crop was bound for export (Ghorbani, 2008). In the same year, second-ranked Greece produced 5.7 t (5,700.0 kg). In decreasing order, Iran, Greece, Morocco, Kashmir, Azerbaijan, Spain, and Italy lead the global harvest. Microscale cultivation occurs in Tasmania, (Courtney, 2002)

China, Mexico, Egypt, France, New Zealand, Israel, Turkey (especially Safranbolu), Central Africa, and California (Hill, 2004; Abdullaev, 2002). Saffron is part of India's cultural identity and even Indian national flag features the color saffron denoting strength and courage. India was world's fourth-biggest importer by importing $14.9 million in saffron in 2016. Increased economic growth has made saffron reachable to more consumers from middle and high-income families. Indian saffron demand averaged annual growth of 23% from 2012–16 (International Trade Centre, 2018). Most of the India region's saffron is grown in the more climatically suitable "Valley of Kashmir," and Kishtawar of Jammu and Kashmir, tied as the next-highest producer, each produced 2.3 t (2,300.0 kg) (Ghorbani, 2008) with few attempts to introduce it in Uttranchal, Himachal Pradesh and Sikkim (Nauriyal, Gupta, & George, 1977; Munshi, 1989; Rana et al., 2003). Even though successful attempts to grow saffron in other parts of J&K state like Kargil (Munshi et al., 2001) or other areas of India such as Uttar Pradesh and Himachal Pradesh have been reported (Dhar & Mir, 1997), almost all saffron production is actually limited to Kashmir. Dastranj, Sepaskhah, & Kamgar-Haghighi (2019) showed that fall and winter seasonal index and annual rainfall index can be exercised to forecast the rain-fed saffron yield when merged with Crocus age and mean highest daily air temperature. After implementation of National Saffron Mission in 2010 by Agriculture Production Department J and K, Saffron farming system a legendary crop of Jammu and Kashmir state was on rise up till 2013, as overall Saffron production of state increased from 9.46 M.T to 16.5 M.T with an increase in average productivity from 2.5 to 4.4 kg/ha (Nehvi and Salwe, 2017). Climatic abnormalities observed in Kashmir over a couple of years (2014–17) has put saffron farming system under great distress leading to reduction in overall production from 16.5 M.T recorded in 2013 to almost 1.5 M.T recorded in 2017 (Nehvi & Yasmin, 2019).

An overview of Saffron producing nations (major and minor) and trading centres (2019) is shown in Fig. 14.2.

Saffron phytochemistry

The first citations of saffron's medicinal use was recorded in inscriptions of Assurbanipal library (668–627 BC) (Ferrence & Bendersky, 2004; Tolner, 2005). In ancient Egypt (3100 BC – 476 AD) saffron was imported from Crete and employed in the treatment of disorders of eye, menstruation and urinary system and to induce labor (Sigerist, 1955; Tolner, 2005)

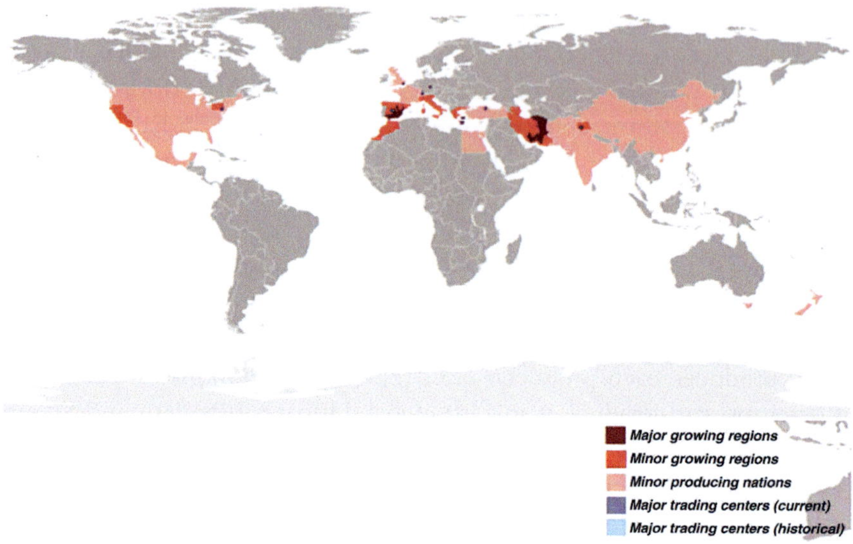

Figure 14.2 Saffron producing regions, producing nations (major and minor) and trading centers (present and past). *(Taken from: https://en.wikipedia.org/wiki/Trade and use of saffron)*

which finds mention in "The Ebers Papyrus." In ancient Greece saffron was utilized as an herbal remedy (Dadkhah Ehtesham et al., 2003). Hippocrates (5–4th century BC), Erasistratus (4–3rd century BC), Diokles (3rd century BC), and Discorides (1st century AD) believed that saffron has styptic and soothing properties and used it for therapeutics of various eye diseases, earache, tooth-ache, ulcers, and erysipelas (Ferrence & Bendersky, 2004; Giaccio, 2004; Tolner, 2005).

Saffron contains more than 150 volatile and aroma-yielding compounds and various non-volatile active compounds many of which are carotenoids. Saffron contains lipophilic and hydrophilic carbohydrates, proteins, amino acids, minerals, mucilage, starch, gums, vitamins (especially riboflavin and thiamine), pigments (crocin, alfa and beta carotenes, mangicrocin, xanthone carotenoid glycosidic conjugate, anthocyanin, zigzantin, lycopene, flavonoids, and zeaxanthin), alkaloids, saponins, safranal (aromatic essence terpene) and picrocrocin (bitter flavor) together with other chemical compounds (Carmona, Zalacain, Sanchez, Novel la, & Alonso, 2006; Samarghandian & Borji, 2014; Singla & Bhat, 2011; Zarinkamar, Tajik, & Soleimanpour, 2011; Abdullaev 1993, Abdullaev 2002; Rios et al., 1996; Winterhalter & Straubinger, 2000, Table 14.1). The main constituents localized in the red stigmatic lobes of the saffron are crocetin and its glucosidic

Table 14.1 USDA Nutrient values per 100g, Saffron

Nutrient	Unit	Value per 100g
Proximates		
Water	g	11.9
Energy	kcal	310
Energy	kJ	1298
Protein	g	11.43
Total lipid (fat)	g	5.85
Ash	g	5.45
Carbohydrate, by difference	g	65.37
Fiber, total dietary	g	3.9
Minerals		
Calcium, Ca	mg	111
Iron, Fe	mg	11.1
Magnesium, Mg	mg	264
Phosphorus, P	mg	252
Potassium, K	mg	1724
Sodium, Na	mg	148
Zinc, Zn	mg	1.09
Copper, Cu	mg	0.328
Manganese, Mn	mg	28.408
Selenium, Se	µg	5.6
Vitamins		
Vitamin C, total ascorbic acid	mg	80.8
Thiamin	mg	0.115
Riboflavin	mg	0.267
Niacin	mg	1.46
Vitamin B-6	mg	1.01
Folate, total	µg	93
Folic acid	µg	0
Folate, food	µg	93
Folate, DFE	µg	93
Vitamin B-12	µg	0
Vitamin A, RAE	µg	27
Retinol	µg	0
Vitamin A, IU	IU	530
Vitamin D (D2 + D3)	µg	0
Vitamin D	IU	0
Lipids		
Fatty acids, total saturated	g	1.586
4:00	g	0
6:00	g	0
8:00	g	0
10:00	g	0
12:00	g	0

(*Continued*)

Table 14.1 USDA Nutrient values per 100g, Saffron (*Cont.*)

Nutrient	Unit	Value per 100g
14:00	g	0.006
16:00	g	1.157
18:00	g	0.247
Fatty acids, total monounsaturated	g	0.429
16:1 undifferentiated	g	0.003
18:1 undifferentiated	g	0.39
20:01	g	0.006
22:1 undifferentiated	g	0
Fatty acids, total polyunsaturated	g	2.067
18:2 undifferentiated	g	0.754
18:3 undifferentiated	g	1.242
18:04	g	0
20:4 undifferentiated	g	0.013
20:5 n-3 (EPA)	g	0
22:5 n-3 (DPA)	g	0.006
22:6 n-3 (DHA)	g	0
Fatty acids, total trans	g	0
Cholesterol	mg	0
Amino Acids		
Other		
Alcohol, ethyl	g	0
Caffeine	mg	0
Theobromine	mg	0
Flavonoids		
Flavonols		
Kaempferol	mg	205.5

Source: United States Department of Agriculture Agricultural Research Service (2018). National Nutrient Database for Standard Reference Legacy Release - 02037, Spices, Saffron. Software v.3.9.5.2_2019-05-07.

derivatives, that is, crocins (pigment), picrocrocin (taste), safranal (odor) (Rubio-Moraga, Trapero-Mozos, Gómez-Gómez, & Ahrazem, 2010) and flavonoids (quercetin and kaempferol) (Pitsikas et al., 2007). The main components of saffron are the carotenoids crocetin (also called- α-crocetin or crocetin I). Its glycosidic forms are digentiobioside (crocin), gentiobioside, glucoside, gentioglucoside and diglucoside; β- crocetin (monoethyl ester), γ- crocetin (dimethylester), transcrocetin isomer, 13-cis-crocetin isomer; α-carotene, β- carotene, lycopene, zeaxanthin and mangiocrocin, a xanthone carotenoid glycosidic conjugate (Fernández, 2004). Sources of strong coloring capacity are glycosyl esters of crocetin carrying up to five glucoses, which are unusual water-soluble carotenoids (Tarantilis, Tsoupras, & Polissiou, 1995). The digentiobiosyl ester of crocetin ($C_{44}H_{64}O_{24}$) is known as crocin (Tarantilis and Polissiou, 1997). The coloring ability and many medicinal properties of saffron are attributed to cis- and trans-crocins, which are glycosidic esters of crocetin, a dicarboxylic apocarotenoid (Winterhalter & Straubinger, 2000). The picrocrocin ($C_{16}H_{26}O_7$) which is a monoterpene aldehyde and its deglycosylated derivate safranal ($C_{10}H_{14}O$) are formed in saffron during drying and storage by hydrolysis of picrocrocin (Tarantilis and Polissiou, 1997; Kanakis, Daferera, Tarantilis, & Polissiou, 2004; Carmona, Zalacain, Salinas, & Alonso, 2007). The principal element which imparts saffron exceptional "bitter" flavor is the glycoside picrocrocin. This bitter tasting substance can be crystallized and produces glucose and the aldehyde safranal by hydrolysis. The main aroma factor in saffron is safranal, which comprises of about 60% of the volatile components of saffron. In fresh saffron, this substance exists as a stable picrocrocin but as a result of heat and with the passage of time, it decomposes releasing the volatile aldehyde, safranal (Abdullaev, 2002, Tarantilis & Polissiou, 1997). The spicy, slightly bitter taste is due to picrocrocin, while the characteristic aroma is ascribed to safranal, a monoterpene aldehyde which is the major volatile component (Carmona et al., 2007). The concentrations of these constituents combine to determine the saffron spice quality, as defined by the International Organization for Standardization (ISO 3632-1:2011). Table 14.1 shows the United States Department of Agriculture (USDA) National Nutrient values for saffron (2018).

Few most important phytochemicals of various parts of Saffron are discussed further

Apocarotenoids in saffron flowers

Apocarotenoids are the products of the oxidative cleavage of carotenoids by specific carotenoid cleavage oxygenases (CCDs) that recognize and specifically cleave one or two double bonds (Walter, Floss, & Strack, 2010). Crocins are widely present in the species of the genus *Crocus* (Rubio-Moraga, Ahrazem, Rambla, Granell, & Gomez Gomez, 2013). Crocin is in fact 8, 8-diapocarotene-8, 8-dioic acid, with this chemical formula: $C_{44}H_{70}O_{28}$ and molecular weight of 976.96 g/mol (Samarghandian & Borji, 2014). It is a diester of crocetin with gentobiose and only water-soluble cis- and trans-carotenoid so, crocin is a peculiar carotenoid used more commonly as a colorant in food and medicine. Crocin is the most important saffron pigment (approx. 80%) and responsible for the golden bright yellow-red color of saffron (Tarantilis et al., 1998). Different types of crocin include trans-crocin 4 (the highest amount), trans-crocin 3, cis crocin 4, trans-crocin 2, 2', and cis-crocin 2 (Caballero-Ortega, Pereda-Miranda, & Abdullaev, 2007); crocin-1 or a-crocin or digentiobioside crocetin, crocin-2 or tricrocin or gentioglucoside crocetin, crocin-3 or gentiobioside crocetin, crocin-4 or glucoside crocetin, crocin-5 or diglucoside crocetin) a-Crocin or Crocin 1 (trans-crocetin di-(b-D-gentiobiosyl) ester) is the most abundant crocin with golden- yellow- orange color blends which can be isolated in pure form and directly crystallized at a melting point of 186°C. a-crocin also may comprise more than 10% of dry saffron's mass (Lage & Cantrell, 2009; Singla & Bhat, 2011). Crocetin (8, 80-diapo-8, 80- carotenoic acid) is a unique lipophilic carotenoid with a multi-unsaturated conjugate olefin acid structure and melting point of 285°C and a molecular weight of 328.4 g/mol (Samarghandian & Borji, 2014). Crocetins (a-crocetin or crocetin I, crocetin II, b-crocetin, g-crocetin) in saffron is formed from the enzymatic cleavage/oxidative degradation of zeaxanthin precursor on the $7 - 8/7 - 8'$ double bonds (Rubio-Moraga et al., 2008; Pfander & Schurtenberger, 1982) after breaking; it gives rise to crocetin (Christodoulou, Kadoglou, Kostomitsopoulos, & V, 2015; Grilli Caiola & Canini, 2010). Many research reports have revealed that 94% of total crocetin of saffron exists as glycosides molecules and the remaining 6% occurs in the form of free crocetin (Habibi & Bagheri, 1989). The resulting products are crocetindial and hydroxyl-β-cyclocitral, which is further glucosylated to form picrocrocin (β-D glucopyranosidehydroxyl- β-cyclocitral), the degradation product of which is the odor-active safranal (Pfander & Schurtenberger, 1982) formed from b-glucosidase action on picrocrocin by dehydration (heating combined with an enzymatic action) during postharvest drying and storage stages (Carmona et al., 2007). Picrocrocin with the chemical formula of $C_{16}H_{26}O_7$ and molecular weight of 330.37 g/mol, is a colorless monoterpene glycoside, precursor of safranal and second most abundant saffron component (approximately 1%–13% of saffron's dry matter), responsible for the bitter flavor of saffron (Samarghandian & Borji, 2014; Pitsikas, 2016; Grilli Caiola & Canini, 2010). Saffranal –a monoterpene aldehyde and a glycon of picrocrocin with a molecular weight of 150. 21 g/mol produces saffron's aroma and is a main component of the saffron's distilled essential oil. Also, aroma and aroma-active compounds of Iranian saffron (*Crocus sativus* L.) were analyzed by GCeMS eolfactometry. The results revealed that the most powerful aroma active compounds were safranal, 4-ketoisophorone and dihydrooxophorone

(Amanpour, Sonmezdag, Kelebek, & Selli, 2015). Accumulation of apocarotenoids in saffron is stigma specific and appears to be developmentally regulated (Rubio-Moraga et al., 2009) Higher levels of saffron apocarotenoids increase as stigmas develop and reach their peak in the red stage, (Rubio-Moraga et al., 2009) and it is likely that the genes involved in their biosynthesis may be stimulated during the process and enhanced in the stigma tissue. Carotenoid biosynthesis, cleavage activities, and the expression of corresponding genes have been studied during the development of the stigma of *C. sativus*. Five CCD genes which belong to two classes of cleavage enzymes were identified from *C. sativus* (Rubio-Moraga et al., 2008; Bouvier, Suire, Mutterer, & Camara, 2003; Ahrazem et al., 2010). From *Crocus sativus* stigmas, 8 glycosides were isolated including a new safranal glycoside, (4R)-4-hydroxy-2,6,6- trimethylcyclohex- 1-enecarbaldehyde 4- *O* -[β- D - glucopyranosyl(1→3)-β- D -glucopyranoside], and a new carotenoid pigment, *trans* - crocetin-1-al 1- *O* -β-gentiobiosyl ester, along with picrocrocin, crocetin mono (β-gentiobiosyl) ester (crocin-3), crocin, crocetin-(-β- D -glucosyl)- (β- gentiobiosyl) ester (crocin-2), kaempferol-7- sophoroside and sophorafl avonoloside (Tung & Shoyama, 2013). The carotenoids like phytoene, phytofluene, tetrahydrolycopene, β-carotene, zeaxanthin and crocetin were isolated from *C. sativus* (Pfander & Schurtenberger, 1982). In *C. sativus* flowers crocetin tri (β-D -glucosyl)-(β- D - gentiobiosyl) ester; crocetin di(β- D -gentiobiosyl) ester; crocetin (β-D -glucosyl)-(β- D -gentiobiosyl) ester; crocetin (β- D -gentiobiosyl) ester; crocetin di (β-D -glucosyl)ester; crocetin (β- D -glucosyl) ester; crocetin;picrocrocin; sinapic acid derivative; sinapicacid, crocusatin B; crocusatin C and safranal were reported by Rios et al. (1996). Montoro et al. (2012) identified crocetin di(β- D - gentiobiosyl) ester; crocetin (β- D- glucosyl)-(β- D -gentiobiosyl) ester; crocetin (β- D - gentiobiosyl) ester; crocetin di(β- D -glucosyl) ester; crocetin (β- D -glucosyl) ester; crocetin; picrocrocin and sinapic acid derivative in *C. sativus* petals and found only sinapic acid derivative in the stamen and flower. Twelve components were isolated from saffron: crocin-1, crocin-2, crocin-3, picrocrocin, acid form of picrocrocin, HTCC-diglycosyl- kaempferol, *trans* -crocin-4, *trans* -crocin-2, *trans* -crocin-3, safranal, crocetin and *cis* -crocin-3 (Abdullaev, 2002). Four new compounds, crocusatins F, G, H and I, together with 21 known compounds, were isolated from an aqueous extract of the stigmas of *Crocus sativus* (Li & Wu, 2002b). Using a two-step, low-pressure liquid chromatography, crocin-1, crocin-2 and crocin-3 were isolated from the hydroalcoholic extract of yellow pigments of saffron (Zhang et al., 2004). As for volatile cleavage products produced in these

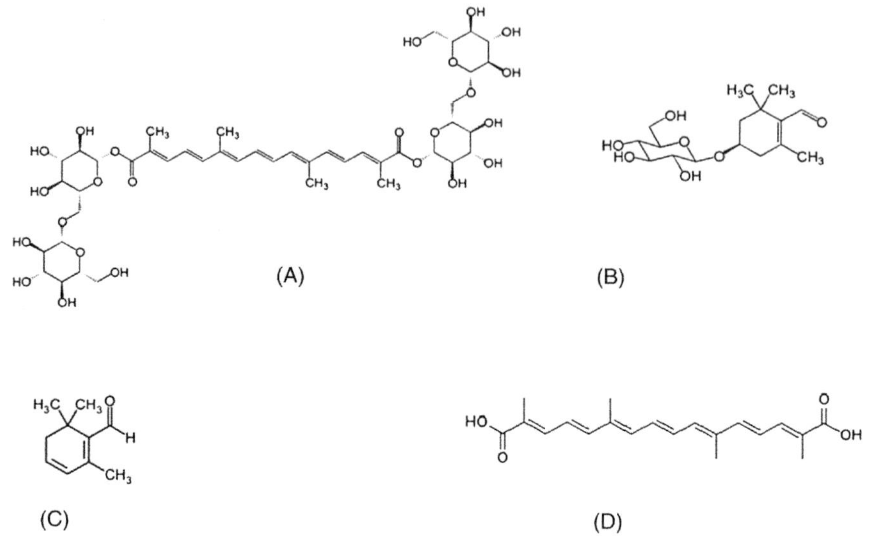

Figure 14.3 Structure of important phytochemicals of C. sativus (A) Crocin (B) Picrocrocin (C) Safranal (D) Crocetin. *https://en.wikipedia.org/wiki/Saffron.*

strains, only C13 β-ionone could be detected, but no C10 apocarotenoids, whereas a $9 - 10/9' - 10'$ cleavage activity for both types of enzymes is predicted. Analysis of stigma volatiles in different stages of flower development indicated high production of β-ionone shortly before and at anthesis (Rubio-Moraga et al., 2008; Rubio-Moraga et al., 2009). The chemical structure of apocarotenoids crocin, picrocrocin, safranal and crocetin of *C. sativus* is drawn in Fig. 14.3.

Flavonoids in saffron flowers

In *Crocus sativus*, flavonols are usually glycosylated at their 3-OH, 7-OH and 4′-OH positions (Rubio-Moraga et al., 2009; Li & Wu, 2002) producing a complex pattern of flavonols. In *C. sativus* stigmas three main glucosides of kaempferol have been identified— kaempferol 7-O-sophoroside, kaempferol 3-O-sophoroside-7-O-glucopyranoside and kaempferol 3, 7, 4′-triglucoside (Moraga et al., 2009) which increased with stigma development. Kaempferol 7- *O* -sophoroside relative levels reached maximum at anthesis stage, while kaempferol 3- *O* -sophoroside- 7- *O* – glucopyranoside and kaempferol 3, 7, 4′-triglucoside, with relative high levels in the scarlet stages. The flavonoid pyrogallol and gallic acid were isolated from saffron stigma (Karimi, Oskoueian, Hendra, & Jaafar, 2010). The quercetin, kaempferol, and galangin were isolated from the fresh saffron petals (Kubo &

Kinst-Hori, 1999). Kaempferol was isolated from saffron petals (Hadizadeh, Khalili, Hosseinzadeh, & Khair-Aldine, 2010). The flavonoids identified in diethyl ether, ethyl acetate and aqueous fractions of *C. sativus* petals included 4′-methyl ether dihydrokaempferol 3- *O* -deoxy-hexoside, taxifolin 7- *O* - hexoside, dihydrokaempferol 3- *O* -hexoside, naringenin 7- *O* -hexoside and naringenin (Termentzi & Kokkalou, 2008). The nine flavonoids identified in methanolic extract of *Crocus sativus* petals included glycosidic derivatives of quercetin and kaempferol as major compounds (1–2), and their methoxylated and acetylated derivatives (Montoro, Tuberoso, Maldini, Cabras, & Pizza, 2008). Some 31 flavonoids were identified in *C. sativus* petal extracts comprising mainly glycosidated and metoxilated derivatives of kaempferol, quercetin, isorhamnetin and tamarixetin (Montoro et al., 2012). Kaempferol and kaempferol-3- *O* -β- D – glucopyranosyl -(1→2)-β- D -glucopyranoside were isolated in the petals (Song, 1990). Nineteen flavonols, kaempferol, quercetin and isorhamnetin glycosides as mono-, di- or triglycosides were isolated from *C. sativus* tepals (Goupy, Vian, Chemat, & Caris-Veyrat, 2013). By contrast, 21 different glycosides of isorhamnetin, kaempferol, myricetin, naringenin, quercetin, tamarixetin, and taxifolin have been identified in tepals (Termentzi & Kokkalou, 2008). The main flavonoid present in saffron tepals is kaempferol-3- O-β-D-glucopyranosyl-(1 – 2)-β-D-glucopyranoside (kaempferol 3-O-β-sophoroside) (Trapero-Mozos et al., 2012), kaempferol and kaempferol glycosides being the most dominant class of flavonoids in this tissue (70% and 90% of the total content of flavonoids) followed by quercetin glycosides (5% – 10%) (Nørbæk et al., 2002). In pollen kaempferide, a methylated kaempferol and the isorhamnetin glycosides isorhamnetin-3, 4′-diglucoside, isorhamnetin- 3-O-robinobioside, and isorhamnetin-3-β-D-glucoside are detected (Li & Wu, 2002). In addition, the anthocyanins delphinidin, petunidin, and malvidin have been identified in tepals, with different sugar substitutions (Nørbæk et al., 2002) and are responsible for their blue color.

Saponins and other phytochemicals of saffron corms

Saponins are a widespread group of plant defense terpenoid compounds with antimicrobial, fungicidal, allelopathic, and insecticidal activities (Sparg, Light, & van Staden, 2004). The name saponin is derived from sapo, soap in Latin, because the surfactant properties produce soap-like foams in aqueous solution. According to the chemical nature of the aglycone (known as sapogenin), the saponins are divided into steroidal and triterpenoid saponins. In saffron, two triterpenic saponins have been identified in the corm, azafrines 1 and 2 (Rubio-Moraga et al., 2011). Azafrine 1 contains 3-O-β- D-glucopyranosiduronic acid echinocystic acid as prosapogenin, and azafrine 2 contains 3-O-β-D-galactopyranosiduronic acid echinocystic acid as prosapogenin. The bidesmosidic saponins were respectively characterized as 3- *O* -β- D -glucopyranosiduronic acid echinocystic acid 28- *O* -β- D -galactopyranosyl-(1→2)-α-l - arabinopyranosyl-(1→2)-[β- D -xylopyranosyl- (1→4)]-α- D -rhamnopyranosyl-(1→2)-[4- *O* -di-α- l - rhamnopyranosyl-3,16-dihydroxy-10-oxohexadecanoyl]- α- D -fucopyranoside and 3- *O* -β- D -galactopyranosiduronic acid echinocystic acid 28- *O* -β- D -galactopyranosyl-(1→2)-α-l -arabinopyranosyl-(1→2)-[β- D -xylopyranosyl- (1→4)]-α- L -rhamnopyranosyl-(1→2)-[4- *O* -di-α-L - rhamnopyranosyl-3,16-dihydroxy-10-oxohexadecanoyl]- β- D -fucopyranoside. Azafrines 1 and 2 are localized in the external part of the corms, suggesting their involvement in plant defense (Rubio-Moraga et al., 2011; Rubio-Moraga et al., 2013). Phenolic compounds detected in saffron corms included catechol, vanillin, salicylic acid, cinnamic acid, *p* -hydroxy benzoic acid, gentisic acid (highest 5.693 μg/g), syringic acid, *p* -coumaric acid, gallic acid (lowest 0.416 μg/g), *t* -ferulic acid and caffeic acid (Esmaeili, Ebrahimzadeh, Abdi, & Sarfarian, 2011). Volatile compounds identified in the corm extract included hexadecanoic acid (33.23 %); octadecadienoic acid, palmitic acid ethyl ester, *n* -tetradecane, 1,3,5-tribenzoylbenzene, *n* –heptadecane, *n* –pentadecane, *n* –catane, diethyltoluamide, *n* –tridecane, *n* –octadecane, and *n* –eicosane (Yu-Zhu et al., 2008). 34 compounds from the volatile oil from *C. sativus* corms include monoterpenoids, sesquiterpenoids, total terpenoids, hydrocarbons, alcohols, aldehydes, esters, acids, sum aliphatics, aldehydes, frans, latones and aromatics (Masuda, Mori, & Miyazawa, 2012).

Saffron therapeutic significance

Although culinary applications of saffron are dominant, saffron has had traditional medicinal applications in many parts of the ancient world for millennia. For example, in Ancient Greece saffron was used as a cure for insomnia as well as for hangovers due to excessive wine drinking. One of the first historical references about the use of saffron comes from Ancient Egypt, where it was known as incense with sedative qualities. Today, the biological compounds present in saffron are associated with certain health benefits. Saffron contains high levels of antioxidants such as alpha and beta-carotenes, as well as to contain many other vitamins and minerals. Research studies have shown that crude saffron extracts have cytotoxic, antitumor, anticarcinogenic and antimutagenic effects (Nair et al., 1991; Abdullaev and Frenkel, 1992a; Abdullaev and Frenkel, 1992b). Saffron helps clear up canquer sores and decreases the pain of teething infants (Rıos et al., 1996Rios, Recio, Giner, & Manez, 1996; Winterhalter & Straubinger, 2000; Abdullaev, 1994, Abdullaev, 2002).

For over 3000 years, saffron has been considered the most expensive spice and is widely used in Persian, Indian, European, Arabian, and Turkish cuisines as a food colorant as well as in traditional medicine for the treatment of some 90 illnesses (Mousavi & Bathaie, 2011; Basker & Negbi, 1983; Abdullaev, 2002). The oldest documents about the edible use of saffron are recorded by Polyen (in the 2nd century BC), a Greek military writer who, in his graduation thesis entitled "Stratagemes" reports a list of foods consumed by the court of the Persian Achemenide dynasty (550–330 BC), which had been carved on a bronze column in front of the kitchen (Abrishami, 2004). Saffron (*Crocus sativus* L.) and its active ingredients have the following effects such as anticancer, antimutagenic, antitumor, antioxidant, antigenotoxic, memory and learning enhancer, antiinflammatory, anticonvulsants, antidepressants, blood pressure regulators, oxygen boosting tissues, bronchodilator, and so on (Abdullaev & Espinosa-Aguirre, 2004; Chryssanthi et al., 2007; Asalgoo, Tat, Sahraei, & Jahromi, 2017; Kianbakht, 2008; Mokhtari Hashtjini, Pirzad Jahromi, Meftahi, Esmacili, & Javidnazar, 2018a) showing that the species can have potential pharmaceutical applications. In addition, laboratory findings have shown that saffron or its active ingredients can significantly reduce oxidative damage in the ischemic kidney (Hosseinzadeh, Sadeghnia, Ziaee, & D, 2005), skeletal muscle (Hosseinzadeh, Modaghegh, & Saffari, 2009), heart (Joukar et al., 2010) and brain (Ochiai et al., 2007). Several preclinical studies have confirmed the antidepressant effects of crocin and crocetin (Shafiee, Arekhi, Omranzadeh, & Sahebkar, 2018). According to Tóth et al. (2019), a meta-analysis revealed that saffron has a significant effect on the severity of depression. The data of randomized, controlled clinical trials hold that saffron is appreciably more efficient than placebo ($g = 0.891$; 95% CI: 0.369–1.412, $p = 0.001$), and non-inferior to tested antidepressant drugs ($g = -0.246$; 95% CI: -0.495– 0.004, $p = 0.053$).

According to Siddiqui et al. (2018) research conducted up to now provides initial support for the use of *Crocus sativus* for the treatment of mild-to-moderate depression. Further research is recommended to increase our understanding about the role and actions of saffron in major depression. Hausenblas, Saha, Dubyak, & Anton (2013) stated that, similar to antidepressants, saffron may exert its antidepressant effect by modulating the levels of certain chemicals in the brain, including serotonin (a mood-elevating neurotransmitter). In Iranian traditional medicine, saffron has been used for obstruction therapy inside the brain to protect it from oxygen deprivation and to treat apoplexy. Notably, it has been demonstrated that topical use of *Crocus sativus* in the boiling water is useful to relieve severe headaches (Mousavi & Bathaie, 2011). One recent study also indicates its potential protective effect against ischemia/reperfusion injury, cerebral edema and decreased infarct volume in a rat model of stroke (Vakili, Einali, & Bandegi, 2014). Effective ingredients of *Crocus sativus*, such as safranal, crocin, picrotoxin and trichrosin, have protective effects against the induced oxidative stress in the global (Zheng, Liu, Wang, & Xu, 2007) and local cerebral ischemia in vitro conditions (Ochiai et al., 2007). Oral administration of saffron seven days before the onset of local cerebral ischemia leads to a reduction in neuronal death and an increase in the antioxidant capacity of the brain in rats (Saleem et al., 2006). Some animal studies have reported the neuroprotective effects of saffron in rats (Deslauriers et al., 2011; Sarshoori, Asadi, & Mohammadi, 2014; Zheng et al., 2007). Asadollahi et al. (2019) confirmed the short and long-term neuroprotective effects of saffron capsules on ischemic stroke in humans. Only the stigmas of the *Crocus sativus* flower along with the style tops are used medicinally, although high dosages (>30 g) can be toxic, dangerous and abortive (Winterhalter & Straubinger, 2000; Azafrán, 2002).

Saffron is credited with various medicinal properties. In small doses, it acts as a mild stimulant and in large doses as an aphrodisiac and narcotic. Since time immemorial, *Crocus sativus* has been used as a drug to treat various human ailments such as asthma, cough, flatulence, stomachic disorder, colic, insomnia, chronic uterine hemorrhage, amenorrhea, smallpox, cardiovascular disorders, bronchospasm, Alzheimer's disease, infertility, carminative, depression, cholesterol, diuretic, febrifuge, stimulant, stomach ailments, retina-degeneration, multiple sclerosis and different kinds of cancers (Abdullaev, 2004; Giaccio, 2004; Chryssanthi et al., 2007; Dagostino et al., 2007; Fernandez, 2007; De-Juan et al., 2009; Gresta et al., 2009; Poma, Fontecchio, Carlucci, & Chichiriccò, 2012; Makri, Ferlemi, Lamari, & Georgakopoulos, 2013; Siracusa et al., 2013; Winterhalter & Straubinger, 2000, Basker & Negbi, 1983; Kirtikar & Basu, 1933). It is also used to regulate menstrual disorders in women. Saffron is used in weakness for rejuvenation. When saffron paste is applied on the forehead, it is said to cure headaches. Some of its components have cytotoxic, anticarcinogenic and antitumor properties (Abdullaev & Frankel, 1999; Fernandez & Escribano, 2000; Abdullaev, 2004; Abdullaev and Espinosa, 2004; Escribano et al., 1999). Saffron is used in the treatment of mild to moderate depression and epilepsy cases (Akhondzadeh et al., 2005; Moshiri et al., 2006; Noorbala, Akhondzadeh, Tahmacebi-Pour, & Jamshidi, 2005; Akhondzadeh et al., 2005). It has also been tested in rats in gastric ailments (Al-Mofleh et al., 2006) and used as a pro-memory agent (Saito, 2004), neurodegenerative and psychiatric disorders (Pitsikas, 2015; Farkhondeh, Samarghandian, Shaterzadeh Yazdi, & Samini, 2018), coronary artery diseases, bronchitis, asthma, diabetes, and cancer (Boskabady & Farkhondeh, 2016). Saffron is known to having anti-mutagenic (mutation preventing), immunomodulating and anti-oxidant like properties (Chang, Wang, Liang, & Kuo, 1964; Abdullaev, 2002; Assimopoulou, Sinakos, & Papageorgiou, 2005), hypolipidaemic, antitussive, antioxidant, anticancer, antidiabetic activities (Rios et al., 1996 ; Abdullaev, 2002 ; Melnyk, Wang, & Marcone, 2010 ; Poma et al., 2012 ; Gutheil, Reed, Ray, Anant, & Dhar, 2012), antihypertensive, anticonvulsant, antitussive, antigenotoxic and cytotoxic effects, cardioprotective, neuroprotective anxiolytic aphrodisiac, antioxidant, antidepressant, antinociceptive, antiinflammatory and relaxant activity (Srivastava, Ahmed, Dixit, & Dharamveer, 2010; Melnyk et al., 2010), improves memory and learning skills, increase blood flow in retina and choroid and is effective-against gastric disorders, cardiovascular disease, insulin resistance, depression, premenstrual syndrome, insomnia and anxiety, prevention and maintenance of cancer due to its antioxidant properties (Abdullaev, 2002; Abdullaev & Espinosa-Aguirre, 2004; Melnyk et al., 2010) and anti-convulsant activity (Zhang, Shoyama, Sugiura, & Saito, 1994).

In herbal medicine, saffron is traditionally used as a nerve sedative, stress-reliever, anti-depressant, aphrodisiac, expectorant, and anti-spasmodic agent (José Bagur et al., 2017; Christodoulou et al., 2015; Hosseini, Razavi, & Hosseinzadeh, 2018; WHO, 2007), neuroprotective, anticonvulsant, anti-depressive, anxiolytic, antioxidant, anti-inflammatory, hypolipidemic, anti-atherogenic, anti-hypertensive, anti-tumour (José Bagur et al., 2017; Christodoulou et al., 2015; Hosseini et al., 2018; Ghaffari & Roshanravan, 2019; Poma et al., 2012). Recently meta-analysis by Pourmasoumi, Hadi, Najafgholizadeh, Kafeshani, & Sahebkar (2019) showed that *Crocus sativus* may be beneficial for several CVD-risk related outcomes (e.g., blood pressure, body weight, waist circumference, and fasting blood glucose levels), suggesting that saffron may have protective effects for multiple systemic conditions related to such CVD risk factors (Pourmasoumi et al., 2019).

Saffron and its constituents have favorable results on CVD risk (Ghaffari & Roshanravan, 2019; Pourmasoumi et al., 2019), endothelial function (Rahiman, Akaberi, Sahebkar, Emami, & Tayarani-Najaran, 2018), inflammatory diseases (Poma et al., 2012), oxidative stress (Rahiman et al., 2018), and glycaemic factors (Pourmasoumi et al., 2019), indicating that *Crocus sativus* may have promising potential as adjunct therapy in both systemic conditions and ocular diseases mainly through its anti-inflammatory and anti-oxidative effects (José Bagur et al., 2017; Hosseini et al., 2018; Ghaffari & Roshanravan, 2019). On the other hand, *Crocus sativus* and its constituents may improve ocular function more indirectly by improving the aforementioned systemic CVD-risk factors through anti-inflammatory and anti-oxidant effects (Ghaffari & Roshanravan, 2019; Poma et al., 2012; Rahiman et al., 2018).

Saffron contains a number of bioactive components which are believed to be largely responsible for its health promoting properties including treating various disorders like asthma, atherosclerosis, painful menstrual periods and even depression; its role as an antioxidant with anti-cancerous and memory enhancing properties; its effectiveness at treating mild to moderate depression; and its high efficiency in lowering the levels of cholesterol and triglycerides in the blood of people suffering from cardiovascular diseases. Studies related to *Crocus sativus* quality are expanding mainly due to the antioxidant properties of this spice and their positive influence on human health (Shahi et al., 2016). Antitumor and cancer-preventive properties are mainly attributed to the high carotenoids content (Bagur et al., 2018). Short-term therapy with saffron capsule showed the same efficacy compared with methylphenidate in the treatment of children with Attention-deficit/hyperactivity disorder (ADHD). So it may attest as a good herbal substitute for ADHD treatment, particularly for the 30% of patients who do not respond to or cannot tolerate stimulants like methylphenidate (Baziar et al., 2019). Saffron supplementation appears to have promising potential as an effective and safe adjunct therapy in certain ocular diseases (Piccardi et al., 2012; Marangoni et al., 2013; Lashay et al., 2016; Riazi et al., 2017; Broadhead et al., 2019; Bonyadi, Yazdani, & Saadat, 2014; Sepahi et al., 2018; Waugh et al., 2018). An increasing number of experimental, animal, and human studies have investigated the effects and mechanistic pathways of these compounds in order to assess their potential therapeutic use in ocular diseases (e.g., in age-related macular degeneration, glaucoma, and diabetic maculopathy) (Heitmar et al., 2019). Heitmar et al. (2019) presents the key findings of published clinical studies that examined the effects of *Crocus sativus* and/or its constituents in the context of ocular disease, as well as an overview of the proposed underlying mechanisms mediating these effects.

The stigmas of *Crocus sativus* have been used in folk medicine to alleviate different health problems (Rios et al., 1996). Crocetin is known to possess pharmaceutical properties among the rest of chemical compounds present in saffron extract. In studies on animal models, it has been shown that crocetin has several pharmacological properties, including antioxidant (Tseng, Chu, Huang, Shiow, & Wang, 1995; Kanakis et al., 2007a), antiinflammatory (Kazi & Qian, 2009), antiatherosclerotic (Zheng, Qian, Sheng, & Wen, 2006), insulin resistance improvement (Sheng et al., 2008), attenuation of physical fatigue (Mizuma et al., 2009), and sleep (Kuratsune, Umigai, Takeno, Kajimoto, & Nakano, 2010). Crocetin also alters the growth of cancer cells by inhibiting replication, inducing apoptosis, and enhancing the antioxidative system (Gutheil et al., 2012). Crocetin has also shown some promising health-promoting effects such as, cardiovascular improver (Wang, Suna, Liub, & Fang, 2014), anti-cancer (Bhandari, 2015), antioxidant activity (Ohba et al., 2016) and antidepressant (Lopresti & Drummond, 2014). Crocetin shows neuroprotective effects (Ahmad et al., 2005; Nam et al., 2010). *Crocus sativus* acts as drug modulators and not only controls myocardium injury induced by Doxorubicin (Chahine et al., 2013) but also attenuates harmful effects of chemotherapeutic drug induced cardiac toxicity (Razavi, Hosseinzadeh, Abnous, & Imenshahidi, 2014) nephro- toxicity (Boroushaki & Sadeghnia, 2009) and cardiac toxicity (Goyal et al., 2010). Crocetin inhibits in vitro amyloid-β aggregation (Papandreou et al., 2006) and a clinical pilot study revealed significant improvements in cognition after the treatment of mild-to-moderate Alzheimer's disease patients with crocetin (Akhondzadeh et al., 2010). Crocins had protective effects on neuronal injury (Ochiai et al., 2007) and attenuated the symptoms of obsessive-compulsive disorder, a common psychiatric disorder defined by the presence of obsessive thoughts and repetitive compulsive actions (Georgiadou, Tarantilis, & Pitsikas, 2012). Safranal, the main constituent of the volatile oil fraction, attenuated cerebral ischemia (Hosseinzadeh & Sadeghnia, 2005) and retinal degeneration (Fernández-Sánchez et al., 2012; Hooshmandi et al., 2011). A recent study demonstrates the potential use of azafrines 1 and 2 saponins in human immunotherapeutic vaccines (Castro-Díaz et al., 2012). Some other pharmacological properties of *Crocus sativus* are listed in Table 14.2.

Table 14.2 Pharmacological properties of *Crocus sativus*.

S.No.	Therapeutic property	References
1	**Antioxidant activity**	Serrano-Díaz et al. (2012); Montoro et al. (2012); Assimopoulou et al. (2005); Li, Lee, & Wu (2004); Termentzi & Kokkalou (2008); Ordoudi, Befani, Nenadis, Koliakos, & Tsimidou (2009); Esmaeili et al. (2011); Karimi et al. (2010); Chen et al. (2008); Goli, Mokhtari, & Rahimmalek (2012); Vatankhah, Niknam, & Ebrahimzadeh (2010); Sánchez-Vioque et al. (2012); Keyhani, Ghamsari, Keyhani, & Hadizadeh (2006)
2	**Anticancer activity**	Abdullaev & Frenkel (1992b); Abdullaev & Frenkel (1992a); Abdullaev (1994); Tarantilis, Morjani, Polissiou, & Manfait (1994); Nair, Kurumboor, & Hasegawa (1995); Escribano, Alonso, Coca-Prados, & Fernández (1996); Escribano, Ríos, & F (1999c); Escribano, Piqueras, Medina, & Rubio, (1999b); Escribano et al., (2000a); Abdullaev, (2002, 2004); Malaekeh-Nikouei et al. (2013); Caballero-Ortega et al. (2004); Tavakkol-Afshari, Brook, & Mousavi (2008); Noureini & Wink (2012); Samarghandian, Boskabady, & Davoodi (2010); Aung et al. (2007); Bajbouj, Schulze-Luehrmann, Diermeier, Amin, & Schneider-Stock (2012); Mousavi, Tavakkol-Afshari, Brook, & Jafari- Anarkooli (2009); Samarghandian, Afshari, & Davoodi (2011); Feizzadeh et al. (2008); Bakshi et al. (2010); Mousavi et al. (2011); Rubio-Moraga et al. (2011); Chryssanthi, Dedes, Karamanos, Cordopatis, & Lamari (2011a); Umigai, Tanaka, Tsuruma, Shimazawa, & Hara (2012); Nair, Pannikar, & Panikkar (1991a); Bakshi et al. (2009); Salomi, Nair, & Panikkar (1991); Das, Chakrabarty, & Das (2004); Garcia-Olmo et al. (1999); Das, Das, & Saha, (2010); Bathaie et al. (2013); Rastgoo et al. (2013)
3	**Cardioprotective activity**	Shen, Qian, Chen, & Wang (2004); Shen & Qian (2006); Khori et al. (2012); Joukar et al. (2010); Sachdeva et al. (2012); Mehdizadeh, Parizadeh, Khooei, Mehri, & Hosseinzadeh, (2013); Joukar, Ghasemipour-Afshar, Sheibani, Naghsh, & Bashiri (2013); Goyal et al. (2010); Bharti, Golechha, Kumari, Siddiqui, & Arya (2011) ; Razavi, Hosseinzadeh, Movassaghi, Imenshahidi, & Abnous (2013b)

4	**Hypotensive activity**	Fatehi, Rashidabady, & Fatehi-Hassanabad, (2003); Boskabady, Shafei, Shakiba, & Sefi di (2008); Imenshahidi, Hosseinzadeh, & Javadpour (2010); Razavi, Hosseinzadeh, Abnous, Motamedshariaty, & Imenshahidi (2013a)
5	**Antigenotoxic/genoprotective activity**	Premkumar, Abraham, Santhiya, Gopinath, & Ramesh, (2001), Premkumar, Abraham, Santhiya, & Ramesh (2003a), Premkumar, Abraham, Santhiya, & Ramesh (2003b), Premkumar, Thirunavukkarasu, Abraham, Santhiya, & Ramesh (2006), Premkumar, Kavitha, Santhiya, Ramesh, & Suwanteerangkul (2004); Hosseinzadeh & Sadeghnia (2007a); Hosseinzadeh, Abootorabi, & Sadeghnia (2008a)
6	**Antihyperlipidemic/ antihypercholesterolemic activity**	Xu, Yu, Gong, & Zhang (2005); Sheng, Qian, Zheng, & Xi, (2006); Asdaq & Inamdar (2010); Tang et al. (2006)
7	**Weight loss activity**	Gout, Bourges, & Paineau-Dubreuil (2010)
8	**Amyloid-beta aggregation inhibitory activity and Alzheimer's disease**	Papandreou et al. (2006); Akhondzadeh et al., (2010a), (2010b); Ebrahim-Habibi, Amininasab, Ebrahim-Habibi, Sabbaghian, & Nemat-Gorgani (2010); Geromichalos et al. (2012)
9	**Neuroprotective and cognitive enhancement activities**	Zhang et al. (1994); Abe, Sugiura, Yamaguchi, Shoyama, & Saito (1999); Soeda et al. (2001), Soeda et al. (2003); Ochiai et al. (2004a,b, 2007); Saleem et al. (2006); Zheng et al. (2007); Ghazavi et al. (2009); Mousavi et al. (2010); Shati et al. (2011); Ahmad et al. (2005); Bie et al. (2011); Vakili et al. (2013); Linardaki et al. (2013); Berger et al. (2011); Lechtenberg et al. (2008); Hosseinzadeh & Sadeghnia (2005); Hosseinzadeh et al. (2008b); Hosseinzadeh et al. (2012); Sadeghnia et al. (2013); Abe & Saito (2000); Saito et al. (2001) ; Howes & Perry (2011); Howes & Houghton (2012); Pitsikas & Sakellaridis (2006); Pitsikas et al. (2007); Asadpour & Sadeghnia (2011); Khalili & Hamzeh (2010); Papandreou et al. (2011); Ghadrdoost et al. (2011); Mehri et al. (2012); Hosseinzadeh & Jahanian (2010); Ghoshooni et al., (2011); Naghibi et al. (2012); Georgiadou et al. (2012)

(*Continued*)

Table 14.2 Pharmacological properties of *Crocus sativus*. (*Cont.*)

S.No.	Therapeutic property	References
10	**Antidepressant activity**	Hosseinzadeh et al. (2004); Akhondzadeh et al. (2004); Akhondzadeh et al. (2005); Akhondzadeh et al. (2007); Noorbala et al. (2005); Moshiri et al. (2006); Wang et al. (2010); Dwyer et al. (2011)
11	**Anxiolytic activity**	Sarris et al. (2013); Pitsikas et al. (2008); Hosseinzadeh & Noraei (2009)
12	**Anticonvulsant activity**	Hosseinzadeh & Khosravan (2002); Hosseinzadeh & Talebzadeh (2005); Hosseinzadeh & Sadeghnia (2007b); Sadeghnia et al. (2008)
13	**Anticataractogenic and ocular protective activities**	Xuan et al. (1999); Laabich et al. (2006); Maccarone et al. (2008); Falsini et al. (2010); Yamauchi et al. (2011); Qi et al. (2013); Ohno et al. (2012); Shukurova & Babaev (2010) ; Fernández-Sánchez et al. (2012); Makri et al. (2013); Ishizuka et al. (2013)
14	**Antinociceptive activity**	Hosseinzadeh & Younesi (2002)
15	**Antiinflammatory activity**	Hosseinzadeh & Younesi (2002); Ma et al. (1998); Xu et al. (2009); Hariri et al. (2010,2011); Moallem et al. (2013); Hemshekhar et al., (2012); Boskabady et al. (2012); Byrami & Boskabady (2012); Byrami et al. (2013); Ding et al. (2013)
16	**Antihyperinsulinemic/ antidiabetic activity**	Xi et al. (2007a); Kang et al. (2012); Rajaei et al. (2013)
17	**Antitussive activity**	Hosseinzadeh & Ghenaati (2006)
18	**Antimicrobial activity**	Motamedi et al. (2010); Pintado et al., 2011; Zheng et al. (2011)
19	**Hepatoprotective activity**	Lin & Wang (1986); Amin et al. (2011); El-Beshbishy et al. (2012)
20	**Nephroprotective activity**	El Daly (1998); Nair et al. (1991b); Hosseinzadeh et al. (2005); Ajami et al. (2010); Naghizadeh et al. (2010), Naghizadeh et al. (2011)
21	**Adaptogenic activity**	Hooshmandi et al. (2011); Halataei et al. (2011); Yang et al. (2011)
22	**Relaxant activity**	Fatehi et al. (2003); Boskabady & Aslani (2006); Nemati et al. (2008); Boskabady et al. (2010); Boskabady et al. (2011a)
23	**Spasmodic activity**	Sadraei et al. (2003)
24	**Analgesic activity**	Amin & Hosseinzadeh (2012)
25	**Effect on menstrual distress**	Agha-Hosseini et al. (2008); Fukui et al. (2011)

26	**Aphrodisiac activity**	Hosseinzadeh et al. (2008c); Heidary et al. (2008); Shamsa et al. (2009); Safarinejad et al. (2010); Safarinejad et al. (2011); Modabbernia et al. (2012); Kashani et al. (2013)
27	**Immunomodulatory activity**	Escribano et al. (1999a); Boskabady et al. (2011b); Bani et al. (2011); Kianbakht & Ghazavi (2011); Castro-Díaz et al. (2012)
28	**Sleep enhancement activity**	Masaki et al. (2012)
29	**Wound healing activity**	Khorasani et al. (2008)
30	**Antiplatelet activity**	Liakopoulou-Kyriakides & Skubas (1990); Liakopoulou-Kyriakides et al. (1985); Jessie & Krishnakantha (2005); Yang et al. (2008); Tsantarliotou et al. (2013); Ayatollahi et al. (2013)
31	**Antivenin activity**	Santhosh et al. (2013a); Santhosh et al. (2013b)
32	**Protective activity against extremity ischemia–reperfusion injury**	Hosseinzadeh et al. (2009)
33	**Protein interaction (binding and polymerization) activity**	Oda et al. (2000); Kakehi et al. (2003); Kanakis et al. (2007a); Kanakis et al. (2007b); Bathaie et al. (2007); Hosseinzadeh et al. (2013); Zarei Jaliani et al. (2013)
34	**Reproductive/estrogenic activity**	Tavana et al. (2012); Chang et al. (1964)
35	**Pharmacokinetic, safety, and toxicity studies**	Chang et al. (1964); Abdullaev (2002); Liu & Qian (2002); Tang et al. (2004); Xi et al. (2007a); Xi et al. (2007b); Ramadan et al. (2012) ; Umigai et al. (2011); Chryssanthi et al. (2011b); Modaghegh et al. (2008); Poma et al. (2012)
36	**Allergy problems**	Feo et al. (1997); Gómez-Gómez et al. (2010a)

Saffron omics

The research paradigm of medicinal plant genome and evolution is evolving, and the use of omics techniques is reshaping the landscape of this dynamic field. Genomics, transcriptomics, proteomics, metabolomics, and other omics platforms generate formidably large data, which cannot be used efficiently in probing plant genome and evolution without the aid of advancing bioinformatics (Hao & Xiao, 2015). High-throughput techniques mainly genome and transcriptome sequencing of medicinal plants could unreveal the biosynthetic pathways of medicinal secondary metabolites (Boutanaev, Moses, & Zi, 2015; Zhang, Gao, Khan, Luo, & Chen, 2014; Hao et al., 2011, Hao, Ma, & Mu, 2012) and their regulatory molecular mechanisms and provides a road map for the molecular breeding of high-yielding cultivars and molecular farming of medicinal transgenic varieties. The genome sequence of any organism provides information of its origin, evolution, inheritable traits, development and physiology and epigenomic mechinery which are basis for deciphering genome diversity and chemodiversity. DNA barcoding techniques (Hao et al., 2012) help in species identification. Karyotypes are determined at metaphase chromosomes while flow cytometry and pulsed-field gel electrophoresis (Hao et al., 2011, Hao et al., 2012) are employed to decipher the ploidy level and genome size. The phylogenomics, pharmacophylogenomics and phylotranscriptomics terms are overlapping with pharmaphylogeny (pharmacophylogeny/ pharmaco phylogenetics) (Hao, Xiao, Liu, Peng, & He, 2014). Phylogenomics (the field of pharmaphylogeny at the omic level) is the marriage of evolutionary biology with genomics, in which genome data is employed for evolutionary reconstructions. The new conceptual framework of phylogenomics finds use in the drug discovery, development and the genomic analysis of the evolutionary history of drug targets (Searls, 2003). The pharmacophylogenomics emphasizes the complete analysis of the evolutionary history of the predominant medicinal plants and the analogy and divergence between molecular phylogeny and chemotaxonomy (Day, Berger, & Hill, 2014, Hao, Gu, Xiao, & Peng, 2013, He, Peng, Dan, Peng, & Xiao, 2014), the orthology and paralogy relationships (Yang & Smith, 2014), the degree and landscape of evolutionary transformations, metabolic pathways and regulatory networks. The phylotranscriptomic approach provides information on the evolution of plant traits and chemodiversity. The link between the plant molecular phylogeny and therapeutic utility has been put forward by Leonti et al. (2013) and Grace, Buerki, & Symonds (2015). The bulky and juicy leaves which symbolize medicinal aloes (Aloeaceae, Liliales) evolved ~10 million years ago and are strongly associated with the molecular phylogeny and correlated to the probability of this species being used for therapeutics (Grace et al., 2015).

Crocus sativus a sterile triploid propagates vegetatively by means of corms (Brighton, 1977; Mathew, 1977). Genetic erosion and sterile non-recombination nature of *C. sativus* account for its limited genetic diversity which is a bottleneck for the genetic improvement of this highly valued crop (Fernández et al., 2011). In *C. sativus*, many researches have been done to obtain seeds (Grilli-Caiola, 1999) and in vitro-regenerated plants (Souret & Weathers, 2000). The omics biology can prove as a bench mark for the genetic modification of this plant. 'Saffron omics', under the initiative of the European Cooperation in Science and Technology (COST), aspires to build up collaborative research exploration on developing 'omic' approaches for the structural organization of saffron genome, Saffron DNA fingerprinting to guard its quality and for improvement of the saffron genetics, saffron chemical fingerprinting, saffron proteomics, saffron transcriptomics and saffron metabolomics (http://www.saffronomics.org/).

Saffron tissue culture

Biotechnological approaches like micropropagation of saffron using direct/indirect shoot induction or plantlet regeneration via somatic embryogenesis and subsequently microcorm production produce huge and disease-free propagating material in minimum time. Saffron and or its phytochemicals like crocin, picrocrocin, crocetin and safranal can be generated by biotechnological approaches. Crocin production in cell cultures is focus of research studies mainly because of its anticancer properties. Biotechnology can cater to the worldwide demand and preservation of this spring-flowering "Golden Condiment" (Ahmad et al., 2014). Saffron is propagated vegetatively by corms wherein only 4–5 corms/mother corm are produced in one growing season. Low multiplication rates and fungal diseases of corms are the bottlenecks for sufficient quality planting material (Kiran, Madhu, Markandey, & Paramvir, 2011). Tissue culture is employed for extensive production of disease-free *Crocus* plants (Ding, Bai, Wu, & Fang, 1981; Chrungoo & Caiola 1987; Ilahi, Jabeen, & Firdous, 1987; Plessner, Ziv, & Negbi, 1990; Fakhrai & Evans 1990; Chen, Wang, Zhao, Yuan, & Wang, 2003; Majourhay, Fernandez, Martınez-Gomez, & Piqueras, 2007; Sheibani, Nemati, Davarinejad, Azghandi, & Habashi, 2007). Ding et al. (1981) pioneered the tissue culture of *Crocus* through regenerated callus and intact plantlets from corm explants in culture media with indole-3-acetic acid and 2, 4 Dichlorophenoxyacetic acids. Ilahi et al. (1987) explained the morphogenesis of saffron tissue culture. Floral organs of 4 *Crocus* species were studied for their proficiency to produce callus by Choob,Vlassova, & B (1994). Shoot development on corm explants was advanced by cytokinins and 2, 4-D while corm formation and growth by ethylene exposure (Plessner et al., 1990). In ovary wall explants, stigma and style-type structures regenerated from the explants (Choob et al., 1994). The continuous darkness initiated shoot primordial formation which elongated on light exposure and regenerated Saffron plantlets with corms (Bhagyalakshmi, 1999). Loskutov et al. (1999) optimized in vitro conditions for stigma-like structure formation in *C. sativus* from half-ovary explants whereas successful stigma and ovary formation took place on culture media containing BA and kinetin (Sano & Himeno, 1987). Style and perianth explants produced stigma-like structures which multiplyed forming upto 100 structures/explant (Ebrahimzadeh Karamian, 2004). Zeng, Yan, Tang, & Chen (2003) documented the amplified crocin production and induction frequency of stigma-like structures from Saffron floral organs by precursor feeding. Chen, Zhao, Wang, Yuan, & Wang (2004) investigated the promoted growth of *C. sativus* cells and the increased crocin production by La3+ and Ce3+ individually and/or their mixture in callus. Somatic embryos and plantlets regenerated from leaf explants on culture of 10 lM BA and 0.5 lM 2, 4-D (Raja, Zaffer, & Wani, 2007) which were utilized for microcorm induction promoted by a half-strength MS medium in addition with 9% sucrose (Raja et al., 2007). The higher concentration of PAC (5 mg/L), BA (0.25 mg/L) and 9% sucrose outcome in relatively large microcorms (Zaffar, Wani, T, & Zeerk, 2004). Primary and secondary hardening guaranteed 100% corm viability (Yasmin, Nehvi, & Wani, 2013).

Saffron genomics

The genome of saffron is not publicly available, although http://www.crocusgenome.org/ provides an online interface to query individual sequences. Inspite of lot of work on tissue culture and hybridization (Rubio-Moraga et al., 2014; Mir et al., 2015), vegetative propagation through corms leads to no or little genetic variation in the form of somatic mutations, segregation distortions, transversions, etc. in a population and heritable changes owing to its sterility (Agayev, Fernandez, & Zarifi, 2009). Research is available on identifying variations in phenotypic and phytochemical traits because of epigenetic variations which urge the immediate demand for developing molecular markers for identification of these variations at molecular level for improvement of saffron (Mir et al., 2015). Classical cytogenetic analyses in the genus Crocus involved chromosome counting and karyotyping of *C. sativus* and *C. cartwrightianus*, as well as chromosome length measurements and centromere positioning using chromatin staining methods such as C banding (Brighton, 1977; Agayev and Zarifi, 2010). So far, molecular cytogenetic studies of Crocus are limited to the diversity of repeat families in ornamental hybrids (Ørgaard, Jacobsen, & Heslop-Harrison, 1995; Frello & Heslop-Harrison, 2000; Frello, Ørgaard, Jacobsen, & Heslop-Harrison, 2004). There is currently no whole genome sequence available for any Crocus species. Only some studies on genome segments have been done e.g., RAPD (Grilli-Caiola et al., 2004), IRAP markers (Alavi-Kia, Mohammadi, Aharizad, & Moghaddam, 2008; Alsayied et al., 2015), Nuclear gene diversity (Grilli Caiola & Canini 2010; Tsaftaris et al., 2011; Harpke et al., 2013) AFLP and SSR (Larsen et al., 2015), etc., for understanding the origin of the triploid *C. sativus*. Despite the existence of different commercial *Crocus* ecotypes and performance of different studies on evaluation of genetic analysis in *Crocus* species based on various molecular markers such as IRAP (Alavi-Kia et al., 2008); RAPD (Beiki et al., 2010); RAPD and SRAP (Keify & Beiki, 2012); RAPD and ISSR (Rubio-Moraga et al., 2009); ISSR (Rubio-Moraga et al., 2010); AFLP (Siracusa et al., 2012; Erol et al., 2014) and SSR (Rubio-Moraga et al., 2009; Nemati et al., 2012), the actual genetic diversity present in *C. sativus* is still an open question. In this study, 27 microsatellite markers were tested to measure level of polymorphism and investigated the genetic relationship and structure among Iranian *Crocus* ecotypes. Nemati et al. (2012) detected a good level of polymorphism by 12 microsatellite markers within 50 Iranian individuals of *Crocus sativus*. Like other studies, we observed a reasonable polymorphism among Iranian *C. sativus* germplasms (Beiki et al., 2010; Keifi & Beiki 2012) which may be due to suitable climatic conditions for growth and development of this valuable crop in Iran. The use of a large number of polymorphic markers is vital for accurate assessment of genetic variation among different *Crocus* species according to their geographical origin and ploidy level, formation of core collection, and construction genetic map. The diploid *Crocus cartwrightianus*, a species occurring in southern mainland Greece and on the Aegean Islands, is the sole progenitor of the saffron crocus. Phylogenetic analyses of nuclear loci and genome-wide DNA polymorphisms together with chloroplast genome comparisons indicate that saffron is genetically most similar to the Attic *C. cartwrightianus* populations. Nemati (2018), postulated that *C.*

sativus is an autopolyploid that originated in southern Attica by combining two different genotypes of *C. cartwrightianus*. For resolving the phylogeny of the Crocus series, Nemati (2018) firstly analyzed two chloroplast (*trn*L-F, *mat*K-*trn*K) and three nuclear (TOPO6, nrDNA ETS and ITS) markers in 53 individuals belonging to all taxa of series *Crocus*.

Busconi et al. (2015) analyzed 112 accessions using Factorial Correspondence Analysis (individual level) of Amplified Fragment Length Polymorphism (AFLP) for investigating the variations at the genetic (33.57 % polymorphic peaks) and methyl-sensitive AFLP for investigating the variations at the epigenetic levels (28 typses of effective epigenotypes). Research investigations are underway to prevent genetic erosion and stimulate genetic variability in Saffron to develop its superior varieties throughout the globe. The variability of morpho-physiological traits in *Crocus sativus* populations coming from Iran and Italy and to evaluate the genetic differences among different origins (Iran, Italy and Spain) with AFLP markers was done. The data show that saffron is an allotriploid species, with the IRAP analysis indicating that the most likely ancestors are *C. cartwrightianus* and *C. pallasii* subsp. *Pallasii* (or close relatives) (Torricelli, Yousefi Javan, Albertini, Venanzoni, & Hosseinzadeh, 2019). *Crocus sativus* survey genome sequencing of repetitive DNA families to develop six-color fluorescent CroSat satellites probes for physical mapping of satellite DNA in combination with rRNA genes (18S-5.8S-25S rRNA and 5S rRNA) for comparative chromosome high resolution FISH karyotype (92 chromosomal positions) revealed 92 cytogenetic landmarks, distributed over the 24 saffron chromosomes and including multiple tagging of chromosomes with up to seven different probes in *C. sativus* and allied species thus concluding the autotriploid origin of saffron solely derived from heterogeneous cytotypes of *C. cartwrightianus* (Schmidt et al., 2019).

Saffron transcriptomics

The transcriptome analysis of stigma of *Crocus sativus* is a prerequisite to present a comprehensive view of the molecular basis of carotenoid/apocarotenoid biosynthesis and corresponding regulatory networks, the gynoecium biology, and the genomic organization (D'Agostino, Pizzichini, Chiusano, & Giuliano, 2007; Baba et al., 2015). Efforts by independent saffron research groups have generated de novo transcriptome assemblies from different tissues of *Crocus sativus*, including leaves, stamens, corm, tepals, and stigmas (105,269 transcripts in leaf, corm, tepal, stamen and stigma (Jain, Srivastava, Verma, Ghangal, & Garg, 2016) and 64,438 transcripts in flowers

(Baba et al., 2015); while 248,099 transcripts in tepals of *Crocus ancyrensis* at two developmental stages (Ahrazem et al., 2018). These transcriptome analyses on Crocus species (Ahrazem et al., 2018; Baba et al., 2015; Jain et al., 2016) have unveiled thousands of transcription factor-coding genes, providing a foundation for investigating their involvement in apocarotenoid metabolism. Transcriptomic and genomic studies on saffron had received much lesser attention when compared to its potential applications in therapeutics and phytochemistry. The reason may be the low or nearly null genetic variability attributed to sterile triploid and vegetative propagation (Piqueras et al., 1999). Previously data was only available from mature stigmas, thus information on the critical stages of apocarotenoid biosynthesis was lacking (Rubio-Moraga et al., 2009).

To understand the molecular basis of apocarotenoid biosynthesis, Jain et al. (2016) performed transcriptome sequencing from five different tissues/organs of *C. sativus*. Gene annotation identified 54% of *Crocus sativus* transcripts could be functionally annotated involved in various biological processes and molecular functions. Transcriptome analysis of *Crocus sativus* revealed the presence of 16,721 SSRs and 3819 transcription factor encoding transcripts. Differential expression of transcripts encoding for transcription factors (MYB, MYB related, WRKY, C2C2-YABBY and bHLH) involved in secondary metabolism was revealed. Results will pave the way for understanding the molecular basis of apocarotenoid biosynthesis and other aspects of stigma development in *C. sativus*. Largest numbers of differentially expressed transcripts were detected in stamen as compared to other tissues. In stamen, at least 4828 transcripts exhibited up-regulation as compared to stigma followed by 4781 as compared to leaf. However, nearly equal numbers of transcripts were up-regulated in stigma as compared to stamen (3829) and leaf (3818). However, least number of transcripts showed differential expression in tepal. The expression profile matrix for all the *C. sativus* transcripts is available at the Saffron Transcriptome web page. Among 2910 up-regulated transcripts in stigma, 92 transcripts were found to encode for TFs representing 37 different families. These TF encoding transcripts revealed substantial differences in their expression patterns among the five tissues with higher expression in stigma. TFs reported to be involved in secondary metabolite biosynthesis, such as Aux/IAA, MYB, MADS, C2C2-YABBY, WRKY, bHLH and SNF2 families, were also represented among the up-regulated transcripts in stigma. Jain et al. (2016) identified the transcripts exhibiting tissue-specific expression. A total of 1075 transcripts exhibited tissue-specific expression, 124 in corm, 161 in tepal, 304 in leaf, 144 in stigma and 342 in stamen. All the genes have been identified encoding for all the enzymes catalyzing different intermediate reactions involved in apocarotenoid biosynthetic pathway from the *C. sativus* transcriptome.

A systematic comparative analysis approach for transcriptomes and crocin data in *C. sativus*, *C. cartwrightianus* and *C. ancyrensis* is presented by Ahrazem et al. (2019) to identify putative transcription factors that may affect apocarotenoid accumulation during stigma development in saffron. Very recently (Ahrazem et al., 2019) sequenced and assembled new transcriptome data from 3 Crocus species in two key stigma developmental stages for apocarotenoid biosynthesis and accumulation. The six transcriptomes obtained perfectly complement previous transcriptomes of saffron tissues (Jain et al., 2016; Baba et al., 2015), providing new data of the earliest developmental stages in stigma development, when crocetin, crocin and picrocrocin biosynthesis begin (Rubio-Moraga et al., 2009). A positive correlation among DXS-CLA1, PDS, ZDS, Z-ISO, CrtISO, LYC-B, BCH-2, CCD2, and UGT74AD2 expression levels and apocarotenoid levels was observed in the three species. Therefore, the formation from phytoene to crocetin and from crocetin to crocins appears to be principally under transcriptional regulation (Ahrazem, Rubio-Moraga, Nebauer, Molina, & Gomez-Gomez, 2015). In emphasizing transcription factors, the data from six transcriptomes was compared in order to identify candidate regulators impacting apocarotenoid accumulation. Information on the spatiotemporal pattern of gene expression/metabolite accumulation can facilitate the understanding of gene function and the generation of a hypothesis for apocarotenoid biosynthesis control at the transcriptional level. 11 TFs belonging to the ARF, bHLH, C2H2, HB, CBF/DREB1, ALFIN, and NF-YC families express in the stigma and correlate with apocarotenoid levels in the three species. Seven of them showed gene expression patterns that suggested a direct participation in the regulatory network influencing apocarotenoid profiles and content, and are involved in a complex network influenced by environmental factors and developmental stages known to regulate carotenoid accumulation in plants. Among these TFs, we found TFs involved in flower and fruit development, retrograde signaling, epigenetic modifications, and light and cold responses.

Nemati (2018), studied gene expression differences in the stigmas of saffron and its closest relative *C. cartwrightianus* by RNA sequencing to know the genetic changes that characterize saffron comparatively its closest relatives, *C. cartwrightianus* from southwest and southeast of Attica. By the identification of the parental origin of saffron in the previous studies, Nemati (2018), have for the first time the chance to compare the transcriptome and transcription activity in the stigmas of wild progenitors and the domesticated saffron crocus through RNA-seq. It will further lead to investigation of evolutionary and developmental mechanisms contributing to the superiority of *C. sativus* over *C. cartwrightianus* in terms of the apocarotenoids production. In this study, Nemati (2018) compared for the first time the transcriptomes of cultivated and wild saffron to understand the molecular basis of apocarotenoid biosynthesis. Nemati (2018) studied the expression profiles of all the transcripts in stigmas of *C. sativus* and the Attic *C. cartwrightianus* and several genes showing differential gene expression were identified. Nemati (2018) further analysed the expression patterns of the putative genes involved in apocarotenoid biosynthesis. These data provide a resource to understand presence and the expression pattern of these important genes in the stigma of wild saffron and are a first step to identify traits and genotypes necessary for future improvement of saffron. Jain et al. (2016) reported 78 genes involved in apocarotenoid biosynthetic pathway in *C. sativus.* Nemati (2018) found seven genes out of 78 (~10%) genes related to apocarotenoid biosynthesis that were expressed differentially between samples. These genes are orthologues of zeaxanthin epoxidase (CsTc006236), UDP-glucosyltransferase (CsTc020060), carotenoid isomerase (CsTc091265), lycopene beta-cyclase (CsTc018497), phytoene synthase (CsTc009491), carotene beta-hydroxylase (CsTc000418) and nine-cis-epoxy carotenoid dioxygenase (CsTc035409). Higher expression diversity in the individuals of *C. cartwrightianus* than in saffron as expected, due to asexual propagation of saffron and the allelic diversity caused by the outcrossing nature of *C. cartwrightianus*. Among these genes, there is only one gene (CsTc020060) down-regulated in all *Crocus sativus* individuals while up-regulated in all *C. cartwrightianus* individuals. This gene was reported to be related to UDP-glucosyltransferase, which is an important enzyme involved in the last step of the apocarotenoid pathway where crocetin is converted to crocin and the pigments in the saffron stigmas are accumulating. This could cause an important difference in metabolite accumulation between wild and cultivated *Crocus staivus*. The hypothesis is that triploidy and the sterility it caused was a way to safeguard a very favourable allele composition (regarding aroma and color of the styles it causes) from being broken up by recombination after it occurred in *C. cartwrightianus*. Selection for genotypes with different color and aroma intensity then could be a way to improve diversity within saffron, either with or without a final step towards triploidisation and sterility to fix new superior gene combinations in certain lineages.

Similar transcriptomic studies on *Crocus sativus* have led to the dissection biosynthetic pathways of carotenoids (Castillo, Fernandez, & Gomez-Gomez, 2005) and flavonoids for characterization of glucosyltransferase (Rubio-Moraga et al., 2009). Additionally, deep transcriptomics analysis has identified carotenoid cleavage dioxygenase (CCD2), a novel dioxygenase which catalyzes the first step of crocin biosynthesis originating from carotenoid zeaxanthin (Frusciante et al., 2014). Although many of the genes involved in the carotenoid/apocarotenoid biosynthesis have been identified and characterized in *Crocus sativus* (Frusciante et al., 2014; Baba et al., 2015; Jain et al., 2016), very little is known about the change in expression of genes in a stage specific manner particularly with regard to apical and axillary bud development. Upregulation of apocarotenoid biosynthesis genes at stages S4 to S8 during the development of stigma indicates tissue specific expression of apocarotenoid genes in *Crocus sativus* (Mir et al., 2012). The results are in conformity with the earlier findings (Gomez-Gomez et al., 2017), where the transcript UGT85U1 increased from yellow stage (floral initiation period) with the highest peak of expression at red stage and anthesis. CsNCED, a regulatory gene encoding the enzyme involved in ABA biosynthesis in *Crocus sativus* showed low expression in all the developmental stages analyzed.

Functional genomics of saffron

The primary objective of functional genomics in *Crocus sativus* is to narrow down the list of candidate genes implicated in the biological processes for production of flavoring compounds and stigma pigments, so that their expression can be enhanced using a transgenic approach and hence improve quality of the saffron stigma. Bioinformatics plays a huge technical function in sequence-level structural characterization of saffron genomic DNA. An important database for saffron has been designed to manage and explore the expressed sequence tags (ESTs) from saffron stigmas (Agostino, Pizzichini, Chiusano, & Giuliano, 2007). Database is the first ever reference collection for the Iridaceae genomics, molecular biology of stigma biogenesis and for the metabolic pathways underlying saffron secondary metabolism (Agostino et al., 2007). This publicly available 5′-EST library of saffron stigma (*Crocus sativus* L., Iridaceae) containing 6202 ESTs (http://www.ncbi.nlm.nih.gov/) has been generated by D'Agostino (D'Agostino et al., 2007). A total of 6768 *Crocus sativus* ESTs (http://www.ncbi.nlm.nih.gov/) are available, since the first set of 6202 high quality ESTs from cDNA library of a saffron stigma were produced by Agostino et al. (2007)

(available at http://www.saffrongenes.org). Agostino et al. (2007) produced 6603 high quality ESTs from a *Crocus sativus* stigma cDNA library and grouped these into 1893 clusters, each cluster corresponding to a different expressed gene. Several contigs have been characterized in saffron genome and based on the presence of tentative consensus sequences categorized into groups of putative function. Cl000944:1 encodes non-heme-β-carotene-hydroxylase and is highly expressed in saffron stigmas (Castillo et al., 2005); Cl000627:1 encodes a putative glucosyltransferase which is very similar to UGTCs2, and glycosylates crocetin in vitro (Rubio-Moraga, Nohales, Perez, & Gomez-Gomez, 2004); Cl001532:1 and Cl001032:1 encoding putative isoprenoid GTases, one of which could represent the still missing enzyme responsible for the glycosylation of picrocrocin; and Cl000348:1 encodes a Myb-like protein with high similarity to LhMyb (*Lilium*), Myb8 (*Gerbera*) (Elomaa et al., 2003) and Myb305 (*Antirrhinium*) (Jackson, Culianez-Macia, Prescott, Roberts, & Martin, 1991) which acts as a putative transcription factor.

A large number of Cytochrome P450 sequences are expressed in saffron stigmas, some at very high levels (Agostino et al., 2007). Mining of an EST database from *Crocus sativus* stigmas has provided further information on potential saffron biosynthetic genes but no candidate for a novel CCD with $7-8/7'-8'$ cleavage specificity (D'Agostino et al., 2007). Analysis of *Crocus sativus* EST collections obtained from the stigma at different developmental stages revealed that CsCCD2 ESTs were more highly represented in libraries obtained from early stages (Frusciante et al., 2014). Recently, homologues of CsCCD2 have been identified in spring Crocus species, which accumulate crocins in stigmas and tepals (Ahrazem et al., 2015). The Expression Sequence Tags identified upto now correspond to floral development (4 MIKC type-II MADS-box cDNAs) (Tsaftaris et al., 2011), markers for detection of adulteration in *Crocus sativus* under trade (Bar-MCA analysis) (Jiang et al., 2014), AS-PCR and SCAR (Shen, Luo, Ding, & Mao, 2007; Torelli, Marieschi, & Bruni, 2014), environmental and pathogenic stresses (Husaini, 2014) and developmental pathways (Álvarez-Ortí et al., 2004). Expression pattern of CsLYC, CsZCD, CsBCH and CSgt-2 was studied in different *Crocus sativus* flower parts and highest expression was found in stigma followed by style and petal (Mir et al., 2012a). The Reverse transcriptase-PCR analysis revealed that CsZCD gene expression followed different patterns during stigma development. CsZCD gene expression reached at its peak in fully developed scarlet stage of stigma. Real time PCR analysis demonstrated a sharp elevation in gene expression from yellow - orange and orange - scarlet stages of stigma development. Increased CsZCD gene expression with the stigma development suggests its regulatory role for stigma development in *Crocus sativus* (Mir et al., 2012b).

Saffron miRNomics

The plant miRNome is ubiquitous, small in size and mighty in nature (Pandita D, 2018) and responsive to abiotic stress conditions (Pandita D, 2019). The most formidable challenge that threatens the existence of *Crocus sativus* industry is biotic and abiotic factors and the climate change. Till date microRNA sequence is not available in *Crocus sativus* but the above-mentioned factors demand the sequencing of biotic and abiotic stress responsive miRNA of saffron for regulation of its genome. Zinati, Shamloo-Dashtpagerdi, & Behpouri (2016) used an EST library from mature saffron stigmas to discover the miRNAs and their targets expressed in *Crocus sativus* stigma and their relation with the genes of carotenoid/apocarotenoid biosynthetic pathways. Then Regulatory functions of these miRNAs were predicted by searching for potential target genes. Finally, network modeling was used to investigate the roles of putative miRNAs in carotenoid/apocarotenoid biosynthesis in Crocus stigma. Two putative microRNAs, miR414 and miR837-5p and their corresponding stemlooped precursors were identified in *Crocus sativus* for the first time. Co-expression network analysis was done which suggested that miR414 and miR837-5p play important roles in *Crocus sativus* metabolic pathways and identified candidate genes with six transcription factors and one protein kinase probably involved in carotenoid/apocarotenoid biosynthetic pathway. The predicted targets of putative miR414 are gb|EX143431.1|EX143431 (Beta-carbonic anhydrase 5), three-transposable element genes AT2G13700.1, AT4G06613.1, AT3G29783.1 and AT4G38480.1 (F20M13_40) (Transducin/WD40 repeat-like superfamily protein). The predicted targets of putative miR837-5p are AT2G15670.1 SEC14 cytosolic factor family protein / phosphoglyceride transfer family protein, AT5G04670.1 (T1E3_30) Enhancer of polycomb-like protein, AT5G48790.1 Domain of unknown function (DUF1995), AT1G52565.1 unknown protein and AT3G42770.1 F-box/RNI-like/FBDlike domains containing protein.The identified miRNAs and their target and co-expressed genes including transcription factors and protein kinase in *Crocus sativus* may help unraveling regulatory networks underlying the carotenoid/apocarotenoid biosynthesis and designing metabolic engineering for enhanced secondary metabolites (Zinati et al., 2016). Guleria, Goswami, & Yadav (2012) predicted three miRNAs of *Crocus staivus*, csa-miR1, csa-miR2 and csa-miR3 by using in silico methods of EST analysis. In addition, predicted targets for respective miRNAs were reported to play roles in regulation of plant growth, disease resistance, senescence, stress responses, mRNA export, protein synthesis and post-translational modifications (Guleria et al., 2012).

Saffron proteomics

To understand the biological role of proteins, their structure and function must be known (Pieper et al., 2006). Proteomics holds promise in categorization of both known and unknown proteins but till now only 312 protein sequence entries are reported in GenBank (http://www.ncbi.nlm.nih.gov/protein/term=Saffron). Therefore, three-dimensional structure can be predicted from this Protein information. Proteomic analysis identified differentially accumulated proteins in somatic embryos of *Crocus sativus* (Sharifi, Ebrahimzadeh, Ghareyazie, Gharechahi, & Vatankhah, 2012). Additionally, lack of validated protein structure information for majority of plants is a major impediment to functional annotation, evolutionary analysis and construction of protein protein interaction networks (Pentony et al., 2012). Inspite of a plethora of tools available for the prediction and visualization of secondary and tertiary structures of proteins, detailed analyses is limited to few plant gene families. UniProt hosts only 98 protein entries for saffron, out of which only 5 have been reviewed and only three crystal structures are available in protein data bank (PDB) (http://www.rcsb.org/pdb/explore/explore.dostructureId=3U8E). Thus, there is an enormous scope for both *in silico* (e.g., generation of 3D-protein models for use in drug screening) and in vitro studies for inferring biochemical function for these *Crocus sativus* proteins.

Saffron metabolomics

Metabolome is a unique compilation of cellular working parts associated with the expression of the sequenced genomes in all living beings including bacteria, plant, animal, etc. Recently, metabolomic analysis has proved to be an incipient tool for functional gene annotation and characterization, particularly for those genes involved in regulatory pathways. Metabolomic studies assist in identifying enzyme substrates and products with no need for going through heterologous expression systems (Beale & Sussman, 2011). Saffron metabolomics has offerd an unbiased, comprehensive qualitative and quantitative synopsis of its metabolites for example crocetin, crocin, picrocrocin, safranal, etc., elucidating their therapeutic and esthetic properties (Ordoudi et al., 2015).

Through chromatography alongwith spectroscopy (UV, IR, NMR) and mass spectrometry (MS) techniques, over 160-volatile compounds have been identified (Assimiadis, Tarantilis, & Polissiou, 1998; Calsteren et al., 1997). Metabolite fingerprinting obtained using ^{1}H NMR spectra and chemo-metrics was used for the authentication of Iranian and Italian *Crocus sativus* (Cagliani, Culeddu, Chessa, & Consonni, 2015; Yilmaz, Nyberg, Molgaard, Asili, & Jaroszewsk, 2010) and adulteration of saffron with certain plant adulterants (Petrakis, Cagliani, Polissiou, & Consonni, 2015). The structural difference in crocetin esters and picrocrocin and differentiation of sugars bound to them by ^{1}H NMR method were well recognized (Ordoudi et al., 2015; Ordoudi & Tsimidou 2004). ^{1}H NMR is an effective tool to manage *Crocus sativus* quality deterioration and offers advantages for characterization of secondary metabolites. Concurrent identification and quantification of metabolites is essential to know the metabolome dynamics in analyzing fluxes and pathways linked with *Crocus staivus*. But, the main challenge remains in identifying discrepancy in biochemical pathways and metabolic networks correlating with the physiology and developmental phenotype of a cell and tissue (Ordoudi et al., 2015; Ordoudi & Tsimidou, 2004). NMR has been used to isolate *Crocus sativus* constituents or saffron extracts for identification purposes (Assimiadis et al., 1998; Straubinger, Jezussek, Waibel, & Winterhalter, 1997). The potential of NMR-based metabolomics in controlling the saffron quality deterioration was unexplored. Several authentic *Crocus sativus* samples ($n = 98$) of known storage history investigated with ^{1}H NMR and chemometrics answered the query of "when" and "why" a saffron sample is no longer "fresh" wherein the insights to the structural changes in crocetin esters and picrocrocin and markers of the quality deterioration i.e.; bound or free forms of glucose and gentiobiose fatty acids spectra were obtained and found that *Crocus sativus* preserve its valuable characteristics for 1-4 years (Ordoudi et al., 2015). Bagri et al. (2017) provided new insights into *Crocus sativus* corm composition and metabolome at various stages of corm development. They identified metabolites of various metabolic pathways for example, tricarboxylic acid cycle (TCA), glycolysis, biosynthesis of amino acids, fatty acids, organic acids and sugars. The majority of known metabolites were detected first at stage3 (62), then S2 (58), S5 (39), S4 (38) and S1 (37). At sprouting process glucose, fructose, and maltose showed strong positive connection with palmitic acid, turanose, oxalic acid, tetronic acid, ethanolamine, linoleic acid, and negative correlation with sitosterol, mannoside, and octadeconoic acid. Their findings recommend that during bud growth and development, carbohydrate metabolism raised only when free amino acid growth enhanced while fatty-acid biosynthesis significantly relied on the carbohydrate metabolism intermediates. Hence, a progressive enhancement in total sugar levels, mainly glucose and fructose during transition from stage 2 - 3 plus a decline in the content of sucrose advocates that reducing monosaccharides because of sucrose breakdown finally reached its maximum relative abundance to begin the sprouting and bud growth in *Crocus staivus*.

Saffron bioinformatics

Few *Crocus sativus* research investigations have applied target deconvolution, reverse screening, protein modelling and docking criteria for the retrospective detection of molecular targets and functional components (Nithya & Sakthisekaran, 2015; Bhattacharjee, Vijayasarathy, Karunakar, & Chatterjee, 2012). Computational biology approaches have been implicated to the nontherapeutic aspects of saffron, for example, building complete metabolic pathways of the bioactive compounds, spatial-temporal expression of genes involved in clonal propagation and quantification of factors. Zeraatkar, Khalili, & Foorginejad (2015), for the first time generated a three-dimensional geometrical model of *Crocus sativus* flower by reverse engineering and laser scanning technology. The mechanical behavior of the flower might play an imperative role in the blueprint of post-harvesting machinery and process. The prediction of the functions and metabolic pathway was connected to the construction of protein interaction networks (PIN) (Guan & Kiss-Toth, 2008; Wetie et al., 2014). Bioinformatic molecular dynamics and docking approaches have been exploited to investigate interactions between secondary metabolites of saffron and transport proteins such as β-lactoglobulin (Sahihi, 2015). Reports on *Crocus sativus* have often highlighted the necessity for refining bioinformatics tools for transcriptomic and genomic data analysis (Fernández, 2004; Husaini et al., 2009; Gómez-Gómez et al., 2010b), but little has been done in this path. Panchangam et al. (2016) investigated a case study of 35 *Crocus sativus* protein sequences selected as a query to search for orthologs (*Oryza sativa* as reference). Sequence similarity searches were done on local FASTA (http://fasta.bioch.virginia.edu) and using BLASTp (http://blast.ncbi.nlm.nih.gov/Blast.cgi) tool against non-redundant protein sequences of *Oryza sativa*. Query proteins 10 out of the 35 selected have orthologs in *Oryza*. The Classification scoring approach discovered that the 15 crocetin-related proteins of *Crocus sativus* have functional protein associations. These were visualized by a protein interaction network where in, interologs of three genes, namely, HMGR (putative 3-hydroxy-3-methylglutaryl-CoA reductase), lycopene cyclase and phytoene synthase were known to be co-expressed. These candidates that are derived from the methods employed in this analysis are concurrent with earlier reports which focused on transcriptome and metabolome experiments. It would be interesting to exploit pull-down assays and computational biology tools which could enhance our knowledge of the carotenoid biosynthetic pathway in *Crocus sativus* and establish other key protein interacting partners.

Although bioinformatics tools have been applied for the prediction and regulation of signaling pathways, a stringent validation using *in vitro* experiments should not be given a miss. An immune-perspective model was premeditatedd with TGFβ (Kahlem & Newfeld, 2009) where booming applications of both fine-scale and network-scale informatics approaches for understanding signaling pathways were reviewed. Similarly, T and B-cell epitopes of Iranian *Crocus sativus* profiling were predicted using bioinformatics tools (Saffari, Mohabatkar, & Mohsenzadeh, 2008). The insilico tools and DNA microarray technology can be functional in locating sources of resistance and agronomically interesting candidate genes for transfer to *Crocus sativus* by appropriate biotechnological tools. Since the C-class MADS-box gene function is vital for both saffron reproductive organ (stamen and carpel) formation, Tsaftaris, Pasentsis, & Polidoros (2005) recently characterized the expression of MADS-box genes in *Crocus sativus* flowers using several molecular biology techniques, bioinformatics tools and database resources. This kind of research studies assist to understand and exploit the molecular mechanisms that control flower development in *Crocus sativus* and in realization of the objective of producing flowers with carpels in place of stamens. This knowledge can find applications in molecular medicine as well, for example, T and B-cell epitopes of *Crocus sativus* of Iranian origin were mapped by bioinformatics tools and the predicted peptides were found useful for vaccine development (Hassan, Babak, & Sasan, 2008).

Saffron metagenomics

Fungal diversity was not only different between roots and corm of *Crocus sativus*. *Zygomycota* was dominant fungal phylum in the rhizosphere whereas *Basidiomycota* was dominant in cormosphere during flowering stage. Whereas in cormosphere *Basidiomycota* was dominant phylum during flowering stage but *Zygomycota* was dominant during dormant stage. In cormosphere, the microorganism group which was dominant at dormant stage was infrequent at flowering stage and vice-versa (*Basidiomycota*: Flowering = 93.2% Dormant = 0.05% and *Zygomycota*: Flowering = 0.8% Dormant = 99.7%).

Rhizopus was dominant in dormant stage but was infrequent in flowering stage (*Rhizopus*: Dormant = 99.7% Flowering = 0.55%). These dynamics is not followed by the bulk soil fungi which were dominated by *Ascomycota* during both stages under study. *Fusarium oxysporum* causes corm rot in *Crocus staivus*. This is the first testimony of the fungal diversity of the root of *Crocus sativus* and first report on the fungi associated with corm of any plant with the temporal and spatial variation in the fungal community structure (Ambardar, Singh, Gowda, & Vakhlu, 2016). The fungi like *Rhizoctonia crocorum, Phoma crocophila* (Madan et al., 1967) *Macrophomina phaseolina* (Thakur, Singh, & Kaul, 1992), *Fusarium moniliforme* var *intermedium* (Dhar, 1992), *Fusarium oxysporum, F. solani, F. equiseti, F. pallidoroseum, Penicillium* sp., *Mucor* sp., (Wani, 2004; Ahmad & Sagar, 2007) and *Sclerotium rolfsii* (Kalha et al., 2007) also infect *Crocus staivus*. RNA-seq based transcriptome of *Crocus sativus* from Jammu and Kashmir by Baba et al. (2015) provides several, yet explored, insights into the metagenome of the plant from that region. YeATS suite from the NCBI and Ensembl databases to enable faster comparisons were used (Chakraborty, 2016) to determine the metagenome from the transcriptome of *Crocus sativus* from Jammu and Kashmir and smaller, yet comprehensive databases were created for -viruses (V-DB), bacterial (B-DB), fungal (F-DB), and plants (mitochondria, chloroplast and ribosomes—CMR-DB). Soybean mosaic virus, a potyvirus, was found to be abundantly expressed in all five tissues analyzed. Leifsonia bacteria, Elizabethkingia bacteria and Staphylococcus bacteria and Mycosphaerella fungi and Pyrenophora fungi were detected. A proper disease-management strategy for *Crocus sativus* can be devised based on this knowledge (Chakraborty, 2016).

Conclusions

The saffron crop is medicinally very potent. With wide interest in growing numbers of therapeutics, there remains a challenge in investigating several noncurative compounds of *Crocus sativus*. This therapeutic potential of the *Crocus sativus* compounds in the form of chemotherapy or radiotherapy can allow us to find novel insights to study impact on diseases. With *Crocus sativus* as a chemical modulator derived from wide number of plant nutrients, employing omics technologies is the need of the hour so as to enhance the potential for drugs through possible anti-disease agents like colorants, stigmas, etc. The Integratome Omics can not only attribute to a better understanding of drug targets but also allow us to consider *Crocus sativus* "ome." The knowledge of origin of *Crocus sativus* offers routes towards resynthesis of *Crocus sativus* by cross breeding tetraploid *C. cartwrightianus* cytotypes with favorable allele combinations which can be selected by flow cytometry or microscopy. Moreover, restoration of fertility in sterile saffron will enable seed propagation, introduction of genetic variability by recombination and selection for *Crocus sativus* breeding. Climatic changes is showing effect on saffron farming system in Kashmir from last few years (2014–17) and has declined saffron production from 16.5 M.T (2013- 1.5 M.T (2017) (Nehvi & Yasmin, 2019). That's why, the information obtained through genomics, transcriptomics, proteomics, miRNomics, and metabolomics of the saffron genes, microRNAs, proteins, and metabolites and other regulatory sequences involved in Apocarotenoid/Carotenoid biosynthetic pathway will enable their genetic editing with CRISPR Cas to boost the yield and required metabolites in *Crocus sativus* future crops.

References

Álvarez-Ortí, M., Gómez, L. G., Rubio, A., Escribano, J., Pardo, J., Jiménez, F., & Fernández, J. A. (2004). Development and gene expression in saffron corms. *Acta Horticulturae (ISHS), 650*, 141–153.

Abdullaev, F. I., & Espinosa-Aguirre, J. J. (2004). Biomedical properties of saffron and its potential use in cancer therapy and chemoprevention trials. *Cancer Detection and Prevention, 28*(6), 426–432.

Abdullaev, F. I., & Frankel, G. D. (1999). Saffron in biological and medical research. In M. Negbi (Ed.), *Saffron:* Crocus sativus *L* (pp. 103–113). Amsterdam, The Netherlands: Harwood Academic Publishers.

Abdullaev, F. I., & Frenkel, G. D. (1992a). Effect of saffron on cell colony formation and cellular nucleic acid and protein synthesis. *Biofactors, 3*(3), 201–204.

Abdullaev, F. I., & Frenkel, G. D. (1992b). The effect of saffron on intracellular DNA, RNA and protein synthesis in malignant and non-malignant human cells. *Biofactors, 4*(1), 43–45.

Abdullaev, F. I., (1993). Biological effects of saffron. Biofactors 4:83–6.

Abdullaev, F. I., & Frenkel, G. D. (1993). Biological effects of saffron. *Biofactors, 4*, 83–86.

Abdullaev, F. I. (1994). Inhibitory effect of crocetin on intracellular nucleic acid and protein synthesis in malignant cells. *Toxicology Letters, 70*(2), 243–251.

Abdullaev, F. I. (2002). Cancer chemopreventive and tumoricidal properties of saffron (*Crocus sativus* L.). *Experimental Biology and Medical Maywood, 227*(1), 20–25.

Abdullaev, F. I. (2004). Antitumour effect of saffron (*Crocus sativus* L.): overview and prospectives. *Acta Horticulturae, 650*, 491–499.

Abe, K., & Saito, H. (2000). Effects of saffron extract and its constituent crocin on learning behaviour and long term potentiation. *Phytotherapy Research, 14*(3), 149–152.

Abe, K., Sugiura, M., Yamaguchi, S., Shoyama, Y., & Saito, H. (1999). Saffron extract prevents acetaldehyde-induced inhibition of long-term potentiation in the rat dentate gyrus in vivo. *Brain Research, 851*(1-2), 287–289.

Abrishami, M. H. (1987). *Understanding of Iranian Saffron* (1st ed.). Tehran: Tous.

Abrishami, M. H. (1997). *Persian Saffron, a Comprehensive Cultural and Agricultural History*. Mashhad: Astan Ghods Razavi Publication.

Abrishami, M. H. (2004). *Saffron, from yesterday till today, an encyclopaedia of its production, trade and use*. Tehran: Amirkabir.

Agayev, Y. M., & Zarifi, E. (2010). Peculiar evolution of saffron (*Crocus sativus* L.): prosperity and decline. *Acta Horticulturae, 850*, 29–34.

Agayev, Y. M. O., Fernandez, J. A., & Zarifi, E. (2009). Clonal selection of saffron (*Crocus sativus* L.): the first optimistic experimental results. *Euphytica, 169*, 81–99.

Agha-Hosseini, M., Kashani, L., Aleyaseen, A., Ghoreishi, A., Rahmanpour, H., Zarrinara, A. R., & Akhondzadeh, S. (2008). *Crocus sativus* L. (saffron) in the treatment of premenstrual syndrome: a double-blind, randomised and placebo- controlled trial. *BJOG, 115*(4), 515–519.

Agostino, N. D., Pizzichini, D., Chiusano, M. L., & Giuliano, G. (2007). An EST database from saffron stigmas. *BMC Plant Biology*, 7, 53.

Ahmad, M., Sagar, V., (2007). Integrated management of corm/tuber rot of saffron and Kalazeera. Horticulture Mini Mission-1, Indian Council for Agricultural Research (ICAR), India, 22 pp.

Ahmad, A. S., Ansari, M. A., Ahmad, M., Saleem, S., Yousuf, S., Hoda, M. N., & Islam, F. (2005). Neuroprotection by crocetin in a hemi-parkinsonian rat model. *Pharmacology and Biochemistry Behavior, 81*(4), 805–813.

Ahmad, M., Zaffar, G., Mir, S.D., Razvi, S.M., Rather, M.A., Mir, M.R. (2011). Saffron (Crocus sativus L.) strategies for enhancing productivity. Research Journal of Medicinal Plant, 5, 630-649.

Ahmad, M., Zaffar, G., Habib, M., Arshid, A., Dar, N. A., & Dar, Z. A. (2014). Saffron (*Crocus sativus* L.) in the light of biotechnological approaches: A review. *Scientific Research and Essays, 9*(2), 13–18. doi: 10.5897/SRE2013.5773 30.

Ahrazem, O., Trapero, A., Gomez, M. D., Rubio-Moraga, A., & Gomez-Gomez, L. (2010). Genomic analysis and gene structure of the plant carotenoid dioxygenase 4 family: a deeper study in *Crocus sativus* and its allies. *Genomics, 96*, 239–250.

Ahrazem, O., Rubio-Moraga, A., Nebauer, S. G., Molina, R.V., & Gomez-Gomez, L. (2015). Saffron: its Phytochemistry, developmental processes, and biotechnological prospects. *Journal of Agriculture and Food Chemistry, 63*(40), 8751–8764.

Ahrazem, O., Argandona, J., Fiore, A., Aguado, C., Lujan, R., Rubio-Moraga, A., Marro, M., Araujo-Andrade, C., Loza-Alvarez, P., Diretto, G., et al. (2018). Transcriptome analysis in tissue sectors with contrasting crocins accumulation provides novel insights into apocarotenoid biosynthesis and regulation during chromoplast biogenesis. *Scientific Reports, 8*(1), 2843.

Ahrazem, O., Argandoña, J., Fiore, A., Rujas, A., Rubio-Moraga, A., Castillo, R., & Gómez-Gómez, L. (2019). Multi-species transcriptome analyses for the regulation of crocins biosynthesis in Crocus. *BMC Genomics, 20*, 320 https://doi.org/10.1186/s12864-019-5666-5.

Ajami, M., Eghtesadi, S., Pazoki-Toroudi, H., Habibey, R., & Ebrahimi, S. A. (2010). Effect of *Crocus sativus* on gentamicin induced nephrotoxicity. *Biology Research, 43*(1), 83–90.

Akhondzadeh, S., Fallah-Pour, H., Afkham, K., Jamshidi, A. H., & Khalighi-Cigaroudi, F. (2004). Comparison of *Crocus sativus* L. and imipramine in the treatment of mild to moderate depression: a pilot double-blind randomized trial. *BMC Complement Alternative Medicine, 4*, 12.

Akhondzadeh, S., Tahmacebi-Pour, N., Noorbala, A. A., Amini, H., Fallah-Pour, H., Jamshidi, A. H., & Khani, M. (2005). *Crocus sativus* L. in the treatment of mild to moderate depression: a double-blind, randomized and placebocontrolled trial. *Phytotherapy Research, 19*(2), 148–151.

Akhondzadeh, B. A., Moshiri, E., Noorbala, A. A., Jamshidi, A. H., Abbasi, S. H., & Akhondzadeh, S. (2007). Comparison of petal of *Crocus sativus* L. and fluoxetine in the treatment of depressed outpatients: a pilot doubleblind randomized trial. *Progress in Neuro-Psychopharmacology and Biological Psychiatry, 31*(2), 439–442.

Akhondzadeh, S., Sabet, M. S., Harirchian, M. H., Togha, M., Cheraghmakani, H., Razeghi, S., Hejazi, S. S. H., Yousefi, M. H., Alimardani, R., Jamshidi, A., Zare, F., & Moradi, A. (2010a). Saffron in the treatment of patients with mild to moderate Alzheimer's disease:

a 16-week, randomized and placebo-controlled trial. *Journal of Clinical Pharmacy and Therapeutics*, *35*(5), 581–588.

Akhondzadeh, S., Shafi ee Sabet, M., Harirchian, M. H., Togha, M., Cheraghmakani, H., Razeghi, S., Hejazi, S. S., Yousefi, M. H., Alimardani, R., Jamshidi, A., Rezazadeh, S. A., Yousefi, A., Zare, F., Moradi, A., & Vossoughi, A. (2010b). A 22-week, multicenter, randomized, double-blind controlled trial of *Crocus sativus* in the treatment of mild to-moderate Alzheimer's disease. *Psychopharmacology (Berl)*, *207*(4), 637–643.

Al-Mofleh, I. A., Alhaider, A. A., Mossa, J. S., Al-Sohaibani, M. O., Qureshi, S., & Rafatullah, S. (2006). Antigastric ulcer studies on 'saffron' *Crocus sativus* L. in rats. *Pakistan Journal of Biological Science*, *9*, 1009–1013.

Alavi-Kia, S. S., Mohammadi, S. A., Aharizad, S., & Moghaddam, M. (2008). Analysis of genetic diversity and phylogenetic relationships in Crocus genus of Iran using inter retrotransposon amplified polymorphism. *Biotechnology & Biotechnological Equations*, *22*, 795–800.

Alsayied, N. F., Fernandez, J. A., Schwarzacher, T., & Heslop-Harrison, J. S. (2015). Diversity and relationships of *Crocus sativus* and its relatives analysed by inter-retroelement amplified polymorphism (IRAP). *Annals of Botany*, *116*(3), 359–368.

Amanpour, A., Sonmezdag, A. S., Kelebek, H., & Selli, S. (2015). GC-MS- olfactometric characterization of the most aroma-active components in a representative aromatic extract from Iranian saffron (*Crocus sativus* L.). *Food Chemistry*, *182*, 251–256.

Ambardar, S., Singh, H. R., Gowda, M., & Vakhlu, J. (2016). Comparative metagenomics reveal phylum level temporal and spatial changes in mycobiome of belowground parts of *Crocus sativus*. *PLoS ONE*, *11*(9), e0163300. doi: 10.1371/journal.pone.0163300.

Amin, B., & Hosseinzadeh, H. (2012). Evaluation of aqueous and ethanolic extracts of saffron, *Crocus sativus* L., and its constituents, safranal and crocin in allodynia and hyperalgesia induced by chronic constriction injury model of neuropathic pain in rats. *Fitoterapia*, *83*(5), 888–895.

Amin, A., Hamza, A. A., Bajbouj, K., Ashraf, S. S., & Daoud, S. (2011). Saffron: a potential candidate for a novel anticancer drug against hepatocellular carcinoma. *Hepatology*, *54*(3), 857–867. https://en.wikipedia.org/wiki/saffron.

Asadollahi, M., Nikdokht, P., Hatef, B., Sadr, S. S., Sahraei, H., & Assarzadegan, F. (2019). Pirzad Jahromi G (2019) Protective properties of the aqueous extract of saffron (Crocus sativus L.) in ischemic stroke, randomized clinical trial. *Journal of Ethnopharmacology* https://doi.org/10.1016/j.jep. 2019.111833.

Asadpour, E., & Sadeghnia, H. R. (2011). P03-173 – effect of safranal, a constituent of *crocus sativus* , on MK-801- induced behavioral and memory deficits in rat. *European Psychiatry*, *26*(Supplement 1), 1342.

Asalgoo, S., Jahromi, G., Meftahi, G., & Sahraei, H. (2015). Posttraumatic stress disorder (ptsd): mechanisms and possible treatments. *Neurophysiology*, *47*(6), 482–489.

Asalgoo, S., Tat, M., Sahraei, H., & Jahromi, G. P. (2017). The psychoactive agent crocin can regulate hypothalamic-pituitary-adrenal axis activity. *Frontiers in Neuroscience*, *11*.

Asdaq, S. M., & Inamdar, M. N. (2010). Potential of *Crocus sativus* (saffron) and its constituent, crocin, as hypolipidemic and antioxidant in rats. *Applied Biochemistry and Biotechnology*, *162*(2), 358–372.

Assimiadis, M. K., Tarantilis, P. A., & Polissiou, M. G. (1998). UV-Vis, FT-Raman and H-NMR spectroscopies of cis-trans carotenoids from saffron (Crocus sativus L.). *Applied Spectroscopy*, *52*, 519–522.

Assimopoulou, A. N., Sinakos, Z., & Papageorgiou, V. P. (2005). Radical scavenging activity of *Crocus sativus* L. extract and its bioactive constituents. *Phytotherapy Research*, *19*(11), 997–1000.

Aung, H. H., Wang, C. Z., Ni, M., Fishbein, A., Mehendale, S. R., Xie, J. T., Shoyama, C. Y., & Yuan, C. S. (2007). Crocin from *Crocus sativus* possesses significant anti-proliferation effects on human colorectal cancer cells. *Experimental Oncology*, *29*(3), 175–180.

Ayatollahi, H., Javan, A. O., Khajedaluee, M., Shahroodian, M., & Hosseinzadeh, H (2013). Effect of *Crocus sativus* L (saffron) on coagulation and anticoagulation systems in healthy volunteers. *Phytotherapy Research*doi: 10.1002/ptr.5021.

Aytekin, A., & Acikgoz, A. O. (2008). Hormone and microorganism treatments in the cultivation of saffron (*Crocus sativus* L.) plants. *Molecules, 13*, 1135–1147.

Azafrán. http://www.herbotecnia.com.ar/exo-azafran.html. 2002 (accessed 18 December 2006).

Baba, S. A., Mohiuddin, T., Basu, S., Swarnkar, M. K., Malik, A. H., Wani, Z. A., Abbas, N., Singh, A. K., & Ashraf, N (2015). Comprehensive transcriptome analysis of *Crocus sativus* for discovery and expression of genes involved in apocarotenoid biosynthesis. *BMC Genomics, 16*, 698 https://doi.org/10.1186/s12864-015-1894-5.

Bagri, J., Yadav, A., Anwar, K., Dkhar, J., Singla-Pareek, S. L., & Pareek, A. (2017). Metabolic shift in sugars and amino acids regulates sprouting in Saffron corm. *Scientific Reports*, 7(1), 11904 https://doi.org/10.1038/s41598-017-10528-2.

Bagur, M., Salinas, G., Jiménez-Monreal, A., Chaouqi, S., Llorens, S., Martinez-Tomé, M., & Alonso, G. (2018). Saffron: an old medicinal plant and a potential novel functional food. *Molecules, 23*, 30.

Bajbouj, K., Schulze-Luehrmann, J., Diermeier, S., Amin, A., & Schneider-Stock, R. (2012). The anticancer effect of saffron in two p53 isogenic colorectal cancer cell lines. *BMC Complement and Alternative Medicine, 12*, 69.

Bakshi, H. A., Sam, S., Feroz, A., Ravesh, Z., Shah, G. A., & Sharma, M. (2009). Crocin from Kashmiri saffron (*Crocus sativus*) induces in vitro and in vivo xenograft growth inhibition of Dalton's lymphoma (DLA) in mice. *Asian Pacific Journal of Cancer Prevention, 10*(5), 887–890.

Bakshi, H., Sam, S., Rozati, R., Sultan, P., Islam, T., Rathore, B., Lone, Z., Sharma, M., Triphati, J., & Saxena, R. C. (2010). DNA fragmentation and cell cycle arrest: a hallmark of apoptosis induced by crocin from Kashmiri saffron in a human pancreatic cancer cell line. *Asian Pacific Journal of Cancer Prevention, 11*(3), 675–679.

Ballabh, B., Chaurasia, O. P., Ahmed, Z., & Singh, S. B. (2008). Traditional medicinal plants of cold desert Ladakhused against kidney and urinary disorders. *Journal of Ethnopharmacology, 118*(2), 331–339.

Bamford, D. (2006). Dyeing for a king. *Journal for Weavers, Spinners and Dyers, 218*, 24–25.

Bani, S., Pandey, A., Agnihotri, V. K., Pathania, V., & Singh, B. (2011). Selective Th2 upregulation by *Crocus sativus:* a nutraceutical spice. *Evidence Based Complement and Alternative Medicine, 2011*, 639862.

Basker, D., & Negbi, M. (1983). Uses of saffron. *Economic Botany, 37*, 228–236.

Bathaie, S. Z., Bolhasani, A., Hoshyar, R., Ranjbar, B., Sabouni, F., & Moosavi-Movahedi, A. A. (2007). Interaction of saffron carotenoids as anticancer compounds with ctDNA, Oligo (dG.dC)15, and Oligo (dA.dT)15. *DNA Cell Biology, 26*(8), 533–540.

Bathaie, S. Z., Miri, H., Mohagheghi, M. A., Mokhtari-Dizaji, M., Shahbazfar, A. A., & Hasanzadeh, H. (2013). Saffron aqueous extract inhibits the chemically-induced gastric cancer progression in the Wistar albino rat. *Iranian Journal of Basic Medical Science, 16*(1), 27–38.

Baziar, S., Aqamolaei, A., Khadem, E., Mortazavi, S. H., Naderi, S., & Sahebolzamani, E. (2019). *Crocus sativus* L. versus methylphenidate in treatment of children with attention-deficit/hyperactivity disorder: a randomized, double-blind pilot study. *Journal of Child and Adolescent Psychopharmacology, 29*, 205–212. doi: 10.1089/cap.2018.0146.

Beale, M. H., & Sussman, M. R. (2011). Metabolomics of *Arabidopsis thaliana. Annals of Plant Reviews, 43*, 157–180.

Behmanesh, A (1959). *The history of ancient Greece.* Tehran: Tehran University publication.

Behnia, M. R. (1991). *Saffron, botang, cultivation and production.* Tehran: Tehran Univerity publication.

Beiki, A. H., Keifi, F., & Mozafari, J. (2010). Genetic differentiation of Crocus species by random amplified polymorphic DNA. *Genetic Engineering and Biotechnology Journal, 18*, 1–10.

Berger, F., Hensel, A., & Nieber, K. (2011). Saffron extract and trans-crocetin inhibit glutamatergic synaptic transmission in rat cortical brain slices. *Neuroscience, 180*, 238–247.

Bhagyalakshmi, N. (1999). Factors influencing direct shoot regeneration from ovary explants of saffron. *Plant Cell Tissue and Organ Culture, 58*, 205–211.

Bhandari, R. P. (2015). *Crocus sativus* L. (saffron) for cancer chemoprevention: a mini review. *Traditional & Complementary Medicine, 5*(2), 81–87.

Bharti, S., Golechha, M., Kumari, S., Siddiqui, K. M., & Arya, D. S. (2011). Akt/GSK-3β/eNOS phosphorylation arbitrates safranal-induced myocardial protection against ischemia- reperfusion injury in rats. *European Journal of Nutrition, 51*(6), 719–727.

Bhattacharjee, B., Vijayasarathy, S., Karunakar, P., & Chatterjee, J. (2012). Comparative reverse screening approach to identify potential anti-neoplastic targets of saffron functional components and binding mode. *Asian Pacific Journal of Cancer Prevention, 13*, 5605–5611.

Bie, X. D., Chen, Y. Q., Zheng, X. S., & Dai, H. B. (2011). The role of crocetin in protection following cerebral contusion and in the enhancement of angiogenesis in rats. *Fitoterapia, 82*(7), 997–1002.

Blois, L. d., & Spek, R. J. v. d. (2005). *An Introduction to the Ancient World*. Tehran: Ghoghnus.

Bonyadi, M. H. J., Yazdani, S., & Saadat, S. (2014). The ocular hypotensive effect of saffron extractin primary open angle glaucoma: a pilot study. *BMC Complement Alternative Medicine, 14*, 399.

Boroushaki, M. T., & Sadeghnia, H. R. (2009). Protective effect of safranal against gentamicin-induced nephrotoxicity in rat. *Iranian Journal of Medical Science, 34*, 285–288.

Boskabady, M. H., & Aslani, M. R. (2006). Relaxant effect of *Crocus sativus* (saffron) on guinea-pig tracheal chains and its possible mechanisms. *Journal of Pharmacy and Pharmacology, 58*(10), 1385–1390.

Boskabady, M. H., & Farkhondeh, T. (2016). Antiinflammatory, antioxidant, and immunomodulatory effects of *Crocus sativus* L. and its Main constituents. *Phytother Res, 30*(7), 1072–1094.

Boskabady, M. H., Shafei, M. N., Shakiba, A., & Sefi di, H. S. (2008). Effect of aqueous-ethanol extract from *Crocus sativus* (saffron) on guinea-pig isolated heart. *Phytotherapy Research, 22*(3), 330–334.

Boskabady, M. H., Rahbardar, M. G., Nemati, H., & Esmaeilzadeh M (2010). Inhibitory effect of *Crocus sativus* (saffron) on histamine (H1) receptors of guinea pig tracheal chains. *Pharmazie, 65*(4), 300–305.

Boskabady, M. H., Rahbardar, M. G., & Jafari, Z. (2011a). The effect of safranal on histamine (H1) receptors of guinea pig tracheal chains. *Fitoterapia, 82*(2), 162–167.

Boskabady, M. H., Seyedhosseini Tamijani, S. M., Rafatpanah, H., Rezaei, A., & Alavinejad, A. (2011b). The effect of *Crocus sativus* extract on human lymphocytes' cytokines and T helper 2/T helper 1 balance. *J Med Food, 14*(12), 1538–1545.

Boskabady, M. H., Tabatabaee, A., & Byrami, G. (2012). The effect of the extract of *Crocus sativus* and its constituent safranal, on lung pathology and lung inflammation of ovalbumin sensitized guinea-pigs. *Phytomedicine, 19*(10), 904–911.

Boutanaev, A. M., Moses, T., Zi, J., et al. (2015). Investigation of terpene diversification across multiple sequenced plant genomes. *Proceedings of National Academy of Science USA, 112*(1), E81-8.

Bouvier, F., Suire, C., Mutterer, J., & Camara, B. (2003). Oxidative remodeling of chromoplast carotenoids: identification of the carotenoid dioxygenase CsCCD and CsZCD genes involved in Crocus secondary metabolite biogenesis. *Plant Cell, 15*, 47–62.

Bowels, E. A. (1952). *Ed. A handbook of Crocus and Colchicum for Gardeners*. London: Bodley Head 222.

Brandizzi, F., Grilli Caiola, M. (1998). Flow cytometric analysis of nuclear DNA in *Crocus sativus* and allies (Iridaceae). *Plant Systematics and Evolution* 211(3): 149–154.

Brighton, C. A. (1977). Cytology of *Crocus sativus* and its allies (Iridaceae). *Plant Systematics and Evolution, 128*, 137–157.

Broadhead, G. K., Grigg, J. R., McCluskey, P., Hong, T., Schlub, T. E., & Chang, A. A. (2019). Saffron Therapy for the treatment of mild/ moderate age-related macular degeneration: a randomized clinical trial. *Graefe's Archives of Clinical Experiment in Ophthalmology, 257*, 31–40.

Busconi, M., Colli, L., Sánchez, R. A., Santaella, M., De-Los-Mozos, P. M., Santana, O., Roldán, M., & Fernández JA (2015). AFLP and MS-AFLP analysis of the variation within saffron crocus (*Crocus sativus* L.) germplasm. *PLoS One, 10*, e0123434.

Byrami, G., & Boskabady, M. H. (2012). The potential effect of the extract of *Crocus sativus* and safranal on the total and differential white blood cells of ovalbumin sensitized guinea pigs. *Research in Pharmaceutical Science*, 7(4), 249–255.

Byrami, G., Boskabady, M. H., Jalali, S., & Farkhondeh, T. (2013). The effect of the extract of *Crocus sativus* on tracheal responsiveness and plasma levels of IL-4, IFN-γ, total NO and nitrite in ovalbumin sensitized guinea-pigs. *Journal of Ethnopharmacology, 147*(2), 530–535.

Caballero-Ortega, H., Pereda-Miranda, R., Riverón-Negrete, L., Hernández, J. M., Medécigo-Ríos, M., Castillo-Villanueva, A., & Abdullaev, F. I. (2004). Chemical composition of saffron (*Crocus sativus* L.) from four countries. *Acta Horticulturae (ISHS), 650*, 321–326.

Caballero-Ortega, H., Pereda-Miranda, R., & Abdullaev, F. I. (2007). HPLC quantification of major active components from 11 different saffron (*Crocus sativus* L.) sources. *Food Chemistry, 100*, 1126–1131.

Cagliani, L. R., Culeddu, N., Chessa, M., & Consonni, R. (2015). NMR investigations for quality assessment of Italian PDO saffron (*Crocus sativus* L.). *Food Control, 50*, 342–348.

Calsteren, M. R., Bissonnette, M. C., Cormier, F., Dufrense, C., Ichi, T., Le BlancCY, et al. (1997). Spectroscopic characterization of crocetin derivatives from *Crocus sativus* and *Gardenia jasminoides*. *Journal of Agriculture and Food Chemistry, 45*, 1055–1061.

Carmona, M., Zalacain, A., Sanchez, A. M., Novel la, J. L., & Alonso, G. L. (2006). Crocetin esters, picrocrocin and its related compounds present in *Crocus sativus* stigmas and Gardenia jasminoides fruits. Tentative identification of seven new compounds by LC-ESI - MS. *Journal of Agricultural and Food Chemistry, 54*, 973–979.

Carmona, M., Zalacain, A., Salinas, M. R., & Alonso, G. L. (2007). A new approach to saffron aroma. *CRC Critical Reviews in Food Science and Nutrition, 47*, 145–159.

Castillo, R., Fernandez, J. A., & Gomez-Gomez, L. (2005). Implications of carotenoid biosynthetic genes in apocarotenoid formation during the stigma development of *Crocus sativus* and its closer relatives. *Plant Physiology, 139*, 674–689.

Castro-Díaz, N., Salaun, B., Perret, R., Sierro, S., Romero, J. F., Fernández, J. A., Rubio-Moraga, A., & Romero, P. (2012). Saponins from the Spanish saffron *Crocus sativus* are efficient adjuvants for protein-based vaccines. *Vaccine, 30*(2), 388–397.

Chahine, N., Hanna, J., Makhlouf, H., Duca, L., Martiny, L., & Chahine, R. (2013). Protective effect of saffron extract against doxorubicin cardiotoxicity in isolated rabbit heart. *Pharmaceutical Biology, 51*, 1564–1571.

Chakraborty, S. (2016) Transcriptome from saffron (Crocus sativus) plants in Jammu and Kashmir reveals abundant soybean mosaic virus transcripts and several putative pathogen bacterial and fungal genera. bioRxiv. doi: http://dx.doi.org/10.1101/079186.

Chang, P. Y., Wang, C. K., Liang, J. D., & Kuo, W. (1964). Studies on the pharmacological action of zang hong hua (*Crocus sativus* L.). I. Effects on uterus and estrous cycle. *Yao Xue Xue Bao, 11*, 94–100.

Chen, S., Wang, X., Zhao, B., Yuan, X., & Wang, Y (2003). Production of crocin using *Crocus sativus* callus by two-stage culture system. *Biotechnology Letters, 25*(15), 1235–1238.

Chen, S., Zhao, B., Wang, X., Yuan, X., & Wang, Y (2004). Promotion of the growth of *Crocus sativus* cells and the production of crocin by rare earth elements. *Biotechnology Letters, 26*, 27–30.

Chen, Y., Zhang, H., Tian, X., Zhao, C., Cai, L., Liu, Y., Jia, L., Yin, H. X., & Chen, C. (2008). Antioxidant potential of crocins and ethanol extracts of *Gardenia jasminoides* Ellis and *Crocus sativus* L.: a relationship investigation between antioxidant activity and crocin contents. *Food Chemistry, 109*(3), 484–492.

Choob, V. V., Vlassova, T. A., & Butenko RG (1994). Callusogenesis and morphogenesis in generative organ culture of the spring flowering species of *Crocus* L. *Russian Journal of Plant Physiology, 41*, 712–716.

Christodoulou, E., Kadoglou, N. P., Kostomitsopoulos, N., & Valsami G (2015). Saffron: a natural product with potential pharmaceutical applications. *Pharmacy and Pharmacology, 67*(12), 1634–1649.

Chrungoo, G., & Caiola, M. G. (1987). In vitro development of parthenocarpic fruits of *Crocus sativus* L. *Plant Cell Tissue and Organ Culture, 11*, 75–78.

Chryssanthi, D. G., Lamari, F. N., Iatrou, G., Pylara, A., Karamanos, N. K., & Cordopatis, P. (2007). Inhibition of breast cancer cell proliferation by style constituents of different *Crocus* species. *Anticancer Research, 27*, 357–362.

Chryssanthi, D. G., Dedes, P. G., Karamanos, N. K., Cordopatis, P., & Lamari, F. N. (2011a). Crocetin inhibits invasiveness of MDA-MB-231 breast cancer cells via downregulation of matrix metalloproteinases. *Planta Medica*, 77(2), 146–151.

Chryssanthi, D. G., Lamari, F. N., Georgakopoulos, C. D., & Cordopatis, P (2011b). A new validated SPE-HPLC method for monitoring crocetin in human plasma– application after saffron tea consumption. *Journal of Pharmaceutical and Biomedical Analysis, 55*(3), 563–568.

Courtney, P. (2002). *Tasmania's Saffron Gold*. Landline: Australian Broadcasting Corp.

D'Agostino, N., Pizzichini, D., Chiusano, M. L., & Giuliano, G. (2007). An EST database from saffron stigmas. *BMC Plant Biology*, 7, 53.

Dadkhah, M., Ehtesham, M., & Fekrat, H. (2003). *Iranian Saffron an unknown jewel*. Tehran: Shahr Ashub Publication.

Dalby A (2002) Ed. *Dangerous Tastes: The Story of Spices*; University of California Press: Berkely, CA; 256.

Das, I., Chakrabarty, R. N., & Das, S. (2004). Saffron can prevent chemically induced skin carcinogenesis in Swiss albino mice. *Asian Pacific Journal of Cancer Prevention, 5*(1), 70–76.

Das, I., Das, S., & Saha, T. (2010). Saffron suppresses oxidative stress in DMBA-induced skin carcinoma: a histopathological study. *Acta Histochemistry, 112*(4), 317–327.

Dastranj, M., Sepaskhah, A. R., & Kamgar-Haghighi, A. A. (2019). Rainfall and its distribution influences on rain-fed saffron yield and economic analysis. *Theoretical and Applied Climatology, 137*, 3139 https://doi.org/10.1007/s00704-019-02804-0.

Day, P. D., Berger, M., Hill, L., et al. (2014). Evolutionary relationships in the medicinally important genus *Fritillaria* L. (Liliaceae). *Molecular Phylogenetics and Evolution, 80*, 11–19.

Deslauriers, A. M., Afkhami-Goli, A., Paul, A. M., Bhat, R. K., Acharjee, S., Ellestad, K. K., Noorbakhsh, F., Michalak, M., & Power, C. (2011). Neuroinflammation and endoplasmic reticulum stress are coregulated by crocin to prevent demyelination and neurodegeneration. *The Journal of Immunology, 187*(9), 4788–4799.

De Juan, J. A., Córcoles, H. L., Muñoz, R. M., & Picornell, M. R. (2009). Yield and yield components of saffron under different cropping systems. *Industrial Crops and Products, 30*(2), 212–219.

Dhar, A. K., & Mir, G. M. (1997). Saffron in Kashmir-VI: A review of distribution and production. *Journal of Herbs Spices and Medicinal Plants*, *4*, 83–90.

Dhar, A.K., (1992). Bio-ecology and control of corm rot of saffron (*Crocus sativus* L.). MSc thesis, Sher-e- Kashmir University of Agricultural Sciences and Technology of Kashmir, India, 109 pp.

Dhar, A. K. (2000). Saffron: biology, utilization, agriculture, production and quality. *Journal of Medicinal and Aromatic Plant Science*, *22*, 355–360.

Ding, B., Bai, S. H., Wu, Y., & Fang, X. P. (1981). Induction of callus and regeneration of plantlets from corms of *Crocus sativus* L. *Acta Botanica Sinica*, *23*, 419–420.

Ding, Q., Zhong, H., Qi, Y., Cheng, Y., Li, W., Yan, S., & Wang, X. (2013). Anti-arthritic effects of crocin in interleukin- 1β- treated articular chondrocytes and cartilage in a rabbit osteoarthritic model. *Inflammation Research*, *62*(1), 17–25.

Douglas, M., & Perry, N. (2003). Growing saffron-the world's most expensive spice. *Crop and Food Research*, *20*, 1–4.

Dwyer, A.V., Whitten, D. L., & Hawrelak, J. A. (2011). Herbal medicines, other than St. John's Wort, in the treatment of depression: a systematic review. *Alternative Medicine Reviews*, *16*(1), 40–49.

Ebrahim-Habibi, M. B., Amininasab, M., Ebrahim-Habibi, A., Sabbaghian, M., & Nemat-Gorgani, M. (2010). Fibrillation of alpha-lactalbumin: effect of crocin and safranal, two natural small molecules from *Crocus sativus*. *Biopolymers*, *93*(10), 854–865.

El Daly, E. S. (1998). Protective effect of cysteine and vitamin E, *Crocus sativus and Nigella sativa* extracts on cisplatin-induced toxicity in rats. *Journal of Pharmacy Belgium*, *53*(2), 87–93 93-95.

El-Beshbishy, H. A., Hassan, M. H., Aly, H. A. A., Doghish, A. S., & Alghaithy AAA (2012). Crocin "saffron" protects against beryllium chloride toxicity in rats through diminution of oxidative stress and enhancing gene expression of antioxidant enzymes. *Ecotoxicology and Environment Safety*, *83*(1), 47–54.

Elomaa, P., Uimari, A., Mehto, M., Albert, V. A., Laitinen, R. A., & Teeri, T. H. (2003). Activation of anthocyanin biosynthesis in *Gerbera hybrida* (Asteraceae) suggests conserved protein-protein and protein- promoter interactions between the anciently diverged monocots and eudicots. *Plant Physiology*, *133*, 1831–1842.

Erol, O., Kaya, H. B., Azik, L., Tuna, M., Can, L., & Tanyolac, M. B. (2014). The genus *Crocus*, series *Crocus* (*Iridaceae*) in Turkey and 2 East Aegean islands: a genetic approach. *Turkish Journal of Biology*, *38*, 48–62.

Escribano, J., Alonso, G. L., Coca-Prados, M., & Fernández, J. A. (1996). Crocin, safranal and picrocrocin from saffron (*Crocus sativus* L.) inhibit the growth of human cancer cells in vitro. *Cancer Letters*, *100*(1), 23–30.

Escribano, J., Díaz-Guerra, M. J., Riese, H. H., Ontañón, J., García-Olmo, D., García-Olmo, D. C., Rubio, A., & Fernández, J. A. (1999a). In vitro activation of macrophages by a novel proteoglycan isolated from corms of Crocus sativus L. *Cancer Letters*, *144*(1), 107–114.

Escribano, J., Piqueras, A., Medina, J., Rubio, A., Alvarez- Ortı M, & Fernández, J. A. (1999b). Production of a cytotoxic proteoglycan using callus culture of saffron corms (*Crocus sativus* L.). *Journal of Biotechnology*, *73*(1), 53–59.

Escribano, J., Ríos, I., & Fernández JA (1999c). Isolation and cytotoxic properties of a novel glycoconjugate from corms of saffron plant (*Crocus sativus* L.). *Biochemistry and Biophysical Acta*, *1426*(1), 217–222.

Escribano, J., Díaz-Guerra, M. J., Riese, H. H., Alvarez, A., Proenza, R., & Fernández, J. A. (2000a). The cytolytic effect of a glycoconjugate extracted from corms of saffron plant (*Crocus sativus*) on human cell lines in culture. *Planta Medica*, *66*(2), 157–162.

Esmaeili, N., Ebrahimzadeh, H., Abdi, K., & Sarfarian, S. (2011). Determination of some phenolic compounds in *Crocus sativus* L. corms and its antioxidant activities study. *Pharmcognosy Magzine*, 7(25), 74–80.

Fakhrai, F., & Evans, P. K. (1990). Mophogenic potential of cultured floral explants of *Crocus sativus* L. for the *in vitro* production of saffron. *Journal of Experimental Botany, 41*, 47–52.

Falsini, B., Piccardi, M., Minnella, A., Savastano, C., Capoluongo, E., Fadda, A., Balestrazzi, E., Maccarone, R., & Bisti, S. (2010). Influence of saffron supplementation on retinal flicker sensitivity in early age-related macular degeneration. *Investigative Ophthalmology and Visual Science, 51*(12), 6118–6124.

Farkhondeh, T., Samarghandian, S., Shaterzadeh Yazdi, H., & Samini, F. (2018). The protective effects of crocin in the management of neurodegenerative diseases: a review. *American Journal of Neurodegenerative Disorder*, 7(1), 1–10.

Fatehi, M., Rashidabady, T., & Fatehi-Hassanabad, Z. (2003). Effects of *Crocus sativus* petals' extract on rat blood pressure and on responses induced by electrical field stimulation in the rat isolated vas deferens and guinea pig ileum. *Journal of Ethnopharmacology, 84*(2–3), 199–203.

Feizzadeh, B., Afshari, J. T., Rakhshandeh, H., Rahimi, A., Brook, A., & Doosti, H. (2008). Cytotoxic effect of saffron stigma aqueous extract on human transitional cell carcinoma and mouse fi broblast. *Urology Journal, 5*(3), 161–167.

Feo, F., Martinez, J., Martinez, A., Galindo, P. A., Cruz, A., Garcia, R., Guerra, F., & Palacios, R. (1997). Occupational allergy in saffron workers. *Allergy, 52*(6), 633–641.

Fernández, J. A., Santana, O., Guardiola, J. L., Molina, R. V., Heslop-Harrison, P., Borbely, G., et al. (2011). The World Saffron and Crocus collection: strategies for establishing, management, characterisation and utilization. *Genetic Resource of Crop Evolution, 58*, 125–137 https://doi.org/10.1007/s10722-010-9601-5.

Fernández, J. A., & Pandalai, S. G. (2004). Biology, biotechnology and biomedicine of saffron. *Recent Research and Development of Plant Science, 2*, 127–159 https://doi.org/10.1007/s10722-010-9601-5.

Fernández-Sánchez, L., Lax, P., Esquiva, G., Martín-Nieto, J., Pinilla, I., & Cuenca, N. (2012). Safranal, a saffron constituent, attenuates retinal degeneration in P23H rats. *PLoS One*, 7(8), e43074.

Fernandez, J. A., & Escribano, M. J. (2000). *Biotecnologia del azafran;*. Cuenca: Ediciones de la Universidad de Castilla La Mancha.

Fernandez, J. A. (2007). Genetic resources of saffron and allies (Crocus spp.). *Acta Horticulture, 739*, 167–185.

Ferrence, S. C., & Bendersky, G. (2004). Therapy with saffron and the goddess at Thera. *Perspective of Biological Medicine, 47*, 199–226.

Fluch, S., Hohl, K., Stierschneider, M., Kopecky, D., & Kaar, B (2010). *Crocus sativus* L. – Molecular evidence on its clonal origin. *Acta Horticulturae, 850*, 41–46.

Frello, S., & Heslop-Harrison, J. S. (2000). Chromosomal variation in *Crocus vernus* Hill (Iridaceae) investigated by *in situ* hybridization of rDNA and a tandemly repeated sequence. *Annals of Botany, 86*(2), 317–322.

Frello, S., Ørgaard, M., Jacobsen, N., & Heslop-Harrison, J. S. (2004). The genomic organization and evolutionary distribution of a tandemly repeated DNA sequence family in the genus *Crocus* (Iridaceae). *Hereditas, 141*(1), 81–88.

Frizzi, G., Miranda, M., Pantani, C., & Tammaro, F. (2007). Allozyme differentiation in four species of the *Crocus cartwrightianus* group and in cultivated saffron (*Crocus sativus*). *Biochemical Systematics and Ecology, 35*, 859–868.

Frusciante, S., Diretto, G., Bruno, M., Ferrante, P., Pietrella, M., Prado-Cabrero, A., & Giuliano, G. (2014). Novel carotenoid cleavage dioxygenase catalyzes the first dedicated step in saffron crocin biosynthesis. *Proceedings of National Academy of Science USA, 111*(33), 12246–12251 https://doi.org/10.1073/pnas. 1404629111.

Frusciante, S., Diretto, G., Bruno, M., Ferrante, P., Pietrella, M., Prado-Cabrero, A., Rubio-Moraga, A., Beyer, P., Gomez-Gomez, L., Al- Babili, S., & Giuliano, G. (2014). Novel carotenoid cleavage dioxygenase catalyzes the first dedicated step in saffron crocin biosynthesis. *Proceedings of National Academy of Science USA, 111*, 12246–12251.

Fukui, H., Toyoshima, K., & Komaki, R. (2011). Psychological and neuroendocrinological effects of odor of saffron (*Crocus sativus*). *Phytomedicine, 18*(8–9), 726–730.

Gómez-Gómez, L., Feo-Brito, F., Rubio-Moraga, A., Galindo, P. A., Prieto, A., & Ahrazem, O. (2010a). Involvement of lipid transfer proteins in saffron hypersensitivity: molecular cloning of the potential allergens. *Journal of Investigating Allergology and Clinical Immunology, 20*(5), 407–412.

Gómez-Gómez, L., Moraga-Rubio, A., & Ahrazem, O (2010b). Understanding carotenoid metabolism in saffron stigmas: unravelling aroma and color formation. *Functional Plant Science and Biotechnology, 4*, 56–63.

Garcia-Olmo, D. C., Riese, H. H., Escribano, J., Ontanon, J., Fernandez, J. A., Atiénzar, M., & García-Olmo, D (1999). Effects of long-term treatment of colon adenocarcinoma with crocin, a carotenoid from saffron (*Crocus sativus* L.): an experimental study in the rat. *Nutrition and Cancer, 35*(2), 120–126.

Georgiadou, G., Tarantilis, P. A., & Pitsikas, N. (2012). Effects of the active constituents of *Crocus sativus* L., crocins, in an animal model of obsessive–compulsive disorder. *Neuroscience Letter, 528*(1), 27–30.

Geromichalos, G. D., Lamari, F. N., Papandreou, M. A., Trafalis, D. T., Margarity, M., Papageorgiou, A., & Sinakos, Z. (2012). Saffron as a source of novel acetylcholinesterase inhibitors: molecular docking and in vitro enzymatic studies. *Journal of Agriculture and Food Chemistry, 60*(24), 6131–6138.

Ghadrdoost, B., Vafaei, A. A., Rashidy-Pour, A., Hajisoltani, R., Bandegi, A. R., Motamedi, F., Haghighi, S., Sameni, H. R., & Pahlvan, S. (2011). Protective effects of saffron extract and its active constituent crocin against oxidative stress and spatial learning and memory deficits induced by chronic stress in rats. *European Journal of Pharmacology, 667*(1–3), 222–229.

Ghaffari, S., & Roshanravan, N. (2019). Saffron; An updated review on biological properties with special focus on cardiovascular effects. *Biomedicine in Pharmacotherapy, 109*, 21–27.

Ghazavi, A., Mosayebi, G., Salehi, H., & Abtahi, H. (2009). Effect of ethanol extract of saffron (*Crocus sativus* L.) on the inhibition of experimental autoimmune encephalomyelitis in C57bl/6 mice. *Pakistan Journal of Biological Science, 12*(9), 690–695.

Ghorbani, M. (2008). The efficiency of saffron's marketing channel in Iran. *World Applied Sciences Journal, 4*(4), 523–527.

Ghoshooni, H., Daryaafzoon, M., Sadeghi-Gharjehdagi, S., Zardooz, H., Sahraei, H., Tehrani, S. P., Noroozzadeh, A., Bahrami-Shenasfandi, F., Kaka, G. H., & Sadraei, S. H. (2011). Saffron (*Crocus sativus*) ethanolic extract and its constituent, safranal, inhibits morphineinduced place preference in mice. *Pakistan Journal of Biological Science, 14*(20), 939–944.

Giaccio, M. (2004). Crocetin from saffron: an active component of an ancient spice. *Critical Reviews of Food Science and Nutrition, 44*, 155–172.

Goli, S. A. H., Mokhtari, F., & Rahimmalek, M. (2012). Phenolic compounds and antioxidant activity from saffron (*Crocus sativus* L.) petal. *Journal of Agricultural Science, 4*(10), 175–181.

Gomez-Gomez, L., Parra-Vega, V., Rivas-Sendra, A., Segui-Simarro, J. M., Molina, R. V., Pallotti, C., Rubio-Moraga, A., Diretto, G., Prieto, A., & Ahrazem, O (2017). Unraveling massive crocins transport and accumulation through proteome and microscopy tools during the development of saffron stigma. *International Journal of Molecular Science, 18*, 76.

Goupy, P., Vian, M. A., Chemat, F., & Caris-Veyrat, C (2013). Identification and quantification of flavonols, anthocyanins and lutein diesters in tepals of *Crocus sativus* by ultra performance liquid chromatography coupled to diode array and ion trap mass spectrometry detections. *Industrial Crop and Research, 44*, 496–510.

Gout, B., Bourges, C., & Paineau-Dubreuil, S (2010). Satiereal, a *Crocus sativus* L extract, reduces snacking and increases satiety in a randomized placebo-controlled study of mildly overweight, healthy women. *Nutrition Research, 30*(5), 305–313.

Goyal, S. N., Arora, S., Sharma, A. K., Joshi, S., Ray, R., Bhatia, J., Kumari, S., & Arya, D. S. (2010). Preventive effect of crocin of *Crocus sativus* on hemodynamic, biochemical, histopathological and ultrastructural alterations in isoproterenol- induced cardiotoxicity in rats. *Phytomedicine, 17*(3–4), 227–232.

Goyal, S. N., Arora, S., Sharma, A. K., Joshi, S., Ray, R., Bhatia, J., Kumari, S., & Arya, D. S. (2010). Preventive effect of crocin of *Crocus sativus* on hemodynamic, biochemical, histopathological and ultrastuctural alterations in isoproterenol-induced cardiotoxicity in rats. *Phytomedicine, 17*, 227–232.

Grace, O. M., Buerki, S., Symonds, M. R., et al. (2015). Evolutionary history and leaf succulence as explanations for medicinal use in aloes and the global popularity of *Aloe vera*. *BMC Evolution Biology, 15*, 29.

Gresta, F., Avola, G., Lombardo, G., Siracusa, L., Ruberto, G. (2009). Analysis of flowering, stigmas yield and qualitative traits of saffron (Crocus sativus L.) as affected by environmental conditions. Scientia Horticulturae 119, 320-324. doi:10.1016/j.scienta.

Grilli Caiola, M., Canini, A. (2010). Looking for saffron's (*Crocus sativus* L.) parents. In: Husaini, A.M., (ed.), Saffron. Functional Plant Science and Biotechnology 4: 1-14.

Grilli-Caiola, M., Caputo, P., & Zaier, R. (2004). RAPD analysis in *Crocus sativus* L. accessions and related *Crocus* species. *Biologia Plantarum, 48*, 375–380.

Grilli-Caiola, M. (1999). Reproduction biology in saffron and its allies. In M. Negbi (Ed.), *Saffron* (pp. 31–44). Harwood Academic Publishers.

Guan, H., & Kiss-Toth, E. (2008). Advanced technologies for studies on protein interactomes. In M. Werther, & H. Seitz (Eds.), *Protein–Protein Interaction* (pp. 1–24). Springer Berlin Heidelberg.

Guleria, p., Goswami, D., & Yadav, K. (2012). Computational identification of miRNAs and their targets from *Crocus sativus* L. *Archives of Biological Science, 64*, 65–70.

Gutheil, W. G., Reed, G., Ray, A., Anant, S., & Dhar, A. (2012). Crocetin: an agent derived from saffron for prevention and therapy for cancer. *Current Pharmaceutical Biotechnology, 13*(1), 173–179.

Habibi, M. B., & Bagheri, B. (1989). *Agriculture processing and chemicals and standards of saffron. Khorasan*. Iran: Iran Science Research Center.

Hadizadeh, F., Khalili, N., Hosseinzadeh, H., & Khair-Aldine, R. (2010). Kaempferol from saffron petals. *Iranian Journal of Pharmaceutical Research, 2*(4), 251–252.

Halataei, B. A., Khosravi, M., Arbabian, S., Sahraei, H., Golmanesh, L., Zardooz, H., Jalili, C., & Ghoshooni, H. (2011). Saffron (*Crocus sativus)* aqueous extract and its constituent crocin reduces stress-induced anorexia in mice. *Phytotherapy Research, 25*(12), 1833–1838.

Hao, & Xiao (2015). Genomics and evolution in traditional medicinal plants: road to a healthier life. *Evolutionary Bioinformatics, 11*, 197–212. doi: 10.4137/EBO.S31326.

Hao da, C., Ge, G., Xiao, P., Zhang, Y., & Yang, L. (2011). The first insight into the tissue specific taxus transcriptome via illumina second generation sequencing. *PLoS One, 6*(6), e21220.

Hao, D., Ma, P., Mu, J., et al. (2012). De novo characterization of the root transcriptome of a traditional Chinese medicinal plant *Polygonum cuspidatum. Science China Life Sciences, 55*(5), 452–466.

Hao, D. C., Xiao, P. G., Peng, Y., Dong, J., & Liu, W. (2012). Evaluation of the chloroplast barcoding markers by mean and smallest interspecific distances. *Pakistan Journal of Botany, 44*(4), 1271–1274.

Hao, D. C., Gu, X. J., Xiao, P. G., & Peng, Y. (2013). Phytochemical and biological research of *Fritillaria* medicine resources. *Chinese Journal of Natural Medicine, 11*(4), 330–344.

Hao, D. C., Xiao, P. G., Liu, M., Peng, Y., & He, C. N. (2014). Pharmaphylogeny vs. pharmacophylogenomics: molecular phylogeny, evolution and drug discovery. *Yao Xue Xue Bao, 49*(10), 1387–1394.

Hao, D. C., Vautrin, S., Song, C., et al. (2015). The first insight into the *Salvia* (*Lamiaceae*) genome via BAC library construction and high-throughput sequencing of target BAC clones. *Pakistan Journal of Botany*, *47*(4), 1347–1357.

Hariri, A. T., Moallem, S. A., Mahmoudi, M., Memar, B., & Hosseinzadeh, H. (2010). Sub-acute effects of diazinon on biochemical indices and specifi c biomarkers in rats: protective effects of crocin and safranal. *Food and Chemical Toxicology*, *48*(10), 2803–2808.

Hariri, A. T., Moallem, S. A., Mahmoudi, M., & Hosseinzadeh, H. (2011). The effect of crocin and safranal, constituents of saffron, against subacute effect of diazinon on hematological and genotoxicity indices in rats. *Phytomedicine*, *18*(6), 499–504.

Harpke, D., Meng, S., Kerndorff, H., Rutten, T., & Blattner, F. R. (2013). Phylogeny of *Crocus* (Iridaceae) based on one chloroplast and two nuclear loci: ancient hybridization and chromosome number evolution. *Molecular Phylogenetics and Evolution*, *66*, 617–627.

Harpke, D., Carta, A., Tomovic´, G., Ranđelović, V., Ranđelovic´, N., Blattner, F. R., & Peruzzi, L. (2015). Phylogeny, karyotype evolution and taxonomy of Crocus ser. Verni (Iridaceae). *Plant Systematics and Evolution*, *301*, 309–325.

Hartwell, J. L. (1982). *Plants used against cancer. A survey*. Lawrence: Quaterman Publications pp. 284-289.

Hassan, M., Babak, S., & Sasan, M. (2008). T and B-cell epitopes prediction of Iranian saffron (*Crocus sativus*) profilin by bioinformatics tools. *Protein & Peptide Letters*, *15*, 280–285.

Hausenblas, H. A., Saha, D., Dubyak, P. J., & Anton, S. D. (2013). Saffron (*Crocus sativus* L.) and major depressive disorder: a meta-analysis of randomized clinical trials. *Journal of Integrated Medicine*, *11*, 377–383.

He, C., Peng, B., Dan, Y., Peng, Y., & Xiao, P. (2014). Chemical taxonomy of tree peony species from China based on root cortex metabolic fingerprinting. *Phytochemistry*, *107*, 69–79.

Heidary, M., Vahhabi, S., Reza Nejadi, J., Delfan, B., Birjandi, M., Kaviani, H., & Givrad, S. (2008). Effect of saffron on semen parameters of infertile men. *Urology Journal*, *5*(4), 255–259.

Heitmar, R., Brown, J.,& Kyrou, I., (2019) Saffron (*Crocus sativus* L.) in ocular diseases: a narrative review of the existing evidence from clinical studies. *Nutrients* 11: 649.

Hemshekhar, M., Santhosh, M. S., Sunitha, K., Thushara, R. M., Kemparaju, K., Rangappa, K. S., & Girish, K. S. (2012). A dietary colorant crocin mitigates arthritis and associated secondary complications by modulating cartilage deteriorating enzymes, inflammatory mediators and antioxidant status. *Biochimie*, *94*(12), 2723–2733.

Hill, T. (2004). *The Contemporary Encyclopedia of Herbs and Spices: Seasonings for the Global Kitchen* (1st ed.). Wiley ISBN 978-0-471-21423-6.

Hooshmandi, Z., Rohani, A. H., Eidi, A., Fatahi, Z., Golmanesh, L., & Sahraei, H. (2011). Reduction of metabolic and behavioral signs of acute stress in male Wistar rats by saffron water extract and its constituent safranal. *Pharmaceutical Biology*, *49*(9), 947–954.

Hosseini, A., Razavi, B. M., & Hosseinzadeh, H. (2018). Saffron (*Crocus sativus*) petal as a new pharmacological target: a review. *Iranian Journal of Basic Medicine Science*, *21*, 1091–1099.

Hosseinzadeh, H., & Ghenaati, J. (2006). Evaluation of the antitussive effect of stigma and petals of saffron (*Crocus sativus*) and its components, safranal and crocin in guinea pigs. *Fitoterapia*, *77*(6), 446–448.

Hosseinzadeh, H., & Jahanian, Z. (2010). Effect of *Crocus sativus* L. (saffron) stigma and its constituents, crocin and safranal, on morphine withdrawal syndrome in mice. *Phytotherapy Research*, *24*(5), 726–730.

Hosseinzadeh, H., & Khosravan, V. (2002). Anticonvulsant effect of *Crocus sativus* L. stigmas aqueous and ethanolic extracts in mice. *Archives of Iranian Medicine*, *5*(1), 44–47.

Hosseinzadeh, H., & Noraei, N. B. (2009). Anxiolytic and hypnotic effect of *Crocus sativus* aqueous extract and its constituents, crocin and safranal, in mice. *Phytotherapy Research*, *23*(6), 768–774.

Hosseinzadeh, H., & Sadeghnia, H. R. (2005). Safranal, a constituent of *Crocus sativus* (saffron), attenuated cerebral ischemia induced oxidative damage in rat hippocampus. *Journal of Pharmacy and Pharmaceutical Science, 8*(3), 394–399.

Hosseinzadeh, H., & Sadeghnia, H. R. (2007a). Effect of safranal, a constituent of *Crocus sativus* (saffron), on methyl methanesulfonate (MMS)-induced DNA damage in mouse organs: an alkaline single-cell gel electrophoresis (comet) assay. *DNA and Cellular Biology, 26*(12), 841–846.

Hosseinzadeh, H., & Sadeghnia, H. R. (2007b). Protective effect of safranal on pentylenetetrazol-induced seizures in the rat: involvement of GABAergic and opioids systems. *Phytomedicine, 14*(4), 256–262.

Hosseinzadeh, H., & Talebzadeh, F. (2005). Anticonvulsant evaluation of safranal and crocin from *Crocus sativus* in mice. *Fitoterapia, 76*(7–8), 722–724.

Hosseinzadeh, H., & Younesi, H. M. (2002). Antinociceptive and anti-inflammatory effects of *Crocus sativus* L. stigma and petal extracts in mice. *BMC Pharmacology, 2*(1), 7.

Hosseinzadeh, H., Karimi, G., & Niapoor, M. (2004). Antidepressant effect of *Crocus sativus* L. stigma extracts and their constituents, crocin and safranal, in mice. *Acta Horticulturae (ISHS), 650*, 435–445.

Hosseinzadeh, H., Sadeghnia, H. R., Ziaee, T., & Danaee A (2005). Protective effect of aqueous saffron extract (*Crocus sativus* L.) and crocin, its active constituent, on renal ischemia-reperfusion-induced oxidative damage in rats. *Journal of Pharmacy and Pharmaceutical Science, 8*(3), 387–393.

Hosseinzadeh, H., Abootorabi, A., & Sadeghnia, H. R. (2008a). Protective effect of *Crocus sativus* stigma extract and crocin (trans-crocin 4) on methyl methane sulfonate induced DNA damage in mice organs. *DNA and Cellular Biology, 27*(12), 657–664.

Hosseinzadeh, H., Sadeghnia, H. R., & Rahimi, A. (2008b). Effect of safranal on extracellular hippocampal levels of glutamate and aspartate during kainic acid treatment in anesthetized rats. *Planta Medica, 74*(12), 1441–1445.

Hosseinzadeh, H., Ziaee, T., & Sadeghi, A. (2008c). The effect of saffron, *Crocus sativus* stigma, extract and its constituents, safranal and crocin on sexual behaviors in normal male rats. *Phytomedicine, 15*(6–7), 491–495.

Hosseinzadeh, H., Modaghegh, M. H., & Saffari, Z. (2009). Crocus sativus L. (Saffron) extract and its active constituents (crocin and safranal) on ischemia-reperfusion in rat skeletal muscle. *Evidence-Based Complement and Alternative Medicine, 6*(3), 343–350.

Hosseinzadeh, H., Sadeghnia, H. R., Ghaeni, F. A., Motamedshariaty, V. S., & Mohajeri, S. A. (2012). Effects of saffron (*Crocus sativus* L.) and its active constituent, crocin, on recognition and spatial memory after chronic cerebral hypoperfusion in rats. *Phytotherapy Research, 26*(3), 381–386.

Hosseinzadeh, H., Mehri, S., Abolhassani, M. M., Ramezani, M., Sahebkar, A., & Abnous, K. (2013). Affinity-based target deconvolution of safranal. *Daru, 21*(1), 25.

Howes, M. J., & Houghton, P. J. (2012). Ethnobotanical treatment strategies against Alzheimer's disease. *Current Alzheimer Research, 9*(1), 67–85.

Howes, M. J., & Perry, E (2011). The role of phytochemicals in the treatment and prevention of dementia. *Drugs Aging, 28*(6), 439–468.

Husaini, A.M., Wani, S.A., Sofi, P., Rather, A.G., Parray, G.A., Shikari, A.B., Mir, J.I. (2009). Bioinformatics for saffron (Crocus sativus L.) improvement. Communications in Biometry and Crop Science, 4(1), 3e8.

Husaini, A.M., Hassan, B., Ghani, M.Y., Teixeira da Silva, J.A., Kirmani, N.A., (2010) Saffron (*Crocus sativus* Kashmirianus) cultivation in Kashmir: Practices and problems. In: Husaini, A.M., (Ed) Saffron. Functional Plant Science and Biotechnology 4 (Special Issue 2), 108–115.

Husaini, A. M. (2014). Challenges of climate change: Omics-based biology of saffron plants and organic agricultural biotechnology for sustainable saffron production. *GM Crops and Food, 5*, 97–105.

Ilahi, I., Jabeen, M., & Firdous, N. (1987). Morphogenesis with saffron tissue cultures. *Journal of Plant Physiology*, *128*, 227–232.

Imenshahidi, M., Hosseinzadeh, H., & Javadpour, Y. (2010). Hypotensive effect of aqueous saffron extract (*Crocus sativus* L.) and its constituents, safranal and crocin, in normotensive and hypertensive rats. *Phytotherapy Research*, *24*(7), 990–994.

International Trade Centre (2018). Afghanistan's National Export Strategy 2018-2022, Saffron Sector. Geneva.

Ishizuka, F., Shimazawa, M., Umigai, N., Ogishima, H., Nakamura, S., Tsuruma, K., & Hara, H. (2013). Crocetin, a carotenoid derivative, inhibits retinal ischemic damage in mice. *European Journal of Pharmacology*, *703*(1–3), 1–10.

ISO 3632-1:2011. Spices—Saffron (Crocus sativus L.). Food Products SC 7, Spices, Culinary Herbs and Condiments; International Organization for Standardization: Geneva, Switzerland, 2011.

Jackson, D., Culianez-Macia, F., Prescott, A. G., Roberts, K., & Martin, C. (1991). Expression patterns of myb genes from Antirrhinum flowers. *Plant Cell*, *3*, 115–125.

Jain, M., Srivastava, P. L., Verma, M., Ghangal, R., & Garg, R. (2016). De novo transcriptome assembly and comprehensive expression profiling in *Crocus sativus* to gain insights into apocarotenoid biosynthesis. *Scientific Reports*, *6*, 22456 https://doi.org/10.1038/srep22456.

Jessie, S. W., & Krishnakantha, T. P. (2005). Inhibition of human platelet aggregation and membrane lipid peroxidation by food spice, saffron. *Molecular and Cellular Biochemistry*, *278*(1–2), 59–63.

Jiang, C., Cao, L., Yuan, Y., Chen, M., Jin, Y., & Huang, L. (2014). Barcoding melting curve analysis for rapid, sensitive, and discriminating authentication of saffron (*Crocus sativus* L.) from its adulterants. *BioMedicine Research International*, 809037.

José Bagur, M., Alonso Salinas, G. L., Jiménez-Monreal, A. M., Chaouqi, S., Llorens, S., Martínez-Tomé, M., & Alonso, G. L. (2017). Saffron: an old medicinal plant and a potential novel functional food. *Molecules*, *23*, 30.

Joukar, S., Najafipour, H., Khaksari, M., Sepehri, G., Shahrokhi, N., Dabiri, S., Gholamhoseinian, A., & Hasanzadeh, S. (2010). The effect of saffron consumption on biochemical and histopathological heart indices of rats with myocardial infarction. *Cardiovascular Toxicology*, *10*(1), 66–71.

Joukar, S., Ghasemipour-Afshar, E., Sheibani, M., Naghsh, N., & Bashiri, A. (2013). Protective effects of saffron (*Crocus sativus*) against lethal ventricular arrhythmias induced by heart reperfusion in rat: a potential anti-arrhythmic agent. *Pharmaceutical Biology*, *51*(7), 836–843.

Kahlem, P., & Newfeld, S. J. (2009). Informatics approaches to understanding TGFβ pathway regulation. *Dev (Cambridge, England)*, *136*, 3729–3740.

Kakehi, K., Kinoshita, M., Oda, Y., & Abdul-Rahman, B. (2003). Lectin from bulbs of *Crocus sativus* recognizing n-linked core glycan: isolation and binding studies using fluorescence polarization. *Methods in Enzymology*, *362*, 512–522.

Kalha, C.S., Gupta, V., Gupta, D. (2007) First report of sclerotial rot of saffron caused by *Sclerotium rolfsii* in India. Plant Dis 91:1203–6. http://dx.doi.org/10.1094/PDIS-91-9-1203B.

Kanakis, C. D., Daferera, D. J., Tarantilis, P. A., & Polissiou, M. G. (2004). Qualitative determination of volatile compounds and qualitative evaluation of safranal and 4-hydroxy-2,6,6-trimethyl-1- cyclohexene-1-carboxaldehyde (HTCC) in Greek saffron. *Journal of Agricultural and Food Chemistry*, *52*, 4515–4521.

Kanakis, C. D., Tarantilis, P. A., Tajmir-Riahi, H. A., & Polissiou, M. G. (2007a). Crocetin, dimethylcrocetin, safranal bind human serum albumin, stability, antioxidative properties. *Journal of Agriculture and Food Chemistry*, *55*(3), 970–977.

Kanakis, C. D., Tarantilis, P. A., Tajmir-Riahi, H. A., & Polissiou, M. G. (2007b). DNA interaction with saffron's secondary metabolites safranal, crocetin, and dimethylcrocetin. *DNA and Cellular Biology*, *26*(1), 63–70.

Kang, C. K., Lee, H.Y., Jung, E. S., Seyedian, R., Jo, M. N., Kim, J. H., Kim, J. S., & Kim, E. Y. (2012). Saffron (*Crocus sativus* L.) increases glucose uptake and insulin sensitivity in muscle cells via multipathway mechanisms. *Food Chemistry, 135*(4), 2350–2358.

Kanth, R. H., Khanday, B. A., & Tabassum, S. (2008). Crop weather relationship for saffron production. In F. A. Nehvi, & S. A. Wani (Eds.), *Saffron Production in Jammu and Kashmir, Directorate of Extension Education* (pp. 170–188). India: SKUAST-K.

Karamian, R. (2004). Plantlet regeneration via somatic embryogenesis in four species of *Crocus*. *Acta Horticulturae, 650*, 253–259.

Karimi, E., Oskoueian, E., Hendra, R., & Jaafar, H. Z. (2010). Evaluation of *Crocus sativus* L. stigma phenolic and flavonoid compounds and its antioxidant activity. *Molecules, 15*(9), 6244–6256.

Kashani, L., Raisi, F., Saroukhani, S., Sohrabi, H., Modabbernia, A., Nasehi, A. A., Jamshidi, A., Ashrafi, M., Mansouri, P., Ghaeli, P., & Akhondzadeh, S. (2013). Saffron for treatment of fluoxetine-induced sexual dysfunction in women: randomized double-blind placebo-controlled study. *Human Psychopharmacology, 28*(1), 54–60.

Katzer, G. (2010) "Saffron (*Crocus sativus* L.)", Gernot Katzer's Spice Pages.

Kazi, H. A., & Qian, Z. (2009). Crocetin reduces TNBS-induced experimental colitis in mice by downregulation of NFkB. *Saudi Journal of Gastroenterology, 15*, 181–187.

Keify, F., & Beiki, A. H. (2012). Exploitation of random amplified polymorphic DNA (RAPD) and sequence-related amplified polymorphism (SRAP) markers for genetic diversity of saffron collection. *Journal of Medicinal Plants Research, 12*, 2761–2768.

Keyhani, E., Ghamsari, L., Keyhani, J., & Hadizadeh, M. (2006). Antioxidant enzymes during hypoxia-anoxia signaling events in *Crocus sativus* L. corm. *Annals of New York Academy of Science, 1091*, 65–75.

Khalili, M., & Hamzeh, F. (2010). Effects of active constituents of *Crocus sativus* L., crocin on streptozocin-induced model of sporadic Alzheimer's disease in male rats. *Iranian Biomedical Journal, 14*(1–2), 59–65.

Khan, I. A. (1996). Cytomorphological studies of saffron (*Crocus sativus* L.). *Indian Journal of Agricultural Research, 30*, 48–52.

Khorasani, G., Hosseinimehr, S. J., Zamani, P., Ghasemi, M., & Ahmadi, A. (2008). The effect of saffron (*Crocus sativus*) extract for healing of second-degree burn wounds in rats. *Keio Journal of Medicine, 57*(4), 190–195.

Khori, V., Alizadeh, A. M., Yazdi, H., Rakhshan, E., Mirabbasi, A., Changizi, S., Mazandarani, M., & Nayebpour, M. (2012). Frequency-dependent electrophysiological remodeling of the AV node by hydroalcohol extract of *Crocus sativus* L. (saffron) during experimental atrial fibrillation: the role of endogenous nitric oxide. *Phytotherapy Research, 26*(6), 826–832.

Kianbakht, S., & Ghazavi, A. (2011). Immunomodulatory effects of saffron: a randomized double-blind placebo controlled clinical trial. *Phytotherapy Research, 25*(12), 1801–1805.

Kianbakht, S. (2008). A systematic review on pharmacology of saffron and its active constituents. *Journal of Medicinal Plants, 4*(28), 1–27.

Kiran, D., Madhu, S., Markandey, S., & Paramvir, S. A. (2011). In vitro cormlet production and growth evaluation under greenhouse conditions in saffron (*Crocus sativus* L.) - A commercially important crop. *Life Sciences, 11*(1), 1–6.

Kirtikar, K.R., Basu, B.D., (1933) Eds. *Indian Medicinal Plants*, Vol. IV; M/s Lalit Mohan Basu, Leader road: Allahabad, India, 2462 pp.

Kubo, I., & Kinst-Hori, I. (1999). Flavonols from saffron flower: tyrosinase inhibitory activity and inhibition mechanism. *Journal of Agriculture and Food Chemistry, 47*(10), 4121–4125.

Kuratsune, H., Umigai, N., Takeno, R., Kajimoto, Y., & Nakano, T. (2010). Effect of crocetin from Gardenia jasminoides Ellis on sleep: a pilot study. *Phytomedicine, 17*, 840–843.

Laabich, A., Vissvesvaran, G. P., Lieu, K. L., Murata, K., McGinn, T. E., Manmoto, C. C., Sinclair, J. R., Karliga, I., Leung, D. W., Fawzi, A., & Kubota, R. (2006). Protective effect of crocin against blue light- and white light-mediated photoreceptor cell death in bovine and primate retinal primary cell culture. *Investigating Ophthalmology and Visual Science, 47*(7), 3156–3163.

Lage, M., & Cantrell, C. L. (2009). Quantification of saffron (*Crocus sativus* L.) metabolites crocins, picrocrocin and safranal for quality determination of the spice grown under different environmental Moroccan conditions. *Scientia Horticulturae, 12*, 366–373.

Larsen, B., Orabi, J., Pedersen, C., & Ørgaard, M. (2015). Large intraspecific genetic variation within the Saffron-*Crocus* group (*Crocus* L., series *Crocus*; Iridaceae). *Plant Systematics and Evolution, 301*(1), 425–437.

Lashay, A., Sadough, G., Ashrafi, E., Lashay, M., Movassat, M., & Akhondzadeh, S. (2016). Short-term outcomes of saffron supplementation in patients with age-related macular degeneration: A double-blind, Placebo-controlled, randomized trial. *Medical Hypothesis Discovery and Innovation of Ophthalmology, 5*, 32.

Lechtenberg, M., Schepmann, D., Niehues, M., Hellenbrand, N., Wünsch, B., & Hensel, A. (2008). Quality and functionality of saffron: quality control, species assortment and affinity of extract and isolated saffron compounds to NMDA and sigma1 (sigma-1) receptors. *Planta Medica, 74*(7), 764–772.

Leonti, M., Cabras, S., Castellanos, M. E., Challenger, A., Gertsch, J., & Casu, L. (2013). Bioprospecting: evolutionary implications from a post-olmec pharmacopoeia and the relevance of widespread taxa. *Journal of Ethnopharmacology, 147*(1), 92–107.

Li, C.Y., & Wu, T. S. (2002). Constituents of the stigmas of *Crocus sativus* and their tyrosinase inhibitory activity. *Journal of Natural Products, 65*(10), 1452–1456.

Li, C.Y., Lee, E. J., & Wu, T. S. (2004). Antityrosinase principles and constituents of the petals of *Crocus sativus*. *Journal of Natural Products, 67*(3), 437–440.

Liakopoulou-Kyriakides, M., & Skubas, A. I. (1990). Characterization of the platelet aggregation inducer and inhibitor isolated from *Crocus sativus*. *Biochemistry International, 22*(1), 103–110.

Liakopoulou-Kyriakides, M., Sinakos, Z., & Kyriakidi, D. A. (1985). A high molecular weight platelet aggregating factor in *Crocus sativus*. *Plant Science, 40*(2), 117–120.

Lin, J. K., & Wang, C. J. (1986). Protection of crocin dyes on the acute hepatic damage induced by aflatoxin B and dimethylnitrosamine in rats. *Carcinogenesis*, 7(4), 595–599.

Linardaki, Z. I., Orkoula, M. G., Kokkosis, A. G., Lamari, F. N., & Margarity, M. (2013). Investigation of the neuroprotective action of saffron (*Crocus sativus* L.) in aluminum exposed adult mice through behavioral, neurobiochemical assessment. *Food and Chemical Toxicology, 52*, 163–170.

Liu, T. Z., & Qian, Z.Y. (2002). Pharmacokinetics of crocetin in rats. *Yao Xue Xue Bao, 37*(5), 367–369.

Lopresti, A. L., & Drummond, P. D. (2014). Saffron (*Crocus sativus*) for depression: A systematic review of clinical studies and examination of underlying antidepressant mechanisms of action. *Human Psychopharmacology, 29*, 517–527.

Loskutov, A.V., Beninger, C. W., Ball, T. M., Hosfield, G. L., Nair, M., & Sink, K. C. (1999). Optimization of in vitro conditions for stigma-like structure production from half-ovary explants of Crocus sativus L. vitro. *Cellular Development in Biological Plant, 35*, 200–205.

Ma, S. P., Zhou, S., Shu, B., & Zhou, J. (1998). Pharmacological studies on *Crocus* glycosides I. Effects on anti-inflammatory and immune function. *Zhongcaoyao, 29*, 536–539.

Maccarone, R., Di Marco, S., & Bisti, S. (2008). Saffron supplement maintains morphology and function after exposure to damaging light in mammalian retina. *Investigating Ophthalmology and Visual Science, 49*(3), 1254–1261.

Madan, C.L., Kapoor, B.M., Gupta, U.S., (1967) Saffron. Economic Botany 20:377-85; http://dx.doi.org/10.1007/BF02904059.

Majourhay, K., Fernandez, J. A., Martınez-Gomez, P., & Piqueras, A. (2007). Enhanced plantlet regeneration from cultured meristems in sprouting buds of saffron corms. *Acta Horticulturae, 739*, 275–278.

Makri, O. E., Ferlemi, A.V., Lamari, F. N., & Georgakopoulos, C. D. (2013). Saffron administration prevents selenite induced cataractogenesis. *Molecular Vision, 19*, 1188–1197.

Malaekeh-Nikouei, B., Mousavi, S. H., Shahsavand, S., Mehri, S., Nassirli, H., & Moallem, S. A. (2013). Assessment of cytotoxic properties of safranal and nanoliposomal safranal in various cancer cell lines. *Phytotherapy Research, 27*, 1868–1873. doi: 10.1002/ptr.4945.

Marangoni, D., Falsini, B., Piccardi, M., Ambrosio, L., Minnella, A. M., Savastano, M. C., Bisti, S., Maccarone, R., Fadda, A., Mello, E., et al. (2013). Functional effect of saffron supplementation and risk genotypes in early age-related macular degeneration: A preliminary report. *Journal of Translational Medicine, 11*, 228.

Masaki, M., Aritake, K., Tanaka, H., Shoyama, Y., Huang, Z. L., & Urade, Y. (2012). Crocin promotes non-rapid eye movement sleep in mice. *Molecular Nutrition and Food Research, 56*(2), 304–308.

Masuda, A., Mori, K., & Miyazawa, M. (2012). Comparative analysis of volatile compounds from corms of *Crocus sativus* and *C. vernus*. *Chemistry of Natural Compounds, 48*(2), 319–321.

Mathew, B. (1977). *Crocus sativus* and its allies (Iridaceae). *Plant Systematics and Evolution, 128*, 89–103.

Mathew, B. (1982). *The Crocus. A Revision of the Genus* Crocus *(Iridaceae)*. Portland: Timber Press.

McGee, H.J. (2004) Ed. *On Food and Cooking: The Science and Lore of the Kitchen*; Press Scribner: New York, NY, 896.

Mehdizadeh, R., Parizadeh, M. R., Khooei, A. R., Mehri, S., & Hosseinzadeh, H. (2013). Cardioprotective effect of saffron extract and safranal in isoproterenol-induced myocardial infarction in Wistar rats. *Iranian Journal of Basic Medicine Science, 16*(1), 56–63.

Mehri, S., Abnous, K., Mousavi, S. H., Shariaty, V. M., & Hosseinzadeh, H. (2012). Neuroprotective effect of crocin on acrylamide-induced cytotoxicity in PC12 cells. *Cell Molecular and Neurobiology, 32*(2), 227–235.

Melnyk, J. P., Wang, S., & Marcone, M. F. (2010). Chemical and biological properties of the world's most expensive spice: Saffron. *Food Research International, 43*(8), 1981–1989.

Mir, J. I., Ahmed, N., Wafai, A. H., & Qadri, R. A. (2012). Relative expression of CsZCD gene and apocarotenoid biosynthesis during stigma development in *Crocus sativus* L. *Physiology of Molecular Biology and Plants, 18*(4), 371–375 https://doi.org/10.1007/s12298-012-0131-9.

Mir JI, Ahmed N, Kumar R, Shafi W, Rashid R, Sheikh MA (2012b) Apocarotenoid gene expression in *in-vitro* developed stigma like structures in (*Crocus sativus* L.): Souvenir and Abstracts 4th International Saffron Symposium. October 22-25:2.8. p. 25.

Mir, J. I., Ahmed, N., Singh, D. B., Khan, M. H., Zffer, S., & Shafi, W. (2015). Breeding and biotechnological opportunities in saffron crop improvement. *African Journal of Agricultural Research, 10*, 1970–1974.

Mizuma, H., Tanaka, M., Nozaki, S., Mizuno, K., Tahara, T., Ataka, S., Sugino, T., Shirai, T., Kajimoto, Y., Kuratsune, H., Kajimoto, O., & Watanabe, Y. (2009). Daily oral administration of crocetin attenuates physical fatigue in human subjects. *Nutrition Research (N. Y., NY, U. S.), 29*, 145–150.

Moallem, S. A., Hariri, A. T., Mahmoudi, M., & Hosseinzadeh, H. (2013). Effect of aqueous extract of *Crocus sativus* L. (saffron) stigma against subacute effect of diazinon on specific biomarkers in rats. *Toxicology and Industrial Health, 30*(2), 141–146.

Modabbernia, A., Sohrabi, H., Nasehi, A. A., Raisi, F., Saroukhani, S., Jamshidi, A., Tabrizi, M., Ashrafi, M., & Akhondzadeh, S. (2012). Effect of saffron on fl uoxetine- induced sexual impairment in men: randomized double- blind placebo controlled trial. *Psychopharmacology (Berl), 223*(4), 381–388.

Modaghegh, M. H., Shahabian, M., Esmaeili, H. A., Rajbai, O., & Hosseinzadeh, H. (2008). Safety evaluation of saffron (*Crocus sativus)* tablets in healthy volunteers. *Phytomedicine, 15*(12), 1032–1037.

Mokhtari Hashtjini, M., Pirzad Jahromi, G., Meftahi, G. H., Esmaeili, D., & Javidnazar, D. (2018a). Aqueous extract of saffron administration along with amygdala deep brain stimulation promoted alleviation of symptoms in post-traumatic stress disorder (PTSD) in rats. *Avicenna Journal of Phytomedicine*, 1–11.

Molina, R., Valero, M., Navarro, Y., Guardiola, J., & García-Luis, A. (2005). Temperature effects on flower formation in saffron (*Crocus sativus* L.). *Scientific Horticulture*, *103*, 361–379.

Montoro, P., Tuberoso, C. I. G., Maldini, M., Cabras, P., & Pizza, C. (2008). Qualitative profi le and quantitative determination of flavonoids from *Crocus sativus* L. petals by LC-MS/M. *Natural Products Community*, *3*(12), 2013–2016.

Montoro, P., Maldini, M., Luciani, L., Tuberoso, C. I., Congiu, F., & Pizza, C. (2012). Radical scavenging activity and LC-MS metabolic profiling of petals, stamens, and flowers of *Crocus sativus* L. *Journal of Food Science*, 77(8), C893–C900.

Moshiri, E., Basti, A. A., Noorbala, A. A., Jamshidi, A. H., Abbasi, S. H., & Akhondzadeh, S. (2006). *Crocus sativus* L.(petal) in the treatment of mild-to-moderate depression: a double-blind, randomized and placebo-controlled trial. *Phytomedicine*, *13*(9–10), 607–611.

Motamedi, H., Darabpour, E., Gholipour, M., & Seyyed Nejad, S. M. (2010). In vitro assay for the anti-Brucella activity of medicinal plants against tetracycline-resistant Brucella melitensis. *Journal of Zhejiang University of Science B*, *11*(7), 506–511.

Mousavi, S. Z., & Bathaie, S. Z. (2011). Historical uses of saffron: Identifying potential new avenues for modern research. *Avicenna Journal of Phytomedicine*, *1*(2), 57–66.

Mousavi, S. H., Tavakkol-Afshari, J., Brook, A., & Jafari- Anarkooli, I. (2009). Role of caspases and Bax protein in saffron-induced apoptosis in MCF-7 cells. *Food Chemistry and Toxicology*, *47*(8), 1909–1913.

Mousavi, S. H., Tayarani, N. Z., & Parsaee, H. (2010). Protective effect of saffron extract and crocin on reactive oxygen species-mediated high glucose-induced toxicity in PC12 cells. *Cell Molecular and Neurobiology*, *30*(2), 185–191.

Mousavi, S. H., Moallem, S. A., Mehri, S., Shahsavand, S., Nassirli, H., & Malaekeh-Nikouei, B. (2011). Improvement of cytotoxic and apoptogenic properties of crocin in cancer cell lines by its nanoliposomal form. *Pharmacology Biology*, *49*(10), 1039–1045.

Munby, L. (1992). *Concise encyclopedia of world history*. Tehran: Elm o Zendegi.

Munshi AM, Wani SA, Tak GM (2002) Improved cultivation practices for saffron. In: Proceedings of Seminar-cum-Workshop on Saffron (Crocus sativus), June 14, 2001. SKUAST-K, India, pp 83-88.

Munshi, A. M. (1989). Economic analysis of saffron under rainfed conditions of Kashmir. *Agricultural Situation in India*, *44*, 379–381.

Mystery solved—biologists explain the genetic origins of the saffron crocus (2019, March 12) retrieved 17 April 2019 from https://phys.org/news/2019-03-mystery-solvedbiologists-geneticsaffron-crocus.html.

Nørbæk, R., Brandt, K., Kvist Nielsen, J., Ørgaard, M., & Jacobsen, N. (2002). Flower pigment composition of Crocus species and cultivars used for a chemotaxonomic investigation. *Biochemical Systematics and Ecology*, *30*, 763.

Naghibi, S. M., Hosseini, M., Khani, F., Rahimi, M., Vafaee, F., Rakhshandeh, H., & Aghaie, A. (2012). Effect of aqueous extract of *Crocus sativus* L. on morphine-induced memory impairment. *Advanced Pharmaceutical Science* Article ID 49436. https://doi.org/10.1155/2012/494367.

Naghizadeh, B., Mansouri, S. M. T., & Mashhadian, N.V. (2010). Crocin attenuates cisplatin-induced renal oxidative stress in rats. *Food Chemistry and Toxicology*, *48*(10), 2650–2655.

Naghizadeh, B., Mansouri, S. M. T., & Mashhadian, N.V. (2011). Evaluation the effect of crocin (the major pigment of *Crocus sativus*) on cisplatin-induced renal toxicity. *Toxicology Letters*, *205*(Supplement), S115.

Nair, S. C., Pannikar, B., & Panikkar, K. R. (1991a). Antitumour activity of saffron (*Crocus sativus*). *Cancer Letters*, *57*(2), 109–114.

Nair, S. C., Salomi, M. J., Panikkar, B., & Panikkar, K. R. (1991b). Modulatory effects of *Crocus sativus* and *Nigella sativa* extracts on cisplatin-induced toxicity in mice. *Journal of Ethnopharmacology*, *31*(1), 75–83.

Nair, S. C., Kurumboor, S. K., & Hasegawa, J. H. (1995). Saffron chemoprevention in biology and medicine: a review. *Cancer Biotherapy*, *10*(4), 257–264.

Nam, K. N., Park, Y. M., Jung, H. J., Lee, J. Y., Min, B. D., Park, S. U., Jung, W. S., Cho, K. H., Park, J. H., Kang, I., Hong, J. W., & Lee, E. H. (2010). Anti-inflammatory effects of crocin and crocetin in rat brain microglial cells. *Eurapean Journal of Pharmacology*, *648*, 110–116.

Nauriyal, J. P., Gupta, R., & George, C. K. (1977). Saffron in India. *Arecanut and Spice Bulletin*, *8*, 59–72.

Negbi, M., & Negbi, O. (2002). The painted plaster floor of the Tell Kabri palace: reflections on saffron domestication in the Aegean Bronze Age. In E. D. Oren, & S. Ahituv (Eds.), *Aharon Kempinski Memorial Volume* (pp. 325–340). Beer-Sheva: Ben-Gurion Univ. Press.

Negbi, M. (1989). Theophrastus on geophytes. *Botanical Journal of the Linnean Society*, *100*, 15–43.

Negbi M (1999) *Saffron cultivation: past, present and future prospects*. In Saffron, Crocus sativus, L., Negbi, M., Ed.; Harwood Academic Publishers: Amsterdam, The Netherlands, 1-18.

Nehvi, F. A., & Yasmin, S. (2019). Saffron (*Crocus sativus* L.) Crop insurance to mitigate ill effects of climate change-a priority of Jammu and Kashmir state. *International Journal of Current Microbiology and Applied Science*, *8*(01), 2972–2984 doi: https://doi.org/10.20546/ijcmas.2019.802.347.

Nehvi, F. A, & Salwee, Y. (2017). Advances in saffron research for integrated development of saffron in Kashmir, India. *Acta horticulturae*, *1184*, 63–68.

Nemati, H., Boskabady, M. H., & AhmadzadefVostakolaei, H. (2008). Stimulatory effect of *Crocus sativus* (saffron) on beta2-adrenoceptors of guinea pig tracheal chains. *Phytomedicine*, *15*(12), 1038–1045.

Nemati, Z., Zeinalabedini, M., Mardi, M., Pirseyediand, S. M., Marashi, S. H., & Nekoui, S. M. K. (2012). Isolation and characterization of a first set of polymorphic microsatellite markers in saffron, Crocus sativus (Iridaceae). *The American Journal of Botany*, *99*, e340–e343.

Nemati, Z., Blattner, F. R., Kerndorff, H., Erol, O., & Harpke, D. (2018). Phylogeny of the saffron Crocus species group, *Crocus* series *Crocus* (Iridaceae). *Molecular Phylogenetics and Evolution*, *127*, 891–897.

Nemati, Z., Harpke, D., Gemicioglu, A., Kerndorff, H., & Blattner, F. R. (2019). Saffron (*Crocus sativus*) is an autotriploid that evolved in Attica (Greece) from wild *Crocus cartwrightianus*. *Molecular Phylogenetics and Evolution*, *136*, 14–20.

Nemati, Z. (2018). The origin of saffron: progenitors, areas and transcriptomics of economic traits. *Dissertation*doi: 10.13140/RG.2.2.16290.89281.

Nithya, G., & Sakthisekaran, D. (2015). In Silico Docking Studies On The Anti-Cancer Effect OfThymoquinone On Interaction With Pten- A Regulator Of Pi3k/ Akt Pathway. *Asian Journal of Pharmaceutical and Clinical Research*, *1*, 192–195.

Noorbala, A. A., Akhondzadeh, S., Tahmacebi-Pour, N., & Jamshidi, A. H. (2005). Hydro-alcoholic extract of *Crocus sativus* L. versus fluoxetine in the treatment of mild to moderate depression: a double-blind, randomized pilot trial. *Journal of Ethnopharmacology*, *97*(2), 281–284.

Noureini, S. K., & Wink, M. (2012). Antiproliferative effects of crocin in HepG2 cells by telomerase inhibition and hTERT down-regulation. *Asian Pacific Journal of Cancer Prevention*, *13*(5), 2305–2309.

Ochiai, T., Ohno, S., Soeda, S., Tanaka, H., Shoyama, Y., & Shimeno, H. (2004a). Crocin prevents the death of rat pheochromyctoma (PC-12) cells by its antioxidant effects stronger than those of α-tocopherol. *Neuroscience Letters*, *362*(1), 61–64.

Ochiai, T., Soeda, S., Ohno, S., Tanaka, H., Shoyama, Y., & Shimeno, H. (2004b). Crocin prevents the death of PC-12 cells through sphingomyelinase-ceramide signaling by increasing glutathione synthesis. *Neurochemistry International*, *44*(5), 321–330.

Ochiai, T., Shimeno, H., Mishima, K., Iwasaki, K., Fujiwara, M., Tanaka, H., Shoyama, Y., Toda, A., Eyanagi, R., & Soeda, S. (2007). Protective effects of carotenoids from saffron on neuronal injury in vitro and in vivo. *Biochim and Biophysics Acta*, *1770*(4), 578–584.

Oda, Y., Nakayama, K., Abdul-Rahman, B., Kinoshita, M., Hashimoto, O., Kawasaki, N., Hayakawa, T., Kakehi, K., Tomiya, N., & Lee, Y. C. (2000). *Crocus sativus* lectin recognizes Man3GlcNAc in the N-glycan core structure. *Journal of Biological Chemistry*, *275*(35), 26772–26779.

Ohba, T., Ishisaka, M., Tsujii, S., Tsuruma, K., Shimazawa, M., Kubo, K., et al. (2016). Crocetin protects ultraviolet A-induced oxidative stress and cell death in skin in vitro and in vivo. *European Journal of Pharmacology*, *789*, 244–253.

Ohno, Y., Nakanishi, T., Umigai, N., Tsuruma, K., Shimazawa, M., & Hara, H. (2012). Oral administration of crocetin prevents inner retinal damage induced by N-methyl-Daspartate in mice. *European Journal of Pharmacology*, *690*(1–3), 84–89.

Ordoudi, S. A., & Tsimidou, M. Z. (2004). Production practices and quality assessment of food crops. In R. Dris, & S. M. Jain (Eds.), *Saffron Quality: Effect of agricultural practices, processing and storage* (pp. 209–260). Netherlands: Kluwer Academic Publ. Dordrecht.

Ordoudi, S. A., Befani, C. D., Nenadis, N., Koliakos, G. G., & Tsimidou, M. Z. (2009). Further examination of antiradical properties of *Crocus sativus* stigmas extract rich in crocins. *Journal of Agriculture and Food Chemistry*, *57*(8), 3080–3086.

Ordoudi, S. A., Cagliani, L. R., Lalou, S., Naziri, E., Tsimidou, M. Z., & Consonni, R. (2015). H NMR-based metabolomics of saffron reveals markers for its quality deterioration. *Food Research International*, *70*, 1–6 http://dx.doi.org/10.1016/j.foodres.2015.01.021.

Ørgaard, M., Jacobsen, N., & Heslop-Harrison, J. S. (1995). The hybrid origin of two cultivars of *Crocus* (Iridaceae) analysed by molecular cytogenetics including genomic Southern and *in situ* hybridization. *Annals of Botany*, *76*(3), 253–262.

Origin of the saffron crocus traced back to Greece (2019) retrieved 17 April 2019 from https://phys.org/news/2019-04-saffron-crocus-greece.html.

Panchangam, S. S., Vahedi, M., Megha, M. J., Kumar, A., Raithatha, K., Suravajhala, P., & Reddy, P. V. V. (2016). Saffron'omics': the challenges of integrating omic technologies. *Avicenna Journal of Phytomedicine*, *6*(6), 604–620.

Pandita, D., (2018). Plant miRNAs: micro structure and macro character research & reviews: Journal of Agriculture and Allied Sciences. 7(1): 83–84.

Pandita, D. (2019). Plant MIRnome: miRNA Biogenesis and Abiotic Stress Response. In M. Hasanuzzaman, K. Hakeem, K. Nahar, & H. Alharby (Eds.), *Plant Abiotic Stress Tolerance*. Cham: Springer https://doi.org/10.1007/978-3-030-06118-0_18.

Papandreou, M. A., Kanakis, C. D., Polissiou, M. G., Efthimiopoulos, S., Cordopatis, P., Margarity, M., & Lamari, F. N. (2006). Inhibitory activity on amyloid-beta aggregation and antioxidant properties of *Crocus sativus* stigmas extract and its crocin constituents. *Journal of Agriculture and Food Chemistry*, *54*(23), 8762–8768.

Papandreou, M. A., Tsachaki, M., Efthimiopoulos, S., Cordopatis, P., Lamari, F. N., & Margarity, M. (2011). Memory enhancing effects of saffron in aged mice are correlated with antioxidant protection. *Behavior Brain Research*, *219*(2), 197–204.

Pentony, M. M., Winters, P., Penfold-Brown, D., Drew, K., Narechania, A., DeSalle, R., Bonneau, R., & Purugganan, M. D. (2012). The Plant Proteome folding project: structure and positive selection in plant protein families. *Genome Biological Evolution*, *4*, 360–371.

Petersen, G., Seberg, O., Thorsøe, S., Jørgensen, T., & Mathew, B. (2008). A phylogeny of the genus *Crocus* (Iridaceae) based on sequence data from five plastid regions. *Taxon*, *57*, 487–499.

Petrakis, E. A., Cagliani, L. R., Polissiou, M. G., & Consonni, R. (2015). Evaluation of saffron (Crocus sativus L.) adulteration with plant adulterants by 1H NMR metabolite fingerprinting. *Food Chemistry*, *173*, 890–896.

Pfander, H., & Schurtenberger, H. (1982). Biosynthesis of C 20 - carotenoids in *Crocus sativus*. *Phytochemistry*, *21*(5), 1039–1042.

Piccardi, M., Marangoni, D., Minnella, A. M., Savastano, M. C., Valentini, P., Ambrosio, L., Capoluongo, E., Maccarone, R., Bisti, S., & Falsini, B. (2012). A Longitudinal follow-up study of saffron supplementation in early age-related macular degeneration: Sustained benefits to central retinal function. *Evid-Based Complement Alternative Medicine*, 429124.

Pieper, U., Eswar, N., Davis, F. P., Braberg, H., Madhusudhan, M. S., Rossi, A., Marti-Renom, M., Karchin, R., Webb, B. M., Eramian, D., Shen, M. Y., Kelly, L., Melo, F., & Sali, A. (2006). MODBASE: a database of annotated comparative protein structure models and associated resources. *Nucleic Acid Research*, *34*, 291–295.

Pintado, C., de Miguel, A., Acevedo, O., Nozal, L., Novella, J. L., & Rafael Rotger, R. (2011). Bactericidal effect of saffron (*Crocus sativus* L.) on Salmonella enterica during storage. *Food Control*, *22*(3–4), 638–642.

Piqueras, A., Han, B. H., Escribano', J., Rubio', C., Hellín, E., & Fernández, J. A. (1999). Development of cormogenic nodules and microcorms by tissue culture, a new tool for the multiplication and genetic improvement of saffron. *Agronomics*, *19*, 603–610.

Pirnia H (2006) Ancient Persia Tehran, Negah.

Pitsikas, N., & Sakellaridis, N. (2006). *Crocus sativus* L. extracts antagonize memory impairments in different behavioural tasks in the rat. *Behavior Brain Research*, *173*(1), 112–115.

Pitsikas, N., Zisopoulou, S., Tarantilis, P. A., Kanakis, C. D., Polissiou, M. G., & Sakellaridis, N. (2007). Effects of the active constituents of *Crocus sativus* L., crocins on recognition and spatial rats' memory. *Behavior and Brain Research*, *183*(2), 141–146.

Pitsikas, N., Boultadakis, A., Georgiadou, G., Tarantilis, P. A., & Sakellaridis, N. (2008). Effects of the active constituents of *Crocus sativus* L., crocins, in an animal model of anxiety. *Phytomedicine*, *15*(12), 1135–1139.

Pitsikas, N. (2015). The effect of *Crocus sativus* L. and its constituents on memory: basic studies and clinical applications. *Evidence Based Complement Alternative Medicine*, 1–7, http://dx.doi.org/10.1155/2015/926284.

Pitsikas, N. (2016). Constituents of saffron (*Crocus sativus* L.) as potential candidates for the treatment of anxiety disorders and schizophrenia. *Molecules*, *21*, 303e314.

Plessner, O., Ziv, M., & Negbi, M. (1990). In vitro corm production in the saffron crocus (*Crocus sativus* L.). *Plant Cell Tissue Organ and Culture*, *20*, 89–94.

Poma, A., Fontecchio, G., Carlucci, G., & Chichiriccò, G. (2012). Anti-inflammatory properties of drugs from saffron *Crocus*. *Antiinflammatory and Antiallergy Agents Medicinal Chemistry*, *11*(1), 37–51.

Pourmasoumi, M., Hadi, A., Najafgholizadeh, A., Kafeshani, M., & Sahebkar, A. (2019). Clinical evidence on the effects of saffron (*Crocus sativus* L.) on cardiovascular risk factors: a systematic review meta-analysis. *Pharmacology Research*, *139*, 348–359.

Premkumar, K., Abraham, S. K., Santhiya, S. T., Gopinath, P. M., & Ramesh, A. (2001). Inhibition of genotoxicity by saffron (*Crocus sativus* L.) in mice. *Drug Chemistry and Toxicology*, *24*(4), 421–428.

Premkumar, K., Abraham, S. K., Santhiya, S. T., & Ramesh, A. (2003a). Inhibitory effects of aqueous crude extract of saffron (*Crocus sativus* L.) on chemical-induced genotoxicity in mice. *Asia Pacific Journal of Clinical Nutrition*, *12*(4), 474–476.

Premkumar, K., Abraham, S. K., Santhiya, S. T., & Ramesh, A. (2003b). Protective effects of saffron (*Crocus sativus* Linn.) on genotoxins-induced oxidative stress in Swiss albino mice. *Phytotherapy Research*, *17*(6), 614–617.

Premkumar, K., Kavitha, S., Santhiya, S.T., Ramesh, A. R., & Suwanteerangkul, J. (2004). Interactive effects of saffron with garlic and curcumin against cyclophosphamide induced genotoxicity in mice. *Asia Pacific Journal of Clinical Nutrition, 13*(3), 292–294.

Premkumar, K., Thirunavukkarasu, C., Abraham, S. K., Santhiya, S.T., & Ramesh, A. (2006). Protective effect of saffron (*Crocus sativus* L.) aqueous extract against genetic damage induced by anti-tumor agents in mice. *Human Experiments and Toxicology, 25*(2), 79–84.

Qi, Y., Chen, L., Zhang, L., Liu, W. B., Chen, X. Y., & Yang, X. G. (2013). Crocin prevents retinal ischaemia/reperfusion injury-induced apoptosis in retinal ganglion cells through the PI3K/AKT signalling pathway. *Experiment Eye Research, 107*, 44–51.

Rahiman, N., Akaberi, M., Sahebkar, A., Emami, S.A., & Tayarani-Najaran, Z. (2018). Protective effects of saffron and its active components against oxidative stress and apoptosis in endothelial cells. *Microvascular Research, 118*, 82–89.

Raja, W., Zaffer, G., & Wani, S. A. (2007). In vitro microcorm formation in saffron (*Crocus sativus* L.). *Acta Horticulturae, 739*, 291–296.

Rajaei, Z., Hadjzadeh, M. A., Nemati, H., Hosseini, M., Ahmadi, M., & Shafiee, S. (2013). Antihyperglycemic and antioxidant activity of crocin in streptozotocin-induced diabetic rats. *Journal of Medicinal Food, 16*(3), 206–210.

Ramadan, A., Soliman, G., Mahmoud, S. S., Nofal, S. M., & Abdel-Rahman, R. F. (2012). Evaluation of the safety and antioxidant activities of *Crocus sativus* and propolis ethanolic extracts. *Journal of Saudi Chemical Society, 16*(1), 13–21.

Rana SS, Sood P, Mondal KK, Thakur KS (2003) Eds. *Technology for saffron cultivation*; Mountain Agricultural Research and Extension Centre publication: Sangla (Kinnaur), HP, India, 20.

Rastgoo, M., Hosseinzadeh, H., Alavizadeh, H., Abbasi, A., Ayati, Z., & Jaafari, M. R. (2013). Antitumor activity of PEGylated nanoliposomes containing crocin in mice bearing C26 colon carcinoma. *Planta Medica, 79*(6), 447–451.

Razavi, B. M., Hosseinzadeh, H., Abnous, K., Motamedshariaty, V. S., & Imenshahidi, M. (2013a). Crocin restores hypotensive effect of subchronic administration of diazinon in rats. *Iranian Journal of Basic Medical Science, 16*(1), 64–72.

Razavi, B. M., Hosseinzadeh, H., Movassaghi, A. R., Imenshahidi, M., & Abnous, K. (2013b). Protective effect of crocin on diazinon induced cardiotoxicity in rats in subchronic exposure. *Chemical Biology Interaction, 203*(3), 547–555.

Razavi, B., Hosseinzadeh, H., Abnous, K., & Imenshahidi, M. (2014). Protective effect of crocin on diazinon induced vascular toxicity in subchronic exposure in rat aorta ex-vivo. *Drug and Chemical Toxicology, 37*, 378–830.

Riazi, A., Panahi, Y., Alishiri, A. A., Hosseini, M. A., Zarchi, A. A. K., & Sahebkar, A. (2017). The impact of saffron (*Crocus sativus*) supplementation on visual function in patients with dry age-related macular degeneration. *Italian Journal of Medicine, 11*, 758.

Rios, J. L., Recio, M. C., Giner, R. M., & Manez, S. (1996). An update review of saffron and its active constituents. *Phytotherapy Research, 10*, 189–193.

Rubio-Moraga, A., Nohales, P. F., Perez, J. A., & Gomez-Gomez, L. (2004). Glucosylation of the saffron apocarotenoid crocetin by a glucosyltransferase isolated from *Crocus sativus* stigmas. *Planta, 219*, 955–966.

Rubio-Moraga, A., Rambla, J. L., Santaella, M., Gomez, M. D., Orzaez, D., Granell, A., & Gomez-Gomez, L. (2008). Cytosolic and plastoglobule-targeted carotenoid dioxygenases from *Crocus sativus* are both involved in β- ionone release. *Journal of Biological Chemistry, 283*, 24816–24825.

Rubio-Moraga A, Castillo-López R, Gómez-Gómez L, Ahrazem O (2009) Saffron is a monomorphic species as revealed by RAPD, ISSR and microsatellite analyses. BMC Research Notes 2: 189. https://doi.org/10.1186/1756-0500-2-189.

Rubio-Moraga, A., Rambla, J. L., Ahrazem, O., Granell, A., & Gomez-Gomez, L. (2009). Metabolite and target transcript analyses during *Crocus sativus* stigma development. *Phytochemistry, 70*(8), 1009–1016.

Rubio-Moraga, A., Trapero-Mozos, A., Gómez-Gómez, L., & Ahrazem, O. (2010). Inter simple sequence repeat markers for molecular characterization of *Crocus cartwrightianus* cv Albus. *Industrial Crop Production, 32*, 147–151.

Rubio-Moraga, A., Gerwig, G. J., Castro-Diaz, N., Jimeno, M. L., Escribano, J., Fernandez, J. A., & Kamerling, J. P. (2011). Triterpenoid saponins from corms of Crocus sativus : localization, extraction and characterization. *Industrial Crop Production, 34*(3), 915–1409.

Rubio-Moraga, A., Ahrazem, O., Rambla, J. L., Granell, A., & Gomez Gomez, L. (2013). Crocins with high levels of sugar conjugation contribute to the yellow colours of early-spring flowering crocus tepals. *PLoS One, 8*, e71946.

Rubio-Moraga, A., Gómez-Gómez, L., Trapero, A., Castro-Díaz, N., & Ahrazem, O. (2013). Saffron corm as a natural source of fungicides: the role of saponins in the underground. *Industrial Crops Production, 49*, 915.

Rubio-Moraga, A., Ahrazem, O., Pérez-Clemente, R. M., Gómez-Cadenas, A., Yoneyama, K., López-Ráez, J. A., Molina, R.V., & Gómez-Gómez, L. (2014). Apical dominance in saffron and the involvement of the branching enzymes CCD7 and CCD8 in the control of bud sprouting. *BMC Plant Biology, 14*, 171.

Sánchez-Vioque, R., Rodríguez-Conde, M. F., Reina-Ureña, J.V., Escolano-Tercero, M. A., Herraiz-Peñalver, D., & Santana-Méridas, O. (2012). In vitro antioxidant and metal chelating properties of corm, tepal and leaf from saffron (*Crocus sativus* L.). *Industrial Crops Production, 39*, 149–153.

Sachdeva, J., Tanwar, V., Golechha, M., Siddiqui, K. M., Nag, T. C., Ray, R., Kumari, S., & Arya, D. S. (2012). *Crocus sativus* L. (saffron) attenuates isoproterenol-induced myocardial injury via preserving cardiac functions and strengthening antioxidant defense system. *Experiment Toxicology and Pathology, 64*(6), 557–564.

Sadeghnia, H. R., Cortez, M. A., Liu, D., Hosseinzadeh, H., & Snead, O. C. (2008). Antiabsence effects of safranal in acute experimental seizure models: EEG and autoradiography. *Journal of Pharmacy and Pharmaceutical Science, 11*(3), 1–14.

Sadeghnia, H. R., Kamkar, M., Assadpour, E., Boroushaki, M. T., & Ghorbani, A. (2013). Protective effect of safranal, a constituent of *Crocus sativus*, on quinolinic acid-induced oxidative damage in rat hippocampus. *Iranian Journal of Basic Medical Science, 16*(1), 73–82.

Sadraei, H., Ghannadi, A., & Takei-bavani, M. (2003). Effects of *Zataria multiflora* and *Carum carvi* essential oils and hydroalcoholic extracts of *Passiflora incarnata, Berberis integerrima* and *Crocus sativus* on rat isolated uterus contractions. *International Journal of Aromatherapy, 13*(2–3), 121–127.

Safarinejad, M. R., Shafi ei, N., & Safarinejad, S. (2010). An open label, randomized, fixed-dose, crossover study comparing efficacy and safety of sildenafil citrate and saffron (*Crocus sativus* Linn.) for treating erectile dysfunction in men naïve to treatment. *International Journal of Impotence Research, 22*(4), 240–250.

Safarinejad, M. R., Shafi ei, N., & Safarinejad, S. (2011). A prospective double-blind randomized placebo-controlled study of the effect of saffron (*Crocus sativus* Linn.) on semen parameters and seminal plasma antioxidant capacity in infertile men with idiopathic oligo astheno terato zoospermia. *Phytotherapy Research, 25*(4), 508–516.

Saffari, B., Mohabatkar, H., & Mohsenzadeh, S. (2008). T and B-cell epitopes prediction of Iranian saffron (*Crocus sativus*) profiling by bioinformatics tools. *Protein Peptide Letters, 15*, 280–285.

Sahihi, M. (2015). In-Silico Study on the Interaction of Saffron Ligands and beta-Lactoglobulin by Molecular Dynamics and Docking Approach. *Journal of Macromolecular Science and Pharmacy*doi: 10.1080/00222348.2015.1125066.

Saito, H., Sugiura, M., Abe, K., Tanaka, H., Morimoto, S., Taura, F., & Shoyama, Y. (2001). Effects of ethanol extract of *Crocus sativus* L. and its components on learning behavior and long-term potentiation. *Study of Natural Products and Chemistry, 25*, 955–969.

Saito, H. (2004). The therapeutic and prophylactic effects of *Crocus sativus* L (saffron) in senile dementia. *Acta Horticulturae, 650*, 407–422.

Saleem, S., Ahmad, M., Ahmad, A. S., Yousuf, S., Ansari, M. A., Khan, M. B., Ishrat, T., & Islam, F. (2006). Effect of Saffron (*Crocus sativus)* on neurobehavioral and neurochemical changes in cerebral ischemia in rats. *Journal of Medicinal Food, 9*(2), 246–253.

Salomi, M. J., Nair, S. C., & Panikkar, K. R. (1991). Inhibitory effects of *Nigella sativa* and saffron (*Crocus sativus*) on chemical carcinogenesis in mice. *Nutrition Cancer, 16*(1), 67–72.

Samarghandian, S., & Borji, A. (2014). Anticarcinogenic effect of saffron (*Crocus sativus* L.) and its ingredients. *Pharmacognosy Research, 6*(2), 99–107.

Samarghandian, S., Boskabady, M. H., & Davoodi, S. (2010). Use of in vitro assays to assess the potential antiproliferative and cytotoxic effects of saffron (*Crocus sativus* L.) in human lung cancer cell line. *Pharmacognosy Magzine, 6*(24), 309–314.

Samarghandian, S., Afshari, J. T., & Davoodi, S. (2011). Suppression of pulmonary tumor promotion and induction of apoptosis by *Crocus sativus* L. extraction. *Applied Biochemistry and Biotechnology, 164*(2), 238–247.

Sampathu, S. R., Shivshankar, S., & Lewis, Y. S. (1984). Saffron (*Crocus sativus* Linn.) cultivation, processing, chemistry and standardization. *CRC Critical Reviews of Food Science and Nutrition, 20*, 123–157.

Sano, K., & Himeno, H. (1987). In vitro proliferation of saffron (*Crocus sativus* L.) stigma. *Plant Cell Tissue and Org Culture, 11*, 159–166.

Santhosh, M. S., Thushara, R. M., Hemshekhar, M., Devaraja, S., Kemparaju, K., & Girish, K. S. (2013a). *Vipera russelli* venom-induced oxidative stress and hematological alterations: amelioration by crocin a dietary colorant. *Cell Biochemical Function, 31*(1), 41–50.

Santhosh, M. S., Thushara, R. M., Hemshekhar, M., Sunitha, K., Devaraja, S., Kemparaju, K., & Girish, K. S. (2013b). Alleviation of viper venom induced platelet apoptosis by crocin (*Crocus sativus)*: implications for thrombocytopenia in viper bites. *Journal of Thrombcytic and Thrombolysis, 36*, 424–432.

Sarris, J., McIntyre, E., & Camfi eld, D. A. (2013). Plant-based medicines for anxiety disorders, Part 1: A review of preclinical studies. *CNS Drugs, 27*(3), 207–219.

Sarshoori, J. R., Asadi, M. H., & Mohammadi, M. T. (2014). Neuroprotective effects of crocin on the histopathological alterations following brain ischemia-reperfusion injury in rat. *Iranian Journal of Basic Medical Sciences, 17*(11), 895.

Schmidt, T., Heitkam, T., Liedtke, S., Schubert, V., & Menzel, G. (2019). Adding color to a century-old enigma: multi-color chromosome identification unravels the autotriploid nature of saffron (*Crocus sativus*) as a hybrid of wild *Crocus cartwrightianus* cytotypes. *New Phytologist*doi: 10.1111/nph.15715.

Searls, D. B. (2003). Pharmacophylogenomics: genes, evolution and drug targets. *Natural Review of Drug Discovery, 2*(8), 613–623.

Sepahi, S., Mohajeri, S. A., Hosseini, S. M., Khodaverdi, E., Shoeibi, N., Namdari, M., & Tabassi, S. A. S. (2018). Effects of crocin on diabetic maculopathy: a placebo-controlled randomized clinical trial. *American Journal of Ophthalmology, 190*, 89–98.

Serrano-Díaz, J., Sánchez, A. M., Maggi, L., Martínez-Tomé, M., García-Diz, L., Murcia, M. A., & Alonso, G. L. (2012). Increasing the applications of *Crocus sativus* flowers as natural antioxidants. *Journal of Food Science*, 77(11), C1162–C1168.

Shabani, R. (2001). *A Short Reviw on Iran History, from Medes to Qajar.* Tehran: Sokhan.

Shafiee, M., Arekhi, S., Omranzadeh, A., & Sahebkar, A. (2018). Saffron in the treatment of depression, anxiety and other mental disorders: Current evidence and potential mechanisms of action. *Journal of Affective Disorder, 227*, 330–337.

Shahi, T., Assadpour, E., & Jafari, S. (2016). Main chemical compounds and pharmacological activities of stigmas and tepals of 'red gold'; saffron. *Trends in Food Science and Technology, 58*, 69–78 http://dx.doi.org/10.1016/j.tifs. 2016.10.010.

Shamsa, A., Hosseinzadeh, H., Molaei, M., Shakeri, M.T., & Rajabi, O. (2009). Evaluation of *Crocus sativus* L. (saffron) on male erectile dysfunction: a pilot study. *Phytomedicine, 16*(8), 690–693.

Sharifi, G., Ebrahimzadeh, H., Ghareyazie, B., Gharechahi, J., & Vatankhah, E. (2012). Identification of differentially accumulated proteins associated with embryogenic and non-embryogeniccalli in saffron (*Crocus sativus* L.). *Proteome Science, 10*, 3–18.

Shati, A. A., Elsaid, F. G., & Hafez, E. E. (2011). Biochemical and molecular aspects of aluminium chloride-induced neurotoxicity in mice and the protective role of *Crocus sativus* L. extraction and honey syrup. *Neuroscience, 175*, 66–74.

Sheibani, M., Nemati, S. H., Davarinejad, G. H., Azghandi, A.V., & Habashi, A. A. (2007). Induction of somatic embryogenesis in saffron using thidiazuron (TDZ). *Acta Horticulturae, 739*, 259–268.

Shen, X. C., & Qian, Z.Y. (2006). Effects of crocetin on antioxidant enzymatic activities in cardiac hypertrophy induced by norepinephrine in rats. *Pharmazie, 61*(4), 348–352.

Shen, X. C., Qian, Z.Y., Chen, Q., & Wang, Y. J. (2004). Protective effect of crocetin on primary culture of cardiac myocyte treated with noradrenaline in vitro. *Yao Xue Xue Bao, 39*(10), 787–791.

Shen, J., Luo, Y. M., Ding, X.Y., & Mao, S. G. (2007). Authentication of *Crocus sativus* L. and its adulterants by rDNA ITS sequences and allele-specific PCR. *Journal of Nanjing Normal University, 30*, 89–92.

Sheng, L., Qian, Z., Zheng, S., & Xi, L. (2006). Mechanism of hypolipidemic effect of crocin in rats: crocin inhibits pancreatic lipase. *European Journal of Pharmacology, 543*(1–3), 116–122.

Sheng, L., Qian, Z., Shi, Y., Yang, L., Xi, L., Zhao, B., Xu, X., & Ji, H. (2008). Crocetin improves the insulin resistance induced by high-fat diet in rats. *British Journal of Pharmacology, 154*, 1016–1024.

Shukurova, P., & Babaev, R. (2010). A study into the effectiveness of the application of saffron extract in ocular pathologies in experiment. *Georgian Medical News, 182*, 38–42.

Siddiqui, M. J., Saleh, M. S. M., Basharuddin, S. N. B., Zamri, S. H., Mohd Najib, M. H., Che Ibrahim, M. Z., et al. (2018). Saffron (*Crocus sativus* L.): As an antidepressant. *Journal of Pharmacy and Biological Science, 10*, 173–180.

Sigerist, H.E. (1955). A history of medicineí, primitive and archaic medicine (2nd ed.). New York: Oxford University Press, 1, 330-490.

Singh, G. C., & Dhar, U. (1976). Origin of Kashmir saffron- a possible clue from weeds. *Science and Culture, 42*, 485–487.

Singla, R. k., & Bhat, G.V. (2011). Crocin: An overview. *Indo Global Journal of Pharmaceutical Sciences, 1*(4), 281–286.

Siracusa, L., Gresta, F., Avola, G., Albertini, E., Raggi, L., Marconi, G., Lombardo, G. M., & Ruberto, G. (2013). Agronomic, chemical and genetic variability of saffron (*Crocus sativus* L.) of different origin by LC-UV-vis-DAD and AFLP analyses. *Genetics Resource Crop Evolution, 60*, 711–721.

Skrubis B. (1990) In The cultivation in Greece of Crocus sativus L, Proceedings of the international conference on saffron (Crocus sativus L.) L'Aquila, Italy, 27-29 October, 1989; Tammaro F, Marra L, Eds. Universita Degil Studi L' Aquila e Academia Italians della Cucina, L'Aquila: Italy, 171-182.

Soeda, S., Ochiai, T., Paopong, L., Tanaka, H., Shoyama, Y., & Shimeno, H. (2001). Crocin suppresses tumor necrosis factor-alpha-induced cell death of neuronally differentiated PC-12 cells. *Life Science, 69*(24), 2887–2898.

Soeda, S., Ochiai, T., Shimeno, H., Saito, H., Abe, K., Sugiura, M., Tanaka, H., Taura, F., Morimoto, S., & Shoyama, Y. (2003). Promising pharmacological actions of croon in *Crocus sativus* on the central nervous system. *Study of Natural Products and Chemistry, 28*(1), 313–329.

Song, C. Q. (1990). Chemical constituents of saffron (*Crocus sativus*). II. The fl avonol compounds of petals. *Zhong Cao Yao, 2*, 439–441.

Souret, F., & Weathers, P. (2000). *Crocus sativus* L. (saffron): cultivation, in vitro culture, secondary metabolite production and phytopharmacognosy, Journal of Herbs. *Spices & Medicinal Plants, 6*, 99–116.

Sparg, S. G., Light, M. E., & van Staden, J. (2004). Biological activities and distribution of plant saponins. *Journal of Ethnopharmacology, 94*, 219–243.

Srivastava, R., Ahmed, H., Dixit, R. K., & Dharamveer, S. S. A. (2010). *Crocus sativus* L.: a comprehensive review. *Pharmacognosy Reviews, 4*(8), 200–208.

Srivastava, R. P. (1963). Cultivation of saffron in India. *Fertilizer News, 8*, 9–16.

Straubinger, M., Jezussek, M., Waibel, R., & Winterhalter, P. (1997). Novel glycosidic constituents from saffron. *Journal of Agricultural and Food Chemistry, 45*, 1678–1681.

Tóth, B., Hegyi, P., Lantos, T., Szakács, Z., Kerémi, B., Varga, G., Tenk, J., Pétervári, E., Balaskó, M., Rumbus, Z., Rakonczay, Z., Bálint, E. R., Kiss, T., & Csupor, D. (2019). The efficacy of saffron in the treatment of mild to moderate depression: a meta-analysis. *Planta Medica, 85*, 24–31 https://doi.org/10.1055/a-0660-9565.

Tang, L., Yan, F., Xu, Y., Rong, F., Li, S., & Chen, F. (2004). Determination of crocin-1 in rabbit plasma and the pharmacokinetics by RP-HPLC. *Yao Xue Xue Bao, 39*(10), 854–856.

Tang, F. T., Qian, Z. Y., Liu, P. Q., Zheng, S. G., He, S. Y., Bao, L. P., & Huang, H. Q. (2006). Crocetin improves endothelium dependent relaxation of thoracic aorta in hypercholesterolemic rabbit by increasing eNOS activity. *Biochemical Pharmacology, 72*(5), 558–565.

Tarantilis, P. A., & Polissiou, M. (1997). Isolation and identification of the aroma constituents of saffron (Crocus sativa). *Journal of Agriculture and Food Chemistry, 45*, 459–462.

Tarantilis, P. A., Morjani, H., Polissiou, M., & Manfait, M. (1994). Inhibition of growth and induction of differentiation of promyelocytic leukemia (HL-60) by carotenoids from *Crocus sativus* L. *Anticancer Research, 14*(5A), 1913–1918.

Tarantilis, P. A., Tsoupras, G., & Polissiou, M. G. (1995). Determination of saffron (*Crocus sativus* L.) components in crude plant extract using high-performance liquid chromatography-UV-visible photodiode-array detection-mass spectrometry. *Journal of Chromatography, 699*, 107–118.

Tarantilis, P. A., Beljebbar, A., Manfait, M., & Polissiou, M. (1998). FT-IR, FT-Raman spectroscopic study of carotenoids from saffron (Crocus sativus L.) and some derivatives. *Spectrochimica Acta Part A, 54*, 651–657.

Tavakkol-Afshari, J., Brook, A., & Mousavi, S. H. (2008). Study of cytotoxic and apoptogenic properties of saffron extract in human cancer cell lines. *Food Chemistry and Toxicology, 46*(11), 3443–3447.

Tavana, S., Eimani, H., Azarnia, M., Shahverdi, A., & Eftekhari-Yazdi, P. (2012). Effects of saffron (*Crocus sativus* L.) aqueous extract on in vitro maturation, fertilization and embryo development of mouse oocytes. *Cell Journal, 13*(4), 259–264.

Termentzi, A., & Kokkalou, E. (2008). LC-DAD-MS (ESI +) analysis and antioxidant capacity of *Crocus sativus* petal extracts. *Plant Medica, 75*(5), 573–581.

Thakur, R. N., Singh, C., & Kaul, B. L. (1992). First report of corm rot in Crocus sativus L. *Indian Phytopathology, 45*, 278–282.

The Royal Horticultural Society (2003) Plants of current interest. http://212.78.71.150/gardens/wisely/archive/wisleypcisept.asp. (Accessed 6 December 2006).

Tolner R (2005) Illustrierte Geschichte der Medizin. Tehran, Encyclopedia of medical history publications, 2, 611-694.

Torelli, A., Marieschi, M., & Bruni, R. (2014). Authentication of saffron (*Crocus sativus* L.) in different processed, retail products by means of SCAR markers. *Food Control, 36*, 126–131.

Torricelli, R., Yousefi Javan, I., Albertini, E., Venanzoni, R., & Hosseinzadeh, Y. G. (2019). Morphological and molecular characterization of Italian, Iranian and Spanish saffron

(Crocus sativus L.) accessions. *Applied Ecology and Environmental Research, 17*(2), 1875–1887 DOI: Http://Dx.Doi.Org/10.15666/Aeer/1702_18751887.

Trapero-Mozos, A., Ahrazem, O., Rubio-Moraga, A., Jimeno, M. L., Gomez, M. D., & Gomez-Gomez, L. (2012). Characterization of a glucosyltransferase enzyme involved in the formation of kaempferol and quercetin sophorosides in Crocus sativus. *Plant Physiology, 159* 1335.

Tsaftaris, A. S., Pasentsis, K., & Polidoros, A. N. (2005). Isolation of a differentially spliced C-type flower specific AG-like MADS-box gene from *Crocus sativus* and characterization of its expression. *Biologia Plantarum, 49*, 499–504.

Tsaftaris, A., Pasentsis, K., Makris, A., Darzentas, N., Polidoros, A., Kalivas, A., & Argiriou, A. (2011). The study of the E-class SEPALLATA3-like MADS-box genes in wild-type and mutant flowers of cultivated saffron (*Crocus sativus* L.) and its putative progenitors. *Journal Plant Physiology, 168*, 1675–1684.

Tsantarliotou, M. P., Poutahidis, T., Markala, D., Kazakos, G., Sapanidou, V., Lavrentiadou, S., Zervos, I., Taitzoglou, I., & Sinakos, Z. (2013). Crocetin administration ameliorates endotoxin-induced disseminated intravascular coagulation in rabbits. *Blood Coagulation Fibrinolysis, 24*(3), 305–310.

Tseng, T. H., Chu, C. Y., Huang, J. M., Shiow, S. J., & Wang, C. J. (1995). Crocetin protects against oxidative damage in rat primary hepatocytes. *Cancer Letters, 97*, 61–67.

Tung, N. H., & Shoyama, Y. (2013). New minor glycoside components from saffron. *Journal of Natural Medicine, 67*(3), 672–676.

Umigai, N., Murakami, K., Ulit, M. V., Antonio, L. S., Shirotori, M., Morikawa, H., & Nakano, T. (2011). The pharmacokinetic profile of crocetin in healthy adult human volunteers after a single oral administration. *Phytomedicine, 18*(7), 575–578.

Umigai, N., Tanaka, J., Tsuruma, K., Shimazawa, M., & Hara, H. (2012). Crocetin, a carotenoid derivative, inhibits VEGF-induced angiogenesis via suppression of p38 phosphorylation. *Current Neurovascular Research, 9*(2), 102–109.

Vakili, A., Einali, M. R., & Bandegi, A. R. (2013). Protective Effect of crocin against cerebral ischemia in a dose- dependent manner in a rat model of ischemic stroke. *Journal of Stroke Cerebrovascular Disease* S1052-3057(12)00345-X.

Vakili, A., Einali, M. R., & Bandegi, A. R. (2014). Protective effect of crocin against cerebral ischemia in a dose-dependent manner in a rat model of ischemic stroke. *Journal of Stroke and Cerebrovascular Diseases, 23*(1), 106–113.

Vatankhah, E., Niknam, V., & Ebrahimzadeh, H. (2010). Activity of antioxidant enzyme during in vitro organogenesis in *Crocus sativus*. *Biology Plant, 54*(3), 509–514.

Vavilov NI (1951) Ed. *The origin, variation, immunity and breeding of cultivated plants*, Translated from Russian by Chester, K.S.; The Ronald Press Company: New York. 364.

Walter, M. H., Floss, D. S., & Strack, D. (2010). Apocarotenoids: hormones, mycorrhizal metabolites and aroma volatiles. *Planta, 232*, 1–17.

Wang, Y., Han, T., Zhu, Y., Zheng, C. J., Ming, Q. L., Rahman, K., & Qin, L. P. (2010). Antidepressant properties of bioactive fractions from the extract of *Crocus sativus* L. *Journal of Natural Medicine, 64*(1), 24–30.

Wang, Y., Suna, J., Liub, C., & Fang, C. (2014). Protective effects of crocetin pretreatment on myocardial injury in an ischemia/reperfusion rat model. *European Journal of Pharmacology, 741*, 290–296.

Wani A (2004) Studies on corm rot of saffron (*Crocus sativus* L.). PhD thesis, Sher-e-Kashmir University of Agricultural Sciences and Technology of Kashmir, India, 108 pp.

Waugh, N., Loveman, E., Colquitt, J., Royle, P., Yeong, J. L., Hoad, G., & Lois, N. (2018). Treatments for dry age-related macular degeneration and Stargardt disease: A systematic review. *Health Technology Assess, 22*, 1–167.

Wetie, A. G. N., Sokolowska, I., Woods, A. G., Roy, U., Deinhardt, K., & Darie, C. C. (2014). Protein–protein interactions: switch from classical methods to proteomics and bioinformatics-based approaches. *Cell Molecular Life Science, 71*, 205–228.

WHO (2007) Monographs on Selected Medicinal Plants; World Health Organization: Geneva, Switzerland, Volume 3. Available online: http://apps.who.int/medicinedocs/en/m/abstract/Js14213e/(accessed on 28 February 2019).

Willard P (2002) Secrets of Saffron: The Vagabond Life of the World's Most Seductive Spice, Beacon Press, ISBN 978-0-8070-5009-5.

Wilson, N. (2006). *Encyclopedia of Ancient Greece*. New York: Routledge, Taylor & Francis Group.

Winterhalter, P., & Straubinger, M. (2000). Saffron-renewed interest in an ancient spice. *Food Reviews International*, *16*, 39–59.

Xi, L., Qian, A.Y., Xu, G. L., Zheng, S. G., Sun, S., Wen, N., Sheng, L., Shi, Y., & Zhang, Y. B. (2007a). Beneficial impact of crocetin, a carotenoid from saffron, on insulin sensitivity in fructose-fed rats. *Journal of Nutrition Biochemistry*, *18*(1), 64–72.

Xi, L., Qian, Z., Du, P., & Fu, J. (2007b). Pharmacokinetic properties of crocin (crocetin digentiobiose ester) following oral administration in rats. *Phytomedicine*, *14*(9), 633–636.

Xu, G. L., Yu, S. Q., Gong, Z. N., & Zhang, S. Q. (2005). Study of the effect of crocin on rat experimental hyperlipidemia and the underlying mechanisms. *Zhongguo Zhong Yao Za Zhi*, *30*(5), 369–372.

Xu, G. L., Li, G., Ma, H. P., Zhong, H., Liu, F., & Ao, G. Z. (2009). Preventive effect of crocin in inflamed animals and in LPS-challenged RAW 2647 cells. *Journal of Agriculture and Food Chemistry*, *57*(18), 8325–8330.

Xuan, B., Zhou, Y. H., Li, N., Min, Z. D., & Chiou, G. C. (1999). Effects of crocin analogs on ocular blood flow and retinal function. *Journal of Ocular Pharmacology and Therapeutics*, *15*(2), 143–152.

Yamauchi, M., Tsuruma, K., Imai, S., Nakanishi, T., Umigai, N., Shimazawa, M., & Hara, H. (2011). Crocetin prevents retinal degeneration induced by oxidative and endoplasmic reticulum stresses via inhibition of caspase activity. *European Journal of Pharmacology*, *650*(1), 110–119.

Yang, Y., & Smith, S. A. (2014). Orthology inference in nonmodel organisms using transcriptomes and low-coverage genomes: improving accuracy and matrix occupancy for phylogenomics. *Molecular Biology of Evolution*, *31*(11), 3081–3092.

Yang, L., Qian, Z., Yang, Y., Sheng, L., Ji, H., Zhou, C., & Kazi, H. A. (2008). Involvement of Ca 2+ in the inhibition by crocetin of platelet activity and thrombosis formation. *Journal of Agriculture and Food Chemistry*, *56*(20), 9429–9433.

Yang, R., Vernon, K., Thomas, A., Morrison, D., Qureshi, N., & Van Way, C. W., 3rd. (2011). Crocetin reduces activation of hepatic apoptotic pathways and improves survival in experimental hemorrhagic shock. *Journal of Parenteral and Enteral Nutrition*, *35*(1), 107–113.

Yasmin, S., & Nehvi, F. A. (2013). Saffron as a valuable spice: a comprehensive review. *African Journal of Agricultural Research*, *8*, 234–242.

Yasmin, S., Nehvi, F. A., & Wani, S. A. (2013). Tissue culture as an alternative for commercial corm production in saffron: a heritage crop of Kashmir. *African Journal of Biotechnology*, *12*(25), 3940–3946.

Yilmaz, A., Nyberg, N. T., Molgaard, P., Asili, J., & Jaroszewsk, J. W. (2010). ^{1}HNMRmetabolic fingerprinting of saffron extracts. *Metabolomics*, *6*, 511–517.

Yu-Zhu, T. -H., Hou, T. T., Hu, Y., Zhang, Q. Y., Rahman, K., & Qin, L. P. (2008). Comparative study of composition of essential oil from stigmas and of extract from corms of *Crocus sativus*. *Chemistry of Natural Compounds*, *44*(5), 666–667.

Zaffar, G., Wani, S. A., Talat Anjum, & Zeerk, N. A. (2004). Colchinicine induced variability in saffron. *Acta Horticulturae*, *650*, 277–280.

Zareena, A. V., Variar, P. S., Cholar, A. S., & Bongirwar, D. F. (2001). Chemical investigation of g-irradiated saffron (*Crocus sativus* L.). *Journal of Agriculture and Food Chemistry*, *49*(2), 687–691.

Zarei Jaliani, H., Riazi, G. H., Ghaffari, S. M., Karima, O., & Rahmani, A. (2013). The effect of the *Crocus sativus* L. carotenoid, crocin, on the polymerization of microtubules, in vitro. *Iranian Journal of Basic Medical Science, 16*(1), 101–107.

Zarinkamar, F., Tajik, S., & Soleimanpour, S. (2011). Effects of altitude on anatomy and concentration of crocin, picrocrocin and safranal in *Crocus sativus* L. *Crop Science, 5*(7), 831–838.

Zeng, Y., Yan, F., Tang, L., & Chen, F. (2003). Increased crocin production and induction frequency of stigma-like-structure from floral organs of *Crocus sativus* by precursor feeding. *Plant Cell Tissue and Organ Culture, 72*, 185–191.

Zeraatkar, M., Khalili, K., & Foorginejad, A. (2015). Studying and generation of saffron flower's 3D solid model. *Procedia Tech, 19*, 62–69.

Zhang, Y., Shoyama, Y., Sugiura, M., & Saito, H. (1994). Effects of *Crocus sativus* L. on the ethanol-induced impairment of passive avoidance performances in mice. *Biology of Pharmaceutical Bulletin, 17*(2), 217–221.

Zhang, H., Zeng, Y., Yan, F., Chen, F., Zhang, X., Liu, M., & Liu, W. (2004). Semi-preparative isolation of crocins from saffron (*Crocus sativus* L.). *Chromatographia, 59*(11–12), 691–696.

Zhang, F., Gao, Q., Khan, G., Luo, K., & Chen, S. (2014). Comparative transcriptome analysis of aboveground and underground tissues of *Rhodiola algida*, an important ethno-medicinal herb endemic to the Qinghai-Tibetan Plateau. *Gene, 553*(2), 90–97.

Zheng, S., Qian, Z., Sheng, L., & Wen, N. (2006). Crocetin attenuates atherosclerosis in hyperlipidemic rabbits through inhibition of LDL oxidation. *Journal of Cardiovascular Pharmacology, 47*, 70–76.

Zheng, Y. Q., Liu, J. X., Wang, J. N., & Xu, L. (2007). Effects of crocin on reperfusion-induced oxidative/nitrative injury to cerebral microvessels after global cerebral ischemia. *Brain Research, 1138*, 86–94.

Zheng, C. J., Li, L., Ma, W. H., Han, T., & Qin, L. P. (2011). Chemical constituents and bioactivities of the liposoluble fraction from different medicinal parts of *Crocus sativus*. *Pharmaceutical Biology, 49*(7), 756–763.

Zinati, Z., Shamloo-Dashtpagerdi, R., & Behpouri, A. (2016). In silico identification of miRNAs and their target genes and analysis of gene co-expression network in saffron (*Crocus sativus* L.) stigma. *Molecular Biology Research Communications, 5*(4), 233–246.

Zohary, D., & Hopf, M., (Eds.), (1994). Domestication of plants in old world (2nd ed.). : Oxford: Clarendon Press.

Zubor, A. A., Suranyi, G., Gyori, Z., Borbely, G., & Prokisch, J. (2004). Molecular biological approach of the systematics of *Crocus sativus* L. and its allies. *Acta Horticulturae, 650*, 85–93.

CHAPTER 15

Exploitation of revered potent medicinal mushroom *Ganoderma lucidum* with particular accent on oncotherapeutics

Nowsheeba Rashid[a], Rouf Ahmad Bhat[b], Nighat Mushtaq[c], Ifra Ashraf[d]

[a]Amity Institute of Food Technology, Amity University, Noida, Uttar Pradesh, India
[b]Department of Environmental Sciences, School of Life Sciences, Sri Pratap College, Cluster University of Srinagar, Jammu and Kashmir, India
[c]Division of Vegetable Science, Sher-e-Kashmir University of Agricultural Sciences and Technology of Kashmir, Srinagar, Jammu and Kashmir, India
[d]College of Agricultural Engineering and Technology, Sher-e-Kashmir University of Agricultural Sciences and Technology of Kashmir, Srinagar, Jammu and Kashmir, India

Introduction

The most promising field of drugs over the precedent decades, is the discovery of innate sources for new bioactive composites providing medicines or direct compounds of significant curative prospective (Newman & Cragg, 2007; Lindequist, Rausch, Füssel, & Hanssen, 2010; Liu et al., 2010; Pan et al., 2010; Xu, Liang, Gao, Zhong, & Liu, 2010; Aly, Debbab, & Proksch, 2011; Debbab, Aly, & Proksch, 2011; Debbab, Aly, & Proksch, 2012). One of the prominent conventional resources of innate bioactive constituents for centuries is different types of mushrooms and they have also been besieged as appreciated therapeutic representatives. Since the succeeding most varied assembly of organisms, it has been hypothesized that multiplicity of fungus (near about 3–5 million varieties) surpasses to that of global plants by an array of enormity (Blackwell, 2011; Dai, 2010). Out of this diverse population, just a portion of every fungal species encompasses description so far, that is, counted as 100,000 and yet lesser amount of it investigated for the manufacturing of pharmacologically essential metabolites. However, until now a number of the majority unbeaten medicines and also some agrochemical fungicides in the marketplace encompass significant developed from secondary metabolites of fungus. These comprise various important derivatives such as certain antibiotics; cephalosporins, penicillins,

Medicinal and Aromatic Plants: Expanding their Horizons through Omics
http://dx.doi.org/10.1016/B978-0-12-819590-1.00015-X

and fusidic acid, bad cholesterol-decreasing compounds like statin derivatives, also immunosuppressive medicines (cyclosporin) and antifungal compounds; echinocandins, griseofulvin, and strobilurins (Jikai, 2002; Li & Vederas, 2009; Aly et al., 2011; Hansen, Jensen, & Carstensen, 2012; Kozlovskii, Zhelifonova, & Antipova, 2013). Also, a few of intoxicating mycotoxins like the ergot alkaloids encompass capitulated medicines to take care of neurological problems, such as mental decline and migraine in the aged people, subsequent to optimization by therapeutic medical chemistry (Hyde, 2001; Jikai, 2002; Li & Vederas, 2009; Zhong & Xiao, 2009; Aly et al., 2011; Mulac, Hüwel, Galla, & Humpf, 2012; Young, 2013).

Ancient herbal medication which fashioned the base of healthiness and well-being right through the human race ever since the most primitive days of existence are still drawing more and more consideration in the perspective of physical condition stipulation and reorganization of healthiness segment. Making record of their scientific, pharmaceutical as well as financially viability is hopeful for global trade, though it differs extensively among countries. For ages a large figure of plant originated conventional arrangement of drugs are being utilized in India and as well in numerous divisions of the globe. Prior to the introduction of present drugs, the conventional organizations of medication were performing a fundamental part in healthcare. As per an approximated estimation, mass global inhabitants, particularly in the emergent countries, till now rely on flora for their primary well-being and healthiness requirements, perhaps for the subsequent grounds:

- Shortage of effortless admittance to drugs of present medicine.
- Admired confidence that herbal medication is without any unfavorable possessions.
- Mainly cost-effective as adjacent to excessive expenditure of the majority of the allopathic drugs.
- Apprehension for the toxicity with medicines.
- The holistic move and belief related to herb-originated drugs.

Conventional drugs have been progressively attaining attention and recognition even among the users of current medication as well (Rajani & Padh, 2000). Medical mushrooms encompass various healing remuneration, chiefly for the reason that they hold a certain quantity of organically dynamic constituents (Bao, Duan, Fang, & Fang, 2001; Petrova, Mahajna, & Denchev, 2005; Chen & Seviour, 2007; Zhang, Cui, Cheung, & Wang, 2007; Lee & Hong, 2011). The specific compounds comprise mostly high-molecular weight composites like proteins, lipids, and polysaccharides. They also comprise a range of intricate metabolites having low-molecular mass with varied concretion of chemical namely, polyketides, alkaloids, terpenoids, and metabolites derivative of nonribosomal peptide synthesis (NRPS) (Erkel & Anke, 2008; Lau, Abdullah, Shuib, & Aminudin, 2012; Wiemann et al., 2012; Gallo, Ferrara, & Perrone, 2013).

Ganodermas

The common name of *Ganoderma lucidum* or *G. lucidum*, in Chinese is Lingzhi. As already discussed, it is a macro fungus which is broadly utilized all the way through centuries for the purpose of common encouragement of fitness and prolonged existence in people of Asian countries. The majority weight of mushrooms is attributed to the water quantity present in them, that is, 90%. The remaining 10% weight of *G. lucidum*, comprises of 26%–28% CHO, 59% rough fiber, 7%–8% rudimentary protein, and 3%–5% crude fat (Mau, Lin, & Chen, 2001). Irrespective of this *G. lucidum* comprises of an extensive range of bioactive components for example polysaccharides, steroids, terpenoids, glycoproteins, and phenols (Boh, Berovic, Zhang, & Zhi-Bin, 2007; Zhou et al., 2007). The chief physiologically vigorous components present in *G. lucidum are* polysaccharides and triterpenes as encompassed by various authors (Boh et al., 2007; Zhou et al., 2007). Natural world is an incredibly excellent resource of numerous therapeutic composites thus; it is required to ascertain intoxicating remediation for microbial/pathogenic infectivity or ailments. Macromycetes are wealthy resources of organically dynamic compounds with a vast diversity of chemical configurations. For that reason, mushrooms may perhaps be constructive for searching novel strong antimicrobial agents (Fig. 15.1).

Figure 15.1 ***Tree with* Ganoderma *on trunk.*** *(Futch et al., 2016)*

Antioxidant property of *Ganodermas*

Basically, antioxidants are known to be those compounds which hinder or satiate free radical responses thus delaying or inhibiting damage of cell (Young & Woodside, 2001). Despite the fact that the defense mechanism of antioxidants varies within the species, the existence of the antioxidant resistance is widespread (Nimse & Pal, 2015). They are considered as best domicile therapy in opposition to a variety of ailments and also play a significant part in thwarting our body from different types of oxidative stresses (Chen, Ju, Li, & Yu, 2012a). Auto-oxidation of free radicals is prevented by these antioxidants; they also suspend the motion of free radical by one or the other mechanisms:

- Per oxidation instigated by the scavenging species
- Chelating metal ions not capable of generating imprudent species are chelated or lipid peroxides are decomposed
- Extinguishing $^{\bullet}O_2^-$ thwarts development of peroxides
- Infringementing self-oxidative sequence response and
- Plummeting restricted O_2 deliberations

Efficiency of an antioxidant relies on a range of feature such as pace constants, activation energy, redox potential, easiness by means of antioxidant is damaged (instability and thermal vulnerability) and also the solubility of an antioxidant. Innate flavonoids and artificial antioxidants (Butylated hydroxyl anisole, Butylated hydroxyl toluene, and propyl gallate) each of the twohold structurally aromatic or phenolic rings. The phenolic rings act by contributing $H^{\bullet}$ to without charge radicals that are produced at the time of oxidation followed by formation of radical intermediates. Resonance delocalization of the electron inside the aromatic structures help in the mitigation of these intermediates. *Ganodermas* are extensively disseminated with antioxidants (Barros, Ferreira, Queiros, Ferreira, & Baptista, 2007). Different species of *Ganoderma* transpire in diverse agro climatic circumstances and are disseminated ubiquitously in the planet. They emerge in diverse sizes, shapes, and colors such as blue/green, white, red, yellow, purple, and black of the crop body. Different varieties of antioxidants are reported to be found in *Ganoderma* (Smina, Mathew, Janardhanan, & Devasagayam, 2011); these are capable of reducing oxidative spoilage by straightforwardly hunting free radicals produced in the cell. Bioactivity is revealed by *Ganoderma lucidium* for various diseases such as bronchitis, hepatopathy, hypercholesterolemia, hypertension, nephritis, neurasthenia, cardiovascular disease, tumorogenic diseases, arthritis, gastric ulcer, chronic hepatitis, AIDS, immunological disorders, scleroderma, and cancer (Sliva, 2003). Nonenzymatic and enzymatic antioxidants vary in their means of action drastically. The action of enzymatic antioxidants is to break down free radicals first and then remove them. The dangerous oxidative products are first converted by antioxidant enzymes into hydrogen peroxide (H_2O_2) and after that to H_2O. This is done in a stepwise procedure in existence of cofactors like Cu, Zn, Mn, and Fe. On the other hand, non-enzymatic antioxidants such as glutathione, vitamin C, plant polyphenol, vitamin E and carotenoids are activated as a result of disruption of free radical chain reactions. The lipid-soluble antioxidants, for example, carotenoids, vitamin E, and lipoic acid are principal components of cell membranes whereas water-soluble antioxidants, for example, vitamin C is present in the cellular fluids like cytoplasmic matrix or cytosol. *Ganoderma* is advantageous to diabetes patients because it includes considerable number of vitamins like B_1, B_2, B_{12}, C, D, and E (Kozarski et al., 2015). It also has high mannitol and less glycemic index, which is in particular advantageous for diabetics. The structural phenols present in *Ganoderma* can be alienated into:

- Phenolic acids such as protochatechuic, rosmarinic, gallic and caffeic acids (Fig. 15.2)

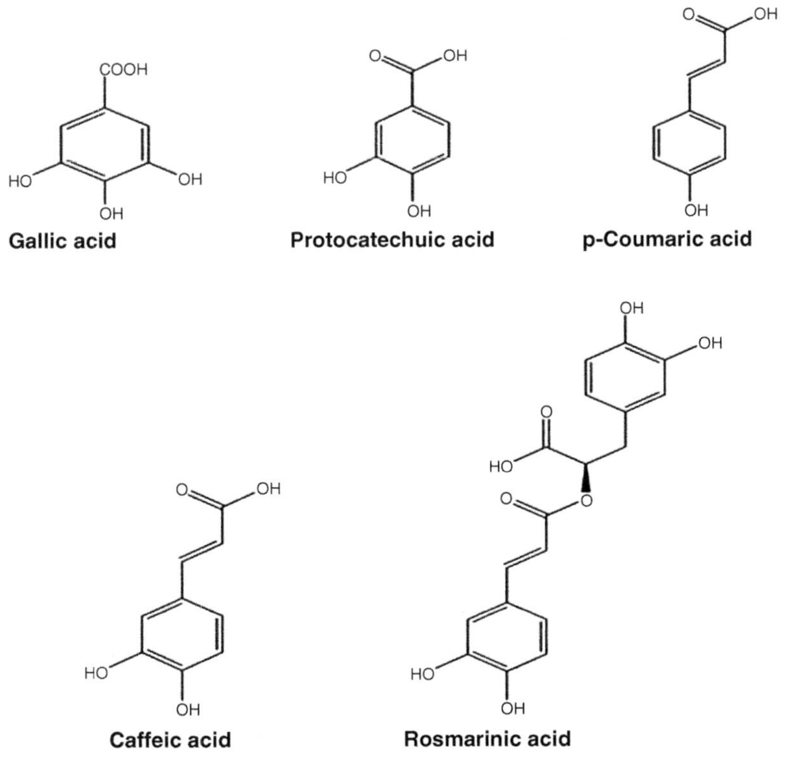

Figure 15.2 ***Representation of phenolic acids.*** *(Shan et al., 2005).*

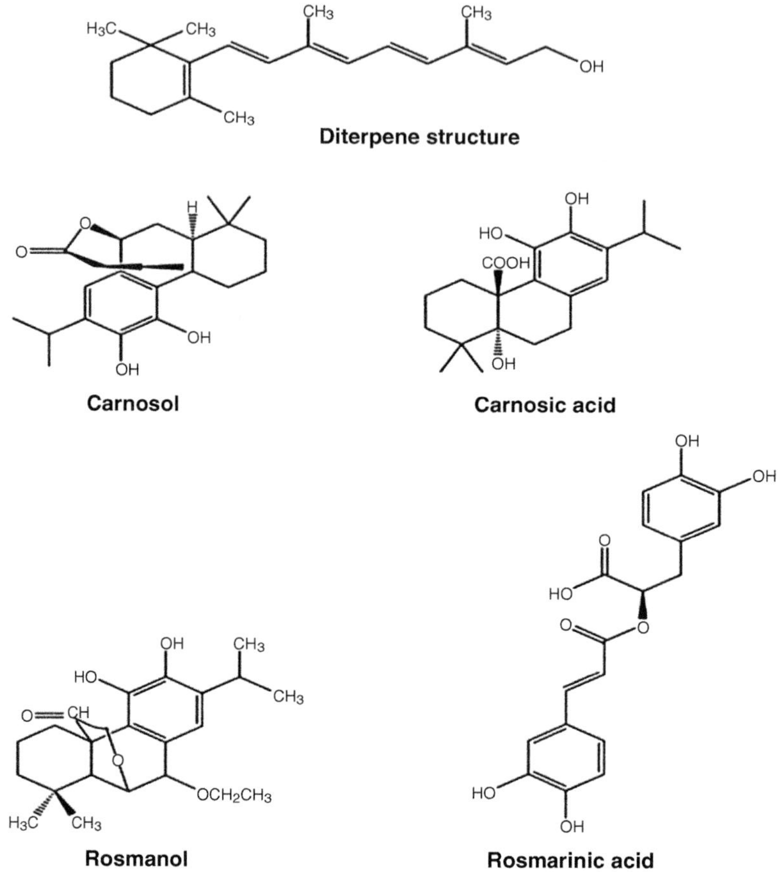

Figure 15.3 ***Representation of phenolic diterpenes.*** *(Shan et al., 2005).*

- Phenolic diterpenes like carnosic acid and carnosol acid (Fig. 15.3)
- Volatile oils like carvacrol, eugenol, menthol, thymol, etc. (Fig. 15.4)
- Flavonoids such as catechin and quercetin (Fig. 15.5) (Shan, Cai, Sun, & Corke, 2005).

Functional commotion straightway depends upon the prototype, that is, both figure and position of open −OH grouping on the flavonoid frame (Lupea, Pop, & Cacig, 2008). Distribution of B-ring prototype is accountable for free radical-scavenging capability of flavonoids ensuing in diverse structures such as myricetin, kaempferol, galangin and quercetin. Flavonoids with only one hydroxyl group with are less effective antioxidants as compared to flavonoids having multiple hydroxyl groups. It is reported that the existence of ortho-3, 4-dihydroxy structure enhances the antioxidative action (Gheldof & Engeseth, 2002). Polyphenols with trihydroxy

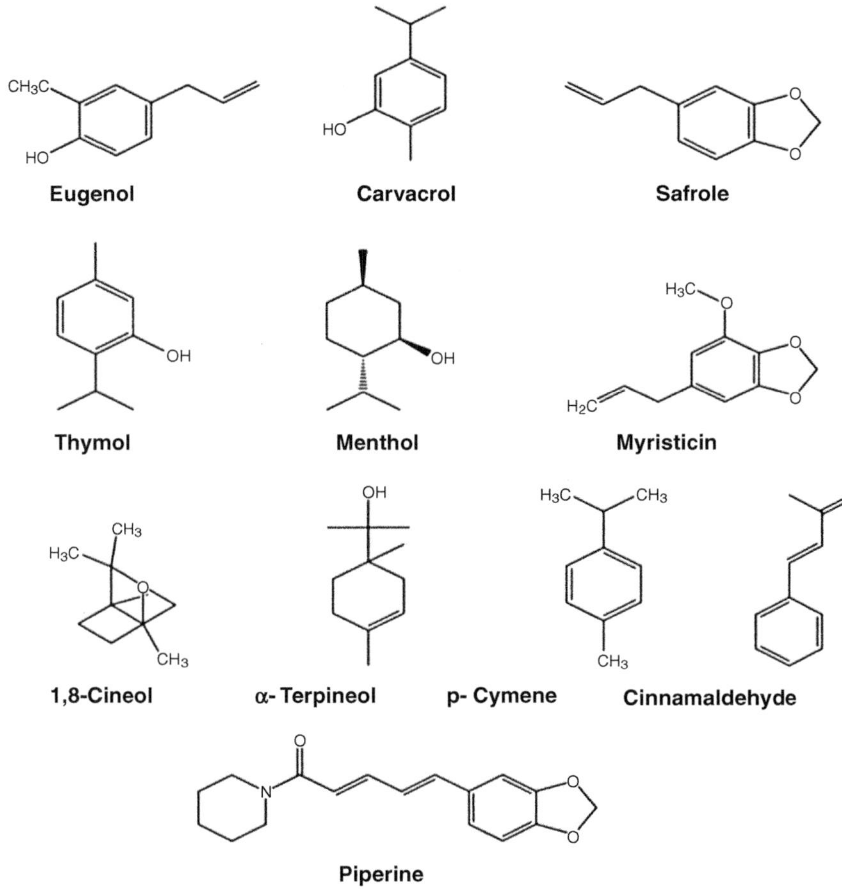

Figure 15.4 ***Representation of volatile oils.*** *(Shan et al., 2005).*

replacement compounds such as gallic acid had simply transitional effectiveness whereas an ortho-dihydroxy replacement compounds like cyanidin, cyanidin-3-glucoside, and protocatechuic acid were the majority efficient. The arrangement of liberated radical-scavenging action of an assembly of polar composites was AA< gallic acid < propyl gallate < coumaric acid < ferulic acid as per studied carried out by (Alamed, Chaiyasit, McClements, & Decker, 2009). Also, the free radical-scavenging commotion of a collection of non-polar complexes was α-tocopherol < tert-butylhydroquinone (TBHQ) < BHT < rosmarinic acid. Proanthocyanidins possessing antioxidant activity takes place in fruits such as apples and cherries, a number of berries like strawberries, blackberries, rosehips, and raspberries, and also in the woof of a lot of plants, foliage (tea), and seeds such as soy,

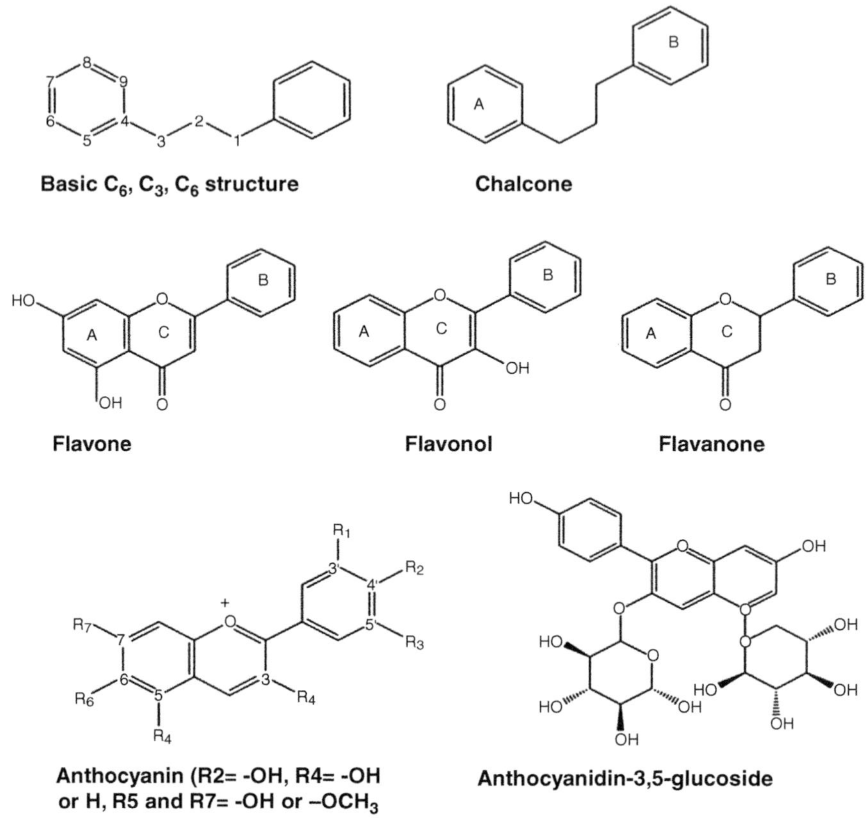

Figure 15.5 ***Representation of flavonoids.*** *(Shan et al., 2005)*

grape, sorghum, and cocoa bean (Dixon, Xie, & Sharma, 2005; Buricova & Reblova, 2008; Bak et al., 2010). The chief vitamin E compound in plant leaves is α-Tocopherol (Fig. 15.6) which is positioned in the chloroplast cover and membranes of thylakoid. α-Tocopherol stops the proliferation of lipid per-oxidation by scavenging lipid peroxyl radicals present in thylakoid membranes, it also neutralizes photosynthesis originated imprudent oxygen species (particularly $^{\bullet}O_2^-$) (Munné-Bosch, 2005).

Antitumor action of *Ganoderma lucidum*

As already quoted *Ganodermas* exhibit an extensive variety of pharmacological behaviors like immunomodulating, anticancer, antitumor, and antiinflammatory effects (Zheng, Jie, Hanchuan, & Moucheng, 2005). Krestin, lentinan, and schizophyllan, are several polysaccharides which acquire

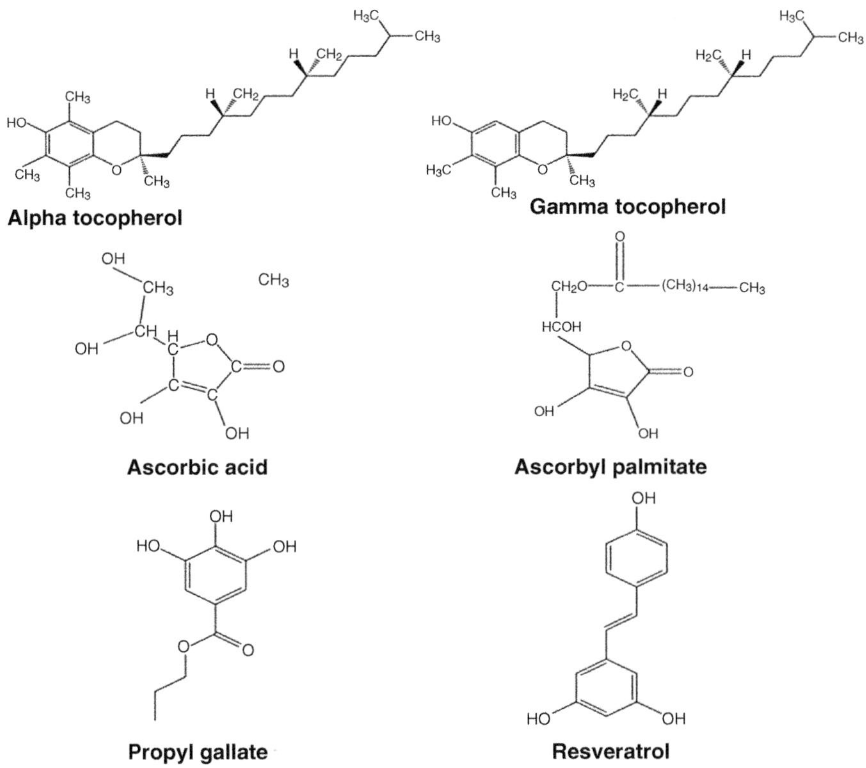

Figure 15.6 ***Representation of α-Tocopherol and derivatives.*** *(Munné-Bosch, 2005).*

pharmaceutical possessions and are present in gandodermas. These polysaccharides are recognized modulators of cellular and humoral resistance. With the commencement of different resistant reactions in the host they exercise antitumor actions (Table 15.1) (Joseph, Sabulal, George, Antony, & Janardhanan, 2011). *Ganoderma lucidum*, as a known fungus grows on decomposed wood, and is a healthy as well as acknowledged resource of curative agents. *G. lucidum*, from past thousands of years has been comprehensively utilized in oriental folk medication to cure different diseases, together with cancer (Stanley, Harvey, Slivova, Jiang, & Sliva, 2005). The chief chemical components of *G. lucidum* are immune modulating polysaccharides and cytotoxic triterpenoids (Paterson, 2006). *G. lucidum* polysaccharide is constituted mainly of β-1, 3 and β -1, 6-D-glucans. The carbohydrate polymer which contains chains of glucose molecules connected with each other by b-glycosidic linkages is β -D-glucan (Hung, Wang, Chen, & Yang, 2008).

Table 15.1 Bioactive constituents having strong anti-tumor characteristics distinguished in *G. lucidum*.

Mushroom	Bioactive multiple	Organic action	Reference
G. lucidum	Ganoderic acid T, Ganoderic acid F 12β-Acetoxy-3,7,11,15,23-pentaoxo-5α-lanosta-8-en-26-oic acid ethyl ester	Antimetastatic effect	Xu et al. (2010); Tang, Gu, & Zhong (2006b)
	4,4,14α-Trimethyl-3,7-dioxo-5α-chol-8-en-24-oic acid 11β-Hydroxy-3,7-dioxo-5α-lanosta-8,24(E)-dien-26-oic acid	Cytotoxic effect	Cheng et al. (2010)
G. lucidum	Ganoderic acid Me (ganoderic derivative)	Antimetastatic outcome, diminishes tumor attack	Chen et al. (2008)
	Ganoderic acid DM	Inhibition of 1. adhesion, migration, and invasion 2. discharge of uPA 3. expression of CDK4	Jiang et al. (2008)
	Lucidenic acids A, B, C, and N	Antiproliferative outcome	Wu et al. (2012)
	Ganoderic acids A, F and H	Inhibition of 1. PMA-induced MMP-9 activity 2. attack	Weng, Chau, Chen, Chen, & Yen (2007)

Polysaccharides of *G. lucidum* stimulate apoptosis, hamper cell propagation and restrain cell relocation of extremely persistent cells of human prostate cancer. The antitumor activity of *G. lucidum* polysaccharides mechanism is advanced all the way through stimulus of host-resistance response (Lu et al., 2004; Zhu & Lin, 2006).

Having least cytotoxicity of ganodermas, their utilization may raise the prospect of immune confrontation and diminished toxicity. It has been found that *G. lucidum* amend various constituents of the resistant system like the antigen-representing cells, T and B lymphocytes, macrophages and NK cells have been found to be modulated by *G. lucidum* ensuing the creation of cytokines, together with interleukins, interferon, and tumor necrosis factor-α (TNF-α) (Lin & Zhang, 2004). The antitumor property of *Ganoderma lucidum* is intervened by cytokines; they are released from triggered macrophages and T-Lymphocytes (Wang et al., 1997). The production of cytokine together with interleukin-6, TNF, interleukin-1, and interferon in which two antitumor cytokines, tumor necrosis feature and interferon, reacted synergistically on the reticence of leukemic-cell development and noticeably induced leukemic cell apoptosis may possibly be potentiated through *Ganoderma lucidum*. The crude Germanium present in *Ganoderma lucidum* could as well contribute to its anti-tumor characteristics. Tang, Liu, Zhao, Wei, & Zhong (2006a) anticipated that GA-T may be an accepted prospective apoptosis inducing agent for severe metastatic lung tumor and it may possibly be as well applicable to take care of additional tumor cell lines.

Antiviral activities of *Ganoderma* sp.

The antiviral properties of ganodermas have been deciphered by their fruiting bodies in particular showing hostility toward protease enzyme present in HIV virus. The chief elements for contending the virus include ganodermanontriol and ganoderiol F which are characterized as anti-HIV-1 agents bearing an repressive strength of 7.8 μg/mL. Various additional compounds having moderate efficiency were also isolated, namely, ganoderiol A, ganoderiol B, ganoderic acid H, ganoderic acid B, ganoderic acid A, and ganoderic acid C1 (El-Mekkawy et al., 1998; El Dine, El Halawany, Ma, & Hattori, 2008).

The IC50 value of many compounds defines the nature and several colossolactone category of triterpenoids for instance colossolactone VIII, colossolactone VII, colossolactone V, colossolactone G, colossolactone A and schisanlactone A were obtained from *G. lucidium* from chloroform extract having IC50 value in the range of 5–39 μg/mL (El Dine et al., 2008). The compounds comprising of ganomycin I, ganomycin B, and farnesyl hydroquinone possess an inhibitory imprint on functional sites of protease enzyme of HIV-1 possessing IC50 values of 1.0 and 7.5 μg/mL, respectively (El Dine et al., 2008). Whereas, triterpenes obtained from *Ganoderma pfeifferi* comprising of applanoxidic acid G, lucidadiol, and ganodermadiol were found to be effective combatant against influenza virus type A having ED50 of more than 0.19 mM, 0.22 mM and 0.22 mM, respectively (Mothana et al., 2003). The IC50 values of diverse triterpenes obtained from *Ganoderma pfeifferi* are in varied ranges like ganoderone C has IC50 value of 2.6 μg/mL, ergosta-7, 22-dien-3α- ol has IC50 value of 0.78 μg/mL and lucialdehyde B has IC50 value of 3.0 μg/mL; and are well-acknowledged for inhibitory effect on the proliferation of Madin Darby canine kidney (MDCK) cells diseased with influenza virus (Niedermeyer et al., 2005). On contrary, other triterpenes like ganoderone A with IC50 value of 0.075 μg/mL, with IC50 value of 0.03 μg/mL and compound 14 with IC50 of 0.03 μg/mL were found to be effectually hostile toward Herpes simplex virus, whereas compounds 21 and 51 were less effectual in contrast (Mothana et al., 2003; Niedermeyer et al., 2005). Triterpenes isolated from *G. lucidium*, lanosta-7, 9, 15; 26-dihydroxy (GLTA), 24-trien-3-one and ganoderic acid Y impose a suppressive effect on enterovirus 71 having IC50 values in the range of 0.16 to 4 μg/mL (Zhang et al., 2014). The details of the antiviral activities of *Ganoderma sp.* have been illustrated in Table 15.2.

Effects on nerves

Because of the "sleep-promoting factor" of *Ganoderma*, they have been usually suggested by herbalists of China and Japan for sleeping disorder (insomnia) (Kenneth, 1992). These mushrooms are given for numerous psychiatric and neurological sufferings in China, together with problems relating the anorexia, muscles and weakness subsequent to prolonged illnesses (Chen & Li, 1993). The dehydrated "mycelium" of *Ganoderma* which is a root-like body that fabricates mushrooms has been established to be greatly valuable in the cure of neuroses originated by "ecological anxiety." Furthermore, patients consuming mycelium product of Ganoderma established noteworthy enhancement in an eight-month study of Alzheimer's disease (Yoon, Eo, Kim, Lee, & Han, 1994).

Ganoderma as Immunomodulator and treatment against cancer

Ganoderma is said to possess antitumor properties by regulating the immune system as has been demonstrated by several pharmacological and

Table 15.2 Detailed antiviral behavior of *Ganoderma* sp.

Ganoderma sp.	Parts/products/compounds	IC50 ($\leq$50 μM)/EC50/ ED50 value	Tested viral strains	Extraction solvent	References
G. colossum	Colossolactone V Colossolactone VII Colossolactone VIII Schisanlactone A Colossolactone A	5–39 μg/mL	HIV 1	$CHCl_3$	El Dine et al. (2008)
G.lucidium	Fruiting bodies	87% clearance of virus	Oral human Papillomavirus	NG	Donatini (2014)
G. colossum	Ganomycin-I & Ganomycin B	7.5 and 1.0 μg/mL	HIV 1	$CHCl_3$	El Dine, El Halawany, Ma, & Hattori (2009)
G.lucidium	Ganoderiol-F & Ganodermanontriol	7.8 μg/mL	HIV-1 protease	MeOH	El-Mekkawy et al. (1998)
G. sinense	Ganoderic acid GS-2, 20-hydroxylucidenic acid N, 20-dehydrolucidenic acid N & ganoderiol F	20 – 40 μM	HIV-1 protease	$CHCl_3$	Sato, Zhang, Ma, & Hattori (2009)
G.lucidium	Carpophores	68–1790 μg/mL-EC50	Herpes Simplex Virus types 1 and 2, Influenza A virus (Flu A) and Vesicular Stomatitis Virus (VSV) Indiana and New Jersey strains	MeOH & H_2O	Eo, Kim, Lee, & Han (2000)

(*Continued*)

Table 15.2 Detailed antiviral behavior of *Ganoderma* sp. (*Cont.*)

Ganoderma sp.	Parts/products/compounds	IC50 (≤50 μM)/EC50/ ED50 value	Tested viral strains	Extraction solvent	References
G.lucidium	Fruiting bodies	96%–100% at 1 103 mol ratio/TPA)	Epstein-Barr virus	MeOH	Iwatsuki et al. (2003)
G.lucidium	Mycelia	IRA(HBsAg, HBeAg) up to 100%	Hepatitis B virus	NG	Li et al. (2006)
G.lucidium	Ganoderic acid	Inhibition of production of HBV surface antigen and HBVe at 8 μg/mL	Hepatitis B virus	H_2O and $CHCl_3$	Li and Wang (2006)
G. pfeiffer	Ganodermadiol, lucidadiol & applanoxidic acid	Influenza ED50(0.19–0.22mmol/L); HSV 1 (0.068 mmol/L for ganodermadiol)	Influenza virus type A and HSV type 1	NG	Mothana et al. (2003)
G. pfeiffer	Ganoderone A, Lucialdehyde B, Ergosta-7,22-dien-3α-ol, Ganoderol A & Ganoderol B	0.03–0.75 μg/mL (IC50)	HSV type 1	CH_2Cl_2	Niedermeyer et al. (2005)
G. pfeiffer	Ganoderone-C, Lucialdehyde B & Ergosta	0.78–2.6 μg/mL(IC50)	Influenza virus type A	CH_2Cl_2	Niedermeyer et al. (2005)
G.lucidium	Lanosta-7,9,24-trien-3-one,15;26-dihydroxy (GLTA), Ganoderic acid Y	0.16–4 μg/mL (IC50)	Enterovirus 71	NG	Zhang et al. (2014)

Source: Basnet, Liu, Bao, & Liu, 2017)

clinical studies (Boh et al., 2007; Cao, Xu, Liu, Huang, & Gu, 2018). Healing properties of *Ganoderma* is because of the presence of immunomodulation fungal proteins, polysaccharides, and triterpenoids. In this chapter, varieties of chemical constituents of Ganoderma and their respective pharmacological effects have been discussed. FIPs consist of proteins extracted from *Ganoderma*. It is also the source for extraction of purified tiny molecular proteins like FIPs. These proteins having anticancer properties are the functional families of *Ganoderma* components. There are four kinds of immunoregulatory proteins, Lingzhi-8 (Lz-8), Fip-gts, GMI, and Fip-gat, obtained and purified from *Ganoderma*. In 1989 Lz-8 an immunodulatory protein from *G. lucidum*, was isolated and cloned. It contains 110 amino acids. The design of Lz-8 resembles to an immunoglobulin which forms noncovalently-linked homodimers with biological activity (Kino et al., 1989). Lz-8 has great healing properties in case of gastric cancer and lung cancers. Liang et al. established that a particular variety of Lz-8 (rLz-8) triggered autophagic cell destruction by accumulation in the endoplasmic reticulum (ER) which caused ER stress and ATF4-CHOP pathway in SGC-7901 human gastric cancer cells (Liang et al., 2012). For treating lung cancers rLz-8 proved to be a helpful chemotherapeutic agent due to the important role of FAK targets in metastasis (Lin & Hsu, 2016). By inducing EGFR mutation or over expression and EGFR dependent processes, Lin et al. found some new anticancer properties of rLz-8 (Lin, Hsu, Sun, Wu, & Tsao, 2017). Another immunomodulatory protein FIP-gts is extracted from *G. tsugae*. Through reverse transcriptase polymerase chain reaction method, the DNA encoding of this protein was isolated from cDNA (Lin, Hung, Hsu, & Lin, 1997). rFIP-gts a recombinant of this protein led to reduction of telomerase catalytic subunit (Liao et al., 2006). It also reduced telomerase activity in lung cancer cells by influencing on nuclear export mechanisms, which in turn might have been mediated by ER stress- induced intracellular calcium level (Liao et al., 2007).

In vivo investigations demonstrated that the development of A549 cells in naked-mice treated with rFIP-gts was altogether slower than those treated with PBS, which affirmed that lung tumor development could be repressed by rFIP-gts (Liao et al., 2008). Additionally, this protein was likewise appeared to influence 217 cervical malignant growth cells. GMI is an immunomodulatory protein cloned from *G. microsporum*. The amino corrosive arrangement of this protein imparted 83% homology to that of FIP-gts (Chiu et al., 2015). In vitro examinations found that GMI restrained the EGF-prompted phosphorylation and initiation of EGFR and AKT pathway kinases in a portion subordinate way (Lin et al., 2010). Hsin et al. discovered that autophagosomal amassing incited autophagic cell demise in a model of GMI treatment, and ATP6V0A1, a subunit of vesicular H + -ATPases, controlled autophagosome lysosome combination. Hsin et al. additionally uncovered that GMI and cisplatin-initiated apoptosis by means of autophagy/caspase-7-subordinate and survivin-and ERCC1-free pathways (Hsin et al., 2012). In vivo examinations proposed that the oral organization of GMI repressed tumor development and actuated autophagy in naked mice that were regulated a subcutaneous infusion of A549 cells (Hsin et al., 2011). Fip-gat is an immunomodulatory protein from *G. atrum* containing 111 amino acids. Xu et al. treated MDA-MB-231 cells with various groupings of recombinant Fip-gat in vitro and found that this protein decreased cell feasibility in a portion subordinate way (Xu et al., 2016). Treatment with FIP-gat set off a critical level of cell cycle capture in the G1/S change and an articulated increment in the apoptotic cell populace (Table 15.3).

Ganoderma lucidum hampering propagation of human prostate

Mushrooms slow down cell movement of extremely persistent individual's prostate cancer cells PC-3 (Sliva et al. 2002). For examining the consequences of *G. lucidum* on propagation of cells, increased deliberations of *G. lucidum* about 0-0.5 mg/mL were used to treat PC-3 cells, followed by the determination of cell growth. It is clearly evident that in a time dependent approach *G. lucidum* clearly subdued propagation of PC-3 cells. Moreover, reticence of growth was as well dosage reliant for the reason that 96 hours of management with 0.125, 0.25, and 0.5 mg/mL of *G. lucidum* repressed production of PC-3 cells by 7.6%, 53.8%, and 79.6%, correspondingly.

Consequently, from the respective data the anti-propagative outcome of *G. lucidum* on extremely persistent human prostate cancer cells is evidently verified (Jiang, Slivova, Valachovicova, Harvey, & Sliva, 2004). PC-3 cells are detained by *G. lucidum* at G2/M stage of the cell cycle. Recently, it was verified that alcoholic excerpt of *G. lucidum* arrests propagation of cells of breast cancer, namely, cell cycle inhibition at Gl phase (Hu, Ahn, Yang, Lee, & Kang, 2002). Thus, (Jiang et al., 2004) observed if the similar mechanism is besides engaged in prostate cancer cells. In their study they treated PC-3 cells with the concentration of 0.5 mg/mL of *G. lucidum* and determined the distribution of cell cycle by flow cytometry following 24 and 48 hours.

Table 15.3 Immunomodulatory of protein constituent extracted from *Ganoderma*.

Resource	Protein	Optimal action concentration dosage	Time interval	Cell shape/Mice	Pharmacological outcome(s)	Reference
G. lucidum	(r)Lz-8	5 μg/mL	12h,4 weeks	A549, CL1-5, H226, LLC1 cells,C57BL/6 mice	Stimulated alteration in epithelial to mesenchymal shift by snooping with cell and focal adhesion kinase (FAK) functions in lung cancer cells.	Lin and Hsu (2016)
		0.5 μg/mL	24h	SGC-7901 cell	Stimulated endoplasmic reticulum stress-mediated autophagic cell death.	Lin et al. (2017)
		1 μg/mL	24h	Human primary and Jurkat T cells	Stimulated IL-2 gene expression via the Src-family protein tyrosine kinase.	Hsu et al. (2008)
		10 μg/mL/7.5mg/kg	48h,18 days	LLC1 cell, C57BL/6 mice	Repressed growth and stimulated apoptosis of lung cancer cells by encouraging epidermal growth factor receptor (EGFR) degradation	Liang et al. (2012)

(Continued)

Table 15.3 Immunomodulatory of protein constituent extracted from *Ganoderma*. (*Cont.*)

Resource	Protein	Optimal action concentration dosage	Time interval	Cell shape/Mice	Pharmacological outcome(s)	Reference
G. tsugae	(r)Fip-gts	0.15 μM	48h	HeLa, SiHa, and Caski cells	Synchronized telomerase in A549 cells.	Liao et al. (2006)
		8 μg/mL	48h	A549, MRC-5 cells	Stimulated inhibition of telomerase activity in lung cancer cells by post-translational alteration of hTERT protein	Liao et al. (2007)
G .microsporum	GMI	1.2 μM, 160 μ g/mouse	48h, 66days	A549, CaLu-1 cells, nude mice	Stimulated-lung cancer cell death by triggering autophagy, but did not provoke apoptotic cell death.	Hsin et al. (2011)
		8 μg/mg	24 h	A549, CCL-185 cells	Displayed an inhibitory effect on EGF-induced migration and invasion	Lin et al. (2010)
G. atrum	(r)Fip-gat	9.96 μg/mL	48 h	MDA-MB-231 cell	Activated important cell cycle capture at the G1/S alteration and marked enhancement in apoptotic cell population.	Xu et al. (2016)

Table 15.4 Impact on cell cycle.

G2/M	G0/G1	S	Time (h)
23.3± 5.5	21.36.8	55.4 ± 4.4	0
36.0± 6.0[a]	20.3 ± 11.6	41.7 ± 6.0[a]	24
48	19.4 ± 5.1	37.2 ± 8.7[a]	43.4 ± 5.2[a]

[a] $p < 0.01$

As clearly depicted in Table 15.4, *G. lucidum* persuades cell cycle seizure in PC-3 cells at G2/M phase. Therefore, the handling of PC-3 cells did not appreciably modify the quantity of cells at G0/G1 stage of the cell cycle, although resulted in an augmentation of 57.1% and 86.3% of the cell populace in G2/M stage prior to 24 and 48 hours, correspondingly. The statistics clearly propose that *G. lucidum* slows down the development of prostate cancer cells by cell cycle seizure at G2/M stage.

Immunomodularity effects of Ganodermal polysaccharides in lung cancer

The inhibitory outcome of *G. lucidum* triterpenes on propogation of cells as well as tumor growth was elucidated by Feng et al. The IC50 of triterpenes was 24.63 μg/mL on A549 cells (Feng et al., 2013). Triterpenes possibly will appreciably hamper tumor enlargement in Lewis tumor carrying mice (i.e., 30, 60, and 120 mg/kg), and the manifestations of invulnerable organs, together with the spleen and thymus, were augmented extraordinarily by the management with triterpenes. In addition, an in vitro examination carried out by Liao et al. establish the fact that Reishi polysaccharide fraction (FMS) supplemented with L-fucose (Fuc) possibly will slow down the development of cancer cells,namely, an enhancement in the antibody intervened cytotoxicity and the diminution of the creation of tumor-related seditious intermediaries, predominantly monocyte chemo attractant protein-1 (MCP-1). In vivo investigations illustrated a noteworthy enhancement in the peritoneal B1 B-cell populace, signifying the FMS intervened anti-glycan IgM creation (Liao et al., 2013). Sun et al. in recent times confirmed that the plasma of lung cancer patients concealed the propagation, CD69 appearance, and perforin and granzyme B creation in lymphocytes upon establishment by PHA (Sun et al., 2014). *G. lucidum* polysaccharides (GLPS) moderately or completely overturn these effects. Additionally, Que et al. recommended that Ganoderic acid 261 Me, a unadulterated lanostane triterpene of *G. lucidum* causative to the indoleamine 2,3-dioxygenase, facilitated in creating a tolerogenic milieu in lung tumors by straightway suggesting T cell apoptosis, slowing down CD8 + T cell commencement, and elevating Treg-mediated immunosuppression (Que et al., 2014).

Effects of *G. lucidum* on inflammatory breast cancer

The mechanization of *G. lucidum* focused on the phosphoinositide-3-kinase (PI3K)/AKT/mammalian objective of rapamycin (mTOR) conduit is demonstrated by Ivette et al. Early conduct of seditious breast cancer (IBC SUM-149) with ganoderma diminishes appearance of mTOR, abridged eIF4G intensity attached with augmented levels of eIF4E confined to 4E-BP also substantial protein synthesis diminution. *G. lucidum* cured serious collective immune-deficient mice introduced with IBC cells for 13 weeks confirmed decrease in development of tumor and weight by 50%. It in addition confirmed diminution in the appearance of p70S6K, E-cadherin, eIF4G, and mTOR and action of extracellular synchronized kinase (ERK1/2). Therefore, it is verified that Reishi restrains production of protein and tumor enlargement by distressing endurance and propagative indication pathways (Suarez-Arroyo et al., 2013).

Remunerations of genome sequencing of *Ganoderma lucidum*

The common biological significance of Ganoderma species, a white rot fungus is the pulverization of woody plants for nutrient mobilization. The lignocelluloses decaying enzymes which are beneficial for bioremediation and bioenergy creation are possessing efficient machineries of ganoderma. Furthermore, the genres contain numerous imperative species that generate pharmacologically dynamic composites utilized in theraupeutic food and drugs. Complete genome sequencing and organized transcriptome scrutinies develop into reasonable approach to recognize genes of an organism with the quick acceptance of DNA sequencing technologies of next-generation. From previous few years, a lot of experiments and researches have been instigated to recognize the hereditary components of numerous Ganoderma species, mainly in diverse strains of *G. lucidum*. Eleven complete genome sequencing schemes for Ganoderma variety was recorded in global catalog in November 2013. Among them three were previously finished with genomes being brought together to elevated eminence. About four transcriptome analyses on different growth phases of Ganoderma species have been executed thus, balancing genome investigation. The base laid for the recognition of genes implicated in organic pathways that are dangerous for consideration of ecology of Ganoderma has been obtained from the information acquired from these analyses. The studies involve the means of pathogenesis, life cycle, the biosynthesis of dynamic mechanism, and cellular progression, etc. A small number of federal resources have been recognized to propagate the plenty of hereditary information suitably accessible and assimilate important data to maintain relative genomic study of Ganoderma species (Kües et al., 2015).

Complete genome sequencing tasks are able to bring numerous reimbursements enumerated as (genus explicit) marker recognition, pathway explication and gene detection. First, the definitive knowledge for the choice of markers proficient for species recognition is offered by the complete genome sequencing. Second, the most affordable, economic, and efficient technique for the detection of vital genes and the explication of biosynthetic path for chemical substances is this complete genome sequencing. Even though, advances like outlining of transcriptome can in addition attain the objective of gene finding. But it may not be competent to discover genes articulated at subordinate points and may not present complete know how of genes. Moreover, this is not going to present the knowledge concerning position of genes on chromosomes. Various chemical compounds present in fungi are created by clustering the genes collectively on DNA. Thus, it can be elucidated that gene co-localization is a successful technique to determine genes implicated in the biosynthesis of meticulous chemical substance (Andersen et al., 2013) also the approaches can be developed for the ultimate combined management of their appearance (Brakhage & Schroeckh, 2011). Third, the proportional examination of genomes out of several species could assist in elucidation of history of origin of important genes (Floudas et al., 2012), for example for the ligninolytic enzymes MnP, LiP, Lac, etc. An appropriate assistance is provided by this information in selection of novel species for approving distinctiveness or for mounting superior approaches to contest pathogens. Complete genome sequencing of Ganoderma spp. turn out to be inexpensive because of this fast growth of DNA sequencing technologies and also about a dozen research ventures are already there (Table 15.5).

Various noteworthy dissimilarities were recognized within the accounted genome assemblies. These variations may be because of numerous reasons. First and foremost, individuality of the species utilized for the purpose of sequencing has not been resolute or decided upon. The result being undecided either the resources utilized are of the identical source or not. The second important consequence is that it recommends the point of conclusion of a genome assemblage hinders the outcome considerably. Even

Table 15.5 Complete genome sequencing for *Ganoderma* in NCBI's Bioproject division (accomplished).

Succession	Strain	Status	Reference
PRJDA61381	*Ganoderma lucidum* strain BCRC 37177	complete genome shotgun sequences obtainable	Huang et al. (2013)
PRJNA68313	*Ganoderma* sp. strain 10597 SS1	Elevated eminence genome assembly obtainable	Binder et al. (2013)
PRJNA77007	*Ganoderma lucidum* strain Xiangnong No. 1	Elevated eminence genome assembly obtainable	Liu et al. (2012)
PRJNA71455	*Ganoderma lucidum* strain G.260125-1	Elevated eminence genome assembly obtainable	Chen et al. (2012b)

though elevated exposure is created by next generation DNA sequencing technologies despite this the shotgun sequencing approach is inadequate to attain high-quality assemblage alone. With the help of the optical mapping technology a substantial map is build which is essential to set up a chromosomal- level assembly. Despite these differences, these assemblies permit us to understand the variety and convolution of genetic makeup of Ganoderma.

Transcriptome investigation of *G. lucidum*

Till the present epoch various studies on transcriptome have been performed for a variety of strains of *G. lucidum* (Table 15.6), even though not any encompass complete molecular species recognition of the substances. Lacuna of the investigations is in respect of technologies engaged and the samples utilized. This is because of the fact that samples were gathered at different times and also at dissimilar developmental stages, for example primordium, fruiting body, and mycelium. For that reason, the outcome could not be effortlessly integrated.

An analysis of *G. lucidum* transcriptome is facing two challenges. First, the resources utilized in different investigations were generally contradictory. This was because the materials for study purposes were collected at diverse time points in cell growth and also from varied cultural media as already discussed. Thus, these entire lacunas were responsible for failure to address particular biological investigations. Upcoming transcriptome investigations ought to

Table 15.6 Evaluation of *G.lucidum* transcriptome contents explained in various studies.

Material/strain(s)	Mycelium	Fruiting body	Primordium	Total	outline	References
MeJA-induced mycelium of *G. lucidum* strain HG	3,910	-	-	3,910	No. of transcript-originated remains	Ren et al. (2013)
16-day-old mycelium and 9-week-old fruiting body of *G. lucidum* strain yw-1	6,439,690	6,416,670	-	12,856,360	No of reads	Yu et al. (2012)
	90	90	-	90	Standard read length (bp)	
	18,892	27,408	-	28,210	No of unigenes	
	498	514	-	507	Standard unigene length (bp)	
	13,332	13,055	13,731	13,144	No. of genes	
50 days old fruiting bodies of a non-specified *G. lucidum* strain	-	-	879	879	No. of ESTs	Luo et al. (2010)
	-	-	600	600	Total no. of exclusive genes	
	-	-	288	288	Standard gene length	
Mono- and dikaryotic *G. lucidum* strains BCRC 36123 and BCRC 37180	1,001	-	-	47,285	ESTs from 5 days dikaryon	Huang et al. (2013)
	6,848	-	-		ESTs from 14 days dikaryon	
	21,547	-	-		ESTs from 18 days dikaryon	
	11,059	-	-		ESTs from 18 days dikaryon	
	6,830	-	-		ESTs from 30 days dikaryon	

reflect on the subsequent issues from the beginning only. For illustration, investigation must be hypothetically using resources from one meticulous, fully sequenced strain, which has been cultivated in consistent environment, like temperature, length of development and medium along with other parameters (Berovic et al., 2003, 2012; Habijanic, Berovic, Boh, Wraber, & Petravic-Tominac, 2013; Wang et al., 2014; Wen et al., 2010). Approach of sequencing as well as bioinformatic data dispensation steps must also be homogenized. For illustration, the complete genome sequences should be utilized as the orientation to gather the transcriptome at whatever time probable.

Ganodermal bioactivity

Various researches have been carried out on bioactivities of rudimentary extorts of glyco-peptides or polysaccharidic, polysaccharides, via *Ganoderma lucidum* (Nie, Zhang, Li, & Xie, 2013). Because of its effective antibacterial, antioxidant, and antitumor behaviors of the glycoproteins, polysaccharides, and polysaccharidic extort elucidated from the fruiting bodies this group has been in particular consideration (Jia et al., 2009; XiaoPing et al., 2009; Shi, Zhang, & Yang, 2013).

With enhancement of the host's resistance arrangement, antitumor polysaccharides put forth their bioactivity typically by means of commencement of the impervious reaction of the multitude (Mizuno, Saito, Nishitoba, & Kawagashi, 1995). The genes are not as much examined as a resource of novel antibacterial agents, although are mostly studied for their remedial use (Gao, Zhou, Huang, & Xu, 2003). Nonetheless, a number of polysaccharides as of *Ganoderma* variety put forth antibacterial action by slowing down the development of microorganisms and, in several cases, by assassinating pathogenic bacteria (Skalicka-Woźniak et al., 2012). Even though being very dynamic as antimicrobials and antioxidants, polysaccharides of *Ganoderma* are typically recognized as antitumor mediators; but, the means of action implicated in their bioactivities are never clearly implicit.

Polysaccharides of *Ganoderma* with non-accounted bioactivity

Ganodermal polysaccharides species with non-accounted bioactivity fit in to water soluble polysaccharides, alkali soluble polysaccharides and/or water unsolvable polysaccharides (Table 15.7). An alkali soluble glucan G-Abut water unsolvable was formerly secluded out of *G. japonicum*. (Ukai, Yokoyama, Hara, & Kiho, 1982; Chen, Zhou, Zhang, Nakamura, & Norisuye, 1998) encompass isolation of water unsolvable glucans, specifically GL4-1 and GL4-2 out of the fruiting body of *G. lucidum*. From the spores of *G. lucidum*

Table 15.7 Polysaccharides secluded out of *Ganoderma* species lacking detailed bioactivity.

Species	Extraction/isolation procedure	Nature of Polysaccharide	Linkage	Sugars composition	References
G. japonicum	Hot dichloromethane and hot methanol Hot water; Dialysis; gelfiltration on Sepharose CL-4B	Alkali-soluble glucan	β-(1→3)-linkedD-glucopyranosyl residueswithside-chainsofsingle, β(1→6)-linkedD-glucopyranosyl groups	Glucose lamin arabiose	Ukai et al. (1982)
G. lucidum	PBS; Ethanol precipitations; 1N NaOH	Water-insoluble glucan	(1→3)-α-D-glucan	Glucose	Chen et al. (1998)
G. lucidum	Hot water followed by ethanol precipitation	Water-soluble polysaccharide	(1→3)-linked-Glc (1→6)-linked-Gal (1→4)-linked-Gal (1→6)-linked-Glc	Glucose galactose	Zhao et al. (2005)
G. lucidum	Deproteinization by Sevag method and frozen-thaw method, fractionated by ultrafiltration and gel chromatographyonCL-6B column.	Heteropolysaccharide	-	Glucose galactose	Zhang et al. (2006)
G. resinaceum	Chloroform–methanol; Hot water; dialysis; Freeze–thawing; ultrafiltration	Water-soluble glucan	(1→3)-linked β-glucan	Glucose mannose galactose xylose	Amaral et al. (2008)

(*Continued*)

Table 15.7 Polysaccharides secluded out of *Ganoderma* species lacking detailed bioactivity. (*Cont.*)

Species	Extraction/isolation procedure	Nature of Polysaccharide	Linkage	Sugars composition	References
G. lucidum	Hot water followed by ethanol precipitation; ultrafiltration	Water-soluble polysaccharide	1,6-disubstituted-agalactopyra nosyl,1,2,6trisubstituted-α-galactopyranosyl, 1,3-disubstituted-β-glucopyranosyl and1,4,6-trisubstituted- β– glucopyranosyl residues	Fucose glucose galactose	Ye et al. (2009)
G. lucidum	Utrasonic/ microwave assisted extraction	Water-soluble neutral polysaccharide	Backbone:1,4-disubstituted-βglucoseopyranoseand1,4,6 trisubstituted- β–glucoseopyranosyl Branchedchains:1,6-disubstituted β-glucopyranosyland1,4disubstituted-β–galactoseopyranosyl	Glucose galactose	Huang et al. (2011)
G. lucidum	Ethyl-acetate; Sevag method; Dialysis	Water soluble Heteropolysaccharides	(1→4)-galactanheteropolysaccharide (1→3)-glucan;1,4,6-glucan, (1→3)galactan,(1→6)-galactan,(1→4) grabinan (1→3)-mannan,and/or(1→4)-xylan linkages	glucose, galactose, mannose, arabinose	Wang et al. (2011)
G. lucidum	Hot-water extraction, graded ethanol precipitation, anionexchange chromatography	Water-soluble glucan	Backbone:1,6-linkedβ-D-Glcp Sidechain:1,4-linked Glcp residues	Glucose	Dong et al. (2012)

Source: Ferreira et al., 2015

a water soluble and small pronged polysaccharide namely SGL-III was secluded (Zhao, Zhang, Yu, Liu, & Zhang, 2005). Out of the sprouting spores of *G. lucidum* impartial, water soluble, hetero-polysaccharide called GLPS3 was secluded (Zhang, Zhao, Wu, Cao, & Zhang, 2006). A was reported in the Basidiocarp of *G. resinaceum* is reported to contain water soluble β-glucan, that is,DESSK5 (Amaral et al., 2008). Also heteropolysaccharide LZ-C-1 a water soluble polysaccharide, was secluded out of *G. lucidum* (Ye et al., 2009). Huang, Li, Li, & Wang (2011) reported that from the fruting bodies of *G. lucidum* a water soluble neutral polysaccharide was secluded. (Wang et al., 2011) reported the isolation of new heteropolysaccharides from GL-1 to GL-5 out of the fruting bodies of *G. lucidum*. Lastly, a fresh neutral as well as water-soluble β-D-glucan namely GLSA50-1B was secluded out of the spores of *G. lucidum* (Dong et al., 2012).

Conclusion

G. lucidum is the renowned conjurer of the entire advanced therapeutic herbs for the reason of its evident medicinal effectiveness and the nonexistence of adverse side-effects and contaminants consequential from utilization. From the present accessible data, namely, in vivo and in vitro studies *G. lucidum* possibly will symbolize a realistic and capable advance for prevention of cancer followed by its treatment. On the other hand, additional investigational, epidemiological, and scientific studies necessitate to be conceded out for the identification of further molecular objectives; resolution of the associations among *G. lucidum* ingestion and risks of cancer; and investigate the most favorable dosage, effectiveness, and protection unaided and in permutation with chemotherapy and/or radiotherapy. In addition to the anti-cancer activity, the attention for *Ganoderma* spp. ascends from their prospective relevance in health care, longevity, bio-energy creation and protection of plants. Due to this reason, genome investigations have to be performed with these budding purposes in psyche. Varieties with meticulous distinctiveness, such as the capability to fabricate a superior assortment of chemical composites or ligninolytic enzymes having elevated-actions ought to be considered initially and genomic knowledge should be incorporated with produced statistics on enzymatic or chemical actions.

References

Alamed, J., Chaiyasit, W., McClements, D. J., & Decker, E. A. (2009). Relationships between free radical scavenging and antioxidant activity in foods. *Journal of Agricultural and Food Chemistry*, *57*(7), 2969–2976.

Aly, A. H., Debbab, A., & Proksch, P. (2011). Fifty years of drug discovery from fungi review. *Fungal Diversity, 50*, 3–19.

Amaral, A. E., Carbonero, E. R., Simão, R. C. G., Kadowaki, M. K., Sassaki, G. L., Osaku, C. A., Gorin, P. A. J., & Iacomini, M. (2008). An unusual water-soluble b-glucan from the basidiocarp of the fungus *Ganoderma resinaceum*. *Carbohydrate Polymers, 72*, 473–478.

Andersen, M. R., Nielsen, J. B., Klitgaard, A., Petersen, L. M., Zachariasen, M., Hansen, T. J., Blicher, L. H., Gotfredsen, C. H., Larsen, T. O., Nielsen, K. F., & Mortensen, U. H. (2013). Accurate prediction of secondary metabolite gene clusters in filamentous fungi. *Proceedings of National Academy of Science USA, 110*, E99–E107.

Bak, I., Lekli, I., Juhasz, B., Varga, E., Varga, B., Gesztelyi, R., Szendrei, L., & Tosaki, A. (2010). Isolation and analysis of bioactive constituents of sour cherry (*Prunus cerasus*) seed kernel: an emerging functional food. *Journal of Medicinal Food, 13*(4), 905–910.

Bao, X., Duan, J., Fang, X., & Fang, J. (2001). Chemical modifications of the (1→ 3)-α-D-glucan from spores of *Ganoderma lucidum* and investigation of their physicochemical properties and immunological activity. *Carbohydrate Research, 336*(2), 127–140.

Barros, L., Ferreira, M. J., Queiros, B., Ferreira, I. C., & Baptista, P. (2007). Total phenols, ascorbic acid, β-carotene and lycopene in Portuguese wild edible mushrooms and their antioxidant activities. *Food Chemistry, 103*(2), 413–419.

Basnet, B. B., Liu, L., Bao, L., & Liu, H. (2017). Current and future perspective on antimicrobial and anti-parasitic activities of *Ganoderma* sp.: an update. *Mycology, 8*(2), 111–124.

Berovic, M., Habijanic, J., Zore, I., Wraber, B., Hodzar, D., Boh, B., & Pohleven, F. (2003). Submerged cultivation of *Ganoderma lucidum* biomass and immunostimulatory effects of fungal polysaccharides. *Journal of Biotechnology, 103*, 77–86.

Binder, M., Justo, A., Riley, R., Salamov, A., Lopez-Giraldez, F., Sjokvist, E., Copeland, A., Foster, B., Sun, H., Larsson, E., Larsson, K. H., Townsend, J., Grigoriev, I.V., & Hibbett, D. S. (2013). Phylogenetic and phylogenomic overview of the polyporales. *Mycologia, 105*, 1350–1373.

Blackwell, M. (2011). The Fungi: 1, 2, 3… 5. 1 million species? *American Journal of Botany, 98*(3), 426–438.

Boh, B., Berovic, M., Zhang, J., & Zhi-Bin, L. (2007). *Ganoderma lucidum* and its pharmaceutically active compounds. *Biotechnology Annual Review, 13*, 265–301.

Brakhage, A. A., & Schroeckh, V. (2011). Fungal secondary metabolites – strategies to activate silent gene clusters. *Fungal and Genetic Biology, 48*, 15–22.

Buricova, L., & Reblova, Z. (2008). Czech medicinal plants as possible sources of antioxidants. *Czech Journal of Food Science, 26*(2), 132–138.

Cao, Y., Xu, X., Liu, S., Huang, L., & Gu, J. (2018). *Ganoderma*: a cancer immunotherapy review. *Frontiers in Pharmacology, 9*, 1217.

Chen, H., Ju, Y., Li, J., & Yu, M. (2012a). Antioxidant activities of polysaccharides from *Lentinus edodes* and their significance for disease prevention. *International Journal of Biological Macromolecules, 50*(1), 214–218.

Chen, J., & Seviour, R. (2007). Medicinal importance of fungal β-(1→ 3),(1→ 6)-glucans. *Mycology Research, 111*(6), 635–652.

Chen, J., Zhou, J., Zhang, L., Nakamura, Y., & Norisuye, T. (1998). Chemical structure of the water-insoluble polysaccharide isolated from the fruiting body of *Ganoderma lucidum*. *Polymer Journal, 30*, 838–842.

Chen, K., & Li, C. (1993). Recent advances in studies on traditional Chinese anti-aging materia medica. *Journal of Traditional and Chinese Medicine, 13*(3), 223–226.

Chen, N. H., Liu, J. W., & Zhong, J. J. (2008). Ganoderic acid Me inhibits tumor invasion through down-regulating matrix metalloproteinases 2/9 gene expression. *Journal of Pharmacology Science, 108*(2), 212–216.

Chen, S., Xu, J., Liu, C., Zhu, Y., Nelson, D. R., Zhou, S., Li, C., Wang, L., Guo, X., Sun, Y., Luo, H., Li, Y., Song, J., Henrissat, B., Levasseur, A., Qian, J., Li, J., Luo, X., Shi, L., He, L.,

Xiang, L., Xu, X., Niu, Y., Li, Q., Han, M.V., Yan, H., Zhang, J., Chen, H., Lv, A., Wang, Z., Liu, M., Schwartz, D. C., & Sun, C. (2012b). Genome sequence of the model medicinal mushroom *Ganoderma lucidum*. *Nature Communnications*, *913*, 1–9.

Cheng, C. R., Yue, Q. X., Wu, Z. Y., Song, X. Y., Tao, S. J., Wu, X. H., Xu, P. P., Liu, X., Guan, S. H., & Guo, D. A. (2010). Cytotoxic triterpenoids from *Ganoderma lucidum*. *Phytochemistry*, *71*(13), 1579–1585.

Chiu, L. Y., Hu, M. E., Yang, T. Y., Hsin, I. L., Ko, J. L., Tsai, K. J., & Sheu, G. T. (2015). Immunomodulatory protein from *Ganoderma microsporum* induces pro-death autophagy through Akt-mTOR-p70S6K pathway inhibition in multidrug resistant lung cancer cells. *PLoS One*, *10*(5), pe0125774.

Dai, Y. C. (2010). Hymenochaetaceae (Basidiomycota) in China. *Fungal Diversity*, *45*(1), 131–343.

Debbab, A., Aly, A. H., & Proksch, P. (2011). Bioactive secondary metabolites from endophytes and associated marine derived fungi. *Fungal Diversity*, *49*(1), 1–12.

Debbab, A., Aly, A. H., & Proksch, P. (2012). Endophytes and associated marine derived fungi—ecological and chemical perspectives. *Fungal Diversity*, *57*(1), 45–83.

Dixon, R. A., Xie, D. Y., & Sharma, S. B. (2005). Proanthocyanidins–a final frontier in flavonoid research? *New Phytologist*, *165*(1), 9–28.

Donatini, B. (2014). Control of oral human papillomavirus (HPV) by medicinal mushrooms, Trametes versicolor and *Ganoderma lucidum*: a preliminary clinical trial. *International Journal of Medicinal Mushrooms*, *16*(5), 497–498.

Dong, Q., Wang, Y., Shi, L., Yao, J., Li, J., Ma, F., & Ding, K. (2012). A novel water-soluble b-D-glucan isolated from the spores of *Ganoderma lucidum*. *Carbohydrate Research*, *353*, 100–105.

El Dine, R. S., El Halawany, A. M., Ma, C. M., & Hattori, M. (2008). Anti-HIV-1 protease activity of lanostane triterpenes from the vietnamese mushroom *Ganoderma colossum*. *Journal of Natural Products*, *71*(6), 1022–1026.

El Dine, R. S., El Halawany, A. M., Ma, C. M., & Hattori, M. (2009). Inhibition of the dimerization and active site of HIV-1 protease by secondary metabolites from the Vietnamese mushroom *Ganoderma colossum*. *Journal of Natural Products*, *72*(11), 2019–2023.

El-Mekkawy, S., Meselhy, M. R., Nakamura, N., Tezuka, Y., Hattori, M., Kakiuchi, N., Shimotohno, K., Kawahata, T., & Otake, T. (1998). Anti-HIV-1 and Anti-HIV-1-Protease substances from *Ganoderma lucidum*. *Phytochemistry*, *49*(6), 1651–1657.

Eo, S. K., Kim, Y. S., Lee, C. K., & Han, S. S. (2000). Possible mode of antiviral activity of acidic protein bound polysaccharide isolated from *Ganoderma lucidum* on herpes simplex viruses. *Journal of Ethnopharmacology*, *72*(3), 475–481.

Erkel, G., & Anke, T. (2008). *Products from Basidiomycetes? Chapter 11. Biotechnology set* (2nd edn.). New York: Wiley pp. 490-526.

Feng, L., Yuan, L., Du, M., Chen, Y., Zhang, M. H., Gu, J. F., He, J. J., Wang, Y., & Cao, W. (2013). Anti-lung cancer activity through enhancement of immunomodulation and induction of cell apoptosis of total triterpenes extracted from *Ganoderma luncidum* (Leyss. ex Fr.) Karst. *Molecules*, *18*(8), 9966–9981.

Ferreira, I. C. F. R., Heleno, S. A., Reis, F. S., Stojkovic, D., João, M., Queiroz, R. P., Helena, M., Vasconcelos, & Sokovic, M. (2015). Chemical features of *Ganoderma* polysaccharides with antioxidant, antitumor and antimicrobial activities. *Phytochemistry*, *114*, 38–55.

Floudas, D., Binder, M., Riley, R., Barry, K., Blanchette, R. A., Henrissat, B., Martínez, A. T., Otillar, R., Spatafora, J. W., Yadav, J. S., Aerts, A., Benoit, I., Boyd, A., Carlson, A., Copeland, A., Coutinho, P. M., de Vries, R. P., Ferreira, P., Findley, K., Foster, B., Gaskell, J., Glotzer, D., Górecki, P., Heitman, J., Hesse, C., Hori, C., Igarashi, K., Jurgens, L. A., Kallen, N., Kersten, P., Kohler, A., Kües, U., Kumar, T. K., Kuo, A., LaButti, K., Larrondo, L. F., Lindquist, E., Ling, A., Lombard, V., Lucas, S., Lundell, T., Martin, R., MyLaughlin, D. J., Morgenstern, I., Morin, E., Murat, C., Nagy, L. G., Nolan, M., Ohm,

R.A., Patyshakuliyeva, A., Rokas, A., Ruiz-Dueñas, F.J., Sabat, G., Salamov, A., Samejima, M., Schmutz, J., Slot, J. C., John, F. S., Stenlid, J., Sun, H., Sun, S., Syed, K., Tsang, A., Wiebenga, A., Young, D., Pisabarro, A., Eastwood, D., Martin, F., Cullen, D., Grigoriev, I.V., & Hibbett, D. S. (2012). The paleozoic origin of enzymatic lignin decomposition reconstructed from 31 fungal genomes. *Science*, *336*, 1715–1719.

Futch, S.H., Graham, J.H., Duncan, L.W., Timmer, L.W. (2016). *Ganoderma* Wood-Rotting Fungi on Citrus Stumps. UF/IFAS Extension HS1019, Available from: http://edis.ifas.ufl.edu.

Gallo, A., Ferrara, M., & Perrone, G. (2013). Phylogenetic study of polyketide synthases and nonribosomal peptide synthetases involved in the biosynthesis of mycotoxins. *Toxins*, *5*(4), 717–742.

Gao, Y., Zhou, S., Huang, M., & Xu, A. (2003). Antibacterial and antiviral value of the genus Ganoderma P. Karst species (Aphyllophoromycetideae): a review. *International Journal of Medicinal Mushroom*, *5*, 235–246.

Gheldof, N., & Engeseth, N. J. (2002). Antioxidant capacity of honeys from various floral sources based on the determination of oxygen radical absorbance capacity and inhibition of in vitro lipoprotein oxidation in human serum samples. *Journal of Agricultural and Food Chemistry*, *50*(10), 3050–3055.

Habijanic, J., Berovic, M., Boh, B., Wraber, B., & Petravic-Tominac, V. (2013). Production of biomass and polysaccharides of Lingzhi or Reishi medicinal mushroom, *Ganoderma lucidum* (W. Curt.:Fr.) P. Karst. (higher Basidiomycetes), by submerged cultivation. *International Journal of Medicinal Mushrooms*, *15*, 81–90.

Hansen, M. B., Jensen, M. L., & Carstensen, B. (2012). Causes of death among diabetic patients in Denmark. *Diabetologia*, *55*(2), 294–302.

Hsin, I. L., Ou, C. C., Wu, T. C., Jan, M. S., Wu, M. F., Chiu, L.Y., Lue, K. H., & Ko, J. L. (2011). GMI, an immunomodulatory protein from *Ganoderma microsporum*, induces autophagy in non-small cell lung cancer cells. *Autophagy*, 7(8), 873–882.

Hsin, I. L., Sheu, G. T., Jan, M. S., Sun, H. L., Wu, T. C., Chiu, L.Y., Lue, K. H., & Ko, J. L. (2012). Inhibition of lysosome degradation on autophagosome formation and responses to GMI, an immunomodulatory protein from *Ganoderma microsporum*. *British Journal of Pharmacology*, *167*(6), 1287–1300.

Hsu, H.Y., Hua, K. F., Wu, W. C., Hsu, J., Weng, S. T., Lin, T. L., Liu, C.Y., Hseu, R. S., & Huang, C.T. (2008). Reishi immune-modulation protein induces interleukin-2 expression via protein kinase-dependent signaling pathways within human T cells. *Journal of Cellular Physiology*, *215*(1), 15–26.

Hu, H., Ahn, N. S., Yang, X., Lee, Y. S., & Kang, K. S. (2002). *Ganoderma lucidum* extract induces cell cycle arrest and apoptosis in MCF-7 human breast cancer cell. *International Journal of Cancer*, *102*(3), 250–253.

Huang, S. Q., Li, J. W., Li, Y. Q., & Wang, Z. (2011). Purification and structural characterization of a new water-soluble neutral polysaccharide GLP-F1-1 from *Ganoderma lucidum*. *International Journal of Biological Macromolecules*, *48*, 165–169.

Huang, Y. H., Wu, H. Y., Wu, K. M., Liu, T. T., Liou, R. F., Tsai, S. F., Shiao, M. S., Ho, L. T., Tzean, S. S., & Yang, U. C. (2013). Generation and analysis of the expressed sequence tags from the mycelium of Ganoderma lucidum. *PLoS One*, *8*, e61127.

Hung, W. T., Wang, S. H., Chen, C. H., & Yang, W. B. (2008). Structure determination of β-glucans from *Ganoderma lucidum* with matrix-assisted laser desorption/ionization (MALDI) mass spectrometry. *Molecules*, *13*(8), 1538–1550.

Hyde, K. D. (2001). Where are the missing fungi? Does Hong Kong have any answers? *Mycology Research*, *105*(12), 1514–1518.

Iwatsuki, K., Akihisa, T., Tokuda, H., Ukiya, M., Oshikubo, M., Kimura, Y., Asano, T., Nomura, A., & Nishino, H. (2003). Lucidenic Acids P and Q, Methyl Lucidenate P, and Other Triterpenoids from the Fungus *Ganoderma lucidum* and Their inhibitory effects on epstein– barr virus activation. *Journal of Natural Products*, *66*(12), 1582–1585.

Jia, J., Zhang, X., Hua, Y. -S., Wua, Y., Wang, O. -Z., Li, N. -N., et al. (2009). Evaluation of in vivo antioxidant activities of *Ganoderma lucidum* polysaccharides in STZ-diabetic rats. *Food Chemistry, 115*, 32–36.

Jiang, J., Grieb, B., Thyagarajan, A., & Sliva, D. (2008). Ganoderic acids suppress growth and invasive behavior of breast cancer cells by modulating AP-1 and NF-κB signaling. *International Journal of Molecular Medicine, 21*(5), 577–584.

Jiang, J., Slivova, V., Valachovicova, T., Harvey, K., & Sliva, D. (2004). *Ganoderma lucidum* inhibits proliferation and induces apoptosis in human prostate cancer cells PC-3. *International Journal of Oncology, 24*(5), 1093–1099.

Jikai, L. (2002). Biologically active substances from mushrooms in Yunnan, China. *Heterocycles, 57*(1), 157–167.

Joseph, S., Sabulal, B., George, V., Antony, K., & Janardhanan, K. (2011). Antitumor and anti-inflammatory activities of polysaccharides isolated from *Ganoderma lucidum*. *Acta Pharmaceutica, 61*(3), 335–342.

Kenneth, J. (1992). *REISHI: Ancient herb for modern times*. Sylvan Press p. 6.

Kino, K., Yamashita, A., Yamaoka, K., Watanabe, J., Tanaka, S., Ko, K., Shimizu, K., & Tsunoo, H. (1989). Isolation and characterization of a new immunomodulatory protein, ling zhi-8 (LZ-8), from *Ganoderma lucidium*. *Journal of Biology and Chemistry, 264*(1), 472–478.

Kozarski, M., Klaus, A., Jakovljevic, D., Todorovic, N., Vunduk, J., Petrovic´, P., Niksic, M., Vrvic, M., & Van Griensven, L. (2015). Antioxidants of edible mushrooms. *Molecules, 20*(10), 19489–19525.

Kozlovskii, A. G., Zhelifonova, V. P., & Antipova, T. V. (2013). Fungi of the genus *Penicillium* as producers of physiologically active compounds. *Applied Biochemistry and Microbiology, 49*(1), 1–10.

Kües, U., Nelson, D. R., Liu, C., Yu, G. J., Zhang, J., Li, J., Wang, X. C., & Sun, H. (2015). Genome analysis of medicinal *Ganoderma* spp. with plant-pathogenic and saprotrophic life-styles. *Phytochemistry, 114*, 18–37.

Lau, C. C., Abdullah, N., Shuib, A. S., & Aminudin, N. (2012). Proteomic analysis of anti-hypertensive proteins in edible mushrooms. *Journal of Agricultural Food and Chemistry, 60*(50), 12341–12348.

Lee, J. S., & Hong, E. K. (2011). Immunostimulating activity of the polysaccharides isolated from *Cordyceps militaris*. *International Immunopharmacology, 11*(9), 1226–1233.

Li, J. W. H., & Vederas, J. C. (2009). Drug discovery and natural products: end of an era or an endless frontier? *Science, 325*(5937), 161–165.

Li, Y., Yang, Y., Fang, L., Zhang, Z., Jin, J., & Zhang, K. (2006). Anti-hepatitis activities in the broth of *Ganoderma lucidum* supplemented with a Chinese herbal medicine. *The American Journal of Chinese Medicine, 34*(02), 341–349.

Li, Y. Q., & Wang, S. F. (2006). Anti-hepatitis B activities of ganoderic acid from *Ganoderma lucidum*. *Biotechnology Letters, 28*(11), 837–841.

Liang, C., Li, H., Zhou, H., Zhang, S., Liu, Z., Zhou, Q., & Sun, F. (2012). Recombinant Lz-8 from *Ganoderma lucidum* induces endoplasmic reticulum stress-mediated autophagic cell death in SGC-7901 human gastric cancer cells. *Oncology Reports, 27*(4), 1079–1089.

Liao, C. H., Hsiao, Y. M., Hsu, C. P., Lin, M. Y., Wang, J. C. H., Huang, Y. L., & Ko, J. L. (2006). Transcriptionally mediated inhibition of telomerase of fungal immunomodulatory protein from *Ganoderma tsugae* in A549 human lung adenocarcinoma cell line. *Molecular Carcinogenetics., 45*(4), 220–229.

Liao, C. H., Hsiao, Y. M., Lin, C. H., Yeh, C. S., Wang, J. C. H., Ni, C. H., Hsu, C. P., & Ko, J. L. (2008). Induction of premature senescence in human lung cancer by fungal immunomodulatory protein from *Ganoderma tsugae*. *Food Chemistry and Toxicology, 46*(5), 1851–1859.

Liao, C. H., Hsiao, Y. M., Sheu, G. T., Chang, J. T., Wang, P. H., Wu, M. F., Shieh, G. J., Hsu, C. P., & Ko, J. L. (2007). Nuclear translocation of telomerase reverse transcriptase and calcium signaling in repression of telomerase activity in human lung cancer cells by fungal immunomodulatory protein from *Ganoderma tsugae*. *Biochemistry and Pharmacology, 74*(10), 1541–1554.

Liao, S. F., Liang, C. H., Ho, M.Y., Hsu, T. L., Tsai, T. I., Hsieh, Y. S.Y., Tsai, C. M., Li, S. T., Cheng, Y.Y., Tsao, S. M., & Lin, T.Y. (2013). Immunization of fucose-containing polysaccharides from Reishi mushroom induces antibodies to tumor-associated Globo H-series epitopes. *Proceedings of National Academy of Science U. S. A., 110*(34), 13809–13814.

Lin, C. H., Sheu, G. T., Lin, Y. W., Yeh, C. S., Huang, Y. H., Lai, Y. C., Chang, J. G., & Ko, J. L. (2010). A new immunomodulatory protein from *Ganoderma microsporum* inhibits epidermal growth factor mediated migration and invasion in A549 lung cancer cells. *Process Biochemistry, 45*(9), 1537–1542.

Lin, T. Y., & Hsu, H. Y. (2016). Ling Zhi-8 reduces lung cancer mobility and metastasis through disruption of focal adhesion and induction of MDM2-mediated Slug degradation. *Cancer Letters, 375*(2), 340–348.

Lin, T.Y., Hsu, H.Y., Sun, W. H., Wu, T. H., & Tsao, S. M. (2017). Induction of Cbl-dependent epidermal growth factor receptor degradation in Ling Zhi-8 suppressed lung cancer. *International Journal of Cancer, 140*(11), 2596–2607.

Lin, W. H., Hung, C. H., Hsu, C. I., & Lin, J.Y. (1997). Dimerization of the N-terminal amphipathic α-helix domain of the fungal immunomodulatory protein from *Ganoderma tsugae* (Fip-gts) defined by a yeast two-hybrid system and site-directed mutagenesis. *Journal of Biology and Chemistry, 272*(32), 20044–20048.

Lin, Z. B., & Zhang, H. N. (2004). Anti-tumor and immunoregulatory activities of *Ganoderma lucidum* and its possible mechanisms. *Acta Pharmacologica Sinica, 25*, 1387–1395.

Lindequist, U., Rausch, R., Füssel, A., & Hanssen, H. P. (2010). Higher fungi in traditional and modern medicine. *Medizinische Monatsschrift fur Pharmazeuten, 33*(2), 40–48.

Liu, D., Gong, J., Dai, W., Kang, X., Huang, Z., Zhang, H. -M., Liu, W., Liu, L., Ma, J., & Xia, Z. (2012). The genome of Ganderma lucidum provide insights into triterpense biosynthesis and wood degradation. *PLoS ONE*, 7, e36146.

Liu, X. T., Winkler, A. L., Schwan, W. R., Volk, T. J., Rott, M. A., & Monte, A. (2010). Antibacterial compounds from mushrooms I: a lanostane-type triterpene and prenylphenol derivatives from *Jahnoporus hirtus* and *Albatrellus flettii* and their activities against *Bacillus cereus* and *Enterococcus faecalis*. *Planta Medica, 76*(02), 182–185.

Lu, Q.Y., Jin, Y. S., Zhang, Q., Zhang, Z., Heber, D., Go, V. L. W., Li, F. P., & Rao, J.Y. (2004). *Ganoderma lucidum* extracts inhibit growth and induce actin polymerization in bladder cancer cells in vitro. *Cancer Letters, 216*(1), 9–20.

Luo, H., Sun, C., Song, J., Lan, J., Li, Y., Li, X., & Chen, S. (2010). Generation and analysis of expressed sequence tags from a cDNA library of the fruiting body of *Ganoderma lucidum*. *Chinese Medicine, 5*(1), 9.

Lupea, A. X., Pop, M., & Cacig, S. (2008). Structure-radical scavenging activity relationships of flavonoids from Ziziphus and Hydrangea extracts. *Revista de Chimime, 59*(3), 309–313.

Mau, J. L., Lin, H. C., & Chen, C. C. (2001). Non-volatile components of several medicinal mushrooms. *Food Research International, 34*(6), 521–526.

Mizuno, T., Saito, H., Nishitoba, T., & Kawagashi, H. (1995). Antitumor active substances from mushrooms. *Food Review International, 11*, 23–61.

Mothana, R. A. A., Ali, N. A., Jansen, R., Wegner, U., Mentel, R., & Lindequist, U. (2003). Antiviral lanostanoid triterpenes from the fungus Ganoderma pfeifferi. *Fitoterapia, 74*(1–2), 177–180.

Mulac, D., Hüwel, S., Galla, H. J., & Humpf, H. U. (2012). Permeability of ergot alkaloids across the blood-brain barrier in vitro and influence on the barrier integrity. *Molecular and Nutrition Food Research, 56*(3), 475–485.

Munné-Bosch, S. (2005). The role of α-tocopherol in plant stress tolerance. *Journal of Plant Physiology, 162*(7), 743–748.

Newman, D. J., & Cragg, G. M. (2007). Natural products as sources of new drugs over the last 25 years. *Journal of Natural Products, 70*(3), 461–477.

Nie, S., Zhang, H., Li, W., & Xie, M. (2013). Current development of polysaccharides from Ganoderma: isolation, structure and bioactivities. *Bioactive Carbohydrates Dietary Fibre, 1*, 10–20.

Niedermeyer, T. H., Lindequist, U., Mentel, R., Gördes, D., Schmidt, E., Thurow, K., & Lalk, M. (2005). Antiviral terpenoid constituents of *Ganoderma pfeifferi*. *Journal of Natural Products, 68*(12), 1728–1731.

Nimse, S. B., & Pal, D. (2015). Free radicals, natural antioxidants, and their reaction mechanisms. *RSC Advances, 5*(35), 27986–28006.

Pan, S.Y., Pan, S., Yu, Z. L., Ma, D. L., Chen, S. B., Fong, W. F., Han, Y. F., & Ko, K. M. (2010). New perspectives on innovative drug discovery: an overview. *Journal of Pharmacology and Pharmaceutical Science, 13*(3), 450–471.

Paterson, R. R. M. (2006). *Ganoderma*–a therapeutic fungal biofactory. *Phytochemistry, 67*(18), 1985–2001.

Petrova, R. D., Mahajna, J. A., & Denchev, C. M. (2005). Potential role of medicinal mushrooms in breast cancer treatment: current knowledge and future perspectives. *International Journal of Medicinal Mushrooms*, 7, 141–155.

Que, Z., Zou, F., Zhang, A., Zheng, Y., Bi, L., Zhong, J., Tian, J., & Liu, J. (2014). Ganoderic acid Me induces the apoptosis of competent T cells and increases the proportion of Treg cells through enhancing the expression and activation of indoleamine 2, 3-dioxygenase in mouse lewis lung cancer cells. *International Immunopharmacology, 23*(1), 192–204.

Rajani, M., Padh, H., (2000). Fourth international symposium on innovations in pharmaceutical sciences and technology, Ahmedabad, India.

Ren, A., Li, M. J., Shi, L., Mu, D. S., Jiang, A. L., Han, Q., & Zhao, M. W. (2013). Profiling and quantifying differential gene transcription provide insights into ganoderic acid biosynthesis in Ganoderma lucidum in response to methyl jasmonate. *PLoS ONE, 8*, e65027.

Sato, N., Zhang, Q., Ma, C. M., & Hattori, M. (2009). Anti-human immunodeficiency virus-1 protease activity of new lanostane-type triterpenoids from *Ganoderma sinense*. *Chemical and Pharmaceutical Bulletin, 57*(10), 1076–1080.

Shan, B., Cai, Y. Z., Sun, M., & Corke, H. (2005). Antioxidant capacity of 26 spice extracts and characterization of their phenolic constituents. *Journal of Agricultural and Food Chemistry, 53*(20), 7749–7759.

Shi, M., Zhang, Z., & Yang, Y. (2013). Antioxidant and immunoregulatory activity of *Ganoderma lucidum* polysaccharide. *Carbohydrate Polymers, 95*, 200–206.

Skalicka-Woźniak, K., Szypowski, J., Łoś, R., Siwulski, M., Sobieralski, K., Głowniak, K., & Malm, A. (2012). Evaluation of polysaccharides content in fruit bodies and their antimicrobial activity of four *Ganoderma lucidum* (W Curt.: Fr.) P. Karst strains cultivated on different wood type substrates. *Acta Societatis Botanicorum Poloniae, 81*, 17–21.

Sliva, D. (2003). *Ganoderma lucidum* (Reishi) in cancer treatment. *Integrative Cancer Therapies, 2*(4), 358–364.

Sliva, D., Labarrere, C., Slivova, V., Sedlak, M., Lloyd, F. P., Jr., & Ho, N. W. (2002). *Ganoderma lucidum* suppresses motility of highly invasive breast and prostate cancer cells. *Biochemistry and Biophysical Research Community, 298*(4), 603–612.

Smina, T. P., Mathew, J., Janardhanan, K. K., & Devasagayam, T. P. A. (2011). Antioxidant activity and toxicity profile of total triterpenes isolated from *Ganoderma lucidum* (Fr.) P. Karst occurring in South India. *Environmental Toxicology and Pharmacology, 32*(3), 438–446.

Stanley, G., Harvey, K., Slivova, V., Jiang, J., & Sliva, D. (2005). *Ganoderma lucidum* suppresses angiogenesis through the inhibition of secretion of VEGF and TGF-β1 from prostate cancer cells. *Biochemical and Biophysical Research Communications, 330*(1), 46–52.

Suarez-Arroyo, I. J., Rosario-Acevedo, R., Aguilar-Perez, A., Clemente, P. L., Cubano, L. A., Serrano, J., Schneider, R. J., & Martínez-Montemayor, M. M. (2013). Anti-tumor effects of *Ganoderma lucidum* (reishi) in inflammatory breast cancer in in vivo and in vitro models. *PloS One, 8*(2), pe57431.

Sun, L. X., Li, W. D., Lin, Z. B., Duan, X. S., Li, X. F., Yang, N., Lan, T. F., Li, M., Sun, Y., Yu, M., & Lu, J. (2014). Protection against lung cancer patient plasma-induced lymphocyte suppression by *Ganoderma lucidum* polysaccharides. *Cellular Physiology in Biochemistry, 33*(2), 289–299.

Tang, W., Liu, J. W., Zhao, W. M., Wei, D. Z., & Zhong, J. J. (2006a). Ganoic acid T from Ganoderma lucidum mycelia induces mitochondria mediated apoptosis in lung cancer cells. *Life Sciences, 80*(3), 205–211.

Tang, W., Gu, T., & Zhong, J. J. (2006b). Separation of targeted ganoderic acids from *Ganoderma lucidum* by reversed phase liquid chromatography with ultraviolet and mass spectrometry detections. *Biochemical Engineering Journal, 32*(3), 205–210.

Ukai, S., Yokoyama, S., Hara, C., & Kiho, T. (1982). Structure of an alkali-soluble polysaccharide from the fruit body of *Ganoderma japonicum* Lloyd. *Carbohydrate Research, 105*, 237–245.

Wang, C. H., Hsieh, S. C., Wang, H. J., Chen, M. L., Lin, B. F., Chiang, B. H., & Lu, T. J. (2014). Concentration variation and molecular characteristics of soluble (1,3;1,6)-b-Dglucans in submerged cultivation products of *Ganoderma lucidum* mycelium. *Journal of Agriculture and Food Chemistry, 62*, 634–641.

Wang, J. G., Ma, Z. C., Zhang, L. N., Fang, Y. P., Jiang, F. T., & Phillips, G. O. (2011). Structure and chain conformation of water-soluble heteropolysaccharides from *Ganoderma lucidum*. *Carbohydrate Polymers, 86*, 844–851.

Wang, S. Y., Hsu, M. L., Hsu, H. C., Lee, S. S., Shiao, M. S., & Ho, C. K. (1997). The anti-tumor effect of *Ganoderma lucidum* is mediated by cytokines released from activated macrophages and T lymphocytes. *International Journal of Cancer, 70*(6), 699–705.

Wen, H., Kang, S., Song, Y., Song, Y., Sung, S. H., & Park, S. (2010). Differentiation of cultivation sources of Ganoderma lucidum by NMR-based metabolomics approach. *Phytochemistry Anals, 21*, 73–79.

Weng, C. J., Chau, C. F., Chen, K. D., Chen, D. H., & Yen, G. C. (2007). The anti-invasive effect of lucidenic acids isolated from a new *Ganoderma lucidum* strain. *Molecular and Nutrition Food Research, 51*(12), 1472–1477.

Wiemann, P., Albermann, S., Niehaus, E. M., Studt, L., Von Bargen, K. W., Brock, N. L., Humpf, H. U., Dickschat, J. S., & Tudzynski, B. (2012). The Sfp-type 4′-phosphopantetheinyl transferase Ppt1 of *Fusarium fujikuroi* controls development, secondary metabolism and pathogenicity. *PLoS One*, 7(5), pe37519.

Wu, G. S., Lu, J. J., Guo, J. J., Li, Y. B., Tan, W., Dang, Y. Y., Zhong, Z. F., Xu, Z. T., Chen, X. P., & Wang, Y. T. (2012). Ganoderic acid DM, a natural triterpenoid, induces DNA damage, G1 cell cycle arrest and apoptosis in human breast cancer cells. *Fitoterapia, 83*(2), 408–414.

XiaoPing, C., Yan, C., Shuibing, L., YouGou, C., JianYun, L., & LanPing, L. (2009). Free radical scavenging of *Ganoderma lucidum* polysaccharides and its effect on antioxidant enzymes and immunity activities in cervical carcinoma rats. *Carbohydrate Polymers, 77*, 389–393.

Xu, H., Kong, Y. Y., Chen, X., Guo, M. Y., Bai, X. H., Lu, Y. J., Li, W., & Zhou, X. W. (2016). Recombinant FIP-gat, a fungal immunomodulatory protein from *Ganoderma atrum*, induces growth inhibition and cell death in breast cancer cells. *Journal of Agriculture and Food Chemistry, 64*(13), 2690–2698.

Xu, K., Liang, X., Gao, F., Zhong, J., & Liu, J. (2010). Antimetastatic effect of ganoderic acid T in vitro through inhibition of cancer cell invasion. *Process Biochemistry, 45*(8), 1261–1267.

Ye, L. B., Zhang, J. S., Yang, Y., Zhou, S., Liu, Y. F., Tang, Q. J., Du, X. J., Chen, H., & Pan, Y. J. (2009). Structural characterisation of a heteropolysaccharide by NMR spectra. *Food Chemistry, 112*, 962–966.

Yoon, S. Y., Eo, S. K., Kim, Y. S., Lee, C. K., & Han, S. S. (1994). Antimicrobial activity of *Ganoderma lucidum* extract alone and in combination with some antibiotics. *Archives of Pharmacal Research, 17*(6), 438–442.

Young, I. S., & Woodside, J. V. (2001). Antioxidants in health and disease. *Journal of Clinical Pathology*, *54*(3), 176–186.

Young, S. N. (2013). Single treatments that have lasting effects: some thoughts on the antidepressant effects of ketamine and botulinum toxin and the anxiolytic effect of psilocybin. *Journal of Psychiatry Neuroscience*, *38*(2), 78–83.

Yu, G. J., Wang, N., Huang, J., Yin, Y. L., Chen, Y. J., Jiang, S., Jin, Y. X., Lan, X. Q., Wong, B. H. C., Liang, Y., & Sun, H. (2012). Deep insight into the Ganoderma lucidum by comprehensive analysis of its transcriptome. *PLOS ONE*, *8*, e44031.

Zhang, L. X., Zhao, G. M., Wu, M. J., Cao, P.Y., & Zhang, L. P. (2006). Purification and characterization of water soluble polysaccharide GLPS 3 from sporderm-unbroken germinting spors of *Ganoderma lucidum*. *Chinese Pharmaceutical Journal*, *41*, 230–232.

Zhang, M., Cui, S. W., Cheung, P. C. K., & Wang, Q. (2007). Antitumor polysaccharides from mushrooms: a review on their isolation process, structural characteristics and antitumor activity. *Trends in Food Science and Technology*, *18*(1), 4–19.

Zhang, W., Tao, J., Yang, X., Yang, Z., Zhang, L., Liu, H., Wu, K., & Wu, J. (2014). Antiviral effects of two *Ganoderma lucidum* triterpenoids against enterovirus 71 infection. *Biochemical and Biophysical Research Communications*, *449*(3), 307–312.

Zhao, G. M., Zhang, L. X., Yu, T. M., Liu, S. M., & Zhang, L. P. (2005). Chemical structure of water soluble polysaccharide from spores of *Ganoderma lucidum*. *Chinese Pharmaceutical Journal*, *41*, 902–904.

Zheng, R., Jie, S., Hanchuan, D., & Moucheng, W. (2005). Characterization and immunomodulating activities of polysaccharide from *Lentinus edodes*. *International Immunopharmacology*, *5*(5), 811–820.

Zhong, J.J., Xiao, J.H., 2009. Secondary metabolites from higher fungi: discovery, bioactivity, and bioproduction. In Biotechnology in China I (79-150). Springer, Berlin, Heidelberg.

Zhou, X., Lin, J., Yin, Y., Zhao, J., Sun, X., & Tang, K. (2007). Ganodermataceae: natural products and their related pharmacological functions. *The American Journal of Chinese Medicine*, *35*(4), 559–574.

Zhu, X., & Lin, Z. (2006). Modulation of cytokines production, granzyme B and perforin in murine CIK cells by *Ganoderma lucidum* polysaccharides. *Carbohydrate Polymers*, *63*(2), 188–197.

About the editors

Dr. Tariq Aftab

Tariq Aftab received his PhD from the Department of Botany, Aligarh Muslim University, India, and is currently an Assistant Professor there. He is the recipient of a prestigious Leibniz-DAAD fellowship from Germany, Raman Fellowship from the Government of India, and Young Scientist Awards from the State Government of Uttar Pradesh (India) and Government of India. After completing his doctorate, he has worked as Research Fellow at National Bureau of Plant Genetic Resources, New Delhi and as Post-doctorate Fellow at Jamia Hamdard, New Delhi, India. Dr. Aftab also worked as Visiting Scientist at Leibniz Institute of Plant Genetics and Crop Plant Research (IPK), Gatersleben, Germany, and in the Department of Plant Biology, Michigan State University, United States. He is a member of various scientific associations from India and abroad.

He has edited 3 books with international publishers, including Springer Nature, and CRC Press (Taylor & Francis Group), coauthored 10 book chapters, and published over 45 research papers in peer-reviewed international journals. His research interests include physiological, proteomic, and molecular studies on medicinal and aromatic plants.

Dr. Khalid Rehman Hakeem

Khalid Rehman Hakeem, PhD is Professor at King Abdulaziz University, Jeddah, Saudi Arabia. After completing his doctorate (Botany; specialization in Plant Ecophysiology and Molecular Biology) from Jamia Hamdard, New Delhi, India, in 2011, he worked as a lecturer at the University of Kashmir, Srinagar, for a short period. Later, he joined Universiti Putra Malaysia, Selangor, Malaysia, and worked there as Post-doctorate Fellow in 2012 and Fellow Researcher (Associate Prof.) from 2013–16. Dr. Hakeem has more than 10 years of teaching and research experience in plant ecophysiology, biotechnology and molecular biology, medicinal plant research, plant-microbe-soil interactions as well as in environmental studies. He is the recipient of several fellowships at both national and international levels; also, he has served as the visiting scientist at Jinan University, Guangzhou, China. Currently, he is involved with a number of international research projects with different government organizations.

So far, Dr. Hakeem has authored and edited more than 36 books with international publishers, including Springer Nature, Academic Press (Elsevier), and CRC Press. He also has to his credit more than 90 research publications in peer-reviewed international journals and 55 book chapters in edited volumes with international publishers.

At present, Dr. Hakeem serves as an editorial board member and reviewer of several high-impact international scientific journals from Elsevier, Springer Nature, Taylor and Francis, Cambridge, and John Wiley Publishers. He is included in the advisory board of Cambridge Scholars Publishing, United Kingdom. He is also a fellow of Plantae group of the American Society of Plant Biologists, member of the World Academy of Sciences, member of the International Society for Development and Sustainability, Japan, and member of Asian Federation of Biotechnology, Korea. Dr. Hakeem has been listed in Marquis Who's Who in the World, since 2014–19. Currently, Dr. Hakeem is engaged in studying the plant processes at ecophysiological as well as molecular levels.

Index

Note: Page numbers followed by "f" indicate figures, "t" indicate tables.

D

CPI Antony Rowe
Eastbourne, UK
May 22, 2020